Secondary Metabolism
and Differentiation in Fungi

MYCOLOGY SERIES

Edited by
Paul A. Lemke
*Department of Botany,
Plant Pathology, and Microbiology
Auburn University
Auburn, Alabama*

Volume 1   Viruses and Plasmids in Fungi, *edited by Paul A. Lemke*

Volume 2   The Fungal Community: Its Organization and Role in the Ecosystem, *edited by Donald T. Wicklow and George C. Carroll*

Volume 3   Fungi Pathogenic for Humans and Animals (in three parts), *edited by Dexter H. Howard*

Volume 4   Fungal Differentiation: A Contemporary Synthesis, *edited by John E. Smith*

Volume 5   Secondary Metabolism and Differentiation in Fungi, *edited by J. W. Bennett and Alex Ciegler*

Other Volumes in Preparation

Secondary Metabolism and Differentiation in Fungi

edited by
J.W. BENNETT
Department of Biology
Tulane University
New Orleans, Louisiana

ALEX CIEGLER
Computer Sciences Corporation
National Space Technology Laboratory Station
Mississippi

MARCEL DEKKER, INC.   New York and Basel

Library of Congress Cataloging in Publication Data

Main entry under title:

Secondary metabolism and differentiation in fungi.

    (Mycology series ; v. 5)
    Includes indexes.
    1. Fungi–Physiology. 2. Fungi–Morphogenesis.
3. Metabolism, Secondary. 4. Fungal metabolites.
I. Bennett, J. W. II. Ciegler, Alex [date].
III. Series.
QK601.S43 1983   589.2'048761   83-10070
ISBN 0-8247-1819-4

COPYRIGHT © 1983 by MARCEL DEKKER, INC. ALL RIGHTS RESERVED

Neither this book nor any part may be reproduced or transmitted in any form or by any means, electronic or mechanical, including photocopying, microfilming, and recording, or by any information storage and retrieval system, without permission in writing from the publisher.

MARCEL DEKKER, INC.
270 Madison Avenue, New York, New York 10016

Current printing (last digit):
10 9 8 7 6 5 4 3 2 1

PRINTED IN THE UNITED STATES OF AMERICA

Series Introduction

Mycology is the study of fungi, that vast assemblage of microorganisms which includes such things as molds, yeasts, and mushrooms. All of us in one way or another are influenced by fungi. Think of it for a moment—the good life without penicillin or a fine wine. Consider further the importance of fungi in the decomposition of wastes and the potential hazards of fungi as pathogens to plants and to humans. Yes, fungi are ubiquitous and important.

Mycologists study fungi either in nature or in the laboratory and at different experimental levels ranging from descriptive to molecular and from basic to applied. Since there are so many fungi and so many ways to study them, mycologists often find it difficult to communicate their results even to other mycologists, much less to other scientists or to society in general.

This Series establishes a niche for publication of works dealing with all aspects of mycology. It is not intended to set the fungi apart, but rather to emphasize the study of fungi and of fungal processes as they relate to mankind and to science in general. Such a series of books is long overdue. It is broadly conceived as to scope, and should include textbooks and manuals as well as original and scholarly research works and monographs.

The scope of the Series will be defined by, and hopefully will help define, progress in mycology.

Paul A. Lemke

Foreword

For too many years, microbial secondary metabolism has been considered a laboratory artifact with no significance in nature. It is indeed peculiar that although biologists accepted the concept that plant secondary metabolites had function, the concept of microbial secondary metabolites, and especially of antibiotics, as ecological effectors has been widely disputed. I believe this bias against function for antibiotics is due to two major facts: (1) antibiotics are of major economic significance and patents on them were much more easily obtained if they were not considered products of nature, and (2) experiments to prove their ecological significance were more often unsuccessful than successful. Although we can dismiss point (1) as being of no scientific significance, point (2) deserves some comment. It is clearly much more difficult to do experiments in natural environments than under controlled conditions in the laboratory. Thus, one should expect that it would be quite difficult to prove or disprove a possible function for a compound in an ecological type of experiment. For this reason, I have been much more impressed with the few successful experiments pointing to a real function of an antibiotic in nature than with the larger number of inconclusive experiments. Broadening our scope from antibiotics to secondary metabolites, there is no doubt that functions exist for these compounds. They serve as antagonistic agents, symbiotic agents, sexual hormones, effectors of sporulation and germination, and metal transporters, etc., and in such roles offer the producing organism an opportunity to survive in the competitive arena of nature.

The editors of the present volume approach secondary metabolism in a very mature way, i.e., as a series of chemical processes correlated with a series of morphological processes, both of which are partners in the extremely important act of differentiation. Although this volume deals with fungi, there is no doubt that it contains messages for those scientists and technologists working with unicellular bacteria and actinomycetes producing secondary metabolites. Even more importantly, it could strike a death blow to the concept that secondary metabolites are without function.

Arnold L. Demain

Preface

Cellular differentiation involves molecular and cytological changes during development. In the fungi, chemical changes have traditionally been studied by natural products chemists and categorized as "secondary metabolites," whereas changes in cellular structure have been studied by biologists and termed "morphogenesis." In this volume we view secondary metabolism and morphogenesis as correlated events in the overall process of cellular differentiation.

Many other books have been written about either secondary metabolism or fungal morphogenesis, usually treating these as independent processes. This illustrates one of the failings of contemporary biological science, wherein there is a tendency to reduce problems to constitutive elements which are then studied as isolated events. The result is an ever-narrowing and myopic view of the entire organism. The sum of the parts no longer equals the whole.

This book does not attempt to be an exhaustive review of previous science or thought. Rather, we have chosen topics that highlight current research, with particular emphasis on genetics, molecular biology, and ecology. It is in this latter field, an area of increasing importance, that we have come to see biological systems not as independent and isolated chemical units, but as entities in an interrelated universal whole.

We asked our authors to indulge in "honorable speculation" and told them we welcomed "responsible hypotheses" in the hope that this volume would serve not only as a reference to what has already been done, but as a guideline for future research. Each chapter presents a unique perspective on fungal differentiation, and each is rich with ideas for future experimentation. We can only hope that our readers will find information and inspiration that will be of use either in their own research or in formulating a more profound view of the biological universe.

J. W. Bennett
Alex Ciegler

Contents

SERIES INTRODUCTION — Paul A. Lemke — iii
FOREWORD — Arnold L. Demain — v
PREFACE — vii
CONTRIBUTORS — xi

1. Differentiation and Secondary Metabolism in Mycelial Fungi — 1
   J. W. Bennett

BIOLOGICAL ASPECTS OF SECONDARY METABOLISM

2. Regulation of Secondary Metabolism and Keys to Its Manipulation — 35
   Stephen W. Drew and David A. Wallis
3. Correlation of Secondary Metabolism and Differentiation — 55
   Iain M. Campbell
4. Comparative Aspects of Secondary Metabolism in Cell Cultures of Green Plants, Animals, and Microorganisms — 73
   Eugene D. Weinberg
5. Biochemistry, Physiology, and Genetics of Carotenogenesis in Fungi — 95
   Manfred Ruddat and Edward D. Garber
6. Evolution and Secondary Pathways — 153
   Hans Zähner, Heidrun Anke, and Timm Anke

MOLECULAR ASPECTS OF MORPHOGENESIS

7. Fungal Nucleic Acids — 175
   Shelby N. Freer
8. Controls for Development and Differentiation of the Dikaryon in Basidiomycetes — 195
   Carlene Allen Raper
9. Hormones and Sexuality in Fungi — 239
   Graham W. Gooday

10	Yeast/Mold Morphogenesis in *Mucor* and *Candida albicans* Jim E. Cutler and Kevin C. Hazen	267
11	The Yeast Genome in Yeast Differentiation Michael Breitenbach and Eva Lachkovics	307

ECOLOGICAL CONSIDERATIONS

12	Phytoalexins H. Grisebach	377
13	Evolution, Ecology, and Mycotoxins: Some Musings Alex Ciegler	429

AUTHOR INDEX	441
SUBJECT INDEX	471

Contributors

Heidrun Anke, Ph.D., Department of Microbiology, Universität Tübingen, Tübingen, Federal Republic of Germany

Timm Anke, Ph.D., Department of Biotechnology, Universität Kaiserslautern, Kaiserslautern, Federal Republic of Germany

J. W. Bennett, Ph.D., Department of Biology, Tulane University, New Orleans, Louisiana

Michael Breitenbach, Ph.D., Institut für Allgemeine Biochemie and Ludwig-Boltzmann-Forschungsstelle für Biochemie, Vienna, Austria

Iain M. Campbell, Ph.D., Department of Biological Sciences, University of Pittsburgh, Pittsburgh, Pennsylvania

Alex Ciegler, Ph.D., Computer Sciences Corporation, National Space Technology Laboratory Station, Mississippi

Jim E. Cutler, Ph.D., Department of Microbiology, Montana State University, Bozeman, Montana

Stephen W. Drew, Ph.D.,* Department of Chemical Engineering, Virginia Polytechnic Institute and State University, Blacksburg, Virginia

Shelby N. Freer, Ph.D., Northern Regional Research Center, Agricultural Research Service, United States Department of Agriculture, Peoria, Illinois

Edward D. Garber, Ph.D., Department of Biology, The University of Chicago, Chicago, Illinois

Graham W. Gooday, Ph.D., Department of Microbiology, University of Aberdeen, Aberdeen, Scotland

H. Grisebach, Ph.D., Biological Institute II, University of Freiburg, Freiburg, Federal Republic of Germany

Present affiliation: Merck and Company, Inc., Rahway, New Jersey

Kevin C. Hazen, Ph.D.,* Department of Microbiology, Montana State University, Bozeman, Montana

Eva Lachkovics, M.Sc.,† Institut für Allgemeine Biochemie and Ludwig-Boltzmann-Forschungsstelle für Biochemie, Vienna, Austria

Carlene Allen Raper, Ph.D., Department of Biological Sciences, Wellesley College, Wellesley, Massachusetts

Manfred Ruddat, Ph.D., Department of Biology, The University of Chicago, Chicago, Illinois

David A. Wallis, Ph.D., Department of Chemical Engineering, Virginia Polytechnic Institute and State University, Blacksburg, Virginia

Eugene D. Weinberg, Ph.D., Department of Microbiology and Medical Sciences Program, Indiana University, Bloomington, Indiana

Hans Zähner, Ph.D., Department of Microbiology, Universität Tübingen, Tübingen, Federal Republic of Germany

---

**Present affiliation*: Washington University School of Medicine, St. Louis, Missouri
†*Present affiliation*: IOCU Regional Office, Penang, Malaysia

Secondary Metabolism and Differentiation in Fungi

1
Differentiation and Secondary Metabolism in Mycelial Fungi

J. W. Bennett / Tulane University, New Orleans, Louisiana

INTRODUCTION

"Differentiation" and "secondary metabolism" are terms which encompass an enormous body of accumulated scientific knowledge and evoke an even larger body of unanswered questions. Neither term is easily defined and many overlapping and sometimes ambiguous definitions have been used by different writers. Table 1 presents a list of pertinent terminology with the definitions that will be employed in this chapter.

Historically, development has been viewed as two separate, although frequently simultaneous, processes: growth and differentiation. A more modern conception might be that there is no such thing as an undifferentiated cell—there are simply changes in states of differentiation. Nevertheless, this arbitrary distinction between undifferentiated and differentiated states is useful for purposes of discussion and will be retained with a recognition of the inherent limitations.

In multicellular plants and animals, after an initial stage of undifferentiated growth, the processes of growth and differentiation occur concomitantly. Indeed, "cellular differentiation is the necessary condition of multicellular life" (Gross, 1968) or, put another way, "in order for organisms to become large, they must divide the labor; the two phenomena are inseparable" (Bonner, 1974, p. 25).

Many aspects of development are common to both plants and animals, but the different nature of the biological systems has profound effects upon the pattern of development. Plant cells generally have rigid cell walls and are not motile. In plants, development is indeterminate, with the growing points remaining permanently embryonic and with specific organ systems exhibiting limited growth; most animals show a distinct embryonic phase which ends when adult structure is achieved. These distinctions are important because most biologists who call themselves "developmentalists" and theorize on the molecular basis of differentiation are trained in the animal sciences. They frequently view

Table 1 Definitions

Major term	Related terms	Definition
Development		Process of growth and differentiation by which the potentialities of a zygote or spore are established; sequence of progressive changes resulting in increased biological complexity.
	Epigenesis	Development involving gradual diversification and differentiation of an initially undifferentiated entity such as a zygote or sphore. Development is an epigenetic phenomenon.
	Morphogenesis	The formation of tissues and organs; the developmental process leading to the characteristic mature form of an organism; those aspects of development related to morphological changes.
Differentiation		Progressive diversification of the structure and function of cells in an organism; acquisition of differences during development.
Growth		An increase in size, especially an increase in the quantity of cellular material; frequently dry weight is used as an equivalent.
Mycelium		The vegetative part (thallus) of a fungus consisting of one or more hyphae; a mass of hyphae.
	Hyphae	Branched filamentous cells; may be septate or aseptate.
Secondary metabolites		Diverse natural products unnecessary for growth, of restricted taxonomic distribution, generally produced during a limited stage of the cell cycle from a few simple precursors derived from primary metabolism; among microorganisms the term "idiolite" may be used as a synonym.

Source: Ainsworth (1971); King (1972); Turner (1971); *Webster's Third* (1972).

differentiation as an ultimately self-limiting process in development whereby differentiation proceeds through a number of significant restrictions in developmental potential (Rutter et al., 1973).

The concept of "pluripotency" is an important one in embryology. Pluripotency is the condition of having a large indeterminate number of possible fates; it is viewed as a property lost during the developmental process in animals. The historical context of "pluripotency" led botanists to adopt the term "totipotency" to express the idea that any somatic, nucleated plant cell could, under the influence of appropriate stimuli, dedifferentiate and regain the ability to act like a zygote and produce an entire new plant (Needham, 1950; Street, 1976; Steward and Mohan Ram, 1961).

Among prokaryotes and unicellular eukaryotes, growth and differentiation are phenomena observed in populations, not simultaneously in single cells. A given cell may divide vegetatively for a number of generations, or it may form a spore. When viewing an entire population, some cells are differentiated, some are not; but when viewing any given cell (organism), it is either vegetative, reproductive, or in transition. Moreover, this transition from a vegetative to a reproductive form is dependent upon the external environment. For this reason, microorganisms have been useful model systems for studying certain forms of differentiation; for example, experimental manipulation of the environment can be used as a "trigger" to initiate the transition between "undifferentiated" and "differentiated" states.

The fungi hold a unique and useful position for experimentalists. Like plants and animals, they are multicellular and eukaryotic; like plants, individual cells are totipotent; and like prokaryotes and unicellular forms, differentiation is generally triggered by changes in the environment. Some of the most elegant experiments on fungal morphogenesis involve the water mold *Blastocladiella emersonii* (Cantino, 1966), the acellular slime molds or Mycomycetes (Alexopoulos, 1962; 1966), and the cellular slime molds or Acrasiales (Ashworth, 1971; Gregg, 1966; Sussman and Brackenbury, 1976). But these are not "mycelial fungi."

What are "mycelial fungi"? Formal fungal taxonomy is by no means settled and a variety of classification schemes have been adopted by different workers. Alexopoulos (1962), Hawker (1966), and others have accepted a general scheme in which the true fungi (Eumycota) are subdivided as follows:

Lower fungi (Phycomycetes). Unicellular or with a mycelium that is generally aseptate.
Higher fungi. The mycelium is regularly septate, with the exception of the yeasts. The higher fungi are usually divided into three classes:

Ascomycetes: Sexual spores (ascospores) borne endogenously.
Basidiomycetes. Sexual spores (basidiospores) born exogenously.
Deuteromycetes. Sexual reproduction lacking. This form class is also called the Fungi Imperfecti.

In the broadest sense, the mycelial fungi encompass all filamentous species of both the lower and the higher fungi. However, perhaps because Vuillemin restricted the term "hyphae" to septate forms (Ainsworth, 1971, p. 281), usage of "mycelial fungi" is often accepted as meaning only the higher fungi. I will follow this limited definition and exclude the Phycomycetes from my discussion. The subject of yeast-mold dimorphism will also be excluded, but it is reviewed by Cutler and Hazen in Chapter 10 of this volume. Breitenback and Lachkovics discuss the role of the yeast genome in differentiation from a contemporary molecular perspective in Chapter 11 of this volume.

MORPHOGENESIS

Spores can be considered as "both a beginning and an end of fungal development" (Ainsworth, 1976, p. 81). Spores possess many attributes in common with the totipotent cells of plants and animals and more attention has been given to the spore than any other fungal structure. Spore morphology and development are basic to fungal systematics and most studies on fungal differentiation involve aspects of spore formation or germination.

"Spore" is a general term for a reproductive structure. In fungi, spores may be produced asexually or sexually, and in many species both forms of reproduction occur. Generally, the expression "vegetative" is associated with the mycelial phase of the fungal life cycle by mycologists. This is in contrast to another common usage of the term which equates "vegetative" to "asexual."

Spores vary in shape, size, ornamentation, origin, modes of liberation, function, and ontogeny. Asexual spores (conidia) have received the most detailed attention by classical mycologists because they provide the basic taxonomic criteria within many groups. The earliest and the majority of studies on spore differentiation are descriptive. Vuillemin was among the first to emphasize spore development as a taxonomic criterion rather than the characteristics of the spores themselves. He distinguished spores which are not separated from the hypae producing them (thallospores) from spores borne upon special hyphae which separate upon maturity (conidiospores) (see Fig. 1).

The extraordinary diversity of spore types in nature has challenged mycologists to find a satisfactory nomenclature. Ainsworth's *Dictionary of the Fungi* (1971) lists 100 spore names. In face of this proliferation, many workers simply

Differentiation in Mycelial Fungi

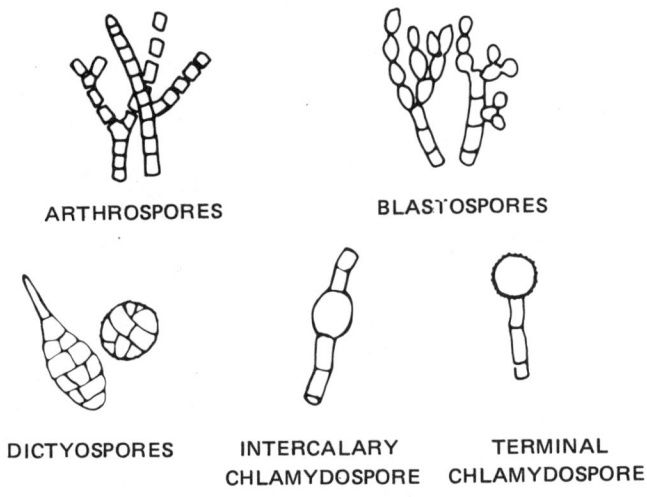

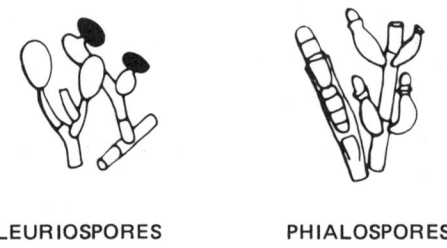

Figure 1 Examples of thallospores and conidiospores; Vuillemin's spore groups (after Tubaki, 1966).

call all asexual fungal spores "conidiospores." This in turn creates a terminological backlash. Vullemin's conidiospores are now designated "phiallospores" or "conidia vera." The major spore types delineated by Tubaki (1966) for the Fungi Imperfecti are illustrated in Fig. 2. An extended treatment of spore types and terminology within this group has been given by Kendrick (1971). Other recent monographs on conidial fungi include Cole and Samson (1979) and Cole and Kendrick (1981).

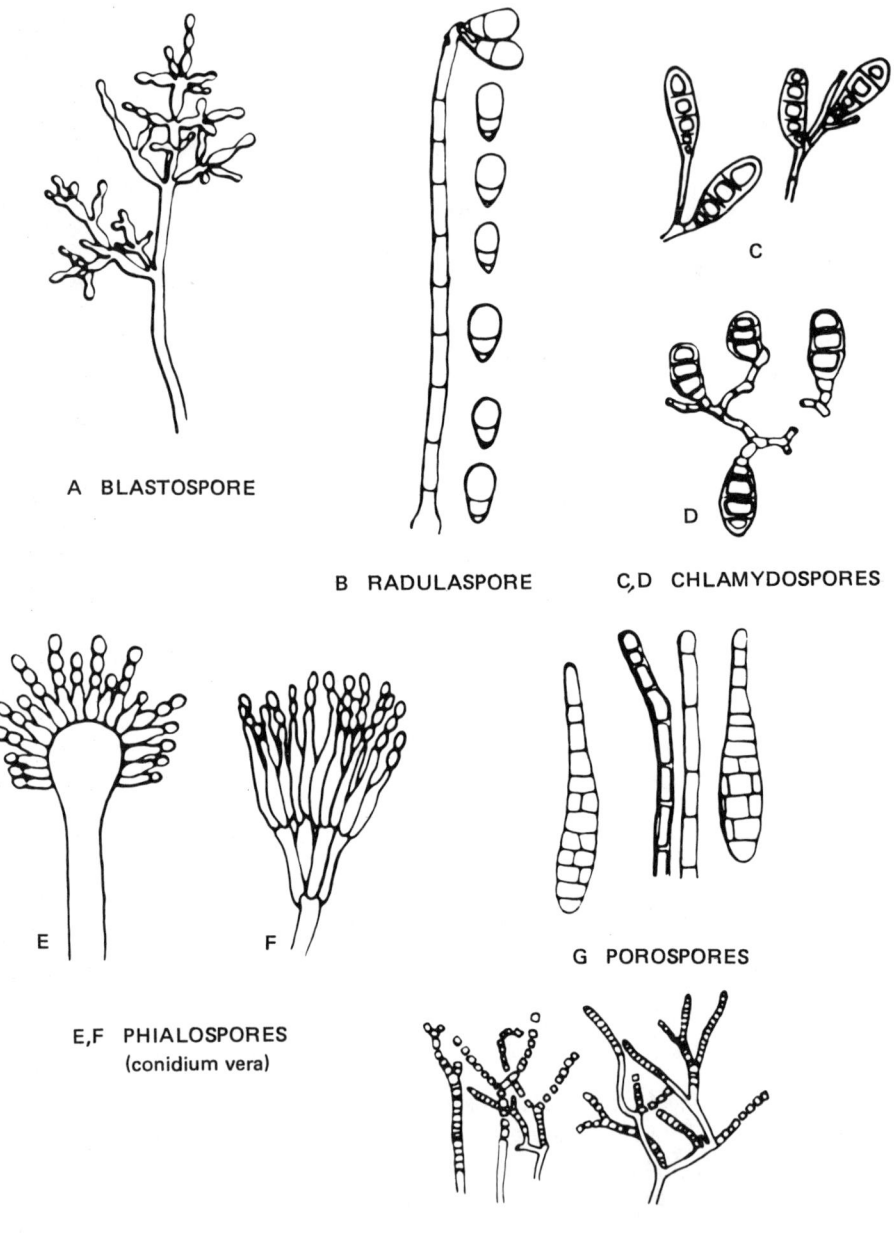

Figure 2

Differentiation in Mycelial Fungi

Sexual

Many Ascomycetes and Basidiomycetes develop macroscopic spore-bearing structures called, respectively, ascocarps and basidiocarps. A great deal of descriptive information is available concerning these structures too. In both forms, the asci or basidia are arranged in a regular manner to form a hymenium. Sterile tissues surround or support the hymenium and give the diversity of form to the entire fruiting body (Burnett, 1968). Some examples of ascocarps and basidiocarps are presented in Fig. 3.

Fungal morphogenesis has been reviewed many times, from many perspectives (Baldwin and Rusch, 1965; Jaffe, 1958; O'Day and Horgen, 1977; Smith and Berry, 1974, 1978; Smith and Galbraith, 1971; Taber, 1966; Weber and Hess, 1976). What follows is a cursory and admittedly idiosyncratic sampling of some of the general observations that have emerged from this enormous literature.

Vegetative Development

Germination of a fungal spore results in formation of a germ tube, followed by hyphal extension, branching, anastomosis, and radial growth of the colony. Early growth proceeds at an exponential rate and the mycelium remains "undifferentiated." As growth continues, anatomically and physiologically distinguishable zones are formed. Growth kinetics of apical branching and the features of normal hyphal morphogenesis have been described in detail (Bull and Trinci, 1977; Burnett and Trinci, 1979; Trinci, 1978).

In addition to the branching patterns and physiological zones apparent in the formation of a fungal colony, specialized hyphal types may be formed. These include rhizoids, haustoria, arthrospores, hyphal traps, and multihyphal structures such as sclerotia and rhizomorphs. The induction of these specialized hyphae is related to the whole problem of hyphal growth, branching, and development. See Butler (1966) for a review of these vegetative multihyphal forms.

Since form and shape are dependent on the rigid outer wall, analysis of this complex macromolecular structure is a prerequisite to understanding fungal

---

Figure 2 Examples of different spore morphologies. (A) Blastospore of *Cladosporium avellaneium* (from Hughes after de Vries, 1953, p. 588), (B) radulaspores of *Brachysporium obovatum* (after Hughes, 1953, p. 598), (C) chlamydospores of *Bactrodesmium spilomeum* (after Hughes, 1953, p. 617), (D) chlamydospores of *Trichocladium opacum* (after Hughes, 1953, p. 610), (E) phialospores from *Aspergillus niveo-glaucus* (after Raper and Fennell, 1965, p. 22), (F) phialospores from *Penicillium wortmanni* (after Raper and Thom, 1968, p. 560), (G) porospores from *Alternaria brassicae* (after Hughes, 1953, p. 636), and (H) arthrospores from *Coremiella ulmariae* (after Hughes, 1953, p. 640).

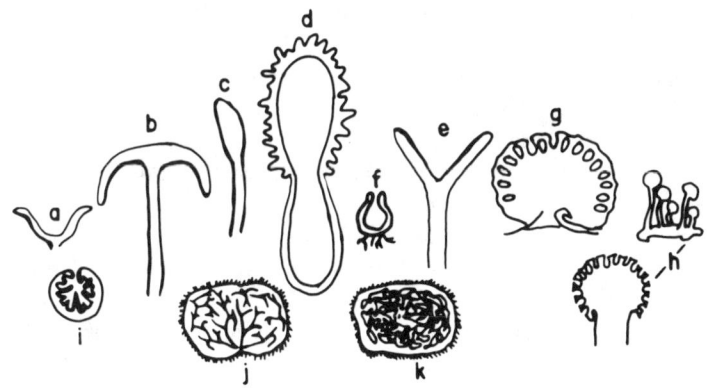

ASCOCARPS

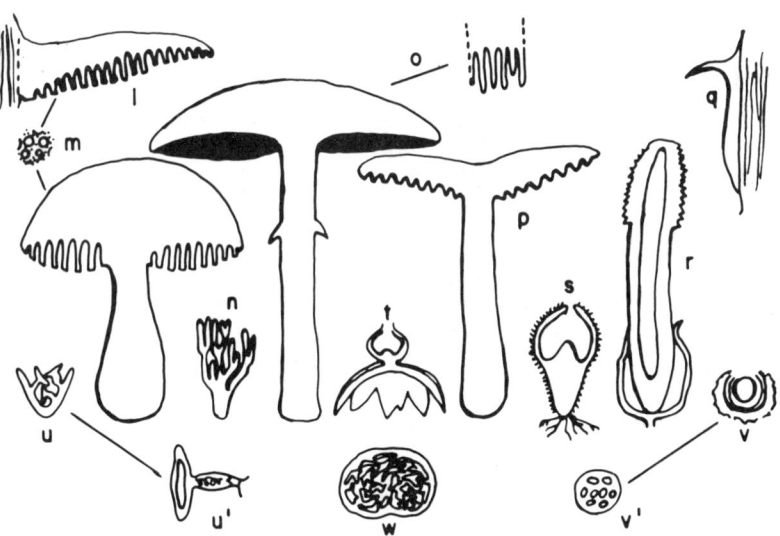

BASIDIOCARPS

Figure 3 Cross sections of selected ascocarps (a-k) and basidiocarps (l-w). (a-3) Discomycetes, (f-h) Pyrenomycetes, (i-k) Tuberales, (m) pore fungus, (n) club fungus, (o) gill fungus, (p) spine fungus, (s) puffball, (t) earthstar, (u) bird's-nest fungus with peridiolum, (u') single peridiolum, magnified, (v) *Sphaerobolus* with projectile (v') containing basidiospores, and (w) *Hymenogaster* (after Burnett, 1968, p. 106).

differentiation. Indeed, Bartnicki-Garcia felt that "morphological development of fungi may be reduced to a question of cell morphogenesis" (Bartnicki-Garcia, 1968, p. 88). He has described cell-wall chemistry in relation to fungal taxonomy and reviewed polymer synthesis, ultrastructure, and cell-wall construction. In septate mycelial fungi, chitin and β-glucans are the major cell-wall components (Bartnicki-Garcia, 1968). A more recent review of the enzymology of hyphal growth has been given by Gooday (1978). Vegetative hyphae grow by deposition of new material at the apices. Chitin synthetase is the best-characterized enzyme involved in the synthesis of hyphal walls; uridine 5'-diphosphate-N-acetylglucosamine is required for chitin formation. Very little is known about glucan synthesis. Because glucans are the major components of cell walls in many species, their biosynthesis deserves more attention.

Fungal protoplasts are an increasingly useful system for the study of cell-wall synthesis and architecture. Not all protoplasts regenerate back to their original morphological form, and regeneration can be affected by the composition of the medium. Two basic patterns for normal protoplast regeneration have been described in liquid media. In one pattern a thick wall is laid down, forming a spherical cell; hyphae subsequently form from this cell. Alternatively, the protoplasts may bulge out on one side and form a "budding chain" from which the hyphae develop. In both cases the wall-rebuilding process begins with the deposition of a microfibrillar skeleton composed of chitin or β-glucan, depending on the species. In the regeneration state of *Schizophyllum commune*, chitin and α-glucan are synthesized; the third polymer, β-glucan, is produced after regeneration is nearly completed and hyphal development has started (Peberdy, 1972, 1978).

The genetic approach to the study of vegetative development is also very useful. In the typical life cycle of the higher basidiomycetes, two functionally distinct mycelial states are found—the homokaryon and the dikaryon. The homokaryon is the form that develops after germination of a single basidiospore, and it contains one nuclear type. The dikaryon forms after fusion of genetically dissimilar homokaryons and contains two genetically different nucleii associated in pairs. Dikaryotic mycelia generally form clamp connections. The conversion of a homokaryon into a dikaryon represents a relatively simple example of hyphal differentiation and has been subjected to detailed genetic analysis. The system is controlled by a series of different genetic elements, termed incompatibility factors, which control nuclear migration, clamp connection formation, and nuclear division. Despite detailed genetic dissection, the molecular action of these incompatibility genes eludes researchers (Raper, 1966; Casselton, 1978). Raper's Chapter 8 of this volume, on controls for development and differentiation of the dikaryon in basidiomycetes, gives a detailed account of current research.

Numerous morphological mutants have been isolated in various species. Commonly the morphology of the entire colony is affected: Hyphae grow faster or slower; the colony is more diffuse or compact; aerial hyphae may be pronounced; and so on (Esser and Kuenen, 1967; Burnett, 1975). Naming these mutants poses problems in that it is often difficult for a reader to get a clear notion of the phenotype. Burnett (1975) wrote:

> Thus, for example, in *Neurospora crassa, bu-botton,* a tight colonial form, is understandable, but is more difficult to visualize *ro-ropy,* with curled hyphae, and impossible to get a clear notion of *sc-scumbo* with "scummy-like" surface growth and no conidia [p. 26].

The biochemical basis of such mutants is rarely understood; a surprising number are inherited cytoplasmically. Among the well-studied examples of extrachromosomal inheritance are *poky* in *N. crassa,* "*vegetative death*" in *Aspergillus glaucus, scenescence* and *barrage* in *Podospora anserina,* and *ragged* in *Aspergillus amstelodami* (Turner, 1978).

Reproductive Development

Reproduction is the process by which new organisms are generated. Most fungi reproduce by spore formation, although vegetative reproduction (new mycelial development from a detached hyphal cell, fragment, or sclerotium) is also possible. The question of whether all hyphae, even specialized, differentiated forms, retain the potential for vegetative reproduction has not been adequately researched. Nevertheless, the general assumption is that all nucleated fungal cells are totipotent.

Mitosis and meiosis provide the criteria for classifying spores into asexual and sexual types, respectively. The initiation of the reproductive phase requires the appropriate genes (sterile mycelial mutants have been observed in numerous species), a certain minimal period of growth, and the correct environmental signals to initiate sporulation. The physiological literature on the effects of environment on reproduction is staggering in its quantity and highly uneven in scientific quality. Hawker (1957, 1966) must be credited as the most successful collator within the field. She concluded that the conditions favoring spore formation are usually of narrower range than those favoring mycelial growth, and that asexual reproduction does not involve any great increase in metabolic activity over that of vegetative growth, whereas sexual reproduction requires a considerable intensification of such activity.

The type and amount of food supply is the most important single factor, and it has been studied more than any other variable. Glucose is usually the most suitable organic substrate, although many other carbohydrates are accepted by most fungi. Nitrogen requirements vary from simple nitrate and ammonium salts to complex organic forms. Trace element requirements tend to be higher for spore formation than for mycelial growth, and narrower ranges of temperature, moisture, pH, and aeration are also the general rule. The effect of light is complex. Some fungi sporulate equally well in darkness and light, others require light, and still others are inhibited by light. Wave lengths in the blue are most effective in stimulating reproductive photoresponses.

Sporulation is usually induced by a decrease in available nutrients. In the laboratory culture this is an inevitable outcome, since the quantity of medium is finite. In nature, food supply is often a limiting factor. Indeed, this simple principle, first formulated by Klebs (1898), remains the only general principle: Reproduction is favored by conditions that limit growth.

Modern workers have attempted to correlate morphology, external physiological triggers, and the underlying biochemical mechanisms. The most definitive studies on single environmental variables have been conducted in pure cultures where all but one variable can be held constant. Mycelia are sampled at various time intervals and the compounds in question (nucleic acids, enzymes, lipids, carbohydrates, or other metabolites) are assayed. This approach was used with great success with the water mold *B. emersonii* (Cantino and Lovett, 1964; Cantino, 1966; Lovett, 1975); subsequent literature has been replete with references to "environmental triggers," "metabolic shifts," and "developmental points of no return." Unfortunately, theorists less able than Cantino, with systems more recalcitrant than *B. emersonii*, have produced a plethora of simplistic models that are inadequate for explaining multidimensional developmental phenomena. This position was described in the introductory chapter of a monograph on developmental mycology by Wright (1978):

> The more we learn about the details of the mechanisms operative in differentiation systems, the more complex will our thoughts become as to the nature of "differentiation." The concept of a "trigger" for "differentiation" will appear naive; a hierarchical scheme of control with the gene at the top or in the centre will become meaningless. The problem is to describe the multitudinous events essential to any differentiation process and to understand the interaction of these events by defining those that are rate-limiting at particular points in time in the living organism. We cannot accomplish this goal simply by breaking open the cells at various times during development and measuring an enzyme or a messenger or a metabolite. Such correlations have marginal relevance to the dynamics of metabolism in the living cell [p. 6].

A far less sophisticated formulation of the same point was given by Massee (1906) over 70 years ago:

The researches of Klebs are especially significant in connection with pure cultures which by some are placed on a par with Caesar's wife. In the future, instead of arguing that such cultures prove incontestably the normal sequence of development of a species, we should only go so far as to say that those show what the particular fungus could do under a special sequence of conditions [p. 9].

Among mycelial species the "particular fungi" that have received the most attention include *Agaricus bisporus* (Hayes and Nair, 1975; Raper et al., 1972), *Aspergillus nidulans* (Clutterbuck, 1974; Smith and Pateman, 1977), *Coprinus radiatus* (Guerdoux, 1974), *N. crassa* (Barratt, 1974; Turian, 1966, 1974, 1978), *P. anserina* (Esser, 1974), and *S. commune* (Niederpreum and Wessles, 1969; Raper, 1966; Raper and Hoffman, 1974).

However, many other species, particularly obligate parasites and those with complex fruiting bodies, cannot even be induced to sporulate in the laboratory. Despite considerable economic incentive to develop a reproducible system of cultivation, ascocarps such as truffles and morels and many delectable mushroom basidiocarps can be collected only in nature. The best we can do is repeat the generalities of the old naturalists: Hot summers and warm, wet autumns favor mushrooms in temperate climates.

Without a doubt morphogenesis is a manifestation of differential gene expression. But how? Molecular biologists have had enormous success in cracking the genetic code and learning about the mechanisms of protein synthesis. We assume that DNA coding must be the basis of the diversity of life as well as of the unity of life. However, our understanding of the mechanics of differential gene expression in multicellular systems remains primitive, even in systems as simple as the mycelial fungi. Our scientific insights are not only limited by the kinds of questions we ask and the technical limits of our experimentation, but by the organisms we choose to study. To use a nonmycological example, although research on *Escherichia coli* can tell us a great deal about the metabolic pathways and genetic control mechanisms in *Homo sapiens,* there are also many things we cannot ever learn from *E. coli,* no matter how perspicacious our questions or how elegant our methodology. There are some things *E. coli* does not do, even with the aid of recombinant DNA technology.

The biological simplicity of *E. coli* or *Blastocladiella* that allows for meaningful experimentation at the biochemical level often excludes their use as tools in understanding the differentiation of complex multicellular systems. I do not wish to echo the sentiments of a certain type of classical biologist who sees the problems of development "in all their elegant complexity and finds himself either unable or unwilling to believe that the biochemist will ever penetrate them with his relatively simple, one-dimensional tools" (Bonner, 1974, p. 3), but neither do I wish to hold the arrogant position of certain molecular biologists

Differentiation in Mycelial Fungi

who assert that we are not asking the right questions, and that self-assembly of macromolecules (á la tobacco mosiac virus) will explain the development of everything from the human liver to the basidiocarp of *Phallus impudicus*. Development is not simply a sequential turning on and off of genes; topography is also important—genetic products must be in the right place at the right time. The relative simplicity of mycelial fungi qualifies them as good candidates for studying such complexities of eukaryotic development.

SECONDARY METABOLISM

"Secondary metabolism" is a term that was coined by plant physiologists. It was conceived and defined in opposition to "primary metabolism." A number of criteria distinguish the two metabolisms; reduced to the barest minimum, the basic difference is that primary metabolism is essential for cellular growth and life, whereas secondary metabolism is not. Primary metabolism includes all the anabolic and catabolic processes that keep an organism alive. Secondary metabolism is anabolic and dispensable. Organisms can and do lose secondary metabolic capability without dying.

Secondary metabolic pathways are unevenly distributed in nature. They are rare in animals, but common in green plants, fungi, and bacteria. Among organic chemists the study of secondary metabolites largely overlaps with the study of "natural products." Among microbiologists, synonyms for "secondary metabolite" include "shunt metabolite" (Foster, 1949), and, more recently, "idiolite" (Walker, 1974).

The "father" of fungal secondary metabolism was Raistrick, whose group at the London School of Hygiene and Tropical Medicine characterized the structures of over 200 mold metabolites (Bentley and Campbell, 1968), Bu'Lock's group at the University of Manchester has made major contributions with respect to the timing of secondary metabolite production in laboratory fermentations. Generally, primary metabolites are produced in association with growth during "trophophase," whereas the production of secondary metabolites occurs only after active growth has ceased, during the "peculiar phase" or "idiophase" (Bu'Lock et al., 1965; Bu'Lock, 1967). However, this distinction is not absolute; changes in growth conditions may alter the time at which secondary metabolism is initiated, and it may accompany growth (Aharonowitz and Demain, 1980).

Secondary metabolites constitute a chemically diverse group, but they are all produced from a few key intermediates of primary metabolism. Consequently, it is convenient to categorize secondary metabolites according to the precursors from which they arise. I have followed the scheme used by Turner (1971). The compounds discussed were selected as examples because they are unique to fungi, of economic importance, widely distributed, or otherwise of particular interest.

I - SECONDARY METABOLITES DERIVED WITHOUT ACETATE

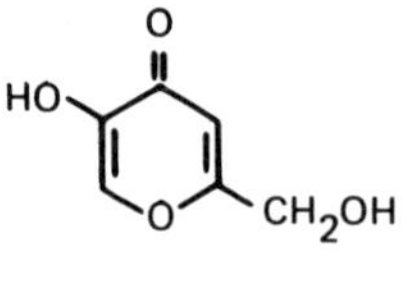

II - SECONDARY METABOLITES FROM FATTY ACIDS

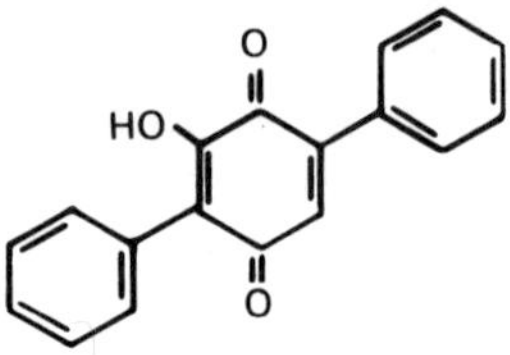

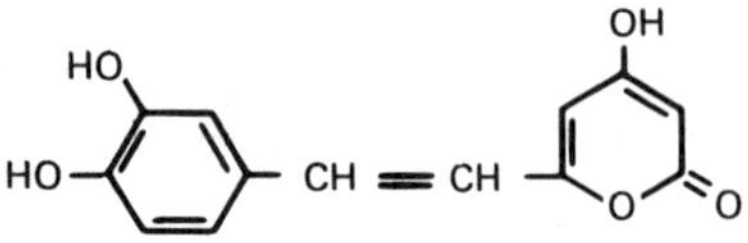

(D) $HOCH_2(C{\equiv}C)_3CH{=}CHCH_2OH$

(E) $HO_2CCH_2CH_2(C{\equiv}C)_2CH{=}CHCO_2H$

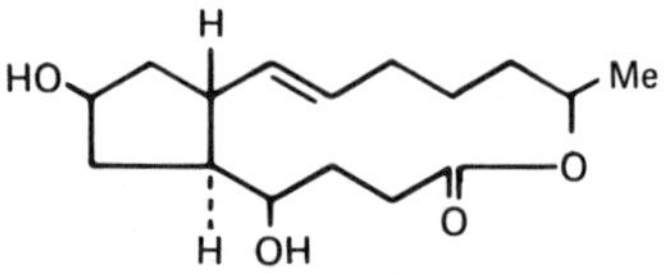

III - POLYKETIDES

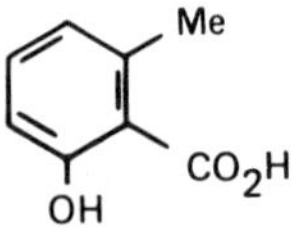

Figure 4 Examples of secondary metabolites (classification after Turner, 1971): (I-A) kojic acid, (I-B), polyporic acid, (I-C) hispidin, (II-D) and (II-E) polyacetylenes, (II-F) brefeldin A, (III-G), 6-methylsalicylic acid, (III-H), mycophenolic acid, (III-I), griseofulvin, (III-J), ochratoxin A, (III-K), aflatoxin B_1, (IV-L), trichlothecin, (IV-M), gibberellin A_3, (IV-N), ergosterol, (IV-O), β-carotene, (V-P), tenuazonic acid, (V-Q), rubratoxin B, (VI-R), psilocin, (VI-S), ergot alkaloids, (VI-T), penicillin G, and (VI-U), cephalosporin C.

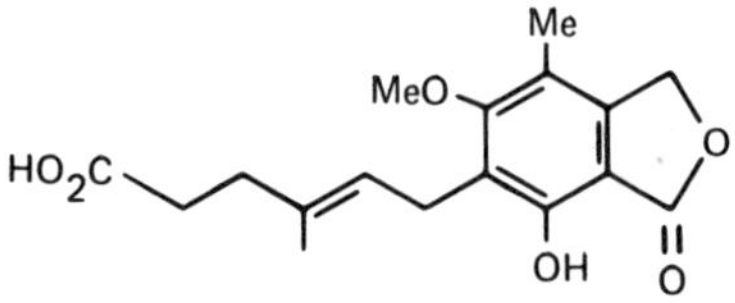

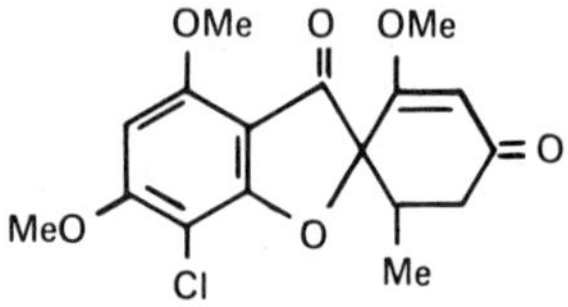

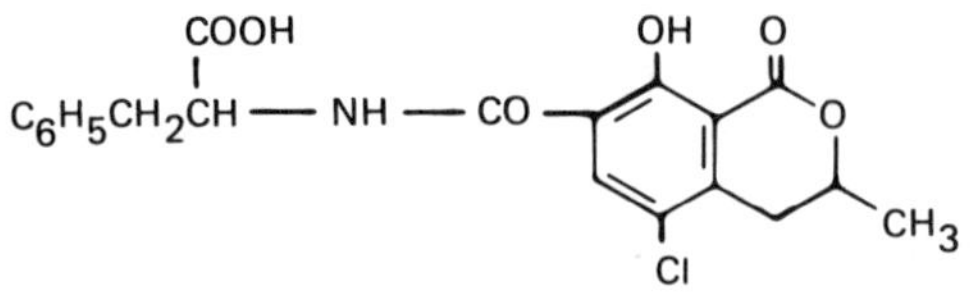

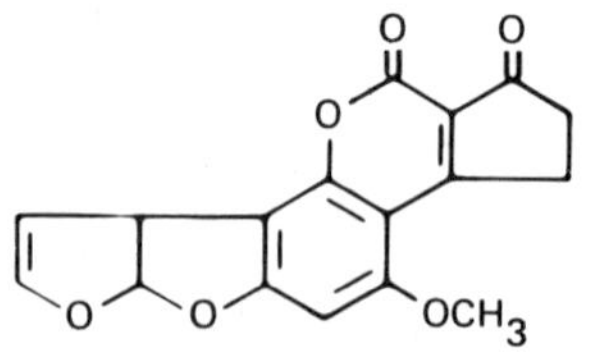

IV - TERPENES AND STEROIDS

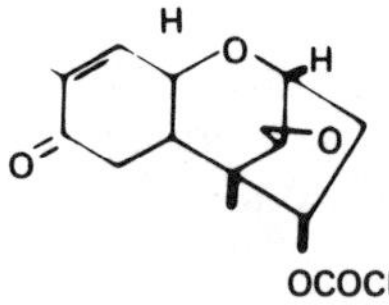

(L)

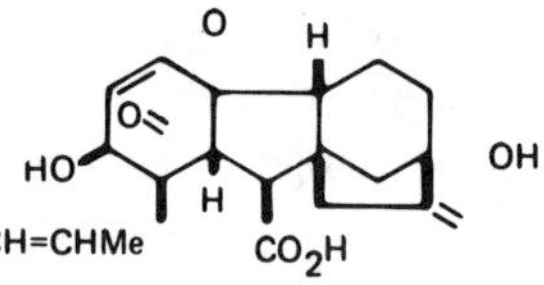

(M)

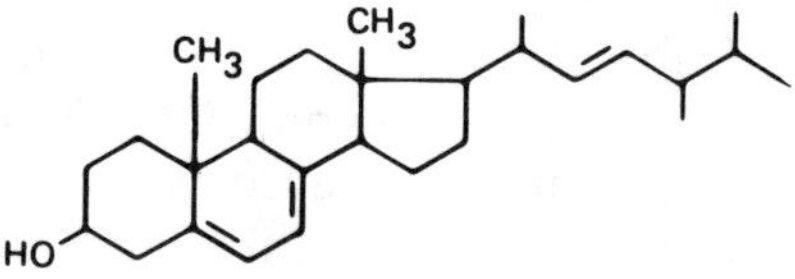

(N)

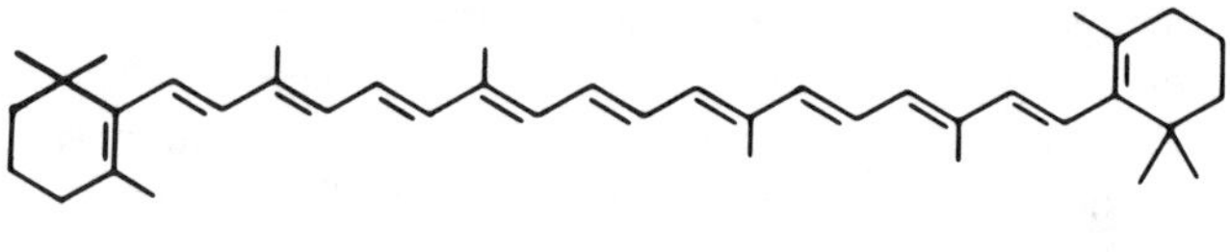

(O)

V - SECONDARY METABOLITES FROM TCA CYCLE

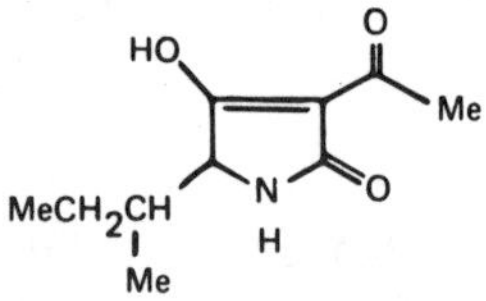

(P)

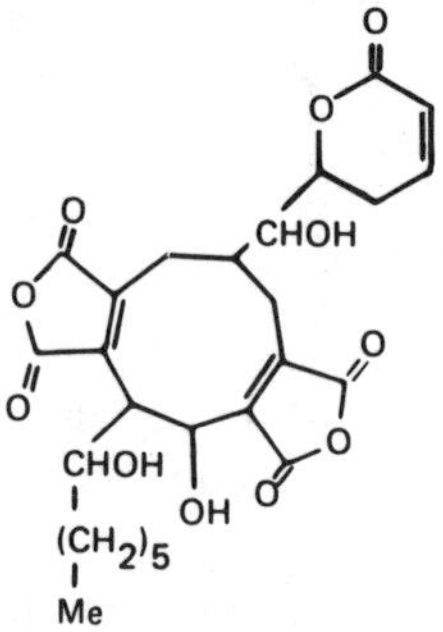

(Q)

VI - SECONDARY METABOLITES DERIVED FROM AMINO ACIDS

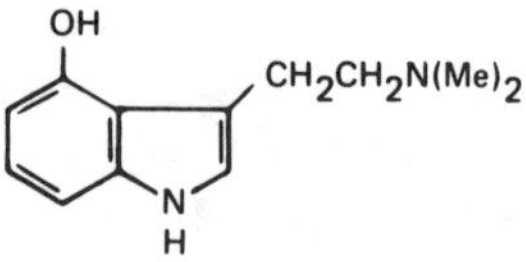

(R)

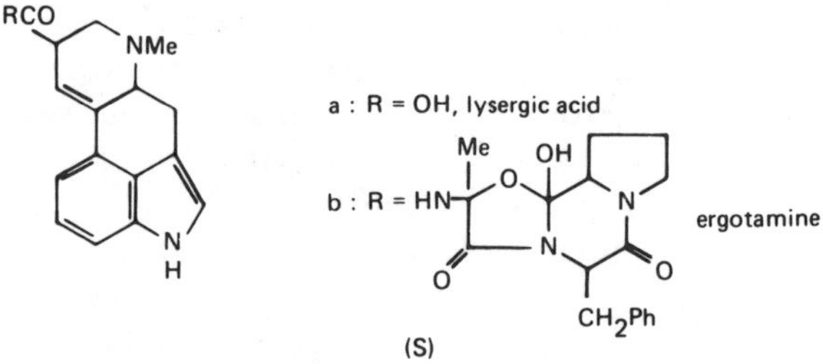

(S)

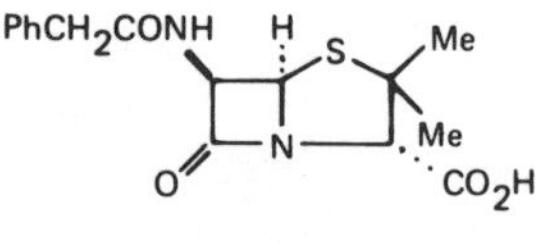

a : R = OH, lysergic acid

b : R = HN— ... ergotamine

(T)

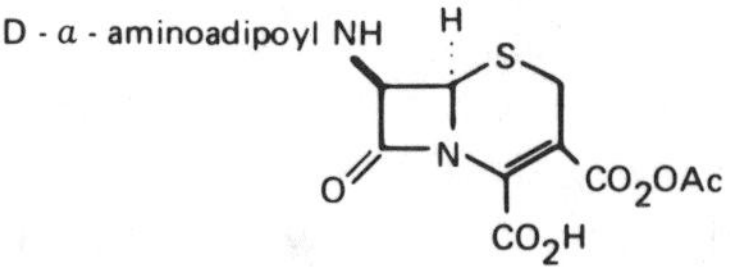

(U)

Secondary Metabolites Derived Without Acetate

This group includes compounds derived directly from glucose, aromatic amino acids, purines, or pyrimidines. The intact carbon skeleton of glucose is commonly a part of the secondary metabolites found in higher plants (glycosides) and many actinomycetes (streptomycin antibiotics), but is uncommon among the fungi. Kojic acid (Fig. 4, I-A), is an exception, and it is produced by numerous *Aspergillus* species.

The shikimic acid pathway yields a number of simple benzene derivatives. Some of these phenolics are common in higher plants and mammals. The diphenylbenzoquinones are restricted to the fungi and, with one exception, have all been isolated from basidiocarps or lichens. Polyporic acid (Fig. 5, I-B) from *Polyporus* species, is an example of this group. Hispidin, another example, (Fig. 4, I-C), is an orange pigment found in the fruiting body of *Polyporus hispidus*. A polymerization of this compound may form a "fungal lignin" responsible for toughening the fruit bodies of *P. hispidus* (Bu'Lock and Smith, 1961; Bu'Lock, 1967).

Secondary Metabolites from Fatty Acids

Along with polyketides, terpenes, and steroids, secondary metabolites from fatty acids are ultimately derived from acetate. In some cases, the functional dividing line between a secondary metabolite and the normal lipid reserves of reproductive structures and of the mycelium is not clear. Both seem to undergo substantial depletion on aging or when the carbon source is insufficient (Brennan and Losel, 1978).

The polyacetylenes are straight-chain compounds containing conjugated acetylenic systems formed by denaturation and chain shortening. Because of their characteristic and usually intense ultraviolet light spectra, they are readily detected in natural products screening programs. Over 400 polyacetylenes are known, largely from higher plants. None have been isolated from bacteria or animals. The 80 or so fungal polyacetylenes are virtually all Basidiomycete products (Packter, 1973; Turner, 1971). The compounds from *Corpinus quadifidus* (Fig. 4, II-D) and *Polyporus anthracophilus* (Fig. 4, II-E) are representative of the group. Bu'Lock (1966) felt that the noteworthy taxonomic distribution of the polyacetylenes, as detected spectroscopically, displays a more accurate picture of the normal range of biosynthetic activity than is provided by a survey based on "a casual property such as antibiotic activity" (Bu'Lock, 1966, p. 144).

Cyclopentanes are derived from fatty acids via cyclization processes. Brefeldin A (Fig. 4, II-F) has been isolated from several *Penicillium* and *Curvularia* species, as well as from *Nectria radiciola* (Turner, 1971). It is structurally similar to the prostaglandins and may arise via a similar mechanism (Bu'Lock and Clay, 1969).

Polyketides

Polyketides are formed by the condensation of an acetyl unit with malonyl units; decarboxylation occurs concomitantly. The term itself derives from an early hypothesis that such compounds come from ketones (Collie, 1907). Birch and Donovan (1953) are responsible for the modern formulation of the hypothesis. Turner (1971) classified polyketides according to the number of C_2 units which contribute to the polyketide chain. Thus he described tetraketides, pentaketides, hexaketides, etc.

The polyketide pathway is used mainly in fungi, and among the fungi is largely found among the Ascomycetes and Fungi Imperfecti. Perhaps because of the limited distribution, and perhaps because this pathway leads almost exclusively to secondary metabolites, very little is known about the detailed mechanisms of polyketide biosynthesis (Turner, 1971).

An example of a simple polyketide, formed with the loss of one oxygen atom from the tetraketide precursor, is 6-methysalicyclic acid (Fig. 4, III-G). This compound was used in the experimental verification of the acetate-malonate hypothesis (Bu'Lock et al., 1962) and in the development of one of the few cell-free systems available for polyketide-type compounds (Light, 1967).

Mycophenolic acid (Fig. 4, III-H) from *Penicillium brevi-compactum* and *Penicillium stoloniferum* is another tetraketide. Here the aromatic ring is derived from acetate, and the side chain from mevalonate. Isolated by Gosio in 1896, mycophenolic acid has the distinction of being the first antibiotic to be purified and crystallized. In addition to antibacterial activity, it also displays antiviral, antitumor, and antipsoriatic properties (Wilson, 1971).

Griseofulvin (Fig. 4, III-I), a heptaketide, was first studied as an antifungal antibiotic from *Penicillium grieseofulvin* and was later reisolated as a hyphal "curling factor" from *Botrytis allii*. Its history and biological properties have been reviewed by Brian (1960).

Ochratoxin A (Fig. 4, III-J) is a chlorine-containing pentaketide originally isolated from *Aspergillus ochraceous* and subsequently from a dozen other species of *Aspergillus* and *Penicillium*. Ochratoxin A induces kidney damage in a number of species in the laboratory; the major veterinary disease entity is a porcine nephropathy of considerable economic importance in Denmark. Epidemiological studies link ochratoxin contamination of foods with endemic Balken nephropathy, a fatal human pathology found in Bulgaria, Rumania, and Yugoslavia (Harwig, 1974; Krogh, 1977).

The best known of the mycotoxins are the family of metabolites from *Aspergillus flavus* and *A. parasiticus* known as aflatoxins. Aflatoxin B_1 (Fig. 4, III-K) is a decaketide which induces both acute and chronic toxicological effects in mammals, and is carcinogenic, mutagenic, and teratogenic in a wide range of experimental systems. Epidemiological studies show an association between estimated aflatoxin contamination of human food supplies and primary liver cancers. Facets of the aflatoxin problem have been reviewed many times (Goldblatt, 1969; Heathcote and Hibbert, 1978).

Terpenes and Steroids

The terpenes and steroids constitute a large family of natural products, many with significant biological activity, which are widely distributed in nature. They are biosynthetically derived from the C_5 "isoprene" unit (isopentenyl pyrophosphate), which is itself formed from acetate via mevalonate. The biosynthesis of isoprenoid compounds has been reviewed extensively (Bu'Lock, 1965; Packter, 1973). Some examples of this group from mycelial fungi include trichothecin (Fig. 4, IV-L), gibberellic acid (Fig. 4, IV-M), ergosterol (Fig. 4, IV-N), and β-carotene (Fig. 4, IV-O).

Trichothecin is one of a family of epoxide-containing sesquiterpenoid mycotoxins which have been found in many imperfect fungi. Trichothecenes elicit a complex clinical syndrome in animals, including vomiting, diarrhea, convulsions, skin irritations, and death, as a result have been implicated in a number of rather obscure human and veterinary syndromes (Mirocha et al., 1977; Carlton and Szczech, 1978). Biochemically, some members of this group are potent inhibitors of eukaryotic protein synthesis (Wei and McLaughlin, 1974). Their greatest recent claim-to-fame has concerned their putative use as agents of biological warfare in Laos and Cambodia. Trichothecenes have been identified in samples of "yellow rain", the substance said to have been sprayed on populations in Southeast Asia (Wade, 1981).

Gibberellic acid was first isolated from *Gibberella fujikuroi,* the perfect stage of *Fusarium moniliforme.* Gibberellins cause "bakanae" or "foolish seedling" disease in rice. Subsequently, the compounds were found endogenously in green plants. They function as important plant growth hormones, promoting shoot but not root growth, and in some species are involved with the initiation of flowering (Hansen, 1967). The biosynthesis of gibberellins in *G. fujikuroi* has been reviewed (Hedden et al., 1978).

Ergosterol is a fungal steroid. First isolated from *Claviceps purpurea,* it has since been recognized as a very common mycological metabolite; yeasts are a particularly good source. Ergosterol can be converted to vitamin D by animals in a nonenzymatic process involving ultraviolet light. The biosynthesis of ergosterol has been studied extensively, and enzymes from fungi have been used in industry to carry out transformations of unnatural sterol substrates (Bu'Lock, 1975).

An even more ubiquitous metabolite is β-carotene, named after the yellow pigment from carrots. It is found in the green tissues of higher plants, and is widespread among the fungi, where it is believed to be involved in certain reproductive processes and photoresponses. β-Carotene is a precursor to vitamin A. Commercially it is used as a coloring agent for margarine and baked goods (Ciegler, 1965; Goodwin, 1952; Isler et al., 1971; Hansen, 1967).

Secondary Metabolites from the Citric Acid Cycle

These compounds arise from the condensation of an intermediate of the citric acid cycle with an acetate-derived chain, usually a fatty acid. Both of the examples presented for this group are lesser known mycotoxins of uncertain economic importance. Tenuazonic acid (Fig. 4, V-P) is derived from acetoacetate and isoleucine, whereas the chemically more complex rubratoxin B (Fig. 4, V-Q) arises from a condensation between acetoacetate and a fatty acid, followed by cyclic dimerization (Turner, 1971). Tenuazonic acid is produced by species of *Alternaria* and by *Phoma sorghina*; rubratoxin has been isolated from *Penicillium rubrum* and *Penicillium purpurogenum*. *Phoma sorghina* has been implicated in a mycotoxicosis, onyalai, a hematological disorder prevalent among black people south of the Sahara (Rabie et al., 1975). Rubratoxin, while toxic to laboratory animals, has not been associated with any naturally occurring mycotoxicoses (Moss, 1971).

Secondary Metabolites from Amino Acids

Turner delimited this group as compounds in which the carbon skeleton is largely or wholly derived from an amino acid, with retention of the nitrogen(s). Psilocin (Fig. 4, VI-R) is a hallucinogenic, indole-containing compound from *Psilocybe aztecorum*, the Mexican magic mushroom, also called "teonanacatl." Native Indian cults utilize the hallucinogenic properties of this and related fungi in religious rituals. Contemporary use, or abuse, of the psilocin and related compounds, is hard to assess. A recent mycological work described current mycological collections, cited the pertinent ethnomycological literature, and commented a bit wistfully, "It seems that the practice of eating hallucinogenic mushrooms among the Indians is diminishing little by little. In San Pedro Nexapa . . . at present the people only know the hallucinogenic fungi as a product to sell, and it seems do not eat them anymore" (Guzman, 1978).

An even larger literature of folklore, pharmacology, mycology, and popular interest surrounds the ergot alkaloids, two of which are pictured in Fig. 4, VI-S (Stoll and Hofmann, 1965; Stadler and Stutz, 1975). Ergot refers to the alkaloids found in the sclerotia of *C. purpurea* infecting rye. Ergotamine is one of the major naturally occurring alkaloids in this group. This compound and related members of the ergot group cause profound vasoconstriction and contraction of smooth muscles. These effects were responsible for the outbreaks of gangrenous ergotism (St. Anthony's fire) which occurred sporadically during the Middle Ages after the consumption of bread made from ergot-contaminated rye flour. Both purified and complex ergot mixtures have been used by midwives, witches, and physicians to induce abortion, hasten labor, and prevent postpartum hemorrhage. Ergotamine has also been prescribed for the relief of migraine.

Lysergic acid diethylamide (LSD) is a chemically modified ergot alkaloid ("acid" in street parlance). Hoffman, the same chemist who characterized psilocin, first observed the hallucinogenic effects of LSD. Because of its so-called psychedelic properties, the compound has been used clinically to facilitate psychotherapy, and even more widely as a recreational drug.

To conclude this rapid survey of secondary metabolites, I have chosen two β-lactam antibiotics, penicillin G (Fig. 4, VI-T) and cephalosporin C (Fig. 4, VI-U). Both exhibit potent antibacterial activity and are much more effective against gram-positive bacteria than against gram-negative bacteria. The penicillin group was the first—and remains the most celebrated and the most important—class of antibiotics.

All penicillins share a basic structure in which a thiazolidine ring is attached to a β-lactam ring. Penicillins have been isolated from numerous species of penicillia, aspergilli, and other filamentous fungi. The commercial fermentation uses high-yielding mutants of *Penicillium chrysogenum*. Several hundred tons of these drugs have been administered to humans in the last 30 years, alleviating untold numbers of bacterial infections; they have been the most successful commercial pharmaceuticals. The success of the penicillins led to the development of many additional antibiotics, including the closely related cephalosporins (Demain, 1966; Meyers et al., 1976) and monobactams (Maugh, 1981).

Function

This brief introduction does not do justice to the splendid diversity and eccentric distribution of fungal secondary metabolites. Weinberg called them "biochemically bizarre" (Weinberg, 1974, p. 70), "an extraordinary bestiary of organic compounds" (Weinberg, 1970, p. 1). Bu'Lock (1961) invoked poetic imagery:

> While the enzymologist's garden is a dream of uniformity, a green meadow where the cycles of Calvin and Krebs tick round in disciplined order, the organic chemist walks in an untidy jungle of uncouthly named extractives, rainbow displays of pigments, where in every bush there lurks the mangled shape of some alkaloid, the exotic perfume of some new turpentine, or some shocking and explosive polyacetylene" [p. 294].

And what do these mangled alkaloids, explosive polyacetylenes, and the like do? Why do fungi make them? We pigeonhole them together in a class called secondary metabolites. Should we? Do they share a common function?

The function of secondary metabolism is a topic of lively debate. The various theories have been compared and contrasted many times (Bu'Lock, 1961; Demain, 1974; Haavik, 1979; Weinberg, 1971). Most modern workers no longer consider them as simple waste or breakdown products, food reserves, or evolutionary relics. A popular theory that has been presented in various modes since Foster's original notion of "shunt metabolism" (Foster, 1949) has it that secondary metabolism provides an orderly disposal of intermediates that accumulate and might turn off certain vital systems, thereby creating a situation of "unbalanced growth." Another theory that gives more emphasis to the process than the product is Bu'Lock's maintenance hypothesis, the view that the operation of secondary metabolism keeps cellular metabolism in working order when growth is not possible (Bu'Lock, 1961). From the perspective of this chapter, the most interesting hypothesis is that secondary metabolites "might function in control of the ordered sequence of events that occurs during differentiation" (Weinberg, 1971, p. 563).

Anatomic differentiation and the initiation of secondary metabolism are phase dependent and often occur simultaneously after active growth has ceased. For example, the peptide antibiotics of bacteria are usually elaborated during sporulation (Katz and Demain, 1977). The phase-specific development of selected polyketides, gibberellins, ergot alkaloids, penicillin, and cephalosporin has been reviewed by Martin and Demain (1978), and a more extended treatment of the subject is available in Luckner (1972), Luckner et al. (1977), and Bu'Lock (1975). Nevertheless, mutational studies indicate that although sporulation and secondary metabolism are generally correlated in the wild type, it is possible to isolate mutants which have lost secondary metabolism yet still sporulate, or mutants that do not sporulate yet still produce secondary metabolites (Katz and Demain, 1977; Sekiguchi and Gaucher, 1977).

The theory that secondary metabolites might serve an ecological advantage has had a few adherents (Brian, 1957), but early work emphasized antibiotics and there was little experimental support for antibiotics serving as agents of "biological warfare" in nature (Woodruff, 1980). Two developments have reopened the issue. The first comes from a field called chemical ecology or ecological biochemistry. Studies on plant-plant and plant-animal interactions show that green plant secondary metabolites demonstrate a variety of functions in defense, reproduction, and dispersal (Gilbert and Raven, 1975; Rosenthal and Janzen, 1979; Swain, 1977). According to these studies, secondary metabolism can only be understood in relationship to the interaction of organisms living, competing, and coevolving with each other.

The second development is the increasing recognition that the pharmacological activity of microbial metabolites is by no means limited to antibiotic and toxigenic reactions. Over 60 distinct pharmacological activities from

"ACTH-like and anabolic" to "ulcerative and vasodilatory" have been listed by Demain (1981). This list reflects the screening protocols used to detect biological activity. It is also anthropocentric; scientists concentrate on compounds that will have clinical, industrial, or agricultural application. Nevertheless, the inference is clear. As a class, microbial secondary metabolites display a wider range of physiological activity than has hitherto been suspected. Examination of the producing fungi in their natural habitats may reveal some interesting ecological interactions. There have been surprisingly few studies, but these few studies are provocative. Two examples will suffice. The Brazilian shrub *Baccharis megapotamica* contains significant concentrations of trichothecenes. The mycotoxins have no apparent ill effect in *B. megapotamica*, but kill tomatoes, peppers, and artichokes. In its native habitat *B. megapotamica* grows in marshes and there appear to be few competitive plants in these areas; the trichothecenes may be important in the ecology of this system (Jarvis et al., 1981).

In natural habitats, most species of mites feed largely or solely on fungi; marked differences in the food preferences of mites exist (Luxton, 1972; Sinha, 1968). Earthworms also eat molds (Cooke and Luxton, 1980). One way microbes may keep from being eaten is by the production of toxins and noxious odors or flavors (Janzen, 1977).

These putative interactions between plants and secondary metabolites, as well as between invertebrates and secondary metabolites, deserve further study. I would remind readers of a basic biological tenet which is often overlooked in discussions of secondary metabolism; Dobzhansky stated it with admirable brevity: "In biology, nothing makes sense except in the light of evolution" (Dobzhansky, 1970, p. 5). Secondary metabolism is a process conducted by living organisms. Evolutionary change occurs when some organisms are more successful than others in passing on their genes to the next generation. The widespread occurrence of secondary metabolism in fungi indicates that species possessing secondary metabolic pathways have been successful in transmitting this capability to their offspring. We are still ignorant of the selective pressures in operation. However, using analogies from green plant secondary metabolism, it would seem plausible that coevolutionary mechanisms may be highly significant. The diversity of chemical structures and biological activities displayed by fungal metabolites suggests that a diversity of ecological functions are not only plausible, but predictable.

SECONDARY METABOLISM AS DIFFERENTIATION

Differentiation is the acquisition of differences, the progressive diversification of cells in an organism. Most biologists immediately think of the acquisition of morphological differences (e.g., morphogenesis) when they think of differentiation; the acquisition of chemical differences is a less obvious facet of the process (Calam, 1979). Yet cells that manufacture secondary metabolites exhibit chemical differentiation; they may or may not concomitantly undergo changes in the composition of cells, and rearrangements of the surface configuration detectable as visible morphological changes. Secondary metabolism is merely one facet of the many possible developmental processes exhibited by differentiating cells. Morphogenesis is differential biochemistry expressed in a three-dimensional array; secondary metabolism is differential biochemistry with or without these cellular rearrangements.

There are many correlation between morphogenesis and secondary metabolism. In general, both processes start after active growth has stopped. Both processes tend to have narrower ranges of trace metal, pH, and temperature requirements than growth. Secondary metabolites occur in chemical families; morphogenesis is displayed in morphological families. Secondary metabolites come from a few simple precursors of primary metabolism; so do cell walls and other building blocks of cellular structure. In both processes we are ignorant of most of the molecular details leading up to the formation of a structure (Bennett, 1981).

Despite these many analogies, secondary metabolism is often viewed as something apart from differentiation, not as part of the differentiation processes. Perhaps this reflects the way in which scientists are trained: Chemists study chemicals; biologists study cells and organisms. Each discipline has its own vocabulary, traditions, approaches, and pet hypotheses. Within chemistry and biology there are subspecialties. Microbiologists working with antibiotics seldom think about what is going on in ecology, molecular biologists frequently ignore taxonomy, and so on. Many students of secondary metabolism are unaware that the chemicals produced by certain differentiated animal cells ("luxury molecules") provide interesting parallels to microbial secondary metabolites.

> Luxury molecules are those molecules, which, though of great utility to the intact organism, are not essential to the viability of the cells synthesizing them. Myosin or fibrinogen, for example, are not essential for the survival of muscle or liver cells. Other luxury molecules would be hemoglobin, insulin, thyroxine, and chondroitin sulfate, as well as their associated mRNA's and repressors and depressors. In contrast, there are those molecules produced by most cells which are essential to the viability of the cell synthesizing them. This category includes molecules such as the cytochromes, hexokinase, the enzymes synthesizing amino acids, the ubiquitous sRNA's and rRNA's, cholesterol, glycogen, etc. From this viewpoint, a better understanding of cell differentiation depends on learning more of how embryonic cells become committed to synthesizing one set of luxury molecules rather than another set [Holzer et al., 1969, pp. 19-20].

Holtzer's projection has been realized. Most of these luxury metabolites are proteins produced in high concentrations by specialized cells. Ovalbumin, globin, and fibroin are cases in point. These specialized molecules and the mRNA's that encode them are ideally suited to analysis by modern biochemical methods: molecular hybridization of nucleic acids, recombinant DNA technology, genetic sequencing methods, and hybridoma production of monoclonal antibodies. The discovery of intervening sequences in the genes for these "luxury molecules" was an unexpected surprise whose significance for eukaryotic development is both profound and enigmatic (Brown, 1981).

The differential activation and expression of genes are the time-honored explanation for how genetic changes occur during cellular differentiation. The modern vocabulary speaks of transcriptional, posttranscriptional, and translational control. Transcriptional control is the principle mechanism for genetic control in prokaryotes. Transcriptional control is now also firmly established for control of the genes for globin, ovalbumin, and fibroin (Brown, 1981). The mechanisms are not known, but may involve direct modification of the DNA. The very conformation of DNA into the right-handed B-DNA or the left handed Z-DNA may have a regulatory role (Nordheim et al., 1981). A major structural modification found in eukaryotic DNA is the methylation of cytosine at the $5'$ position. DNA methylation may negatively control transcription. Large differences have been observed from tissue to tissue for certain "luxury function" genes, suggesting a relationship between decreased methylation and induction of differentiation (Ehrlich and Wang, 1981). Z-DNA is more stable in the methylated state (Nordheim et al., 1981).

Without question, luxury molecules and their association with the recent revolution in the biochemical analysis of DNA and DNA function have important implications for fungal secondary metabolism and morphogenesis. But before carrying the analogy between mammalian luxury metabolites and microbial secondary metabolites too far, I must emphasize an important difference: Luxury molecules are direct genetic products (proteins), whereas secondary metabolites are not. Secondary metabolites are a diverse group of low molecular weight compounds produced by enzymes; the enzymes are the direct genetic products. Even the peptide antibiotics are not manufactured in the normal fashion off a RNA template on a ribosome, but are synthesized by a series of enzymatic reactions (Katz and Demain, 1977).

The same criteria apply to the building blocks of cellular differentiation. Chitin, cellulose, and other nonproteinaceous structural components of cell walls and cell membranes are indirect genetic products. They are produced by enzymatic interactions; as noted earlier, topography is a critical variable.

There is a long distance between the gene, its product, and the final phenotype of a differenitated cell. The morphological structure and the secondary metabolite both represent a level of complexity that our current biochemical methods of DNA analysis cannot decipher. Yet these technologies provide us with the first major probes for understanding the molecular biology of differentiation. The current state of the art is beautifully summarized by one of its most prominent practitioners:

> By these means, as assortment of genes involved in some complex process will be identified and sequenced. Their RNA and protein produces will be characterized. The molecular details of how mutants in these genes affect the process will be determined. We can imagine the paradoxical situation of having in hand a set of fully sequenced genes with completely characterized products all known to have important and interrelated developmental functions but without the slightest notion of what these products do in the cell. Can it be that we will learn all about how a group of genes is controlled in development before we gain any insight into what is being controlled? I, frankly, think the answer is yes. The undeveloped step, expressed in chemical terms, is how can we go from the sequence of a gene or the sequence of a protein to function in a living cell. I am not dismayed by this deficiency in our methodology. After all, a few years ago we could not imagine how we could ever isolate a gene [Brown, 1981, p. 674].

On that optimistic note, I challenge all lovers of molds and metabolites to tackle the ancient riddle of differentiation.

ACKNOWLEDGMENTS

My thanks to John Bu'Lock, Ronald Cape, Arnold Demain, and Eivind Lillehoj for helpful insights concerning differentiation and/or secondary metabolism; to Nancy Meadow for illustrations; and to Bill, Jack, Daniel, and Mark Bennett for their encouragement and patience. Research in this laboratory has been supported by funds from the U.S. Department of Agriculture, Cooperative Agreement 58-7B30-0-216.

REFERENCES

Aharonowitz, Y., and Demain, A. L. (1980). Thoughts on secondary metabolism. Biotechnol. Bioeng. *22* (Suppl. 1):5-9.

Ainsworth, G. C. (1971). *Ainsworth and Bisby's Dictionary of the Fungi*, 6th ed. Commonwealth Mycological Institute, Kew, Surrey.

Ainsworth, G. C. (1976). *Introduction to the History of Mycology*. Cambridge University, Cambridge.

Alexopoulos, C. J. (1962). *Introductory Mycology*, 2nd ed. Wiley, New York.

Alexopoulos, C. J. (1966). Morphogenesis in the Myxomycetes. *The Fungi*, Vol. 2, G. C. Ainsworth and A. S. Sussman (Eds.). Academic, New York, pp. 211-233.
Ashworth, J. M. (1971). Cell development in the cellular slime mould *Dictyostelium discoideum*. Symp. Soc. Exp. Biol. *25*:27-49.
Baldwin, H. H., and Rusch, H. P. (1965). The chemistry of differentiation in lower organisms. Annu. Rev. Biochem. *34*:565-594.
Barratt, R. W. (1974). *Neurospora crassa*. In *Handbook of Genetics*, Vol. 1, R. C. King (Ed.). Plenum, New York, pp. 511-529.
Bartnicki-Garcia, S. (1968). Cell wall chemistry, morphogenesis, and taxonomy of fungi. Annu. Rev. Microbiol. *22*:87-108.
Bennett, J. W. (1981). Genetic perspectives on polyketides, productivity, parasexuality, protoplasts, and plasmids. *Advances in Biotechnology III. Fermentation Products* C. Vezina and K. Singh (Eds.). Pergamon, Toronto, pp. 409-415.
Bentley, R., and Campbell, I. M. (1968). Secondary metabolism of fungi. In *Comprehensive Biochemistry, Metabolism of Cyclic Compounds*, M. Florkin and E. H. Stotz (Eds.). Elsevier, Amsterdam, pp. 415-487.
Birch, A. J., and Donovan, F. W. (1953). Studies in relation to biosynthesis I. Some possible routes to derivatives or orcinol and phloroglucinol. Aus. J. Chem. *6*:361-378.
Bonner, J. T. (1974). *On Development*. Harvard University, Cambridge.
Brennan, P. J., and Losel, D. M. (1978). Physiology of fungal lipids: Selected topics. Adv. Microb. Physiol. *17*:47-179.
Brian, P. W. (1957). The ecological significance of antibiotic production. Soc. Gen. Microbiol. Symp. *7*:168-188.
Brian, P. W. (1960). Griseofulvin. Trans. Br. Mycol. Soc. *43*:1-13.
Brown, D. T. (1981). Gene expression in eukaryotes. Science *211*:667-674.
Bull, A. T., and Trinci, A. P. J. (1977). The physiology and metabolic control of fungal growth. Adv. Microb. Physiol. *15*:1-84.
Bu'Lock, J. D. (1961). Intermediary metabolism and antibiotic synthesis. Annu. Rev. Appl. Microbiol. *3*:293-342.
Bu'Lock, J. D. (1965). *The Biosynthesis of Natural Products*. McGraw-Hill, London.
Bu'Lock, J. D. (1966). Biosynthesis of polyacetylenes in fungi. In *Biosynthesis of Antibiotics*, Vol. 1, J. F. Snell (Ed.). Academic, New York, pp. 141-157.
Bu'Lock, J. D. (1967). *Essays in Biosynthesis and Microbial Development*. Wiley, New York.
Bu'Lock, J. D. (1975). Secondary metabolism in fungi and its relationship to growth and development. In *The Filamentous Fungi*, Vol. 1, J. E. Smith and Dr. R. Berry (Eds.). Wiley, New York, pp. 33-58.
Bu'Lock, J. D., and Clay, P. T. (1969). Fatty acid cyclization in the biosynthesis of brefeldin-A: A new route to some fungal metabolites. *1969*:237-238.
Bu'Lock, J. D., and Smith, H. G. (1961). A fungus pigment of novel type and the nature of fungus "lignin." Experientia *17*:553-554.

Bu'Lock, J. D., Hamilton, D., Hulme, M. A., Powell, A. J., Smalley, H. M., Shepherd, D., and Smith, G. N. (1965). Metabolic development and secondary biosynthesis in *Penicillium urticae*. Can. J. Microbiol. *11*:765-778.

Bu'Lock, J. D., Smalley, H. M., and Smith, G. N. (1962). Malonate as a biosynthetic intermediate in *Penicillium urticae*. J. Biol. Chem. *237*:1778-1780.

Burnett, J. H. (1968). *Fundamentals of Mycology*. Edward Arnold, London.

Burnett, J. H. (1975). *Mycogenetics. An Introduction to the General Genetics of Fungi*. Wiley, New York.

Burnett, J. H., and Trinci, A. P. J. (Eds.) (1979). *Fungal Walls and Hyphal Growth, Papers from a Symposium. British Mycological Society Symposium 2*. Cambridge University, New York.

Butler, G. M. (1966). Vegetative structures. In *The Fungi: An Advanced Treatise*, G. C. Ainsworth and A. S. Sussman (Eds.). Academic, New York, pp. 83-112.

Calam, C. T. (1979). Secondary metabolism as an expression of microbial growth and development. Folia Microbiol. *24*:276-285.

Cantino, E. C. (1966). Morphogenesis in aquatic fungi. In *The Fungi*, Vol. 2, G. C. Ainsworth and A. S. Sussman (Eds.). Academic, New York, pp. 283-337.

Cantino, E. C., and Lovett, J. S. (1964). Non-filamentous aquatic fungi: Model systems for biochemical studies of morphological differentiation. Adv. Morphog. *3*:33-93.

Carlton, W. W., and Szczech, G. M. (1978). Mycotoxicosis in laboratory animals. Moude. In *Mycyotoxic Fungi, Mycotoxins, Mycotoxicoses*, Vol. 2, T. D. Wyllie and L. G. Morehouse (Eds.). Marcel Dekker, New York, pp. 373-406.

Casselton, L. A. (1978). Dikaryon formation in higher basidiomycetes. In *The Filamentous Fungi*, Vol. 3, J. E. Smith and D. R. Berry (Eds.). Wiley, New York, pp. 275-297.

Ciegler, A. (1965). Microbial carotenogenesis. Adv. Appl. Microbiol. *7*:1-34.

Clutterbuck, A. J. (1974). *Aspergillus nidulans*. In *Handbook of Genetics*, Vol. 1, R. C. King (Ed.). Plenum, New York.

Cole, G. T., and Kendrick, B. (1981). *Biology of Conidial Fungi*, Vols. 1 and 2. Academic, New York.

Cole, G. T., and Samson, R. A. (1979). *Patterns of Development in Conidial Fungi*. Pitmans, London.

Collie, J. N. (1907). Derivatives of multiple ketene group. Chem. Soc. J. Trans. *91*:1806-1813.

Cooke, A., and Luxton, M. (1980). Effect of microbes on food selection by *Lubricus terrestris*. Rev. Ecol. Biol. Soc. *17*:365-370.

Demain, A. L. (1966). Biosynthesis of penicillins and cephalosporins. In *Biosynthesis of Antibiotics*, Vol. 1, J. F. Snell (Ed.). Academic, New York, pp. 29-94.

Demain, A. L. (1974). How do antibiotic-producing microorganisms avoid suicide? Ann. N. Y. Acad. Sci. *235*:601-612.

Demain, A. L. (1981). Industrial microbiology. Science *214*:987-995.

Dobzhansky, T. (1970). *Genetics of the Evolutionary Process*. Columbia University, New York.

Ehrlich, M., and Wang, Y.-H. (1981). 5-Methylcytosine in eukaryotic DNA. Science *212*:1350-1357.

Esser, K. (1974). *Podospora anserina*. In *Handbook of Genetics*, Vol. 1, R. C. King (Ed.). Plenum, New York, pp. 531-551.

Esser, K., and Kuenen, R. (1967). *Genetics of Fungi* (translated by E. Steiner). Springer-Verlag, New York.

Foster, J. W. (1949). *Chemical Activities of the Fungi*. Academic, New York.

Gilbert, L. E., and Raven, P. H. (Eds.) (1975). *Coevolution of Animals and Plants*. University of Texas, Austin.

Goldblatt, L. A. (Ed.) (1969). *Aflatoxin. Scientific Background, Control, and Implications*. Academic, New York.

Gooday, G. W. (1978). The enzymology of hyphal growth. In *The Filamentous Fungi*, Vol. 3, J. E. Smith and D. R. Berry (Eds.). Wiley, New York, pp. 51-77.

Goodwin, T. W. (1952). *The Comparative Biochemistry of the Carotenoids*. Chapman and Hall, London.

Gregg, J. H. (1966). Organization and synthesis in the cellular slime molds. In *The Fungi*, Vol. 2, G. C. Ainsworth and A. S. Sussman (Eds.). Academic, New York, pp. 235-281.

Gross, P. R. (1968). Biochemistry of differentiation. Annu. Rev. Biochem. *37*: 631-660.

Guerdoux, J. L. (1974). *Coprinus*. In *Handbook of Genetics*, Vol. 1, R. C. King, (Ed.). Plenum, New York, pp. 627-636.

Guzman, G. (1978). Variation, distribution, ethnomycological data and relationships of *Psilocybe aztecorum*, a Mexican hallucinogenic mushroom. Mycologia *70*:385-396.

Haavik, H. I. (1979). On the physiological meaning of secondary metabolism. Folia Microbiol. *24*:365-367.

Hansen, A. M. (1967). Microbial production of pigments and vitamins. In *Microbial Technology*, H. J. Peppler (Ed.). Reinhold, Amsterdam, pp. 222-250.

Harwig, J. (1974). Ochratoxin A and related metabolites. In *Mycotoxins*, I. F. H. Purchase (Ed.). Elsevier, Amsterdam, pp. 345-368.

Hawker, L. E. (1957). *The Physiology of Reproduction in Fungi*. Cambridge University, Cambridge.

Hawker, L. E. (1966). *Fungi, an Introduction*. Hutchinson, London.

Hayes, W. A., and Nair, N. G. (1975). The cultivation of *Agaricus bisporus* and other edible mushrooms. In *The Filamentous Fungi*, Vol. 1, J. E. Smith and D. R. Berry (Eds.). Wiley, New York, pp. 212-248.

Heathcote, J. G., and Hibbert, J. R. (1978). *Aflatoxins: Chemical and Biological Aspects*. Elsevier, Amsterdam.

Hedden, P., Macmillan, J., and Phinney, B. O. (1978). The metabolism of the gibberellins. Annu. Rev. Plant Physiol. *29*:149-192.

Holtzer, H., Bishcoff, R., Chacko, S. (1969). Activities of the cell surface during myogenesis and chondrogenesis. In *Cellular Recognition Developmental*

Immunology Workshop, Richard T. Smith and Robert A. Good (Eds.). Appleton-Century-Crofts, New York, pp. 19-25.
Hughes, S. J. (1953). Conidiophores, conidia, and classification. Can. J. Bot. *31*:577-659.
Isler, O., Gutmann, H., and Solms, U. (Eds.) (1971). *Carotenoids.* Halsted, New York.
Jaffe, L. F. (1958). Morphogenesis in lower plants. Annu. Rev. Plant. Physiol. *9*:359-383.
Janzen, D. H. (1977). Why fruits rot, seeds mold, and meat spoils. Am. Nat. *111*:691-713.
Jarvis, B. B., Midiwo, J. O., Tuthill, D., and Bean, G. A. (1981). Interaction between the antibiotic trichothecenes and the higher plant *Baccharis megapotamica.* Science *214*:460-462.
Katz, E., and Demain, A. L. (1977). The peptide antibiotics of Bacillus: Chemistry, biogenesis, and possible functions. Bacteriol. Rev. *41*:449-474.
Kendrick, B. (1971). *Taxonomy of Fungi Imperfecti.* University of Toronto, Toronto.
King, R. D. (1972). *A Dictionary of Genetics.* Oxford University, New York.
Klebs, G. (1898). Zur Physiologie der Pflanzen einiger Pilze. Jahrb. Wiss. Bot. *32*:1-70.
Krogh, P. (1977). Ochratoxins. pp. 486-498. In *Mycotoxins in Human and Animal Health,* J. V. Rodricks, C. W. Hesseltine, and M. A. Mehlman (Eds.). Pathotox, Park Forest, Ill.
Light, R. J. (1967). The biosynthesis of 6-methylsalicylic acid. J. Biol. Chem. *242*:1880-1886.
Lovett, J. S. (1975). Growth and differentiation in the water mold *Blastocladiella emersonii*: Cytodifferentiation and the role of ribonucleic acid and protein synthesis. Bacteriol. Rev. *39*:345-404.
Luckner, M. (1972). *Secondary Metabolism in Plants and Animals.* Academic, New York.
Luckner, M., Nover, L., and Bohm, H. (1977). *Secondary Metabolism and Cell Differentiation.* Springer-Verlag, Heidelberg.
Luxton, M. (1972). Studies on the oribatid mites of a Danish beech wood soil. Pedobiologia *12*:434-463.
Martin, J. F., and Demain, A. L. (1978). Fungal development and metabolite formation. In *The Filamentous Fungi,* Vol. 3, J. E. Smith and D. R. Berry (Eds.). Wiley, New York, pp. 426-450.
Massee, G. (1906). *Textbook of Fungi.* Duckworth, London.
Maugh, T. H. (1981). A new wave of antibiotics builds. Science *214*:1225-1228.
Meyers, F. H., Jawets, E., and Goldfien, A. (1976). *Review of Medical Pharmacology.* Lange Medical Publications, Los Altos, Calif.
Mirocha, C. J., Pathres, S. V., and Christensen, C. M. (1977). Chemistry of *Fusarium* and *Stachybotrys.* In *Mycotoxic Fungi, Mycotoxins, Mycotoxicoses, An Encyclopedic Handbook,* Vol. 1, T. D. Wyllie and L. G. Morehouse (Eds.). Marcel Dekker, New York, pp. 365-420.

Moss, M. O. (1971). The rubratoxins, toxic metabolites of *Penicillium rubrum*. In *Microbial Toxins*, Vol. 6, A. Ciegler, S. Kadis and S. J. Ajl (Eds.). Academic, New York, pp. 381-407.

Needham, J. (1950). *Biochemistry and Morphogenesis*. Cambridge Univeristy, Cambridge.

Niederpreum, D. J., and Wessles, J. C. H. (1969). Cytodifferentiation and morphogenesis in *Schizophyllum commune*. Bacteriol. Rev. *33*:505-535.

Nordheim, A., Pardou, M. L., Lafer, E. M., Moller, A., Stollar, D. D., and Rich, A. (1981). Antibodies to left-handed Z-DNA bind to interband regions of *Drosophila* polytene chromsomes. Nature *294*:417-422.

O'Day, D. H., and Horgen, P. A. (Eds.). (1977). *Eucaryotic Microbes as Model Developmental Systems*. Marcel Dekker, New York.

Packter, N. M. (1973). *Biosynthesis of Acetate-Derived Compounds*. Wiley, London.

Peberdy, J. F. (1972). Protoplasts from fungi. Sci. Prog. *60*:73-86.

Peberdy, J. F. (1978). Protoplasts and their development. In *The Filamentous Fungi*, Vol. 3, J. E. Smith and D. R. Berry (Eds.). Wiley, New York, pp. 119-131.

Rabie, C. J., Van Rensburg, S. J., VanDerWatt, J. J., and Lubben, A. (1975). Onyalae—The possible involvement of a mycotoxin produced by *Phoma sorghina* in the aetiology. S. Afr. Med. J. *59*:1647-1650.

Raper, C. A., Raper, J. R., and Miller, R. D. (1972). Genetic analysis of the life cycle of *Agaricus bisporus*. Mycologia *64*:1088-1117.

Raper, J. R. (1966). *Genetics of Sexuality in Higher Fungi*. Ronald Press, New York.

Raper, J. R., and Hoffman, R. M. (1974). *Schizophyllum commune*. In *Handbook of Genetics*, Vol. 1, R. C. King (Ed.). Plenum, New York, pp. 597-626.

Raper, K. B., and Fennell, D. I. (1965). *The Genus Aspergillus*. Williams and Wilkins, Baltimore.

Raper, K. B., and Thom, C. (1968). *A Manual of the Penicillia*. Hafner, New York.

Rosenthal, G. A., and Janzen, D. H. (Eds.) (1979). *Herbivores. Their Interaction with Secondary Plant Metabolites*. Academic, New York.

Rutter, W. J., Pictet, R. L., and Morris, P. W. (1973). Toward molecular mechanisms of developmental processes. Annu. Rev. Biochem. *42*:601-646.

Sekiguchi, J., and Gaucher, M. (1977). Conidiogenesis and secondary metabolism in *Penicillium urticae*. Appl. Environ. Microbiol. *33*:147-158.

Sinha, R. N. (1968). Adaptive significance of mycophagy in stored-product arthropoda. Evolution *22*:785-798.

Smith, J. E., and Berry, D. R. (1974). *An Introduction to the Biochemistry of Fungal Development*. Academic, London.

Smith, J. E., and Berry, D. R. (Eds.) (1978). *The Filamentous Fungi. Developmental Mycology*, Vol. 3, Wiley, New York.

Smith, J. E., and Galbraith, J. C. (1971). Biochemical and physiological aspects of differentiation in the fungi. Adv. Microb. Phys. *5*:45-134.

Smith, J. E., and Pateman, J. A. (1977). *Genetics and Physiology of Aspergillus.* Academic, London.

Stadler, P. A., and Stutz, P. (1975). The alkaloids. Chem. Physiol. *15*:1-40.

Steward, F. C., and Mohan Ram, H. Y. (1961). Determining factors in cell growth: Some implications for morphogenesis in plants. Adv. Morphog. *1*: 189-265.

Stoll, A., and Hofmann, A. (1965). The alkaloids. Chem. Physiol. *8*:725-783.

Street, H. E. (1976). Experimental embryogenesis—The totipotency of cultured plant cells. In *The Developmental Biology of Plants and Animals*, C. F. Graham and P. F. Wareing (Eds.). Saunders, Philadelphia, pp. 73-95.

Sussman, M., and Brackenbury, R. (1976). Biochemical and molecular-genetic aspects of cellular slime mold development. Annu. Rev. Plant Physiol. *27*: 229-265.

Swain, T. (1977). Secondary compounds as protective agents. Annu. Rev. Plant Physiol. *28*:479-501.

Taber, W. A. (1966). Morphogenesis in basidiomycetes. In *The Fungi*, Vol. 2, G. C. Ainsworth and A. S. Sussman (Eds.). Academic, New York, pp. 387-412.

Trinci, A. P. J. (1978). The duplication cycle and vegetative development in moulds. In *The Filamentous Fungi*, J. E. Smith and D. R. Berry (Eds.). Wiley, New York, pp. 132-163.

Tubaki, K. (1966). Sporulating structures in Fungi Imperfecti. In *The Fungi*, Vol. 2, G. C. Ainsworth and A. S. Sussman (Eds.). Academic, New Yokr, pp. 113-131.

Turian, G. (1966). Morphogenesis and ascomycetes. In *The Fungi*, Vol. 2, G. C. Ainsworth and A. S. Sussman (Eds.). Academic, New York, pp. 339-385.

Turian, G. (1974). Sporogenesis in fungi. Annu. Rev. Phytopathol. *12*:129-137.

Turian, G. (1978). Sexual morphogenesis in the Ascomycetes. In *The Filamentous Fungi*, Vol. 3, J. E. Smith and D. R. Berry (Eds.). Wiley, New York, pp. 315-333.

Turner, G. (1978). Cytoplasmic inheritance and senescence. In *The Filamentous Fungi*, Vol. 3, J. E. Smith and D. R. Berry (Eds.). Wiley, New York, pp. 406-425.

Turner, W. B. (1971). *Fungal Metabolites.* Academic, London.

Wade, N. (1981). Yellow rain and the cloud of chemical war. Science *214*:1008-1009.

Walker, J. B. (1974). Biosynthesis of the monoguanidinated inositol moiety of bluensomycin, a possible evolutionary precursor of streptomycin. J. Biol. Chem. *249*:2397-2404.

Weber, D. J., and Hess, W. M. (Ed.) (1976). *The Fungal Spore, Form and Function.* Wiley, New York.

Webster's Third New International Dictionary of the English Language, Unabridged. Merriam, Springfield, Mass.

Wei, D., and McLaughlin, C. S. (1974). Structure-function relationship in the 12, 13-epoxytrichothecenes, novel inhibitors of protein synthesis. Biochem. Biophys. Res. Commun. *57*:838-844.

Weinberg, E. D. (1970). Biosynthesis of secondary metabolites: Roles of trace metals. Adv. Microb. Physiol. *4*:1-44.

Weinberg, E. D. (1971). Secondary metabolism: Raison d'être. Perspect. Biol. Med. *14*:565-577.

Weinberg, E. D. (1974). Secondary metabolism: Control by temperature and inorganic phosphate. Dev. Ind. Microb. *15*:70-81.

Wilson, B. J. (1971). Miscellaneous *Penicillium* toxins. In *Microbial Toxins*, Vol. 6, A. Ciegler, S. Kadis, and S. J. Ajl (Eds.). Academic, New York, pp. 459-521.

Woodruff, H. B. (1980). Natural products from microorganisms. Science *208*: 1225-1229.

Wright, B. E. (1978). Concepts of differentiation. In *The Filamentous Fungi*, Vol. 3, J. E. Smith and D. R. Berry (Eds.). Wiley, New York, pp. 1-7.

Biological Aspects of Secondary Metabolism

2
Regulation of Secondary Metabolism and Keys to Its Manipulation

Stephen W. Drew* and David A. Wallis / Virginia Polytechnic Institute and State University, Blacksburg, Virginia

INTRODUCTION

Two broad categories of metabolism exist in microorganisms and plants restricted to certain taxonomic groups. Primary metabolism involves catabolic, amphibolic, and anabolic pathways to build low molecular weight intermediates, provide the energy for chemical reaction, and assemble the biosynthetic precursors into the essential macromolecules of cellular structure and function. Life cannot exist without primary metabolism, nor could it exist in a competitive environment without a very fine balancing of the efficiencies of the functions of primary metabolism, that is, metabolic regulation. Certain taxonomic groups, including fungi and actinomycetes, also possess metabolic pathways for the biosynthesis of compounds that seem to serve no essential role in the growth of the producing organism. These secondary pathways lead to secondary metabolites, sometimes termed idiolites. Their synthesis depends upon precursors and energy generated through primary metabolism and, as such, primary and secondary metabolism are intimately related in complex ways.

The term "idiolite" arose from observations of batch cultivation of secondary metabolite-producing organisms. Synthesis of secondary metabolites in batch culture often occurs during a distinct production phase (idiophase) that may be quite separate from the growth phase (trophophase). While it seems clear that secondary metabolites do not play a primary role in the growth of the producing organism, they probably confer survival advantages in nature. Some secondary metabolite formation may allow the organism to compete more effectively with other life forms. Although secondary metabolites appear to be nonessential for growth of the producing organism, some of them have a negative impact on the producing organism when exposed to it during a rapid growth

---

Present affiliation: Merck and Company, Inc., Rahway, New Jersey

period. Regulation of secondary metabolism, in these cases by delaying synthesis until after growth, may allow the organism to gain a selective advantage without the disadvantage of autotoxic effects.

Most secondary metabolites are produced as families of closely related compounds. Their chemical structure and activities span a wide range of possibilities, including antibiotics, ergot alkaloids, naphthalenes, nucleosides, certain peptides (often unusual in structure), phenazines, quinolines, terpenoids, and certain complex plant growth factors. Since relatively few microbial types produce the majority of secondary metabolite families [a single strain of *Micromonospora* produces 48 different aminocyclitol antibiotics (Berdy, 1974)], very different secondary metabolites can be produced by a single microbial type [*Streptomyces griseus* and *Bacillus subtilis* each produce more than 50 different antibiotics.] Some of these compounds are of major commercial importance. Production of antibiotics on a global basis reached nearly 25,000 metric tons in 1980, with the majority of tonnage accounted for by the β-lactam antibiotics. The production of economically important metabolites, such as antibiotics, by microbial fermentation is one of the major activities of the biological process industry.

Much of our knowledge of secondary metabolism and its regulation comes from the study of commercially important microorganisms. Another major source of insight has been the studies of cultures which produce toxic secondary metabolites. The penultimate goal of much of this study by microbiologists and biochemical engineers has been to manipulate the production of secondary metabolites. Since the regulation of secondary metabolic pathways is interrelated in complex ways to primary metabolic regulation, adequate understanding of these relationships is necessary for optimal design of secondary metabolite fermentations. The orientation of this chapter is toward understanding these relationships.

Both genetic mutation and manipulation of the culture environment are the primary means of influencing the rate and extent of secondary metabolite production by affecting the culture's physiological state. Malek (1976) defined "physiological state" as "a genetically defined set of metabolic activities of cultures and their integrated physiological unity, with their components heirarchically structured according to their significance in metabolic and genetic processes, with a clear dependence on the history of individual cells and populations." In other words, any gentic or environmental manipulation to enhance secondary metabolite production will affect the physiological state of the culture. Since some of these effects are undoubtedly undesirable, we seek manipulations which are as specific as possible in achieving the desired metabolic alteration. The relationship between primary and secondary metabolism will be presented in the light of this quest for very specific manipulations.

Each relationship between primary and secondary metabolism will be discussed as it was discovered, as a relationship between culture condition and synthesis of a particular compound. It should be noted that although each relationship is discussed individually, several regulatory mechanisms are often involved in the regulation of a single biosynthetic pathway. For example, the synthesis of cephamycins is subject to control by primary precursor supply, high levels of phosphate, as well as carbon and nitrogen catabolite regulation. Since broad classes of metabolic control are manifest in diverse microbial groups, observations taken from bacterial and actinomycete systems may be generally applicable to fungal systems, within the limitations of their respective biochemistries. We develop examples of metabolic control in specific biochemical pathways, yet one must account for differences between prokaryotic and eukaryotic metabolism in the specific cross-application of the control theory.

TROPHOPHASE-IDIOPHASE RELATIONSHIP

One of the first general observations made concerning the formation of secondary metabolites was the presence of a distinct production phase (idiophase) which was apparently dissociated from the active growth phase (trophophase). Secondary metabolism in batch fermentation is often not expressed until the growth phase approaches completion. This trophophase-idiophase relationship has been observed to occur in a large number of secondary metabolite fermentations (Demain, 1972). Although the separation of trophophase and idiophase is often quite distinct in unicellular bacteria, the distinction between phases is less clear in filamentous microorganisms such as the actinomycetes and fungi. The dry cell weight of filamentous microorganisms may increase long after the nucleic acid content of a culture has reached a stationary point. When plotted against dry cell weight, secondary metabolism appears to occur during the growth phase, whereas its comparison with nucleic acid content may show a clear dissociation of the growth (increase in nucleic acid) and production (increase in secondary metabolite) phases.

In some cases, secondary metabolites fail to appear during rapid growth because the enzymes responsible for their biosynthesis are repressed. One example is the production of gramicidin S, which takes place late in the fermentation cycle during batch cultivation of *Bacillus brevis*. Matteo and co-workers (1976) studied the formation of the two enzymes responsible for the synthesis of this antibiotic. Both enzymes are produced in the latter part of the logarithmic growth phase. After reaching the maximum level of specific activity, the enzymes rapidly disappear as the culture proceeds into the stationary phase. Thus the enzymes appear to be synthesized only during a short part of the cell cycle. These workers found that both gramicidin S synthetases were produced under carbon, nitrogen, sulfur, or phosphorus limitation in chemostat studies. The enzyme production was low at dilution rates of 0.45-0.5 hr^{-1} (i.e., high specific growth rates) and increased as the dilution rate was lowered. In other words, a high growth rate appears to be incompatible with gramicidin S synthetase production, and hence with production of the antibiotib itself. A low specific growth rate, regardless of which nutrient was growth limiting, allowed enzyme synthesis. Absence of growth was not a requirement for enzyme formation.

Similar results were found for penicillin production by *Penicillium chrysogenum* (Pirt and Righelato, 1967). It was found that the specific rate of penicillin production in a glucose-limited chemostat was constant at specific growth rates between 0.014 and 0.086 hr^{-1}. However, if growth was stopped by limiting the glucose supply only to that necessary for maintenance requirements, penicillin production dropped to zero. Glucose supplied in excess of that necessary for maintenance requirements inhibited the decay of penicillin synthesis. It was found that a critical specific growth rate between 0.009 and 0.014 hr^{-1} was necessary to completely inhibit decay of the penicillin synthesis capability. Clearly, maintenance of antibiotic synthesis required an energy level that could not be met in a system so limited by its environment as to establish a state of no net growth.

Another example of apparent interaction between secondary metabolism and growth rate occurs in the production of gibberellin and bikaverin by *Gibberella fujikuroi* (Bu'Lock et al., 1974). In batch cultivation, the maximum specific growth rate occurs at 18 hr. Bikaverin reached its maximum concentration 40 hr into the fermentation, and gibberellin concentration was maximum at 70 hr. It was found that the bikaverin production rate was maximal at a specific growth rate of 0.05 hr^{-1} and the gibberellin production rate was maximal at 0.01 hr^{-1} in a chemostat. At the lower specific growth rate, bikaverin was produced only negligibly. Hence, the different secondary metabolic pathways are affected differently by the conditions imposed by the same specific growth rate.

Although the broad regime of the specific growth rates that will allow secondary metabolism suggests sweeping control, such as the restriction of energy pathways or precursor pools, there are examples of more specific effects of rapid growth. The repression of enzymes uniquely associated with secondary metabolism has been documented for streptomycin biosynthesis by Walker and his students (see the review by Walker, 1975). Gallo and Katz (1972) have shown that phenoxazinone synthetase and the enzyme responsible for linkage of the two 4-methyl-3-hydroxyanathranilic acid pentapeptides of actinomycin are repressed during the growth phase of *Streptomyces antibioticus*. Table 1 summarizes those systems in which secondary metabolism-specific enzymes are known to be

Table 1 Enzymes Specific to Secondary Metabolism That Are Repressed During Part or All of Trophophase

Enzyme	Secondary metabolite
Amidinotransferase	Streptomycin
Streptidine kinase	Streptomycin
Phenoxazinone synthase	Actinomycin
Gramicidin S synthetase	Gramicidin S
Bacitracin synthetase	Bacitracin
Tyrocidine synthetase	Tyrocidine
Candicidin synthetase	Candicidin
N-Acetylkanamycin amidohydrolase	Kanamycin
O-Dimethylpuromycin-O-methyltransferase	Puromycin
Cephalosporin C acetylhydrolase	Cephalosporin C
Neomycin phosphatase(s)	Neomycin
"Penicillin synthetase"[a]	Penicillin

[a]Glucose regulates [C^{14}] valine incorporation into penicillin.
Source: G. Revilla and J. F. Martin (unpublished data), and Walker (1975).

repressed during much or all of the growth phase of the producing organism. Many of these systems are described in further detail below.

These studies demonstrate that there is not only an upper limit on the specific growth rate at which secondary metabolism is expressed, but also a minimum specific growth rate below which secondary metabolism cannot be maintained. Continuous cultivation studies have shown that while a trophophase-idiophase relationship exists, the two phases can indeed occur simultaneously. Although the specific regulatory mechanisms which are responsible for these phenomena may not be known, this information is certainly of value in fermentation process design.

Fermentation processes for the production of these antibiotics must seemingly take one of two forms. The form in common use today is to operate the fermentation process with a distinct cell growth phase and a secondary metabolite production phase. This is the case in the use of fed batch or extended batch cultivation. This same approach could apply to continuous cultivation processes if multiple fermenters in series are used, or if fermenters with plug flow characteristics are developed. If a single stirred tank fermenter is desirable, one is limited to operation at low dilution rates.

Control of the specific growth rate as a means of controlling secondary metabolism is a "rough" tool for process control. The production of the enzymes responsible for both primary and secondary metabolism is affected. Thus maximizing or prolonging the production of a secondary metabolite through limitation of the specific growth rate has widespread effects on the overall physiological state of the culture. A more specific means of enhancing secondary metabolism is desirable, since this will reduce the occurrence of any physiological "side effects" which may be detrimental to the production of the secondary metabolite.

CARBON CATABOLITE REGULATION

The suppression of biosynthetic enzymes under conditions of rapid utilization of a carbon source is often implicated in the regulation of secondary metabolism. For example, glucose is often an excellent substrate for microbial growth, yet rapid glucose utilization (i.e., rapid growth) and expression of secondary metabolism have been found to be mutually exclusive in many cases. If glucose supply to the culture is limited, secondary metabolism is found to occur. Thus it appears that the rate at which a carbon source is utilized, and not its chemical identity, most directly affects regulation. Carbon catabolite regulation has been found to occur in the production of puromycin, coumermycin, neomycin, enniatin, bacitracin, mitomycin, prodigiosin, siomycin, indolmycin, ergot alkaloids, violacein, cephalosporins, actinomycin, and penicillins.

Catabolite Repression

Gallo and Katz (1972) conducted one of the first studies to attribute the suppression of secondary metabolism to the repression of a specific enzyme. *Streptomyces antibioticus,* which produces the antibiotic actinomycin, was cultivated in a batch fermentation using a medium composed of 0.1% glucose and 1.0% galactose. The authors found a delay in the production of the antibiotic until the glucose supply was depleted. If additional glucose was added prior to the initiation of actinomycin biosynthesis (i.e., during the growth phase), antibiotic production was severely depressed. They found that the formation of phenoxazinone synthetase, an enzyme unique to actinomycin synthesis, was repressed by glucose. Similarly, other carbon sources that supported rapid growth also suppressed production of the antibiotic. Galactose, on which the organism grows slowly, did not repress the formation of phenoxazinone synthetase or the synthesis of actinomycin.

Pogell and colleagues (1976) observed a similar repression of puromycin synthesis by *Streptomyces alboniger.* Formation of O-dimethylpuromycin-O-methyltransferase, presumably the final enzyme of puromycin synthesis, appears to be subject to glucose-mediated catabolite repression. In batch cultivation, the activity of this enzyme was low during the early stages of growth and increased during the later stages of growth, prior to antibiotic appearance. When glucose was added just prior to the rapid increase in enzyme activity, enzyme production was severely repressed and puromycin production did not occur.

In the case of novobiocin production by *Streptomyces niveus*, citrate causes catabolite repression while glucose does not (Kominek, 1972). In this case, it is citrate and not glucose which is the preferred substrate for growth. In batch cultivation, using a medium containing both citrate and glucose, novobiocin synthesis is delayed until the citrate has been depleted. Antibiotic production begins concurrently with glucose utilization, after a diauxic lag.

Another example of carbon catabolite repression in which the specific enzyme involved has been identified is the repression of kanamycin synthesis (Satoh et al., 1976). The final enzyme involved in this pathway, N-acetylkanamycin amidohydrolase, is repressed by several rapidly utilized carbon sources. Lactose and fructose exert the severest effects, while glucose, mannose, and maltose repress enzyme synthesis to a lesser degree. It was also reported that addition of cyclic adenosine 5'-monophosphate (cAMP) relieved glucose repression. Since intracellular pool levels of cAMP are depleted during rapid utilization of a carbon substrate, this finding is compatible with the other results. Hence the actual effector of repression might more directly involve low cAMP levels than high concentrations of a carbon catabolite. It should be noted that addition of cAMP, cyclic guanosine 5'-monophosphate (cGMP), or their dibutyryl derivatives did not reverse glucose repression of puromycin synthesis, by *S. alboniger*. In that case, repression was partially relieved by adenine, and to a lesser extent by adenosine and guanosine.

Other examples of carbon catabolite repression have been found in the interconversion of compounds within a "family" of antibiotics. The formation of mannosidostreptomycinase, an enzyme which converts mannosidostreptomycin to streptomycin, is repressed during rapid glucose utilization (Demain and Inamine, 1970). Glucose has also been reported to repress the formation of cephalosporin C acetylhydrolase (Hinnen and Nuesch, 1976). Aharonowitz and Demain (1978) have reported catabolite regulation of cephamycin biosynthesis by glycerol in *Streptomyces clavuligerus*.

Catabolite Inhibition

Although carbon catabolite inhibition has not been conclusively proven for secondary metabolism, the biosynthesis of some secondary metabolites does appear to be regulated by enzyme inhibition rather than repression. The addition of glucose to microbial cultures which are actively producing neomycin (Majumdar and Majumdar, 1971) siomycin (Kimura, 1967), or penicillin (Demain, 1972) *rapidly* inhibits antibiotic synthesis. Since the cells were actively producing secondary metabolite when glucose was added, the enzymes of the biosynthetic pathway must have been present; hence the regulatory mechanism appears to be inhibition. It is possible that instead of inhibition, the regulatory mechanism observed is that of repression of the synthesis of an enzyme which is being rapidly degraded and resynthesized. Kennel (1977) studied the production of cephalosporin C and penicillin N by resting cells of *Cephalosporium acremonium*. He found that rapidly utilized sugars (glucose, fructose, maltose) resulted in a lower yeild of both antibiotics than did more slowly metabolized sugars (sucrose and galactose). No evidence of catabolite repression could be found, and a type of catabolite inhibition of the preformed enzymes by a combination of glucose and ammonia was detected. It is unlikely that this phenomenon could be the result of the repression of the synthesis of an enzyme which is being rapidly degraded and resynthesized, because a resting cell culture probably could not maintain such an enzyme system.

Identification of Carbon Catabolite Regulation

The idiosyncracies of fermentation can often indicate the presence of catabolite regulation. Carbon catabolite regulation might be indicated by the trophophase-idiophase relationship discussed above. This would be the case for batch fermentation in which growth is slowed by depletion of the primary carbon source. If the specific regulatory mechanism is repression, addition of the primary carbon source after the initiation of secondary metabolism might have little immediate effect on metabolite production. As discussed above, the exception to this would be the case in which the enzyme is rapidly degraded and resynthesized during idiophase. Similar results would be seen if the regulatory mechanism was enzyme inhibition. The addition of a carbon source (i.e., inhibitor) during the idiophase would rapidly reduce the activity of the enzyme involved in the biosynthesis, thus reducing secondary metabolite formation.

NITROGEN CATABOLITE REGULATION

The regulation of secondary metabolism by nitrogen catabolites has been given little discussion in the literature. However, indications of this mode of regulation, which are quite similar to those found for carbon catabolite regulation, have existed for many years. In 1978, Dulaney found proline to be the best nitrogen source for streptomycin production. Since this amino acid is utilized very slowly as the sole nitrogen source, this situation appears analogous to that of slowly utilizable carbon sources. The usefulness of complex nitrogen sources such as soybean meal, cottonseed flour, or casein digests in antibiotic fermentations instead of ammonia is probably due to their slow breakdown to amino acids and ammonia. This slow breakdown is the result of enzymatic hydrolysis

Regulation of Secondary Metabolism 43

and may result in autofeeding of nitrogen at a rate too slow to allow nitrogen catabolite regulation. Obviously, an imbalance in protease activity could have drastic effects on such systems.

Nitrogen catabolite regulation appears to be involved in the production of phenolic secondary metabolites, such as trihydroxytoluene by *Aspergillus fumigatus* (Ward and Packter, 1974). These compounds are produced in batch cultivation only after the nitrogen source has been consumed. The addition of an ammonium salt completely inhibited their production without affecting either the growth rate or the pH of the medium. Other examples of secondary metabolism which appear to be subject to nitrogen catabolite regulation are bikaverin formation by *G. fujikuroi* (Bu'Lock et al., 1974) and the production of ergot alkaloids by *Claviceps purpurea* (Kybal et al., 1976).

The effects of various nitrogen sources on cephalosporin production by *S. clavuligerus* have been studied in detail (Aharonowitz, 1979; Aharonowitz and Demain, 1979). Asparagine was found to be the best nitrogen source, supporting both growth and antibiotic production. If ammonium chloride was used as the sole nitrogen source, cell growth was supported; however, no synthesis could be detected. In other studies, cephalosporin production in an asparagine-glycerol medium was evaluated with respect to various ammonium chloride concentrations. Maximum production of the antibiotic occurred at a concentration of approximately 5 mM ammonium chloride. Minimum cephalosporin production occurred at an ammonium chloride concentration of approximately 20 mM, at which point the culture was no longer nitrogen limited. These results seem analogous to some of those determined for carbon catabolite regulation, in that it appears to be the rate of utilization of a nitrogen source, and not its identity, which most directly relates to the regulatory mechanism.

The effect of adding excess ammonium chloride to a batch culture of different times was also investigated by these researchers. Cephalosporin was detected after 72 hr of cultivation in the asparagine-glycerol medium. When excess ammonium chloride was added at time 0, that is, at the same time as inoculation, cephalosporin synthesis was severely reduced. Addition of ammonium chloride after 24 hr of growth resulted in a delay of cephalosporin synthesis and a 50% reduction in the amount of antibiotic produced. When ammonium chloride was added 48 hr into the fermentation, there was essentially no effect on cephalosporin synthesis. These results exclude direct inhibition of the biosynthetic enzymes as the regulatory mechanism. Repression of the formation of these enzymes appears to be a possibility; however, this remains to be conclusively proven.

Aharonowitz (1979) also reported increases in extracellular protease and glutamine synthetase activities which corresponded to increased cephalosporin synthesis. When excess ammonium chloride was present, no increase in the activity of these enzymes was detected. Such increases should correspond to increases in the supply of amino acids. Since some amino acids are precursors to cephalosporin biosynthesis, this suggests that the supply of precursor molecules may be the rate-limiting factor in cephalosporin production. Therefore conditions which favor amino acid catabolism (e.g., low ammonia concentration) should also favor cephalosporin production.

Identification of Nitrogen Catabolite Regulation

Identification of nitrogen catabolite regulation of secondary metabolism is analogous to identification of carbon catabolite regulation. The trophophase-idiophase relationship found in batch cultures could be explained by nitrogen catabolite regulation if growth is slowed as a result of depletion of the primary nitrogen source. The results discussed above for trihydroxytoluene production by *A. fumigatus* supported this idea. However, the work of Aharonowitz and Demain (1979) indicates that a more complex regulatory mechanism may be involved. Since the degree of reduction was correlated with how early in the fermentation excess ammonium chloride was added, if the regulatory mechanism is one of simple repression, the enzymes must be synthesized early in the growth phase. Since cephalosporin is not detectable until late in the fermentation, this seems unlikely. Hence a more complex mechanism appears to be involved.

PHOSPHATE REGULATION

For many years it has been known that inorganic phosphate concentrations which are optimal for microbial growth often inhibit secondary metabolism (Weinberg, 1974). Phosphate regulation has been shown to be involved in the production of chlorotetracycline, streptomycin, candicidin, ergot alkaloids, neomycin, vancomycin, polymyxin, bacitracit, gramicidin S, cephamycin, bikaverin, and corynecin. Several mechanism might be involved in phosphate regulation, and these have been reviewed by Martin (1976). One of the most important cellular activities in which phosphate is involved is the storage of energy. It is tempting to attribute all phosphate regulation to the energy state of the cell or to the concentration of a particular phosphonucleotide (e.g., adenosine triphosphate). In some cases, phosphonucleotide concentrations or energy charges (as described by Atkinson, 1969) were correlated with secondary metabolites synthesis; however, in other examples of secondary metabolism, other modes of action seem more probable explanations of phosphate regulation.

Curdova and co-workers (1976) investigated the effect of phosphate concentration on chlorotetracycline production by *Streptomyces aureofaciens*. In batch cultivation, it was found that chlorotetracycline production did not begin until the phosphate in the medium had been depleted. Similarly, increased concentions of phosphate reduced antibiotic production, while also increasing intracellular pool levels of adenosine triphosphate (ATP), adenosine diphosphate (ADP), and AMP. Comparisons of the intracellular adenylate levels of a high antibiotic production strain to those of a low production strain yielded similar results. The adenylate levels in the low production strain were an order of magnitude higher than those in the high production strain. It is interesting to note that the energy charge defined as

$$\frac{[ATP] + (1/2)[ADP]}{[ATP] + [ADP] + [AMP]}$$

was similar for both strains. These results demonstrate a correlation between low intracellular adenylate pool levels and a high level of chlorotetracycline production. Energy charge appears to have little effect on the level of antibiotic production. The level of anhydrotetracycline hydratase, an inducible enzyme of the chlorotetracycline biosynthetic pathway, was greatly reduced under conditions of excess phosphate. This result suggests that the mode of action of increased adenylate levels within the cell is to repress the formation of this enzyme. However, this remains to be proven conclusively. Since the enzyme is known to be inducible, the low enzyme levels might be the result of suppression of the production of the inducer compound due to high phosphate levels.

Other investigations with chlorotetracycline production by *S. aureofaciens* suggest a closer and more complex interaction between primary and secondary metabolism (Hostalek et al., 1979). The suppression of chlorotetracycline production by high concentrations of extracellular phosphate and/or intracellular adenylate might be due to the suppression or stimulation of a primary metabolic pathway. This could be the suppression of malonylcoenzyme A formation, the stimulation of lipogenesis, or the stimulation of the tricarboxylic acid cycle.

Malonyl-CoA is the precursor for both chlorotetracycline synthesis and lipogenesis. Malonyl-CoA can be formed from three pathways (Fig. 1). In one pathway, acetyl-CoA is the precursor of malonyl-CoA, and in the other two pathways oxaloacetate is converted to malonyl-CoA. Phosphoenolpyruvate is converted to pyruvate via the enzyme pyruvate kinase, and pyruvate is converted to acetyl-CoA via pyruvate dehydrogenase. Acetyl-CoA can then be directly converted to malonyl-CoA via acetylcarboxylase. If acetyl-CoA enters the tricarboxcyclic acid cycle, it may directly be converted to malonyl-CoA via the conversion of oxaloacetate, which is an intermediate in the tricarboxylic acid cycle. The other pathway for malonyl-CoA production involves the direct conversion of phosphoenolpyruvate to oxaloacetate via the enzyme phosphoenolpyruvate carboxylase.

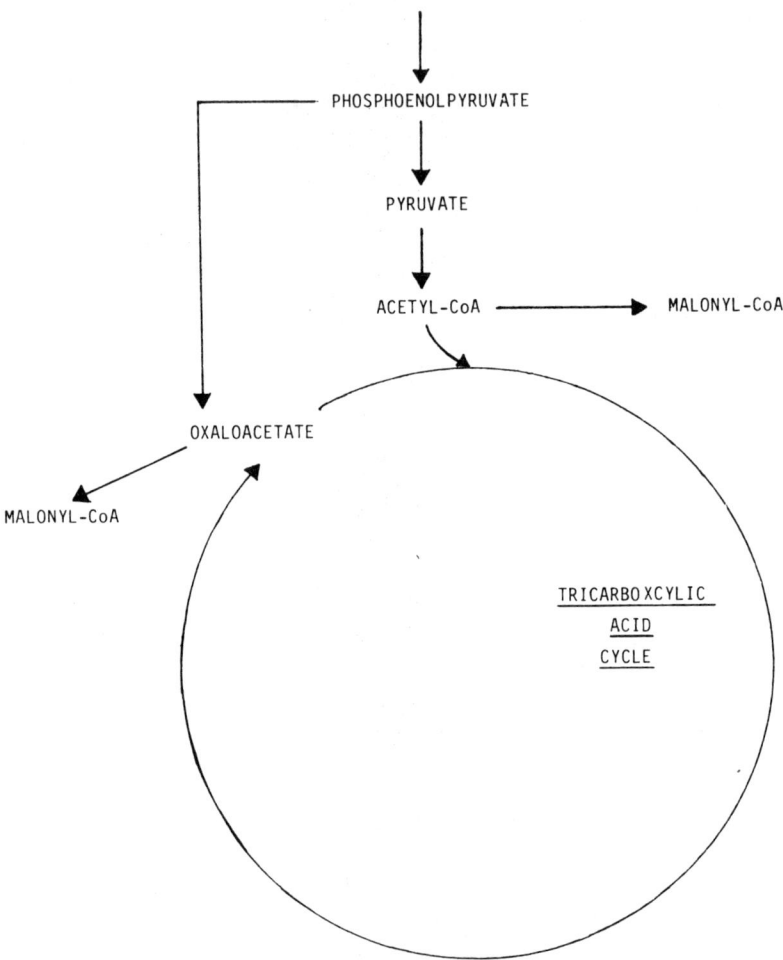

Figure 1 Enzymes specific to secondary metabolism that are repressed during part or all of trophophase.

Both acetyl-CoA carboxylase activity and lipogenesis peak at the onset of chlorotetracycline production and decrease as production of the antibiotic increases. Also, the levels of pyruvate kinase and pyruvate dehydrogenase (the enzymes responsible for the synthesis of acetyl-CoA) are low during antibiotic production. These results were seen in both the high and low antibiotic production strains. Since the phosphoenolpyruvate carboxylase activity of the high producer increased with increased antibiotic production, these results were interpreted as proof that the malonyl-CoA used for chlorotetracycline synthesis was derived directly from the conversion of phosphoenolpyruvate. However, the activity of phosphoenolpyruvate carboxylase drops prior to the onset of antibiotic production and does not begin to increase until 20 hr after antibiotic production has begun. Consequently, it is unlikely that this mechanism could be responsible for the supply of malonyl-CoA for chlorotetracycline synthesis.

Acetyl-CoA carboxylase activity peaks just prior to chlorotetracycline synthesis, and hence intracellular pool levels of acetyl-CoA and malonyl-CoA are probably at high levels. If lipogenesis were suppressed (e.g., feedback inhibition), increased synthesis of chlorotetracycline could relieve the accumulation of these compounds. The difference between the low and high antibiotic production strains might not be due to the difference in the levels of phosphoenolpyruvate carboxylase, but to a difference in the activity of the enzymes of the tricarboxylic acid cycle. Recently, Shantha and Murthy (1981) have shown that competing primary pathways interacting with the tricarboxcyclic acid cycle can control aflatoxin biosynthesis.

Hostalek et al. (1979) reported that the levels of activity of the tricarboxylic acid cycle enzymes in the higher production strain were lower than those in the low producer strain. Thus an excess of malonyl-CoA or acetyl-CoA could be more rapidly utilized via the tricarboxylic acid cycle in the low antibiotic producer than in the high antibiotic production strain. Therefore chlorotetracycline synthesis might be a means for the high antibiotic production strain to rid itself of excess malonyl-CoA or acetyl-CoA should lipogenesis be suppressed. This hypothesis is supported by the fact that peak levels of acetyl-CoA carboxylase and lipogenesis are about 50% higher in the high antibiotic production strain than in the low production strain. Whether either of the hypotheses presented above are correct or whether the biosynthetic enzymes of the chlorotetracycline pathway are more directly suppressed by excess phosphate still remains to be proven.

As with chlorotetracycline production, canicidin production by *S. griseus* must be carried out under phosphate-limited conditions. Liu and co-workers (1975) demonstrated a simultaneous enhancement of growth and suppression of antibiotic production by addition of inorganic phosphate to the medium. Martin and Demain (1976) found a similar suppression of antibiotic synthesis using a short-duration resting cell system of *S. griseus*. They found that addition of inorganic phosphate caused only a slight change in the energy charge of the cells. However, ATP concentrations were three to four times that normally seen during antibiotic synthesis (Martin and Demain, 1977a). The addition of ATP, ADP, AMP, or GMP had the same effect on candicidin production as did addition of inorganic phosphate (Martin and Demain, 1977b). Since it has been shown that phosphate is cleaved from AMP during microbial uptake (Martin et al., 1978), these results suggest that phosphate is the initiator of the regulation of candicidin synthesis. This is supported by the finding that addition of ribonucleosides or adenine had no effect on candicidin production (Martin et al., 1978). In the same study, AMP and ATP were added at different times during the course of the fermentation. In all cases, candicidin production was immediately suppressed, regardless of when the AMP or ATP was added. These results indicate that the relationship between phosphate and the regulation of candicidin synthesis is more complex than direct repression or inhibition of the biosynthetic enzymes. As for the case of chlorotetracycline production, the actual mode of regulation by phosphate may be the suppression of a primary metabolic pathway which supplies precursors for candicidin synthesis (propionate or methyl malonyl-CoA). Another possible explanation would the the stimulation of a primary metabolic pathway which would compete with candicidin synthesis for precursors, thus lowering the precursor supply for candicidin synthesis.

Another example of phosphate regulation, the regulation of streptomycin synthesis, can be explained by the inhibition of a biosynthetic enzyme. At least three of the reactions involved in the synthesis of streptomycin, including the final reaction which converts streptomycin phosphate to streptomycin, involve a phosphate-cleaving reaction (Walker, 1971). The enzyme responsible for the conversion of streptomycin phosphate to streptomycin has been shown to be feedback inhibited by high levels of phosphate (Walker and Walker, 1971). This finding explains why cultivation under conditions of high phosphate levels leads to the accumulation of streptomycin phosphate instead of streptomycin (Miller and Walker, 1970).

FEEDBACK REGULATION

It is known that chloramphenicol, rastomycin, ergot alkaloids, virginiamycin, indolmycin, cycloheximide, mycophenolic acid, aurodoxin, and penicillin feedback-regulate their own biosynthesis. However, in many other cases, feedback regulation of secondary metabolism occurs through the action of a primary metabolite. This mode of regulation involves the limitation of the supply of precursor (primary metabolite) to the biosynthetic pathway of the secondary metabolite. Two types of feedback regulation by primary metabolites will be discussed below. The first type involves feedback regulation by a primary metabolite on the metabolic pathway of its own formation (linear pathway feedback); the other type involves cases in which the primary metabolite is an intermediate in a further primary metabolic pathway, as well as the precursor to the secondary metabolic pathway (branched pathway feedback). In these areas, the end product of the primary pathway may have a regulatory effect on the pathway responsible for the synthesis of the common precursor (primary metabolite).

Linear Pathway Feedback

One example in which a primary metabolite precursor feedback regulated its own synthesis is the production of penicillin by *P. chrysogenum*. The amino acid valine is a direct precursor of the dihydrothiazine ring in penicillin biosynthesis. The first enzyme of valine synthesis, acetohydroxy acid synthetase, is inhibited by high intracellular concentrations of valine (Goulden and Chattaway, 1968). It was also shown that in a superior penicillin-producing strain, this enzyme was less sensitive to valine feedback inhibition than in poor penicillin-producing strains. In other words, the mutational deregulation of valine production increased precursor supply, thus increasing antibiotic production.

A similar situation apparently exists for pyrrolnitrin production by *Pseudomonas aureofaciens*. The cultivation of early strains of this organism required the addition of tryptophan (a precursor to pyrrolnitrin synthesis) to the medium in order to achieve antibiotic production. Mutants which were selected for resistance to the tryptophan analogs, 5-fluorotryptophan and 6-fluorotryptophan, no longer required tryptophan supplementation for antibiotic production (Elander et al., 1971). Tryptophan formation in these mutants appears to be insensitive to tryptophan feedback regulation, and the amino acid is probably overproduced.

Branched Pathway and Feedback Regulation

Many biosynthetic pathways of secondary metabolism share intermediates with primary metabolic pathways. One example of this is penicillin production by *P. chrysogenum*. In this case, the common precursor is α-aminoadipic acid. This compound is converted to lysine via a primary metabolic pathway, or to penicillin via a secondary metabolic pathway. The addition of lysine to a culture of *P. chrysogenum* has been shown to reduce penicillin production (Demain, 1957), whereas addition of α-aminoadipic acid will relieve this inhibition (Somerson et al., 1961). In the latter study it was also shown that in the absence lysine, α-aminoadipic acid stimulates the formation of penicillin. Thus the regulatory effect of lysine on penicillin production appears to be the result of feedback inhibition on the synthesis of α-aminoadipic acid. Lysine has been shown to inhibit the activity of homocitrate synthetase, an enzyme in the biosynthetic pathway of α-aminoadipic acid (Masurekar and Demain, 1972). Also, the addition of homocitrate, which is the product of the reaction catalyzed by homocitrate synthetase, relieves lysine inhibition of penicillin synthesis. This indicates that accumulation of lysine within the cell directly inhibits homocitrate production, which restricts the amount of α-aminoadipic acid formed. Thus the amount of lysine or penicillin which can be synthesized is restricted.

Branched pathway regulation may occur by a totally different mechanism and be affected in a complex way. In this case, which involves candicidin synthesis by *S. griseus,* chorismic acid plays a role which α-aminoadipic acid plays in penicillin. Chorismic acid is a direct precursor of anthranilic acid, which is converted via a primary metabolic pathway to tryptophan. Chorismic acid is also a precursor of p-aminobenzoic acid, which is an intermediate in candicidin synthesis. Tryptophan and anthranilic acid have been shown to strongly reduce candicidin production by *S. griseus* (Liu et al., 1972; Martin et al., 1979). If the mode of regulation of these compounds were similar to that of lysine inhibition of penicillin formation, the addition of p-aminobenzoic acid should relieve the inhibition. Martin et al. (1979) found that addition of p-aminobenzoic acid had no effect on relieving suppression of candicidin production, even though it was readily taken up by the cells. Therefore a "cross-pathway feedback" mechanism was proposed to explain this regulatory phenomenon. In addition to inhibiting candicidin synthesis, the addition of tryptophan or anthranilic acid was found to stimulate protein synthesis. Thus the suppression of candicidin synthesis by accumulation of tryptophan would increase the supply of precursors for the synthesis of aromatic amino acids, which would be used in protein synthesis. This explanation is based on a control logic which is different from that usually proposed for metabolic regulations. As intracellular pool levels of tryptophan increase, candicidin formation would be suppressed, thus increasing the level of tryptophan still further. Conversely, if tryptophan levels drop as a result of stimulated protein synthesis, the suppression of candicidin synthesis will be relieved. Thus there is an increased utilization of the common precursor, reducing the amount of tryptophan which will be produced and causing the level of this compound to drop even further. The logic of this "cross-pathway feedback" mechanism of regulation appears to lead to an unstable control of tryptophan levels. More information is needed in order to explain the complete mechanism of tryptophan regulation on candicidin synthesis.

INDUCTION

All of the interactions between primary and secondary metabolism regulation which were discussed above were of a negative nature, that is, secondary metabolite formation was suppressed. In some cases, the enzymes of the secondary metabolic pathway appear to be induced by the addition of primary metabolites to the growth medium. One example of this phenomenon is the effect of tryptophan on alkaloid synthesis by Claviceps SD58. If tryptophan is absent from the medium, little alkaloid is produced, while supplementation with tryptophan stimulates alkaloid synthesis. Since the effect of tryptophan might be due to its role as a precursor of alkaloid synthesis, Robbers and Floss (1970) tested the effect of addition of tryptophan analogs on alkaloid production. Although these compounds cannot serve as alkaloid precursors, alkaloid production was still enhanced. If the culture is grown in a medium which contains tryptophan, and after the growth phase is completed the mycelia are transferred to a medium lacking tryptophan, alkaloid is still synthesized at a high rate (Robbers et al., 1972). These results indicate that tryptophan induces the synthesis of an enzyme which is not rapidly degraded and resynthesized during idiophase. It was noted that inducer must be added within 12 hr of inoculation in order to achieve maximum levels of alkaloid production. Addition of tryptophan at 12-24 hr after inoculation yielded production rates lower than the maximum. If the inducer was added later than 48 hr after inoculation, alkaloid production was about one-third of the maximum level. The results support the hypothesis that tryptophan enhances alkaloid production by inducing the synthesis of a biosynthetic enzyme. Krupinski et al. (1976) found that dimethylallyltryptophan synthetase, which is believed to be the first enzyme of alkaloid synthesis, was indeed induced by the presence of tryptophan.

It seems that a true inducer must be added during trophophase for the maximum enhancement of the secondary metabolite being produced, efforts must be made to differentiate between induction and an increase in precursor supply. The use of nonmetabolizable structural analogs to the precursor/inducer should have the same effect of enhancing secondary metabolite formation. In addition, increased intracellular pool levels of the inducer should be detectable prior to the onset of increased secondary metabolism.

REFERENCES

Aharonowitz, Y. A. (1979). Regulatory interrelationships of nitrogen metabolism and cephalosporin biosynthesis. In *Genetics of Industrial Microorganisms*, O. K. Sabek and A. I. Laskin (Eds.). American Society of Microbiology, Washington.

Aharonowitz, Y. A., and Demain, A. L. (1978). Carbon catabolite regulation of cephalosporin production in *Streptomyces clavuligerus*. Antimicrob. Agents Chemother. *14*:159-164.

Aharonowitz, Y. A., and Demain, A. L. (1979). Nitrogen nutrition and regulation of cephalosporin production in *Streptomyces clavuligerus*. Can. J. Microb. *25*:61-67.

Atkinson, D. E. (1969). Regulation of enzyme function. Annu. Rev. Microbiol. *23*:47-68.

Berdy, J. (1974). Recent development of antibiotic research and classification of antibiotics according to chemical structure. Adv. Appl. Microbiol. *18*:309-406.

Bu'Lock, J. D., Detory, R. W., Hostalek, Z., and Munim-al-Shakarchi, A. (1974). Regulation of secondary biosynthesis in *Gibberella fujikuroi*. Trans. Br. Mycol. Soc. *62*:377-389.

Curdova, E., Kremen, A., Vanek, Z., and Hostalek, Z. (1976). Regulation and biosynthesis of secondary metabolites. Folia Microbiol. *21*:481-487.

Demain, A. L. (1957). Inhibition of penicillin formation by lysine. Arch. Biochem. Biophys. *67*:244-245.

Demain, A. L. (1972). Cellular and environmental factors affecting the synthesis and excretion of metabolites. J. Appl. Chem. Biotechnol. *22*:345-362.

Demain, A. L., and Inamine, E. (1970). Biochemistry and regulation of streptomycin and mannosidostreptomycinase (α-D-mannosidase) formation. Bacterial Rev. *34*:1-19.

Dulaney, E. L. (1978). Observations on *Streptomyces griseus*. II. Nitrogen sources for growth and streptomycin production. J. Bacteriol. *56*:305-313.

Elander, R. P., Mabe, J. A., Hamill, R. L., and Gorman, M. (1971). Folia Microb. *16*:721-725.

Gallo, M., and Katz, E. (1972). Regulation of secondary metabolite biosynthesis —Catabolite repression of phenoxazinone synthetase and actinomycin formation by glucose. J. Bacteriol. *109*:659-667.

Goulden, S. A., and Chattaway, F. W. (1968). Lysine control of α-aminoadipic acid and penicillin synthesis in *Penicillium chrysogenum*. Biochem. J. *110*: 55P-56P.

Hinnen, A., and Nuesch, J. (1976). Enzymatic hydrolysis of *cephalosporin C acremonium*. Antimicrob. Agents Chemother. *9*:824-830.

Hostalek, Z., Behal, V., Curdova, E., and Jechova, V. (1979). Specific primary pathways supplying secondary biosynthesis. In *Genetics of Industrial Microorganisms*, O. K. Sabek and A. I. Laskin (Eds.). American Society of Microbiology, Washington, p. 225.

Kennel, Y. M. (1977). Carbon source regulation of antibiotic biosynthesis in *Cephalosporin acremonium*. Ph.D. thesis, Massachusetts Institute of Technology, Cambridge.

Kimura, A. (1967). Biochemical studies on *Streptomyces siogaensis*, 2. Mechanism of inhibitory effect of glucose on siomycin formation. Agric. Biol. Chem. *31*:845-852.

Kominek, L. A. (1972). Biosynthesis of of novobiocin by *Streptomyces niveus*. Antimicrob. Agents Chemother. *1*:123-134.

Krupinski, V. M., Robbers, J. E., and Floss, H. G. (1976). Physiological study of ergot induction of alkaloid synthesis by tryptophan at enzymatic level. J. Bacteriol. *125*:158-165.

Kybal, J., Kleinerova, E., and Bulant, V. (1976). Ergot alkaloids. VI. Nitrogen metabolism during the development of sclerotium of *Clavieceps purpurea*. Folia Microb. *21*:474-480.

Liu, C. M., McDaniel, L. E., and Schaffner, C. R. (1972). Studies on candicidin biogenesis. J. Antibiot. *25*:116-212.
Liu, C. M., McDaniel, L. E., and Schaffner, C. P. (1975). Factors affecting production of candicidin. Antimicrob. Agents Chemother. *7*:196-202.
Majumdar, M. K., and Majumdar, S. K. (1971). Synthesis of neomycin by washed mycelium of *Streptomyces fradiae* and some physiological considerations. Folia Microb. *122*:397-404.
Malek, I. (1976). Physiological state of continuously grown microbial cultures. In *Continuous Culture 6: Applications and New Fields*, A. C. R. Dean et al. (Eds.). pp. 31-39.
Martin, J. F. (1976). Control of antibiotic synthesis. In *Microbiology 1976*, D. Schlessinger (Ed.). American Society of Microbiology, Washington.
Martin, J. F., and Demain, A. L. (1976). Control of phosphate of candicidin production. Biochem. Biophys. Res. Commun. *71*:1103-1109.
Martin, J. F., and Demain, A. L. (1977a). Effect of exogenous nucleotides on the candicidin fermentation. Can. J. Microbiol. *23*:1334-1399.
Martin, J. F., and Demain, A. L. (1977b). Cleavage of adenosine 5'-monophosphate during uptake by *Streptomyces griseus*. J. Bacteriol. *132*:590-595.
Martin, J. F., Loras, P., and Demain, A. L. (1978). Adenylate energy charge during phosphate-mediated control of antibiotic synthesis. Biochem. Biophys. Res. Commun. *83*:822-828.
Martin, J. F., Naharro, G., Loras, P., and Villanueva, J. R. (1979). Isolation of mutants deregulated in phosphate control of candicidin biosynthesis. J. Antibiot. *32*:600-606.
Masurekar, P. S., and Demain, A. L. (1972). Lysine control of penicillin biosynthesis. Can. J. Microbiol. *18*:1045-1048.
Matteo, C. C., Cooney, C. L., and Demain, A. L. (1976). Production of gramicidin S synthesis by *Bacillus brevis* in continuous culture. J. Gen. Microbiol. *96*:415-422.
Miller, A. L., and Walker, J. B. (1970). Accumulation of streptomycin phosphate in cultures of streptomycin producers grwon on a high phosphate medium. J. Bacteriol. *104*:8-12.
Pirt, S. J., and Righelato, R. C. (1967). Effect of growth rate on the synthesis of penicillin by *Penicillium chrysogenum* in batch and chemostat cultures. Appl. Microbiol. *15*:1284-1290.
Pogell, B. M., Sankaran, L., Redshaw, P. A., and McCann, P. A. (1976). Regulation of antibiotic biosynthesis and differentiation in Streptomyces. In *Microbiology 1976*, D. Schlessinger (Ed.). American Society of Microbiology, Washington, pp. 543-547.
Robbers, J. E., and Floss, H. G. (1970). Physiological studies on ergot—Influence of 5-methyltryptophan on alkaloid biosynthesis and incorporation of tryptophan analogues into protein. J. Pham. Sci. *59*:702-703.
Robbers, J. E., Robertson, L. W., Hornemann, K. M., Jindras, A., and Floss, H. G. (1972). Physiological studies on ergot—Further studies on induction of

alkaloid synthesis by tryptophan and its inhibition by phosphate. J. Bacteriol. *112*:791-796.

Satoh, A., Ogawa, H., and Satomura, Y. (1976). Role and regulation mechanism of kanamycin acetyltransferase in kanamycin biosynthesis. Agric. Biol. Chem. *40*:191-196.

Shantha, T., and Murthy, V. S. (1981). Influence of tricarboxylic acid cycle intermediates and related metabolites on the biosynthesis of aflatoxin by resting cells of *Aspergillus flavus*. Appl. Environ. Microbiol. *42*:758-761.

Somerson, N. L., Demain, A. L., and Nunheimer, T. D. (1961). Reversal of lysine inhibition of penicillin production by α-aminoadipic or adipic acid. Arch. Biochem. Biophys. *93*:238-241.

Walker, J. B. (1971). Enzymatic reaction involved in streptomycin biosynthesis and metabolism. Lloydia *34*:363-371.

Walker, J. B. (1975). L-Arginine:Inosamine-phosphate amidinotransferase. Methods Enzymol. *43*:451-458.

Walker, M. S., and Walker, J. B. (1971). Streptomycin biosynthesis—Separation and substrate specificites of phosphatases acing on guanidinodeoxy scyllo-insitol phosphate and streptomycin phosphate. J. Biol. Chem. *246*:7034-7040.

Ward, A. C., and Packter, N. M. (1974). Relationship between fatty acid and phenol synthesis in *Aspergillus fumigatus*. Eur. J. Biochem. *46*:323-333.

Weinberg, E. D. (1974). Secondary metabolism, regulation by trace elements. Dev. Ind. Microbiol. *15*:70-81.

3
Correlation of Secondary Metabolism and Differentiation

Iain M. Campbell / University of Pittsburgh, Pittsburgh, Pennsylvania

> No question is settled
> Until it is settled right.
> E. M. Wilcox, *Settle the Question Right*

INTRODUCTION

The question of fungal secondary metabolism's role in the producing organism remains unsettled despite decades of study. Many proposals have been advanced (Demain, 1974; Campbell et al., 1982), but the available experimental evidence fails to confirm any proposal as the "right settlement."

The bulk of that evidence has been obtained by studying how the onset and maintenance of secondary metabolism is correlated with the development of submerged liquid cultures of fungi. A more limited but ever-increasing body of data is emerging from secondary metabolism-colony development correlation studies on liquid or solid media.* By and large, the focus of all this work has been at the organismic level. Moreover, since much of it was conducted toward ends other than the elucidation of the function question, the experimental record lacks integration and cohesion.

Successful settlement of the complex question of the function of fungal secondary metabolism will require a much more integrated and systematic approach. While more information can and must be gleaned from experiments pitched at the organismic level, cellular and organism community/ecological

---

*For the purposes of this article, the terms "culture" and "colony," when used to qualify the word "development," will refer exclusively and restrictively to the process that occurs in the submerged liquid growth format (culture development) or in the surface liquid or solid growth formats (colony development). In accord with normal usage and when it does not grammatically qualify "development," the word "culture" will be used loosely to connote the act, method, or outcome of growing a microorganism.

level studies are also called for. This article deals with the cellular level approach —the correlation of fungal secondary metabolism with cell differentiation. Very little, if any, real cellular level work has been done to date, although some laboratories are coming close. The objective of this article is to introduce the topic, outline its foundation and frame of reference, and suggest why its pursuit might help settle, finally, the tantalizing question; Why do fungi engage in secondary metabolism?

Cellular differentiation is the process whereby a cell alters its genetic expression, normally in a programmed and irreversible manner, such that its morphology and/or biochemical/physiological competence or that of its immediate progeny is similarly altered. Any attempt to relate this process to the phenomenon of fungal secondary metabolism presumes the following:

1. Secondary metabolism is an attribute of one or more specific fungal cell phenotypes.
2. These cell phenotypes can be identified.
3. Methods can be devised whereby experimentally significant numbers of fungal cells can be made to differentiate from a nonproducing state to a secondary metabolite-producing state in a synchronous manner.

As we will show, there is reasonable evidence for believing that the first two prerequisites are satisfied. The evidence derives, appropriately enough, from correlation studies involving culture and colony development—methods for which the cellular level approach is both the natural extension and the essential complement. Not only do these studies establish the feasibility of cellular level work, but they also identify the cell phenotypes required for studying prerequisite (3). Progress toward a satisfying prerequisite (3) is discussed on pp. 65-67. On pp. 67-68, we conclude with a brief analysis of what the future of secondary metabolism-cellular differentiation correlation studies might be and where such studies might lead.

EXPERIMENTAL PARAMETERS REQUIRED FOR CORRELATION

Before reviewing the data on which the fledgling topic of secondary metabolism-cellular differentiation correlation is based, it is appropriate to deal in a general way with the types of experimental data that are needed.

Monitoring the onset, maintenance, and, if appropriate, termination of secondary metabolism in a population of fungal cells is a relatively simple matter. Provided that the inoculum does not contain the metabolite, onset of production can be monitored simply by assaying the metabolite level associated with the cell population. Spectrophotometric, chromatographic, and bio- or radio-immunological assays are all equally good methods for the task. Although often less quantitative, cytological methods are also appropriate. The overall status of secondary metabolism, however, is likely best monitored by means of short-duration in vivo isotope incorporation experiments. In our experience assay of such experiments with radiogas and/or radioliquid chromatographic technologies, often with the assistance of coupled mass spectrometry, has proved particularly valuable (Doerfler, et al., 1978; Campbell, 1979). Such technologies have the additional advantage of being able to detect and monitor the isotopic content of intermediates in the biosynthetic pathway (Doerfler et al., 1980). This is a particularly valuable capability for localizing the cell types of a mixed population that are involved in the synthesis.

In vitro enzyme assay of cell population extracts can also be used to monitor the status of secondary metabolism, but the generally underdeveloped state of secondary metabolism enzymology makes this currently a less useful methodology. Moreover, since the cellular organization is totally disrupted on preparing the in vitro sample, great care has to be taken to control for the possibility that regulatory systems that could have been in operation when the cell was intact are modified positively or negatively by the time the assay is performed.

Localizing pools of secondary metabolites at the cellular level is a key requirement of cellular level correlation. Cytoimmunological methods have been used for this purpose in the past (e.g., Kurylowicz, 1977; Lawellin et al., 1977). Microspectrophotometry should also prove to be a valuable technique. We are currently exploring the feasibility of using mass spectrometry to detect secondary metabolites in single cells or in small numbers of cells of the same phenotype. In the selected ion mode and, if need be, with negative ions/chemical ionization, mass spectrometry can operate at the femtogram level (10^{-15} g, equivalent to approximately a million molecules of a metabolite of molecular weight 500 or a 50 mM concentration in a single cylindrical cell 1 μm in diameter by 10 μm long).

Monitoring cellular differentiation is a more demanding task. Much of the work with culture and colony development that will be discussed relied simply on the frequent assay of such gross parameters as medium nutrient levels, medium pH, biomass weight, and total cellular protein. Rates of RNA, DNA, and total protein synthesis have been used, as have the in vitro activities of key enzymes of primary metabolism. Morphology at the microscopic level can also be a legitimate and valuable index of the progress of culture development; it is likely the best index of colony development.

Many of these parameters are appropriate for monitoring cellular differentiation in synchronous population of cells. The most valuable parameters in this regard are those which are themselves cellular, to wit: rates of macromolecular synthesis, enzyme activities, and the various aspects of cell morphology revealed by light and electron microscopy. Total protein two-dimensional electrophoretic profiles (O'Farrell, 1975; O'Farrell et al., 1977) and total small-molecule metabolic profiles (Doerfler et al., 1978) are parameters that will likely be used increasingly as indices of states of cellular differentiation in future work.

CONTRIBUTION OF SECONDARY METABOLISM-SUBMERGED LIQUID CULTURE DEVELOPMENT CORRELATION STUDIES TO THE CELLULAR LEVEL APPROACH

Much has been written regarding the so-called phasing of batch-mode submerged liquid cultures of fungi. The phenomenon was first described in work on the penicillin fermentation (Koffler et al., 1945). It received its most detailed definition from Borrow et al. (1961, 1964), who were working on the gibberellic acid fermentation. Its most enduring terminology was coined by Bu'Lock et al. (1965); trophophase for the preproduction growth phase, idiophase for the postgrowth production phase. History will likely record that too much functional significance has been attributed to the phasing phenomenon and the conflicting instances where secondary metabolism is idiophasic or trophophasic. Be that as it may; the fact remains that of the secondary metabolites that can be produced in the submerged liquid format, most appear quite reproducibly at specific points in the lifetime of a batch-mode submerged liquid culture (see Martin and Demain, 1978, for a more detailed discussion). In some instances biosynthesis begins early, for example, mycophenolic acid production by *Penicillium brevicompactum* (Doerfler et al., 1978); in other instances it begins late, for example, gibberellic acid production by *Gibberella fujikuroi* (Bu'Lock et al., 1974). Occasionally the time of biosynthesis onset can be modified by medium manipulation, for example, ergot alkaloid production by *Claviceps paspali* (Brar et al., 1968), or by strain selection, for example, 6-methylsalicylic acid production by *Penicillium urticae* (Light, 1967a). In some cases the time of onset can be tightly correlated with an index of culture development, for example, medium nitrogen levels in the patulin-*Penicillium patulum* system (Grootwassink and Gaucher, 1980); in other cases tight correlation has not been detected, for example, enniatin production by *Fusarium sambucinum* (Audhya and Russell, 1975). Notwithstanding these divergences, the clear impression is that fungi express their capability to produce secondary metabolites as part of a prescribed species-associated development program. If this proves to be the case, unique phenotypes should be found to be responsible for secondary metabolite activity [recall prerequisite (1) on p. 56].

A thin web of additional data supports the position that unique phenotypes are indeed involved in secondary metabolite production in submerged liquid culture. The data are of three types. The first type involves the use of inhibitors of macromolecular synthesis and the effect that these substances have on that characteristic point in the lifetime of a batch-mode submerged culture when the secondary metabolite normally appears. In several systems addition of inhibitors of protein synthesis, for example, cycloheximide and p-fluorophenylalanine, or of RNA synthesis, for example, actinomycin D, 5-fluorouracil, when added in an appropriate concentration at an appropriate time, inhibit the scheduled appearance of the secondary metabolite. The best-studied example is the 6-methylsalicyclic acid-patulin-*P. patulum* system (Light, 1967b; Bu'Lock et al., 1969; Grootwassink and Gaucher, 1980). The results indicate that transcription and translation are required for the secondary metabolite trait to be expressed. This is a necessary but not sufficient condition for claiming that a process of differentiation connects the nonproducing state with the producing one. Transcription and translation are involved, for instance, in simple cellular adaptation to utilize a new carbon source. More work has to be done to define phenotypically the producing and nonproducing states of a submerged liquid culture. There is good reason for believing this can be done. Thus Bu'Lock et al. (1965) have shown that the 6-methylsalicylic acid nonproducing state of *P. patulum* differs from the producing state in regard to the relative flux carried by the hexose monophosphate and glycolytic pathways; it remains to be seen if the switch in that flux is also delayed by macromolecular synthesis inhibitors.

The second type of evidence that supports the claim that recognizable phenotypes are associated with expression of the secondary metabolite trait in submerged liquid culture is morphological. In several instances microscopic monitoring of culture development reveals parallel production of secondary metabolites and distinct cell types or cell aggregates: for example, pigmented fat-containing cells: gibberellic acid in *G. fujikuroi* (Borrow et al., 1961), arthrospores: β-lactam antibiotics in *Cephalosporium acremonium* (Nash and Huber, 1971), and chlamydospores: ergot alkaloids in *Claviceps purpurea* (Vorisek et al., 1974).

The third type of evidence derives from continuous culture work. This culture technique has the potential to not only produce very homogeneous populations of cells, but also allows modification of their growth rate and/or nutritional regimen with facility. Unfortunately, continuous culture methods have not been applied to the secondary metabolism-culture development correlation issue as extensively or intensively as they might have been. It is not always clear in any given experiment where the result of culture manipulation is adaptation and where it is true development; nevertheless, two types of findings are clearly beyond dispute. First, specific growth rate ranges do correspond reproducibly to secondary metabolite-producing states (e.g., Pirt and Righelato, 1967; Bu'Lock et al., 1974); outside these ranges significant production does not occur. Second, ranges of specific growth rate also correspond to induction of conidiation (e.g., Righelato et al., 1968); outside these ranges of growth rate conidiation does not occur. Since the second process involves development, it is tempting to presume that the first also does and that markedly different growth rates correspond, generally, to different phenotypes. This is likely a correct presumption, but its correctness has not been rigorously proven. It is interesting to note that Righelato et al. (1968) found no significant morphological difference in cells of *Penicillium chrysogenum* grown at a full range of specific growth rates.

Overall, the contribution of submerged culture development correlation is to *suggest* that secondary metabolism is the attribute of specific fungal cell phenotypes.

CONTRIBUTION OF SECONDARY METABOLISM-SURFACE LIQUID COLONY DEVELOPMENT CORRELATION STUDIES TO THE CELLULAR LEVEL APPROACH

A major problem encountered in correlation work with submerged liquid cultures is that cell morphology—one of the most reliable indices of cellular differentiation—can be severely distorted. The action of the gyroshaker or the fermenter impeller can wreak havoc on fungal cell morphology. Morphological integrity is, however, preserved if the fungus is allowed to grow as a mat on the surface of a growth medium. Luckner, Nover, and their colleagues have used this system with benzodiazapine (cyclopenin and cyclopenol) alkaloid and quinoline (viridicatin and viridicatol) alkaloid production by *Penicillium cyclopium*. Their results support the notion that secondary metabolism is associated with specific cell phenotypes in a fungal colony and that indeed secondary metabolism is a component of the normal development program of many fungi (for an overview, see Nover and Luckner, 1974; Luckner et al., 1976; Luckner et al., 1977; Martin and Demain, 1978).

The correlation work of Luckner, Nover, and their colleagues can be summarized as follows:

1. Morphologically, the development cycle of *P. cyclopium* can be divided into three stages:
   Spore germination (0-12 hr)
   Vigorous formation of mycelium (12-48/60 hr)
   Hyphal specialization—the formation of penicilli and the process of conidiation (48/60 hr onward)
   The latter two stages were termed, respectively, trophophase and idiophase in accord with the nomenclature of submerged liquid culture. The idiophase was induced by medium dilution at hour 48.

2. Induction of idiophase altered three gross indices of colony development. Mycelia weight tended toward a plateau and the rate of both net protein and net nucleic acid synthesis dropped off.
3. Benzodiazepine alkaloids first appeared in the growth medium around hour 60 and were correlated with the conidial number, an invertase activity, a phenol oxidase activity, and the production of the green pigment characteristic of mature *P. cyclopium* spores.
4. Cyclopenase, the enzyme that effects the benzodiazepine → quinoline alkaloid transformation, was induced some 30 hr after conidiation, but little of its products was found through 216 hr.

From these studies it was construed that benzodiazepine alkaloid formation was the property of cells of the conidial head/conidiospores; that is, it was clearly associated with specific cell phenotypes. Based on the behavior of cyclopenase, the quinoline alkaloids were likely similarly associated.

These conclusions have been amply confirmed. Work with inhibitors of macromolecular synthesis (fluorouracil and cycloheximide) has shown that benzodiazepine alkaloid biosynthesis, as judged by the activity of one of the enzymes in the biosynthetic pathway (cyclopeptine dehydrogenase) and the total alkaloid content, can be inhibited by colony treatment at the induction of idiophase. Moreover, conidiation and the invertase activity (but not the cyclopenase activity) were also reduced by the drugs (El Kousy et al., 1975). It has also been demonstrated that cyclopenase is located exclusively on the inner surface of the conidiospore membrane; its access to the supplies of its substrates that invest the outer surface of conidia is prevented by an energy-requiring membrane-associated exclusion process (Roos and Luckner, 1977).

It is unfortunate that the work of the Martin Luther University group has not had a more immediate impact on the fungal secondary metabolism field. The work clearly associates one aspect of this phenomenon, benzodiazepine alkaloid formation by *P. cyclopium,* with a specific group of differentiated cells. The work is clearly seminal, suggesting that the benzodiazepine and quinoline alkaloids might have a role in conidium maturation and/or eventual germination. Moreover, the facts that benzodiazepine alkaloid formation in *P. cyclopium* cannot be demonstrated in submerged liquid culture (Luckner et al., 1976) and that under the more "natural" conditions of surface culture, cyclopenase and its "normal substrates" are segregated (Roos and Luckner, 1977) may have suggested some caution in relying too heavily on the results of submerged liquid culture in discussing possible secondary metabolite functions.

Our own group tried to apply the same basic surface liquid culture approach to mycophenolic acid production by *P. brevicompactum* (Nulton and Campbell, 1977). It quickly became apparent, however, that mycophenolic acid production occurred in what Luckner and Nover would have termed the trophophase. In pursuing this difference, we resorted to solid culture and, finding it superior, abandoned the liquid culture experiments.

CONTRIBUTION OF SECONDARY METABOLISM-SOLID COLONY DEVELOPMENT CORRELATION STUDIES TO THE CELLULAR LEVEL APPROACH

Solid culture on agar is used extensively in fungal taxonomy, in preserving cultures, and in preparing inocula for liquid culture. It has been used only infrequently in the direct study of fungal secondary metabolism, the general feeling being that the agar adds an unnecessary separatory complication, that additions of isotope tracer to the colony would be difficult, and that the growth of a fungus on solid substrates, particularly if started from a point inoculum, is exceedingly complex (Yanagita and Kogane, 1962). We explored anew the potential of solid culture for three reasons. First, for most of the fungi that engage in secondary metabolism, solid culture is a closer laboratory approach to their natural growth circumstance than is liquid culture. There are reasons to believe that liquid culture in general and submerged liquid culture in particular can perturb the normal metabolic and physiological development of fungi and mask functionally significant information (Campbell et al., 1981). Since we are primarily using correlation methods in search of functional information, solid culture seemed more appropriate than liquid culture. Second, continual microscopic examination of the developing culture allows a morphological index of development to be kept; routine microscopy is more easily accomplished with solid cultures on Petri dishes than with liquid cultures in Erlenmeyer flasks or in fermentors. Third, genetic methods required in certain aspects of the work are geared more conveniently to the solid culture format.

It has been our experience that solid cultures are even easier to handle than liquid cultures. Drugs and isotope tracers can be sprayed onto the cultures in aqueous solution in a controlled, safe, and economical way. Colony morphology can be monitored continually, often without breaking sterility (the Nikon X4 objective has a 22-mm working distance). Provided inocula are spread uniformly over the medium surface, rather than there being a point inoculation, growth is rapid and development over the plate surface is reasonably synchronous. Standard 10-cm Petri plates containing 20 ml of medium provide enough biomass for most analytical purposes. With solid culture, micromanipulation is easily accomplished, making isolation of single intact conidial chains and single penicilli or conidiophores reasonably easy. Micromanipulation also allows drugs to be administered to specific cells, for example, conidiophores. Solid culture on Petri plates is easily scaled down to culture on "welled" microscopic slides or scaled up to culture on 1 X 1-m cake pans.

One disadvantage of solid culture—penetration of vegetative hyphae into the growth medium, with resulting difficulty in cleanly separating the biomass from the growth medium—was resolved by growing the fungus on a sheet of dialysis membrane placed on the agar surface ("over" culture). To the best of our knowledge this procedure was first introduced by Yanagita and Kogane (1962); it has been used, with variation, by several others (e.g., Nover and Muller, 1975).

Our correlation studies using the solid culture format have focused on two fungi, *P. brevicompactum* and *P. patulum*. The former produces mycophenolic acid, the Raistrick phenols, the brevianamides A and B (a diastereoisomeric pair), the pebrolides, compactin, and asperphenamate (Campbell et al., 1981); the latter produces 6-methylsalicylic acid (6-MSA), patulin, a series of biosynthetic intermediates between 6-MSA and patulin, and griseofulvin (Neway and Gaucher, 1981). Both organisms also produce ergosterol—a fungal metabolite we arbitrarily consider to be secondary rather than primary (Hendrix, 1970). Information on 6-methylsalicylic acid production by *P. patulum* and on mycophenolic acid, brevianamides A and B, asperphenamate, and ergosterol production by *P. brevicompactum* is discussed in this article.

The results of our studies are quickly summarized here (for fuller details, see Bartman et al., 1981; Peace et al., 1981; Bird et al., 1981; Bird and Campbell, 1982):

1. 6-Methylsalicylic acid. Formed by *P. patulum* only after an aerial mycelium has begun forming; onset of biosynthesis precedes penicillus formation; uncertainty as to whether biosynthesis occurs in vegetative hyphae, in aerial hyphae, or both; bulk of metabolite is excreted into the growth substrate.
2. Mycophenolic acid. Formed by *P. brevicompactum* only after an aerial mycelium has begun forming; onset of biosynthesis precedes penicillus formation; biosynthesis likely takes place in vegetative hyphae; bulk of metabolite is excreted into the growth substrate.
3. Brevianamides A/B. Formed by *P. brevicompactum* only after conidiation is underway; biosynthesis takes place in conidial heads; bulk of metabolite is located in upper reaches of conidiophore and penicillus.
4. Asperphenamate. Formed by *P. brevicompactum* only after aerial mycelium has begun forming; onset of biosynthesis precedes onset of mycophenolic acid biosynthesis and penicillus formation; biosynthesis likely takes place in aerial hyphae; majority of metabolite located in aerial hyphae.
5. Ergosterol. Formed by *P. brevicompactum* only after aerial mycelium has begun forming; onset of biosynthesis precedes onset of mycophenolic acid biosynthesis and penicillus formation; uncertainty regarding site of synthesis; metabolite distributed between vegetative and aerial hyphae.

Although our biosynthesis studies are not yet complete, these results provide a cogent argument for specific instances of secondary metabolism being associated with specific cell types in the developing colony. Other supportive evidence is available. When penicillia or aspergilli (and presumably other filamentous fungi) are grown between two layers of dialysis membrane on an agar surface ("between" culture, based on work by Clutterbuck and Raper, 1966), no aerial mycelium is formed. These between cultures appear to be exclusively cells of the vegetative phenotype. They are metabolically active, synthesizing fatty acids and protein vigorously. As far as the current work is concerned, however, their most important characteristic is that between cultures of wild-type *P. patulum* or *P. brevicompactum* do not synthesize any of the secondary metabolites listed above. If the upper layer of dialysis membrane in a between culture is stripped off, the biomass that remains on the lower membrane quickly gives rise to aerial hyphae which eventually form penicilli and conidiate. These "stripped between" cultures produce the secondary metabolites in the appropriate time sequence.

Further supportive evidence for the association of mycophenolic acid, asperphenamate, and ergosterol production with colonies of *P. brevicompactum* that possess aerial hyphae but no penicilli, and for the association of brevianamides A and B with colonies of the same fungus that possesses actively conidiating penicilli comes from work with two groups of *P. brevicompactum* developmental mutants (A. T. Remaley et al., unpublished work). Members of the first group form aerial hyphae, but never give rise to penicilli. Every member of this group studied so far is mycophenolic acid plus, brevianamide minus, and ergosterol plus; possession of the asperphenamate characteristic is variable. The second group is characterized loosely as giving rise on minimal medium to colonies that have immature penicilli which form no conidia or that possess seemingly mature penicilli which form no more than 10 conidia. On a complete medium, wild-type morphology is observed. This is a heterogeneous group of mutants most likely impaired (but not completely defective) in the uptake or synthesis of key cell components. As far as the correlation results are concerned, the key finding is that on minimal medium mutants of the second group produce only traces (1% normal) of the brevianamides. On complete medium normal levels of brevianamides A and B appear, that is, the return of normal levels of brevianamide is correlated with the return of normal conidiation.

Our solid culture correlation work indicates that secondary metabolism is attributable to specific cell types in the colony and that these types are relatively easy to identify. Two other interesting observations can be made. First, the brevianamide production/localization patterns in *P. brevicompactum* resemble quite closely those of the benzodiazepines in *P. cyclopium*. Second, the close similarity of the mycophenolic acid-*P. brevicompactum* and the 6-methylsalicylic acid-*P. patulum* correlation patterns in solid culture is in contrast to their divergent nature in submerged culture (see Doerfler et al., 1978; Grootwassink and Gaucher, 1980). Is this further evidence for the potentially perturbing influence of submerged liquid culture?

SECONDARY METABOLISM-CELLULAR DIFFERENTIATION CORRELATION: THE PRESENT STATE OF AFFAIRS

On p. 56, three conditions were listed for secondary metabolism correlation studies at the cellular level to have a future. The first two—attributability of the secondary metabolic trait to specific phenotypes and the ability to identify these phenotypes—have been demonstrated to hold true in a limited number of cases. A considerable volume of additional work has to be done before generality can be claimed for the first precondition. Regarding the second precondition, even at this early stage it seems reasonable to assume that identification will be possible.

The third condition—that experimental systems can be devised in which homogeneous populations of the cell phenotypes that engage in secondary metabolism can be derived in a synchronous manner from the cell phenotypes that are their natural precursors—is more demanding, and for one special reason. The work that has been done to date on colony development correlation, limited though it is, suggests that the cell types that engage in secondary metabolism (aerial hyphae, cells of the penicilli, maturing conidia, arthrospores, sclerotial cells, etc.) are those that normally exist in a nonaqueous environment. This being the case, we doubt the legitimacy of deriving the secondary metabolite-producing phenotype in any form of submerged liquid culture. Attempting to study in a churning, bubbling brew the structural, biochemical, and physiological properties of a cell phenotype that normally exists at rest in an aerial environment has to be considered, at first view, ill advised. This is not to say that submerged liquid culture methods, particularly those run in the continuous mode, will not have a role to play in generating nonproducing progenitor phenotypes (see below).

Excluding submerged liquid cultures leaves us with surface liquid and solid cultures. Within the confines of these fungal growth techniques, methods have to be devised such that a homogeneous cell population can be studied as it differentiates synchronously from a state characteristic of the nonproducing phenotype/progenitor to a state characteristic of the producing phenotype and, if appropriate, to a state characteristic of a subsequent nonproducing phenotype. Access has to be available to metabolic, physiological, and structural indices of the differentiation. The following systems seem worthy of consideration:

1. *Between → stripped between culture transition.* When the upper dialysis membrane is stripped off a between culture of *P. patulum* or *P. brevicompactum*, aerial hyphae form quickly, as do secondary metabolites such as 6-methylsalicylic acid, mycophenolic acid, and asperphenamate. Preliminary work shows that cyloheximide blocks *both* aerial hypha formation and secondary metabolite biosynthesis. Recent evidence indicates that several other penicillia and several aspergilli behave similarly. We think this system has a good prospect of being a good model for cellular differentiation correlation studies. It is easy to establish hundreds of Petri plates in the between state at time zero and monitor the passage of the between to the stripped between phenotype in any desired degree of metabolic, physiological, or structural detail.

2. *Submerged → emerged (surface) culture transition.* Nover and Muller (1975) developed a procedure whereby biomass from a submerged culture was quickly transferred to a dialysis membrane under which a liquid medium flowed. The culture could be observed microscopically and the liquid medium could have its composition varied and could be collected for analysis. Following transfer, the "idiophase" characteristics of benzodiazepine alkaloid formation and conidiation in *P. cyclopium* were quickly expressed. Nover and Muller used batch-mode culture pellets to stock the dialysis membrane and followed the development both under normal conditions and following cycloheximide treatment. This procedure has some potential in approaching those secondary metabolism-cellular differentiation correlation problems involving metabolites whose production is associated with the cells of the conidial heads (penicillus, aspergillum, etc.). A particularly valuable modification of this approach would stock the dialysis membrane with the extremely homogeneous batch-mode cell preparation that Gaucher devised (Neway and Gaucher, 1981) or with preparations from continuous culture. Those modification would require prior detailed characterization of the phenotype of the cells placed on the membrane (hopefully in terms of the properties of the vegetative and aerial phenotypes of solid culture). This having been done, a very powerful vehicle for the study of cellular differentiation would exist.

3. *Conditional mutants.* Temperature-sensitive developmental mutants or, possibly, selected members of the "nutrient-sensitive" mutants we have found for *P. brevicompactum* (mutants of the second group, p. 64) offer a general solution to the problem, albeit with a less favorable "signal-to-noise" ratio than methods (1) and (2). If a temperature-sensitive mutant that was blocked in conidiation at the nonpermissive temperature were transferred to the permissive temperature, correlation of the process of conidiation with any associated secondary metabolic event would be possible. The characterization-correlation information would have to be extracted from the background signals deriving from the rest of the fungal colony. Our preliminary exploration of the temperature-sensitive

option in *P. brevicompactum* has been thwarted so far because *P. brevicompactum* wild type grows poorly at temperatures above 28°C.

4. *Microcycle conidiation.* This procedure induces conidia to develop conidiophores and conidia by bypassing the vegetative hyphal stage (Anderson and Smith, 1971). Most of the work that has been done on this phenomenon has utilized submerged liquid culture (e.g., Sekiguchi et al., 1975a,b,c). If it has relevance in solid culture, microcycle conidiation may be a valuable tool with which to approach secondary metabolism-cellular differentiation correlation studies (Pazoutova et al., 1978). At this point in time we believe that mycophenolic acid is formed in vegetative cells of *P. brevicompactum* on receipt of a signal from newly emerged aerial hyphae. It would be intriguing to determine whether microcycle preparations that nominally lacked vegetative cells produced mycophenolic acid.

The methods listed above have focused on cellular differentiation associated with asexual spore formation. Other methods appropriate to the study of sclerotium formation, the sexual spore formation process, and so on, will need to be devised. We have also excluded spore germination. This is clearly an example of cellular differentiation and one that the *products* of secondary metabolism might influence. Since, however, there is no evidence that secondary metabolite production and spore germination run in parallel, correlation studies would not be appropriate.

SECONDARY METABOLISM-CELLULAR DIFFERENTIATION CORRELATION AND THE FUNGAL SECONDARY METABOLISM FUNCTION QUESTION

The previous sections have established that secondary metabolism-cellular differentiation correlation is possible and have suggested how such studies might be undertaken. The following questions now arise: What practical value will such correlation studies have and how will they help settle the fungal secondary metabolism function question? In response, it must be remembered that secondary metabolism correlation studies at any level are a means to an end rather than an end in themselves. They permit secondary metabolism to be viewed within the detailed context of culture development, colony development, or cellular differentiation. From that viewpoint, function hypotheses can be formulated and experiments to evaluate them designed. A contrived example illustrates the general strategy. Assume that the onset of production of the secondary metabolite A is correlated with the development of conidial heads in a *Penicillium* species. If no conidial heads are formed, no A is produced. Metabolite A is synthesized exclusively in sterigmatal cells, but is found both in sterigmata and immature conidia. Assume further that key enzymes involved in the production of A arise anew when sterigmatal cells are formed from metulae and that mutant colonies that lack A produce morphologically abnormal conidia in smaller than usual numbers. A reasonable hypothesis might be that A was one of the building blocks of the spore wall—a possibility that could be tested, for example, by electron microscopy of conidia in various stages of ripening from wild-type and A-deficient mutant strains of the *Penicillium*.

Three further comments are appropriate. First, in the above example cellular and colony level data were used in concert; in all likelihood this will be usual practice. Second, correlational studies that have a major cellular component will tend to be more applicable to elucidating functions that are themselves cellular. Functional proposals, such as those of Brian (1957) and Zähner (1978), that have significance exclusively at the ecological level will not be well served by cellular level correlation studies. Third and finally, correlation studies will have at least one intermediate benefit. As more of this type of work is done on an ever-increasing number of fungi, patterns of association of secondary metabolites with various common cell types could well emerge. Our information base is still minute, but it is nevertheless intriguing that the two phenolic acids (6-methylsalicylic acid and mycophenolic acid) we have studied are both excreted into the medium, while two amino acid dimers (the brevianamides and the *P. cyclopium* benzodiazepine alkaloids) are both associated with conidial head formation. Patterns such as these, if maintained as the information base is increased, will provide valuable collateral information on which to found hypotheses regarding the why and wherefore of the intriguing fungal phenomenon of secondary metabolism.

ACKNOWLEDGMENTS

The author is very pleased to acknowledge the sterling contributions that his co-workers have made in the last 7 years to the design and execution of those research activities of his laboratory that are described in this article. These co-workers are C. D. Bartman, B. A. Bird, B. N. Davis, J. A. Decker, D. L. Doerfler, F. J. Gottlieb, C. P. Nulton, J. N. Peace, and A. T. Remaley. Without them, nothing would have been achieved. Helpful discussions with M. Sussman are also acknowledged, as is the financial support of the U.S. Public Health Service (GM-25592), the National Science Foundation (PCM 78-03852), and the Gulf Oil Foundation.

REFERENCES

Anderson, J. G., and Smith, J. E. (1971). The production of conidiophores and conidia by newly germinated conidia of *Aspergillus niger* (microcycle conidiation). J. Gen. Microbiol. *69*:185-197.
Audhya, T. K., and Russell, D. W. (1975). Enniatin production by *Fusarum sambucinum*: Primary, secondary and unitary metabolism. J. Gen. Microbiol. *86*:327-333.
Bartman, C. D., Doerfler, D. L., Bird, B. A., Remaley, A. T., Peace, J. N., and Campbell, I. M. (1981). Mycophenolic acid production by *Penicillium brevicompactum* on solid media. Appl. Environ. Microbiol. *41*:729-736.
Bird, B. A., and Campbell, I. M. (1982). Disposition of mycophenolic acid, brevianamide A, asperphenamate and ergosterol in solid cultures of *Penicillium brevicompactum*. Environ. App. Microbiol. *43*:345-348.
Bird, B. A., Remaley, A. T., and Campbell, I. M. (1981). In solid cultures of *Penicillium brevicompactum*, brevianamides A and B are formed only after conidiation has begun. Appl. Environ. Microbiol. *42*:521-525.
Borrow, A., Jefferys, E. G., Kessell, R. H. J., Lloyd, E. C., Lloyd, P. B., and Nixon, I. S. (1961). The metabolism of *Gibberella fujikuroi* in stirred culture. Can. J. Microbiol. *7*:227-276.
Borrow, A., Brown, S., Jefferys, E. G., Kessell, R. H. J., Lloyd, E. C., Lloyd, P. B., Rothwell, A., Rothwell, B., and Swait, J. C. (1964). The kinetics of metabolism of *Gibberella fujikuroi* in stirred culture. Can. J. Microbiol. *10*: 407-444.
Brar, S. S., Giam, C. S., and Taber, W. A. (1968). Patterns of *in vitro* ergot alkaloid production by *Claviceps paspali* and their association with different growth rates. Mycologia *60*:806-826.
Brian, P. W. (1957). The ecological significance of antibiotic production. *Symp. Soc. Gen. Microbiol. 7:168-188.*
Bu'Lock, J. D., Hamilton, D., Hulme, M. A., Powell, A. J., Smalley, H. M., Shepherd, D., and Smith, G. N. (1965). Metabolic development and secondary biosynthesis in *Penicillium urticae*. Can. J. Microbiol. *11*:765-778.
Bu'Lock, J. D., Shepherd, D., and Winstanley, D. J. (1969). Regulation of 6-methylsalicylate and patulin synthesis in *Penicillium urticae*. Can. J. Microbiol. *15*:279-287.
Bu'Lock, J. D., Detroy, R. W., Hostalek, Z., and Munim-al-Shakarchi, A. (1974). Regulation of secondary biosynthesis in *Gibberella fujikuroi*. Trans. Br. Mycol. Soc. *62*:377-384.
Campbell, I. M. (1979). Radiogas chromatography-mass spectrometry: Current status and future prospects. Anal. Chem. *51*:1012A-1021A.
Campbell, I. M., Bartman, C. D., Doerfler, D. L., Bird, B. A., and Remaley, A. T. (1981). The role of secondary metabolism in *Penicillium brevicompactum*. Adv. Biotechnol.: *3*:9-14.
Campbell, I. M., Doerfler, D. L., Bird, B. A., Remaley, A. T., Rosato, L. M., and Davis, B. N. (1982). Secondary metabolism and colony development in solid

cultures of *Penicillium brevicompactum* and *Penicillium patulum*. In *Overproduction of Microbial Products*, V. Krumphanzl, B. Sityta, and Z. Vaněk (Eds.). Academic, London, pp. 141-157.

Clutterbuck, A. J., and Raper, J. A. (1966). A direct determination of nuclear distribution in heterokaryons of *Aspergillus nidulans*. Genet. Res. 7:185-194.

Demain, A. L. (1974). How do antibiotic-producing microorganisms avoid suicide? Ann. N.Y. Acad. Sci. 235:601-612.

Doerfler, D. L., Nulton, C. P., Bartman, C. D., Gottlieb, F. J., and Campbell, I. M. (1978). Spore germination, colony development and secondary metabolism in *Penicillium brevicompactum*: A radiogas chromatographic and morphological study. Can. J. Microbiol. 24:1490-1501.

Doerfler, D. L., Ernst, L. A., and Campbell, I. M. (1980). Biosynthetic intermediates en route to mycophenolic acid in *Penicillium brevicompactum*. J. Chem. Soc. Chem. Commun. 329-330.

El Kousy, S., Pfeiffer, E., Ininger, G., Roos, W., Nover, L., and Luckner, M. (1975). Influence of inhibitors of gene expression on processes of cell specialization during the idiophase development of *Penicillium cyclopium* Wrestling. Biochem. Physiol. Pflanz. 168:79-85.

Grootwassink, J. W. D., and Gaucher, G. M. (1980). *De novo* biosynthesis of secondary metabolism enzymes in homogeneous cultures of *Penicillium urticae*. J. Bacteriol. 141:443-455.

Hendrix, J. W. (1970). Sterols in growth and reproduction of fungi. Annu. Rev. Phytopathol. 8:111-130.

Koffler, H., Emerson, R. L., Perlman, D., and Burris, R. H. (1945). Chemical changes in submerged penicillin fermentations. J. Bacteriol. 50:517-548.

Kurylowicz, W. (1977). The site of antibiotic accumulation in *Streptomycetes* and *Penicillium chrysogenum*. Acta Microbiol. Acad. Sci. Hung.24:263-271.

Lawellin, D. W., Grant, D. W., Joyce, B. K. (1977). Aflatoxin localization by the enzyme-linked immunocytochemical technique. Appl. Environ. Microbiol. 34:88-93.

Light, R. J. (1967a). The biosynthesis of 6-methylsalicylic acid. Crude enzyme systems from early and late producing strains of *Penicillium patulum*. J. Biol. Chem. 242:1880-1886.

Light, R. J. (1967b). Effects of cycloheximide and amino acid analogues on biosynthesis of 6-methylsalicylic acid in *Penicillium patulum*. Arch. Biochem. Biophys. 122:494-500.

Luckner, M., Mothes, K., and Nover, L. (Eds.). (1976). *Secondary Metabolism and Coevolution. Nova Acta Leopoldina Supplement 7*. Deutsche Akademie Naturforscher Leopoldina, Halle.

Luckner, M., Nover, L., and Bohm, H. (1977). *Secondary Metabolism and Cell Differentiation*. Springer-Verlag, Berlin.

Martin, J. F., and Demain, A. L. (1978). Fungal development and metabolite formation. In *The Filamentous Fungi*, Vol. 3, J. E. Smith and D. R. Berry (Eds.). Wiley, New York, pp. 426-450.

Nash, C. H., Huber, F. M. (1971). Antibiotic synthesis and morphological differentiation of *Cephalosporium acremonium*. Appl. Microbiol. *22*:6-10.
Neway, J., and Gaucher, G. M. (1981). Intrinsic limitations on the continued production of the antibiotic patulin by *Penicillium urticae*. Can. J. Microbiol. *27*:206-215.
Nover, L., and Luckner, M. (1974). Expressions of secondary metabolism as part of the differentiation process during the idiophase development of *Penicillium cyclopium* Wrestling. Biochem. Physiol. Pflanz. *166*:293-305.
Nover, L., and Muller, W. (1975). Influence of cycloheximide on the expression of alkaloid metabolism in partly synchronized emerged cultures of *Penicillium cyclopium* Wrestling. FEBS Lett. *50*:17-20.
Nulton, C. P., and Campbell, I. M. (1977). Mycophenolic acid is produced during balanced growth of *Penicillium brevicompactum*. Can. J. Microbiol. *23*:20-27.
O'Farrell, P. H. (1975). High resolution two-dimensional electrophoresis of proteins. J. Biol. Chem. *250*:4007-4021.
O'Farrell, P. Z., Goodman, H. M., O'Farrell, P. H. (1977). High resolution dimensional electrophoresis of basic as well as acidic proteins. Cell *12*:1133-1142.
Pazoutova, S., Rehacek, Z., and Pokorny, V. (1978). Microcycle sporulation in the *Claviceps purpurea* 244. Folia Microbiol. *23*:376-378.
Peace, J. N., Bartman, C. D., Doerfler, D. L., and Campbell, I. M. (1981). 6-Methylsalicylic acid production in solid cultures of *Penicillium patulum* occurs only when an aerial mycelium is present. Appl. Environ. Microbiol. *41*:1407-1412.
Pirt, S. J., and Righelato, R. C. (1967). Effect of growth rate on the synthesis of penicillin by *Penicillium chrysogenum* in batch and chemostat culture. Appl. Microbiol. *15*:1284-1290.
Righelato, R. C., Trinci, A. P. J., Pirt, S. J., and Peat, A. (1968). The influence of maintenance energy and growth rate on the metabolic activity, morphology and conidiation of *Penicillium chrysogenum*. J. Gen. Microbiol. *50*: 399-412.
Roos, W., and Luckner, M. (1977). ATP-dependent permeability of membrane barriers in *Penicillium cyclopium* Wrestling. Biochem. Physiol. Pflanz. *171*:127-138.
Sekiguchi, J., Gaucher, G. M., and Costerton, J. W. (1975a). Microcycle conidiation in *Penicillium urticae*: An ultrastructural investigation of spherical spore growth. Can. J. Microbiol. *21*:2048-2058.
Sekiguchi, J., Gaucher, G. M., and Costerton, J. W. (1975b). Microcycle conidiation in *Penicillium urticae*: An ultrastructural investigation of conidial germination and outgrowth. Can. J. Microbiol. *21*:2059-2068.
Sekiguchi, J., Gaucher, G. M., and Costerton, J. W. (1975c). Microcycle conidiation in *Penicillium urticae*: An ultrastructural investigation of conidiogenesis. Can. J. Microbiol. *21*:2069-2083.

Vorisek, J., Ludvik, J., and Rehacek, Z. (1974). Morphogenesis and ultrastructure of *Claviceps purpurea* during submerged alkaloid formation. J. Bacteriol. *120*:1401-1408.
Yanagita, T., and Kogane, F. (1962). Growth and cytochemical differentiation of mold colonies. J. Gen. Appl. Microbiol. *8*:201-213.
Zähner, H. (1978). The search for new secondary metabolites. In *Antibiotics and Other Secondary Metabolites. Biosynthesis and Production. FEMS Symposium No. 5*, R. Hutter, T. Leisinger, J. Nuesch, and W. Wehri (Eds.). Academic, London, pp. 1-17.

4   Comparative Aspects of Secondary Metabolism in Cell Cultures of Green Plants, Animals, and Microorganisms

Eugene D. Weinberg / Indiana University, Bloomington, Indiana

> Since the behavior of cells outside their own tissue or organ system may differ perceptively from their behavior within the system, unknown or unsuspected capacities may not be revealed unless cells are released from the controls of the system and presented with new histogenetic cues.
>
> *H. Eagle*
>
> Genes define, permit, and sustain biosynthetic reactions; they do not initiate them in the absence of exogenous cues.
>
> *H. Holzer, R. Bischoff, and S. Chacko*
>
> The coordinate gene expression in secondary metabolism under the influence of intrinsic and extrinsic factors reflects a general phenomenon of developmental biology.
>
> *K. Mothes*

INTRODUCTION

The scientific study of secondary metabolism in microbial cell cultures essentially began in 1922, when, at the suggestion of the pioneer biochemist Sir Frederick Gowland Hopkins, Harold Raistrick at the London School of Hygiene and Tropical Medicine initiated his classic study of the biochemistry of fungi. Within the subesquent score of years, the field matured sufficiently to permit, in the 1940s and 1950s, explosive development of industrial production of such important agents as vitamins, antibiotics, steroid hormones, and enzymes. In sharp contrast, secondary metabolism in cell cultures of green (i.e., higher) plants and animals has been studied for only the past one and two decades, respectively.

Thus many of the principles and concepts established for microbial secondary metabolism are not yet clearly associated with the process in plant and animal cells. Moreover, workers engaged in either plant or animal cell research generally do not appear to be aware of either the other group or the considerable amount of knowledge in the area of secondary metabolism that has been generated by microbiologists. This chapter will examine selected aspects of secondary metabolism to determine if, indeed, concepts derived from the study of cells of either plants, animals, or micororganisms are shared with or could be extended to one or both of the other two groups.

SECONDARY PRODUCTS OF PLANT AND ANIMAL CELLS

That cells contained in the various organs of intact plants can produce secondary metabolites has long been appreciated. Such products, like microbial secondary substances, consist of a diverse chemical range of small and large compounds which are generally not essential for growth of the intact plant. The secondary substances are irregular or sporadic in occurrence; they can be useful but not entirely dependable as taxonomic markers. The materials usually are secluded spatially and/or are chemically altered so as to not poison producer cells (Whittaker, 1970), Literally thousands of compounds have been described, including at least 4500 alkaloids, and undoubtedly tens of thousands remain to be discovered. In most cases, only the major components of a given structural class have been examined in any one species of green plant (Swain, 1977).

Upon more intensive search, the number of known plant secondary compounds increases dramatically. For example, detailed investigation of *Vinca rosea*, a source of antitumor indole alkaloids, revealed over 100 of these compounds in addition to its full complement of secondary lipids, sterols, carotenoids, phenolic acids, and flavonoids. Indeed, the total number of all secondary products in plants may well equal the 400,000 known species (Swain, 1977).

Secondary metabolites are distributed unevenly throughout the plant in space and time. Seeds often contain high concentrations of secondary compounds; the amounts decrease quickly after germination and early seedling development. Quantities in various plant tissues often fluctuate diurnally; in some cases, large changes occur in periods as brief as 1 hr. Concentrations may also be altered by exposure of the plant to variations in soil, climate, and air pollution, as well as to herbivores and, of course, to microbial plant pathogens (Swain, 1977).

With some exceptions, secondary plant products can be categorized into five major groups: phenylpropanes, acetogenins, terpenoids, steroids, and alkaloids (Whittaker and Feeny, 1971). As with microbial secondary metabolites, the great diversity of higher plant secondary products arises from surprisingly few starting materials, chief of which are acetic acid and a few common amino acids. Examples of plant products are listed in Table 1 (Swain 1977; Bohm, 1980).

Many secondary substances of plants are released into the environment by (1) rainwash or fog drip from leaf surfaces and glands, (2) volatilization from leaves, (3) excretion or exudation from roots, and (4) decay of either above- or belowground plant parts (Whittaker and Feeny, 1971). Some secondary compounds exert ecological functions while either associated with or disassociated from the plant: for example, deterrant action of alkaloids and tannins toward herbivorous vertebrates, of flavonoids toward invertebrates, of acetylenes to-

Table 1 Some Classes of Organic Compounds in Which Plant Secondary Products Are Found

Class	Example	Class	Example
Acetylenes	Wyerone	Phenolic acids	Vanillic acid
Alkaloids	Lupanine	Polyketides	Hircinol
Amino acids	Canavanine	Polysaccharides	Acylated polysaccharides
Carotenoids	β-Carotene		
Cyanogenic glucosides	Linamarin	Proteins	Lectins
Flavonoids	Naringenin	Quinones	Juglone
Glucosinolates	Sinigrin	Sesquiterpenoids	Paniculide B
Lignans	Excelsin	Steroids	Ecdysones
Lipids	Waxes	Terpenes	Glaucolide A

ward microbial pathogens, or of quinones toward plant competitors; or attractant action of some terpenes toward useful invertebrates.

As is the case in microbial secondary metabolism, some known plant compounds do not as yet have an assigned ecological function. It has been suggested for microbial (Weinberg, 1971) and plant (Whittaker, 1970) secondary metabolism that originally all the compounds were formed as "shunt" metabolites and that those which, by chance, possessed biological utility were then selected for retention and exploitation by species that were evolving in a particular environment. Levin (1976) has suggested that "the diversity of secondary compounds is an evolutionary product, indefinite and indeterminate, and subject to self-augmentation through time."

Some compounds appear to be needed in the primary metabolism of one type of organism, but may be secondary metabolites for another. Such plant hormones as cytokinins, gibberellins, and auxins can be produced as secondary substances by nonplant forms of life. For example, high cytokinin activity has been observed in (1) virulent, plasmid-containing strains of *Corynebacterium fascians* and *Agrobacterium tumefacians,* (2) *Rhizopogon roseolus,* and (3) gall-forming insects (Murai et al., 1980). Plant cell DNA that codes for the synthesis of these hormones may have been acquired from the host plants by the microbial or invertebrate cells; overproduction of the hormones by nonplant organisms could then disrupt balanced plant cell growth to enhance establishment of the parasitic forms.

As a corollary, host plant cells might be able to acquire microbial or invertebrate DNA; an example is that of *A. tumefacians* Ti plasmids that induce plant tumor cell growth and synthesis of octopine or nopaline. These unusual amino acid derivatives, secondary substances for the producer plant cells, can then be used as a source of carbon and nitrogen by oncogenic, but not nononcogenic, strains of *A. tumefacians* (Zambryski et al., 1980). A possible example of plant acquisition of invertebrate DNA is observed in the production of phytoecdysones. In some cases, the concentration of these molting hormones is higher in the plant than in the insect (Levin, 1976). These terpenoid plant secondary products can cause severe derangement of insect growth and metamorphosis. Alternatively, identical gene products may have evolved in parallel in plants and insects, or the invertebrates might have acquired the DNA from plants and then utilized the genetic product as a hormone.

Human chlorionic gonadotropin consists of two subunits coded by two genes. Some, but not all, bacteria that have been isolated from human tumors synthesize detectable amounts of the protein (Acevedo et al., 1978; Backus and Affronti, 1981). Bacterial isolates from nonmalignant tissues fail to produce the hormone. Conceivably, the relevant genes could be present in the bacteria as a result of conservation or as an example of convergent evolution. However, inasmuch as the genes appear to be expressed only in bacterial strains that earlier had been associated with tumor cells, the DNA may possibly have been acquired from the host.

Secondary metabolites produced by animal cells appear to be considerably fewer and less diverse than those of either plants or microorganisms. Woodruff (1966) suggested that animal cells have developed sufficient specialization and interdependence so that excess synthesis of shunt metabolites is not needed to retain cell viability. In place of the ecological functions provided by some higher plant and microbial secondary substances, animals may have substituted behavioal mechanisms (Swain, 1977).

Nevertheless, within the animal kingdom, a fairly large number of secondary products are produced by invertebrates (see Eisner, 1970); fewer are found in chordates. Of the animal secondary metabolites, most have a recognized function, structural, coordinative, or defensive. Secondary compounds of animal cells have been termed "luxury" metabolites because they do not appear to be needed for the growth and survival of the producer cells (Holtzer et al., 1969); however, use of this term has been discouraged (Rutter et al., 1973) inasmuch as the compounds generally are "necessities" for either the structure, tissue coordination, or defense of the intact animal. Examples of compounds that can be produced in vertebrate animal cell culture and which appear not to be needed, at least in the quantity synthesized, for the growth or survival of the producer cells are listed in Table 2 (Green and Todaro, 1967; Rutter et al., 1973; Cox and King, 1975). These products are functional within the intact animal to a much greater extent than the substances in Table 1 are useful within the intact plant. Moreover, Table 2 contains a more diverse mix of substances (hormones, enzymes, antibodies, structural compounds, etc.) than Table 1. The unifying theme of Table 2 is that each example is a specialized product which is synthesized in large quantities by animal cells that are differentiated rather than replicating.

Table 2 Examples of Secondary Products Produced in Vertebrate Animal Cell Culture

Cell type	Products
Adrenocortical tumor	Steroids
Amnion	Hyaluronic acid
Chondrocyte	Collagen, chondroitin sulfate
Choriocarcinoma	Human chorionic gonadotropin
Endothelial	Von Willebrand factor
Erythroblast	Hemoglobin
Fibroblast	Collagen, chondroitin sulfate, hyaluronic acid, interferon
Granulocyte	Macrophage and granulocyte development inhibitor
Hepatic (avian)	Phosvitin
Hepatic (parenchymal)	Tyrosine transaminase
Hepatoma (parenchymal)	Tyrosine transaminase, ornithine transaminase, histidase
Lens	Crystallin
Leydig (tumor)	Steroids
Lymphoblast	Ig and T cell mediators
Lymphoma	IgG, IgM, IgA
Macrophage	Macrophage and granulocyte development inhibitor
Mammary gland	Casein
Mast (tumor)	Histamine, serotonin
Melanocyte	Melanin, dopaoxidase, tyrosinase
Melanoma	Melanin, dopaoxidase, tyrosinase
Muscle (cardiac)	Myosin, creatine kinases
Muscle (smooth)	Elastin
Muscle (striated)	Actomyosin, creatine kinase
Neuroblastoma	Axone formation
Osteoblast	Collagen, sulfated polysaccharides, alkaline phosphatase
Pancreatic β cells	Insulin
Parathyroid (tumor)	Parathormone
Pineal (virus transformed)	Hydroxyindole-O-methyl transferase
Pituitary (tumor)	Growth hormone, prolactin
Yolk sac (teratoma)	Hyaline (basement membrane), collagen

SECONDARY METABOLISM IN PLANT AND ANIMAL CELL CULTURE

Plant cell production of secondary substances is generally studied in suspension rather than in callus cultures. Suspensions consist of cells that are single or in short chains and which can be transferred by pipette. The cells need not be derived from organs in the plant than contain the product in nature. In an aerated batch suspension culture, the generation time is between 1 and 2 days; usually about three to five generations are attained. Advantages of cell suspension culture for secondary product formation over the intact plant include (1) homogeneous cell population, (2) freedom from contaminating microbial and insect organisms, (3) ease of manipulation and assay of cell growth, and (4) in some cases as much as a 10-fold increase in product yield over that obtained in the plant (Tabata, 1977; Bohm, 1980). Moreover, like microbial systems, plant cell cultures can be employed to convert compounds to more useful substances; for example, glucosylation of phenolics or racemization of nicotine. Occasionally novel metabolites, hitherto undiscovered in the intact plant, are produced in cell culture (Tabata, 1977).

Problems that may be encountered in the use of plant cell suspension or callus cultures for secondary metabolism include (1) decay in product yield upon repeated subculture, (2) accumulation of intermediates rather than the final desired product, and (3) inability to identify exogenous precursors that might enhance yield.

Animal cell production of secondary substances is generally studied in cultures in 60-mm Petri dishes rather than in suspension. Cells must usually be derived from tissues in the animal that contain the product in nature. The differentiated functions can be expressed by cells of both strains and lines, although in some cases, the biosynthesis can be accomplished only by cells of a line. By starting with tumors originating from the desired cell type, especially transplantable tumors, established lines can frequently be developed that retain the differentiated function. Moreover, the much higher plating efficiency of a line over a strain is very convenient. Care must be taken to ensure that the line has not been comtaminated with foreign cells (Green and Todaro, 1967).

As with plant cell cultures, about three to five generations are customarily obtained during a 6-10 day period. Problems encountered in animal cell production of secondary metabolites include (1) decay in product yield upon repeated subculture, (2) overgrowth by other cell types from either the original tissue or laboratory contaminants, and (3) incomplete knowledge concerning special requirements for expression of the differentiated function (discussed below).

In both plant and animal cell cultures, as is true for microbial systems (Weinberg, 1970), the bulk of the secondary product is formed by cells that have recently stopped dividing (Green and Todaro, 1967; Gresser, 1977). In most instances, several cell divisions appear to be necessary just prior to overt cytodifferentiation and initiation of synthesis of secondary metabolites. Perhaps cell division is needed to enable (1) the replacement of chromosomal regulatory proteins, (2) the assumption of a new chromosomal conformation, or (3) the replacement of cell membrane determinants (Rutter et al., 1973). These authors suggested that "growth and differentiative functions may not be incompatible but may be inversely regulated by the same effectors." Generally, a small quantity of secondary product is formed during the period of maximum cell replication (Fig. 1). It is not clear to what extent synthesis of the small amount might be occurring only in those cells whose G_1 phase had become so lengthened that they have essentially lost, at least temporarily, the ability to undergo mitosis (Konigsberg, 1975).

In contrast to the production of secondary metabolites, viral replication proceeds better in dividing than in nondividing cells. It is perhaps more than a coincidence that interferon inhibits animal cell division and viral replication but stimulates such secondary cell functions as the synthesis of H-2 antigen in mouse mammary tumor cells (Gresser, 1977).

In some cases in either plant or animal cell cultures, secondary compounds are formed in the absence of recognizable morphological differentiation. In *Pinus sylvestris,* resin accumulates in a resin duct in the intact plant, whereas in culture, anatomical development is not needed for product formation. In contrast, to obtain saikosaponins in culture, the cells must be induced to differentiate into root (Tabata, 1977). In such animals systems as erythroblasts, lenticular cells, or keratin-synthesizing epidermal cells, the nuclei degenerate and disappear prior to product formation (Green and Todaro, 1967). In myogenesis in culture, the cell plasmalemma fuses to produce multinucleated myotubes (Holtzer et al., 1969). In the adipose differentiation, fibroblasts become spherical adipocytes while undergoing a significant decrease in the content of actin fibers. Simultaneously, the conversion from the preadipocyte to adipocyte involves a very large shift in the protein composition of the cell; at least 40% of the total soluable adipocyte protein is specifically related to the differentiation (Spiegelmann and Green, 1980). When fibroblasts are most actively engaged in collagen synthesis (stationary phase), their cytoplasm is structurally differentiated for secretion of the product (Green and Todaro, 1967).

Plant and animal cell culture research workers, like microbiologists, have examined a variety of nutritional and environmental factors for possible modulation of product yield. Generally the sets of evaluated factors have little overlap; that is, each of the three groups of investigators has its own array of test items.

In plant cell suspension of callus systems, such natural and synthetic hormones as kinetin, 2,4-dichlorophenoxyacetic acid (2,4-D), and α-naphthalenacetic acid (NAA) are routinely tested. In general, the quantities employed have little effect on vegetative cell yield, but markedly suppress the amount of sec-

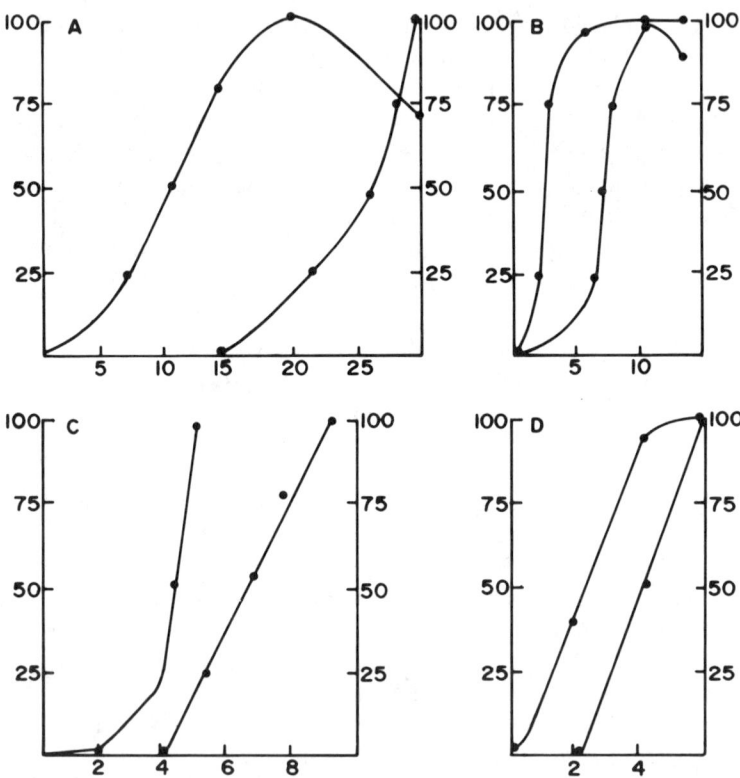

Figure 1 Typical examples of the kinetics of the growth and production of secondary metabolites in cell cultures of green plants and animals [ordinate = percentage of maximal yield; abscissa = days of incubation; left curves = cell dry weight (A,B) and cell number (C,D); right curves = serpentine (A), anthraquinones (B), collagen (C), and H-2 antigen (D)]. (A) *Cantharanthus roseus*, (B) *Morinda citrifolia*, (C) mouse fibroblast, and (D) murine lymphoma. [Redrawn from data in (A) Fig. 10 of Zenk et al., 1977; (B) Fig. 1, Wagner and Vogelmann, 1977; (C) Fig. 1, Green and Goldberg, 1964; and (D) Fig. 1, Cikes, 1970.]

ondary product (Table 3) (Shio and Ohta, 1973; Misawa et al., 1974; Tabata, 1977; Zenk et al., 1977; Bohm, 1980; Kinnersly and Dougall, 1980). Quantities of kinetin greater than 0.5 μm were observed to increase the size of cell aggregates to over 170 μm, whereas cells that produced very high amounts of anthocyanin were contained in aggregates smaller than 63 μm (Kinnersly and Dougall, 1980). In some instances, a small amount of plant hormone enhanced secondary product yield without altering cell yield; for example, 0.5 μm of 2,4-D stimulated

Table 3 Examples of the Effect of Plant Hormones on Cell and Product Yields

Wild carrot cell suspension culture[a]			Tobacco callus culture[b]		
Kinetin (μM)	Cell dry weight (mg)	Anthocyanin/ml (A_{530} nm)	α-NAA (μM)	Cell dry weight (mg)	Nicotine (%)
None	1.98 ± 0.06	2.34 ± 0.06	0.5	342	0.33
0.5	2.12 ± 0.06	1.64 ± 0.08	5.0	316	0.12
5.0	1.98 ± 0.05	0.88 ± 0.04	50	299	0.00

[a]Kinnersly and Dougall (1980).
[b]Shio and Ohta (1973).

synthesis of diosgenin (Tabata, 1977). In cell cultures of *Nicotinia tabacum* 2,4- 2,4-D inhibits activity of putrescine-N-methyl transferase, so that putrescine, instead of being converted to nicotine, remains free to combine with p-coumaric acid to yield p-coumaroylputrescine. Likewise, in *Lithospermum* sp., 2,4-D blocks metabolism of geranylhydroquinone, thus preventing accumulation of shikonin (Tabata, 1977).

As is the case for microbial fermentations, it is not simple to find the most effective precursor(s) for increasing the content of final product. Sometimes a precursor must be added daily in small amounts to avoid injury to the cells. Formation of indole alkaloids (e.g., serpentine and ajmalicine) was enhanced threefold by inclusion of L- but not D-tryptophan (Zenk et al., 1977). Unlike the situation in *Claviceps* sp., tryptophan derivatives were inactive.

In addition to examination of the effect of hormones on plant cell secondary metabolism, various wavelengths of light have been tested. White or blue, but not red, light suppressed yield of terpenoids and shikonin (Tabata, 1977). Blue as well as ultraviolet light induced synthesis of phenylalanine ammonialyase (PAL), which catalyzes deamination of phenylalanine to cinnamic acid (Matsumoto et al., 1973; Schroder et al., 1976). Phenylalanine ammonialyase is the first in a sequence of at least 10 enzymes that ultimately results in flavone products. Ultraviolet (UV) light has also increased the yields of carotenoids, flavonoids, polyphenols, and plastoquinones (Tabata, 1977); the increase in carotenoid synthesis may in turn protect the cells from photodestruction.

In cells exposed to UV light just prior to entry into stationary phase, increased amounts of PAL are synthesized after a 2-2½ hr lag. The inducing effect of UV light is limited to only a few hours; the enzyme attains a peak amount 12-27 hr after induction and has a half-life of 10 hr (Schroder et al., 1976).

In contrast to work in microbial cell cultures, very few plant culture studies have been even partially concerned with examining possible modulation of the yield of secondary products by temperature, Eh, pH, or mineral element constituents of the basal medium. Tabata (1977), however, did observe that in *Datura innoxia* anaerobiasis enhanced glycosylation of phenolic compounds sixfold, and in sycamore cell suspension cultures an increase in the ratio of carbon to nitrogen in the medium resulted in increased formation of catechol tannins. The quantity of iron, but not manganese, needed for bud initiation in excised cotyledons of *Lactuca sativa* was greater than that required for growth of the callus cells (Doerschug and Miller, 1967). Reduction of the phosphate concentration from 5.0 to 1.25 mM had no effect on the amount of cell growth of *Hydrangea macrophylla*, but permitted a threefold increase in the yield of polyphenols (Suzuki et al., 1981).

In animal cell synthesis of secondary metabolites, even when highly active clones of the desired differentiated cell type can be obtained in pure culture, the expression of the differentiated function may have special requirements. One or more of the commonly known animal hormones may be required for maximum product yield. The protein hormones apparently exert their effects at the cell membrane. Of interest in this regard are observations that the number of membrane binding sites for insulin increases two- to ninefold in fibroblasts in stationary phase, as compared with cells in growth phase (Thomopoulus et al., 1976; Reed et al., 1977).

Steroid hormones, on the other hand, bind to receptor proteins in the cytosol; the complex is then translocated to the nucleus, where its presence results in altered template activity so that increased and/or decreased levels of specific proteins are produced (Baxter and Rousseau, 1979). Dexamethasone (DX) is more active in multiplying than in resting cells (Tomkins et al., 1969; Gerschenson et al., 1970). The hormone induces synthesis of tyrosine aminotransferase in the latter portion of the G_1 phase and during the S phase of the cell cycle. Once induced, the enzymes can be synthesized in all phases of the cycle. In hepatoma cells, DX can induce, with a 2-hr lag, tyrosine α-ketoglutarate transaminase in either the growth or the stationary phase; continuous presence of 10-μm hormone is required (Thompson et al., 1966). As predicted, formation of the enzyme is blocked by inhibitors of RNA and protein synthesis.

Green and Todaro (1967) noted that the amount of change in enzyme activity induced by hormones or substrates is usually less than 15-fold, whereas the degree of concentration achieved in "irreversible" differentiation (e.g., to collagen or hemoglobin synthesis) may be as great as four to six orders of magnitude. Moreover, some highly specialized differentiated cells such as the erythrocyte, in which 90% of the cell protein is hemoglobin, lose the ability to return to a replicative phase.

In some cases, a combination of hormones is required for expression of the differentiated function. Mammary gland epithelial cells require hydrocortisone plus prolactin plus insulin in order to form casein (Stockdale and Topper, 1966); the action of the latter two is suppressed by zinc (Rillema, 1979). Stimulation of the fibroblast differentiation to adipocytes by DX is enhanced by insulin (Hiragun et al., 1980). On the other hand, albumin production by hepatocytes was enhanced by either hydrocortisone or insulin (Leffert et al., 1978). Glucocorticoids enhance zinc uptake in hepatocytes presumably by altering one or more components of the transport system. When a threshhold quantity of intracellular zinc is exceeded, the metal induces or increases the synthesis of metallothionein mRNA (Cousins and Failla, 1980).

In the differentiation of fibroblasts to adipocytes, insulin and biotin are needed to induce formation of the mRNAs of some, but not all, of the proteins whose rate of synthesis is increased (Spiegelmann and Green, 1980). Enzymes whose production is dependent on these exogenous factors include glyceraldehyde phosphate dehydrogenase, lactate dehydrogenase, and, to a small extent, aldolase. Virtually insensitive to insulin and biotin is the formation of fatty acid synthetase and glycerophosphate dehydrogenase. The latter enzyme, needed to form glycerol-3-phosphate, the substrate for fatty acylation, increases over 2000-fold during adipose conversion. Much or all of the increase is due to an enhanced amount of the corresponding translatable mRNA (Spiegelmann and Green, 1980). Accompanying the differentiation to adipocytes are changes in the biosynthesis of at least 30 cytoplasmic proteins, 24 membrane proteins, and 9 nonhistone chromosome-associated proteins (Cryer, 1980).

In addition to insulin and biotin, adipogenic factor, as yet uncharacterized, is required for conversion of mouse fibroblasts to adipocytes. The factor is present in high amounts in fetal calf serum, less in bovine serum, and is absent in feline serum. Possibly the latter might contain a factor active for feline but not murine cells. Adult human and porcine sera contained amounts of the factor comparable to that of bovine serum, whereas equine serum was less active (Kuri-Harcuch and Green, 1978).

In contrast to its stimulating effect on the adipose differentiation, serum contains a factor(s) that suppresses axon formation by neuroblastoma cells (Seeds et al., 1970) and collagen synthesis by avian tendon cells (Schwarz et al., 1976). Quantities of serum greater than 0.5% are active; in the axon differentiation, the factor is heat stable and present in both the α_1 and α_4 globulin fractions and, to a lesser extent, in the β-globulin fraction. However, in many cell lines, differentiated functions are equally well expressed in both serum-supplemented and serum-free, hormone-supplemented media (Barnes and Sato, 1980). Unlike the suppressing effect of serum, collagen synthesis is enhanced by anaerobiasis (Schwarz et al., 1976) and ascorbic acid (Schwarz and Bissell, 1977).

Table 4 Examples of Differentiating Systems That Are Altered by
5-Bromodeoxyuridine

Cell type	Product or process[a]
Repression	
Amnion cells	Hyaluronic acid
Chondrocytes	Chondroitin sulfate
Fibroblasts	Conversion to adipocytes[b]
Erythroblasts	Hemoglobin
Hepatocytes (avian)	Phosvitin
Hepatoma cells	Tyrosine aminotransferase
Lymphocytes (primed)	Antibodies
Mammary gland cells	Casein, α-lactalbumin
Melanoma cells	Melanin, plasminogen activator[c]
Myoblasts	Myotube formation, myosin
Pancreatic acinar cells	Exocrine enzymes
Pancreatic β cells	Insulin
Induction	
Lymphoid cells	Epstein-Barr virus
Lymphoblastoid cells	
(primed by Sendai virus)	Interferon[d]
Mammary carcinoma cells	Alkaline phosphatase
Neuroblastoma cells	Neurons
Pancreatic acinar cells	Alkaline phosphatase
Pituitary tumor cells	Prolactin[e]

[a]References are cited in Table 2 of Rutter et al. (1973), except for those cited here.
[b]Green and Meuth (1974); Russell (1979).
[c]Silagi (1976).
[d]Baker et al. (1979).
[e]Biswas et al. (1977).

Such polar solvents as sodium butyrate (1000 μm), dimethylsulfoxide (DMSO), and N,N-dimethylformamide induce differentiated functions in animal cells, presumably by alteration of cell membranes. For instance, the solvents increased creatine kinase activity 5- to 20-fold in murine rhabdomyosarcoma cells (Dexter et al., 1981). Likewise, DMSO increased by two- to sixfold the number of transferrin binding sites on Friend erythroleukemia cells (Hu et al., 1977; Glass et al., 1978; Yeoh and Morgan, 1979).

A compound frequently tested in research on animal cell culture secondary metabolism is 5-bromodeoxyuridine (5-BrdU). This substance is readily incorporated into DNA in place of thymidine by all cellular and in vitro DNA-synthesizing systems. Quantities that exert little influence on total DNA, RNA, or protein synthesis, on the overall rate of cell proliferation, or on viability can, during a restricted period in the culture cycle, generally inhibit differentiation (Rutter et al., 1973). Moreover, in melanoma cultures, the compound causes the cells to become flattened, more adhesive, and sensitive to density-dependent inhibition, as well as lose tumorgenicity (Silagi, 1976). The effect of 5-BrdU is specifically and competitively inhibited by thymidine.

In some systems, 5-BrdU induces rather than represses specialized processes (Table 4). Of the several hypotheses concerning the ability of the compound to inhibit differentiative processes, the most attractive is that poly[d(A-BrdU)] binds repressor proteins more tightly than does poly[d(A-T)] (Rutter et al., 1973). However, the actual molecular mechanisms of selectivity of either repression or induction as well as of the restricted period of sensitivity to 5-BrdU have not yet been established.

PERSPECTIVES AND CONCLUSIONS

It is clear that many aspects of secondary metabolism and differentiation are shared by cells of plants, animals, and microorganisms. An outstanding example is the dependence of the process on the developmental stage of the tissue or batch culture. Furthermore, in each of the three organismic groups, a highly programmed set of genes must be derepressed to code for de novo synthesis of mRNA and protein, while some genes that are needed during growth must be repressed. In each group of organisms, the quantity of specialized product formed by the differentiating cell is under both genetic and environmental influences. An intriguing matter is that of understanding the mechanism(s) whereby nutritional and environmental messages control the expression of genes whose products are secondary metabolites (see Weinberg, 1978).

Each organismic group shares the problem of "safe" disposal of the secondary product; in many cases, of course, the substances are employed as structural units in the differentiating cell; in others, the secondary product is packaged in a storage compartment or, in many instances, secreted into the environment. In each of the three groups of organisms, many of the secondary metabolites have recognized specialized functions, structural, coordinative, or defensive. Despite the apparent lability in the qualitative and quantitative formation of secondary products, their presence can be quite useful in taxonomy. Indeed, many of the distinctive features that assist in the identification of species and varieties are provided by secondary metabolites.

Table 5 Secondary Metabolites of Differentiated Cell Forms Whose Synthesis Requires a Finer Adjustment of Zinc Concentration Than Is Necessary for Maximal Vegetative Growth of the Producer Fungus[a]

Aflatoxin	Gentisyl alcohol	Napthazarin	Sclerotia (*Whetzelinia*)
Cynodontin	Griseofulvin	Oogonia (*Pythium*)	Vesicles (*Puccinia*)
Ergotamine	Lysergic acid	Patulin	Yeast → mycelia (*Candida*)
Fusaric acid	Malformin	Penicillin	Zearalenone

[a]References to each item are cited in Table 4 of Failla and Weinberg (1980). An additional reference to the role of zinc in yeast → mycelia differentiation is Bedell and Soll (1979).

Probably because of their relatively recent entry into the field of synthesis of secondary products in culture, plant and animal cell biologists have made little use of the knowledge gained in each other's area or in the more established field of microbial secondary metabolism. For example, little attention has been given to the possible effects of light on animal cell differentiation or of 5-BrdU on plant cell differentiation. Indeed, these entities should be more intensively examined as well in microbial cell culture differentiations.

Most astonishing from the perspective of microbiology is the relative lack of attention given by plant and animal cell culture research workers to varying such components of the environment as minerals, pH, Eh, and temperature at the time of initiation of the phase of induction of secondary metabolism and differentiation. In microorganisms, the permissible range of concentrations of specific transition series metal(s) and inorganic phosphate, as well as the range of pH, Eh, and temperature, is considerably narrower for expression of the genome of secondary metabolism than that tolerated for growth of the producer cells. The mediating factors appear to operate at the level of transcription, but may also function at the levels of translation and activation of synthetases of secondary metabolism (Weinberg, 1977, 1978).

A few observations of the potential importance of transition series metals in secondary metabolism and differentiation are presented below. For example, the concentration of zinc has long been known to be of critical importance in the morphological and biochemical differentiation of fungi (Table 5), and cyclical uptake of the metal has been observed in both *Candida utilis* (Failla and Weinberg, 1980) and mouse fibroblast cells (Schwarz and Matrone, 1975). In batch cultures of *C. utilis*, "activated" zinc transport occurs in late exponential phase; the K_m of the high-affinity transport system remains similar to that in the early and midexponential phase (2.0 and 1.8 mM, respectively), but the V_{max} increases 17-fold to 3.65 nmol Zn^{2+}/min mg dry wt (Failla and Weinberg, 1980).

Zinc metalloenzymes of both prokaryotic and eukaryotic organisms represent a diverse group of catalytic proteins (oxidoreductases, transferases, hydrolases, lyases, an isomerase, and a ligase) and are required for nucleic acid, protein, carbohydrate, and lipid metabolism (Failla and Weinberg, 1980). Sea urchin DNA and RNA polymerases I, II, and III, wheat germ RNA polymerase II, rat liver RNA polymerases I and II, as well as animal virus reverse transcriptases are zinc dependent. Tightly bound zinc ions present in these polymerases might be required for template binding, as well as for the initiation and elongation of the newly synthesized chain. In *Euglena gracilis,* zinc is required for the structural integrity of ribosomes.

Rat liver fructose-1,6-bisphosphatase is inhibited by 0.3 μM zinc and activated by 10 μM zinc at neutral pH. When present at the low concentration, zinc binds solely to a high-affinity site and inhibits enzyme activation by either magnesium or manganese in a noncompetitive manner (Tejwani et al., 1976). When the zinc concentration is elevated, the metal replaces magnesium or manganese from the lower-affinity activation site and the inhibitory effects of zinc are abolished. Might similar allosteric zinc-binding proteins have a role in regulating secondary metabolic processes of plant, animal, or microbial cells? The concentrations of many precursor substances are altered as cells begin to reach stationary phase. What effects do altered levels of low molecular weight zinc-complexing agents (e.g., histidine, cysteine, and nucleotides) have on the intracellular distribution of the metal and on secondary metabolic synthetase formation and activity?

Among other transition series metals whose effects should be examined on secondary metabolism and differentiation in plant, animal, and microbial cells is copper. The human plasma tripeptide glycyl-L-histidyl-L-lysine (GHL) combined with equimolar copper and about 1/5 M iron stimulates adhesiveness and intercellular attachment of hepatoma cells (Pickart and Thaler, 1980; Pickart et al., 1980). The authors have proposed that GHL acts as a copper transport factor and mediates some basic biochemical function common to many types of cells and organisms. They pointed out that, among other actions, 0.05-0.5 μM GHL enhances the growth of hepatoma and thyroid follicular cells, as well as the growth and differentiation of neurones.

Specific concentrations of copper can be critical for the formation of such fungal products as penicillin (Koffler et al., 1947) and *Candida albicans* mycelia (Vaughn and Weinberg, 1978). The metal activates enzymatic oxidation of polyphenols to quinones and can thus be important in secondary metabolism in the cells of plants, animals, and microorganisms. Plant cell basal media generally contain about 0.01 μM copper; presumably this low amount is likewise present as background in many media formulations used in animal cell and microbial cultures. A range of concentrations of copper between 0.01 and 1.0 μM ± GHL should be supplied to cells at various times during the growth and early stationary phase to identify possible mediation of secondary metabolism and differentiation. It will be of considerable interest to determine if GHL or an alternate peptide might function as a copper transport factor in fungi and cells of higher plants. Additionally, might cells vary during the batch culture cycle in regard to their ability to bind and transport the GHL-copper-iron complex?

Additional transition series metals that should be screened in a wide range of concentrations for possible effects on secondary metabolism and differentiation include vanadium, chromium, manganese, iron, and molybdenum. Iron is especially intriguing. For example, in *Aspergillus nidulans,* a change has been observed in iron availability in the mycelia during differentiation. The authors proposed that at that stage in the batch culture cycle, iron or an iron-binding component might regulate genetic activity at either (or all) of the levels of chromosome structure, transcription, and translation (Hall and Axelrod, 1978). To obtain germination of conidia of *Colletotrichum musae,* conidial iron must be either removed from the cells or transported to an intracellular noninhibiting site (Harper et al., 1980).

Synergism of iron and ascorbic acid has been reported in the production of chromosome aberrations in cultured Chinese hamster ovary cells (Stich et al., 1979). In mammalian (and perhaps all eukaryotic) cells, iron is present in nucleoli during interphase, but is associated with chromosomes during mitosis (Robbins and Pederson, 1970). The metal is required for activity of ribonucleotide reductase, which catalyzes the first unique step in DNA synthesis and provides the cell with a balanced supply of the four deoxyribonucleotides (Thelander and Reichard, 1979).

Abnormal amounts of intracellular metals, including iron, have been proposed to induce neoplasia by a continuous derepression of genes coding for metal ligands or by causing conformational variation in reiterative DNA sequences to modify the binding of repressors and polymerases (Guille et al., 1979). In either normal or neoplastic cells, genetic reiteration may be required during differentiation because of the prodigious demand for a particular genetic product that cannot be met quickly by the transcription capacity of a single gene (Rutter et al., 1973).

Neoplastic cells have an increased capacity to bind transferrin (see Weinberg, 1981). Additionally, malignant and transformed cells might differ from normal cells in their ability to synthesize or use low molecular weight iron-binding compounds in acquisition or intracellular use of the metal. A SV_{40} virus-transformed mutant strain of BALB/3T3 fibroblasts produced a highly specific iron-binding peptide of approximately 1600 daltons that stimulates iron uptake and DNA synthesis (Fernandez-Pol, 1978). Production of the siderophore-like factor is transformation dependent. In contrast, iron deprivation induced synthesis of membrane procollagen proteins by normal, but not virus-transformed, rat kidney cells (Fernandez-Pol, 1980). From these and numerous other observations made on the roles of iron and the problems of its uptake (see Weinberg, 1970, 1978, 1980; Doerschug and Miller, 1967), we can predict that availability of the metal at the appropriate time and intracellular site should be a key factor in the outcome of differentiating systems in cells of plants, animals, and microorganisms.

In conclusion, secondary metabolism and differentiation in cells of higher plants, animals, and microorganisms share many features. It has finally become possible, in each of the three groups of organisms, to rigorously study the process in homogeneous cell cultures. Thus principles and concepts observed in one group of organisms can be extended in theory and sometimes in practice to each of the two other groups. Accordingly, we can begin to anticipate rapid advances in understanding the mechanisms of regulation of the process both within and between members of the three groups. As a corollary, we can expect the emergence of applied spin-offs of benefit to agriculture, medicine, and our environment.

ACKNOWLEDGMENTS

My sincere thanks to Catherine Olsheski Andrews, whose thoughtful M.A. thesis awakened my interest in "luxury" metabolites, and to Professor Mark L. Failla for valuable advice and continuing encouragement.

REFERENCES

Acevedo, H. F., Slifkin, M., Pouchet, G. R., and Pardo, M. (1978). Immunohistochemical localization of a choriogonadotropin-like protein in bacteria isolated from cancer patients. Cancer *41*:1217-1229.

Backus, B. T., and Affronti, L. F. (1981). Tumor-associated bacteria capable of producing a human choriogonadotropin-like substance. Infect. Immun. *32*: 1211-1215.

Baker, P. N., Bradshaw, T. K., Morser, J., and Burke, D. C. (1979). The effect of 5-bromodeoxyuridine on interferon production in human cells. J. Gen. Virol. *45*:177-184.

Barnes, D., and Sato, G. (1980). Methods for growth of cultured cells in serum-free medium. Anal. Biochem. *102*:255-270.

Baxter, J. D., and Rousseau, G. G. (1979). Glucocorticoid hormone action: An overview. In *Glucocorticoid Hormone Action,* J. D. Baxter and G. G. Rousseau (Eds.). Springer-Verlag, Berlin, pp. 1-24.

Bedell, G. W., and Soll, D. R. (1979). Effects of low concentrations of zinc on the growth and dimorphism of *Candida albicans*: Evidence for zinc-resistant and sensitive pathways for mycelium formation. Infect. Immun. *26*:348-354.

Biswas, D. K., Lyons, J., and Tashjian, A. H., Jr. (1977). Induction of prolactin synthesis in rat pituitary cells. Cell *11*:431-439.

Bohm, H. (1980). The formation of secondary metabolites in plant tissue and cell cultures. Int. Rev. Cytol. Suppl. *11B*:183-208.

Cikes, M. (1970). Relationship between growth rate, cell volume, cell cycle kinetics, and antigenic properties of cultured murine lymphoma cells. J. Nat. Cancer Inst. *45*:979-988.

Cousins, R. J., and Failla, M. L. (1980). Cellular and molecular aspects of mammalian zinc metabolism and homeostasis. In *Zinc in the Environment, Part 2: Health Effects*, J. O. Nriagu (Ed.). Wiley, New York, pp. 121-135.

Cox, R. P., and King, J. C. (1975). Gene expression in cultured mammalian cells. Int. Rev. Cytol. *43*:281-351.

Cryer, A. (1980). Adiforyte histogenesis. Trends Biochem. Sci. *5*:196-198.

Dexter, D. L., Konieczny, S. F., Laurence, J. B., Shaffer, M., Mitchell, P., and Coleman, J. R. (1981). Induction by butyrate of differentiated properties in cloned murine rhabdomyosarcoma cells. Differentiation *18*:115-122.

Doerschug, M. R., and Miller, C. O. (1967). Chemical control of adventitious organ formation in *Lactuca sativa* explants. Am. J. Bot. *54*:410-413.

Eagle, H. (1963). Population density and the nutrition of cultured mammalian cells. In *The General Physiology of the Cells*, D. Mazia and A. Tyler (Eds.). McGraw-Hill, New York, pp. 151-170.

Eisner, T. (1970). Chemical defense against predation in insects. In *Chemical Ecology*, E. Sondheimer and J. B. Simeone (Eds.). Academic, London, pp. 157-217.

Failla, M. L., and Weinberg, E. D. (1980). Zinc transport and metabolism by microorganisms. In *Zinc in the Environment, Part 2: Health Effects*, J. O. Nriagu (Ed.). Wiley, New York, pp. 439-465.

Fernandez-Pol, J. A. (1978). Isolation and characterization of a siderophore-like growth factor from mutants of SV_{40}-transformed cells adapted to picolinic acid. Cell *14*:489-499.

Fernandez-Pol, J. A. (1980). Induction of two transformation-sensitive membrane polypeptides in normal rat kidney cells by iron deprivation. Cancer Res. *40*:786-795.

Gerschenson, L. E., Anderson, M., Molson, J., and Okigaki, T. (1970). Tyrosine transaminase induction by dexamethasone in a new rat liver cell line. Science *170*:859-861.

Glass, J., Nunez, M. T., Fischer, S., and Robinson, S. H. (1978). Transferrin receptors, iron transport and ferritin metabolism in Friend erythroleukemia cells. Biochim. Biophys. Acta *542*:154-162.

Green, H., and Goldberg, B. (1964). Collagen and cell protein synthesis by an established mammalian fibroblast line. Nature *204*:347-349.

Green, H., and Meuth, M. (1974). An established preadipose cell line and its differentiation in culture. Cell *3*:127-133.

Green, H., and Todaro, G. J. (1967). The mammalian cell as differentiated microorganism. Annu. Rev. Microbiol. *21*:574-600.

Gresser, I. (1977). On the varied biologic effects of interferon. Cell. Immunol. *34*:406-415.
Guille, E., Grisvard, J., and Sissoeff, I. (1979). Implications of reiteractive DNA-metal ion complexes in the induction and development of neoplastic cells. Biol. Trace Elem. Res. *1*:299-311.
Hall, N. E. L., and Axelrod, D. E. (1978). Sporulation competence in *Aspergillus nidulans*: A role of iron in development. Cell Differ. 7:73-82.
Harper, D. B., Swinburne, T. R., Moore, S. K., Brown, A. E., and Graham, H. (1980). A role for iron in germination of conidia of *Collectotrichum musae*. J. Gen. Microbiol. *121*:169-174.
Hiragun, A., Sato, M., and Mitsui, H. (1980). Establishment of a clonal cell line that differentiates into adipose cells in vitro. In Vitro *16*:685-693.
Holtzer, H., Bischoff, R., and Chacko, S. (1969). Activities of the cell surface during myogenesis and chondrogenesis. In *Cellular Recognition—1969*, R. T. Smith and R. A. Good (Eds.). Appleton-Century-Crofts, New York, pp. 19-25.
Hu, H. -Y. Y., Gardner, J., and Aisen, P. (1977). Inducibility of transferrin receptors on Friend erythroleukemia cells. Science *197*:559-561.
Kinnersly, A. M., and Dougall, D. K. (1980). Increase in anthocyanin yield from wild-carrot cell cultures by a selection system based on cell-aggregate size. Planta *149*:200-204.
Koffler, H., Knight, S. G., and Frazier, W. C. (1947). The effect of certain mineral elements on the production of penicillin in shake flasks. J. Bacteriol. *53*:115-123.
Konigsberg, I. R. (1975). The culture environment and its control of myogenesis. In *Regulation of Cell Proliferation and Differentiation*, W. W. Nichols and D. G. Murphy (Eds.). Plenum, New York, pp. 105-137.
Kuri-Harcuch, W., and Green, H. (1978). Adipose conversion of 3T3 cells depends on a serum factor. Proc. Nat. Acad. Sci. USA *75*:6107-6109.
Leffert, H. L., Koch, K. S., Rubaclava, B., Sell, S., Moran, T., and Boorstein, R. (1978). Hepatocyte growth control: In vitro approach to problems of liver regeneration and function. Nat. Cancer Inst. Monogr. *48*:87-101.
Levin, D. A. (1976). The chemical defenses of plants to pathogens and herbivores. Annu. Rev. Ecol. Syst. 7:121-159.
Matsumoto, T., Nishida, K., Noguchi, M., and Tanaki, E. (1973). Some factors affecting the anthocyanin formation by *Populus* cells in suspension culture. Agric. Biol. Chem. *37*:561-567.
Misawa, M., Sakato, K., Tanaka, H., Hayashi, M., and Samejima, H. (1974). Production of physiologically active substances by plant cell suspension cultures. In *Tissue Culture and Plant Science. 1974*. H. E. Street (Ed.). Academic, London, pp. 405-432.
Mothes, K. (1980). Introduction. In *Secondary Plant Products*, E. A. Bell and B. V. Charlwood (Eds.). Springer-Verlag, Berlin, pp. 1-10.

Murai, N., Skoog, F., Doyle, M. E., and Hanson, R. S. (1980). Relationships between cytokinin production, presence of plasmids, and fasciation caused by strains of *Corynebacterium fascians*. Proc. Nat. Acad. Sci. USA 77: 619-623.

Pickart, L., and Thaler, M. M. (1980). Growth-modulating tripeptide (glycyl-histidyllysine): Association with copper and iron in plasma, and stimulation of adhesiveness and growth of hepatoma cells in culture by tripeptide-metal ion complexes. J. Cell. Physiol. 102:129-139.

Pickart, L., Freedman, J. H., Loker, W. J., Peisach, J., Perkins, C. M., Stenkamp, R. E., and Weinstein, B. (1980). Growth-modulating plasma tripeptide may function by facilitating copper uptake into cells. Nature 288:715-717.

Reed, B. C., Kaufman, S. H., Mackall, J. C., Student, A. K., and Land, M. D. (1977). Alterations in insulin binding accompanying differentiation of 3T3-L1 preadipocytes. Proc. Nat. Acad. Sci. USA 74:4876-4880.

Rillema, J. A. (1979). Effect of zinc ions on the actions of prolactin on RNA and casein synthesis in mouse mammary gland explants. Proc. Soc. Exp. Biol. Med. 162:464-466.

Robbins, E., and Pederson, T. (1970). Iron: Its intracellular location and possible role in cell division. Proc. Nat. Acad. Sci. USA 66:1244-1251.

Russell, T. R. (1979). Differentiation of 3T3-L2 fibroblasts into adipose cells in bromodeoxyuridine-suppressed cultures. Proc. Nat. Acad. Sci. USA 76: 4451-4454.

Rutter, W. J., Pictet, R. L., and Morris, P. W. (1973). Toward molecular mechanisms of development processes. Annu. Rev. Biochem. 42:601-646.

Schroder, J., Betz, B., and Hahlbrook, K. (1976). Light-induced enzyme synthesis in cell suspension cultures of *Petroselinum hortense*. Eur. J. Biochem. 67:527-541.

Schwartz, F. J., and Matrone, G. (1975). Methodological studies on the uptake of zinc by 3T3 cells. Proc. Soc. Exp. Biol. Med. 149:888-892.

Schwarz, R. I., and Bissell, M. J. (1977). Dependence of the differentiated state on the cellular environment; modulation of collagen synthesis in tendon cells. Proc. Nat. Acad. Sci. USA 74:4453-4457.

Schwarz, R. I., Colarusso, L., and Doty, P. (1976). Maintenance of differentiation in primary cultures of avian tendon cells. Exp. Cell. Res. 102:63-71.

Seeds, N. W., Gilman, A. G., Amano, T., and Nirenberg, M. W. (1970). Regulation of axon formation by clonal lines of a neural tumor. Proc. Nat. Acad. Sci. USA 66:160-167.

Shio, I., and Ohta, S. (1973). Nicotine production by tobacco callus tissues and effect of plant growth regulators. Agric. Biol. Chem. 37:1857-1864.

Silagi, S. (1976). Effects of 5-bromodeoxyuridine on tumorigenicity, immunogenicity, virus production, plasminogen activator, and melanogenesis of mouse melanoma cells. Int. Rev. Cytol. 45:65-111.

Spiegelmann, B. M., and Green, H. (1980). Control of specific protein biosynthesis during the adipose conversion of 3T3 cells. J. Biol. Chem. 255:8811-8818.

Stich, H. F., Wei, L., and Whiting, R. F. (1979). Enhancement of the chromosome-damaging action of ascorbate by transition metals. Cancer Res. *39*: 4145-4151.

Stockdale, F. E., and Topper, Y. J. (1966). The role of DNA synthesis and mitosis in hormone-dependent differentiation. Proc. Nat. Acad. Sci. USA *56*:1283-1289.

Suzuki, H., Matsumoto, T., Kisaki, T., and Noguchi, M. (1981). Influences of cultural conditions on polyphenol formation and growth of amacha cells (*Hydrangea macrophylla*). Agric. Biol. Chem. *45*:1067-1077.

Swain, T., (1977). Secondary compounds as protective agents. Annu. Rev. Plant Physiol. *28*:479-501.

Tabata, M. (1977). Recent advances in the production of medicinal substances by plant cell cultures. In *Plant Tissue Culture and Its Biotechnological Applications*, W. Barz, E. Reinhard, and M. H. Zenk (Eds.). Springer-Verlag, Berlin, pp. 3-16.

Tejwani, G. A., Pedrosa, R. O., Pontremoli, S., and Horecker, B. L. (1976). 1,6-Bisphosphatase of rat liver. Proc. Nat. Acad. Sci. USA *73*:2692-2697.

Thelander, L., and Reichard, P. (1979). Reduction of ribonucleotides. Annu. Rev. Biochem. *48*:133-158.

Thomopoulus, P., Roth, J., Lovelace, E., and Postan, I. (1976). Insulin receptors in normal and transformed fibroblasts: Relationship to growth and transformation. Cell *8*:417-423.

Thompson, E. B., Tomkins, G. M., and Curran, J. F. (1966). Induction of tyrosine α-ketoglutarate transaminase by steroid hormones in a newly established tissue culture line. Proc. Nat. Acad. Sci. USA *56*:296-303.

Tomkins, G. M., Gelehrter, T. D., Granner, D., Martin, D., Jr., Samuels, H. H., and Thompson, E. B. (1969). Control of specific gene expression in higher organisms. Science *166*:1474-1480.

Vaughn, V., and Weinberg, E. D. (1978). Copper inhibition of dimorphism in *Candida albicans*. Mycopathologia *64*:39-42.

Wagner, F., and Vogelmann, H. (1977). Cultivation of plant tissue cultures in bioreactors and formation of secondary metabolites. In *Plant Tissue Culture and Its Biotechnological Application*, W. Barz, E. Reinhard, and M. H. Zenk (Eds.). Springer-Verlag, Berlin, pp. 245-252.

Weinberg, E. D. (1970). Biosynthesis of secondary metabolites: Role of trace metals. Adv. Microb. Physiol. *4*:1-44.

Weinberg, E. D. (1971). Secondary metabolism: Raison d'être. Perspect. Biol. Med. *14*:565-577.

Weinberg, E. D. (1977). Mineral element control of microbial secondary metabolims. In *Microorganisms and Minerals*, E. D. Weinberg (Ed.). Marcel Dekker, New York, pp. 289-316.

Weinberg, E. D. (1978). Secondary metabolism: Regulation by phosphate and trace elements. Folia Microbiol. *23*:496-504.

Weinberg, E. D. (1981). Iron and neoplasia. Biol. Trace Elem. Res. *3*:55-80.

Whittaker, R. H. (1970). The biochemical ecology of higher plants. In *Chemical Ecology*, E. Sondheimer and J. B. Simeone (Eds.). Academic, London, pp. 43-70.

Whittaker, R. H., and Feeny, P. P. (1971). Allelochemics: Chemical interactions between species. Science *171*:757-770.

Woodruff, H. B. (1966). The physiology of antibiotic production: The role of the producing microorganism. Symp. Soc. Gen. Microbiol. *16*:22-46.

Yeoh, G. C. T., and Morgan, E. H. (1979). DMSO induction of transferrin receptors on Friend erythroleukemia cells. Cell Differ. *8*:331-343.

Zambryski, P., Holsters, M., Kruger, K., Depicker, A., Schell, J., van Montague, M., and Goodman, H. M. (1980). Tumor DNA structure in plant cells transformed by *Agrobacterium tumefacians*. Science *209*:1385-1391.

Zenk, M. H., El-Shagi, H., Arens, H., Stockigt, J., Weiler, E. W., Deus, B. (1977). Formation of the indole alkaloids serpentine and ajmalicine in cell suspension cultures of *Catharanthus roseus*. In *Plant Tissue Culture and Its Biotechnological Applications*, W. Barz, E. Reinhard, and M. H. Zenk (Eds.). Springer-Verlag, Berlin, pp. 27-43.

5
Biochemistry, Physiology, and Genetics of Carotenogenesis in Fungi

Manfred Ruddat and Edward D. Garber / The University of Chicago, Chicago, Illinois

INTRODUCTION

Carotenoids are bright yellow to red pigments with a widespread distribution in all photosynthetic and in many nonphotosynthetic organisms. Only bacteria, fungi, and plants, however, synthesize carotenoids de novo. Animals obtain carotenoids from their diet and as a rule modify the carotenoid molecule oxidatively. The most important metabolic product is vitamin A, a derivative of β-carotene.

The vast majority of the more than 400 naturally occurring, identified carotenoids are tetraterpenoids with a highly branched C_{40} skeleton corresponding to eight isoprene units, which reflects their mode of biosynthesis. The arrangement of the isoprene units at the center of the molecule is characteristically reversed. The two central methyl groups are in 1,6 positions relative to each other, while all of the other nonterminal methyl groups are separated by five carbon atoms. A second distinguishing feature of the carotenoids is the array of conjugated double bonds which are responsible for light absorption. The chain of conjugated double bonds is mainly or entirely the chromophore.

Not all of the carotenoids are confined to the C_{40} skeleton. Some have fewer than 40 carbon atoms and because of their close structural relationship are defined as apocarotenoids. Neurosporoxanthin (4'-apo-β-caroten-4'-oic acid),* a C_{35} carotenoid acid, has been detected in fungi; crocetin (8,8'-diapocarotene-8,8'-dioic acid) and bixin (methyl hydrogen 9'-cis-6,6'-diapocarotene-6,6'-dioate) are common C_{20} carotenoids in plants. Carotenoids with C_{45} and C_{50} skeletons or with 30 carbon atoms have been identified in nonphotosynthesizing bacteria.

---

*The commonly accepted trivial name of "carotenoids" will be used, accompanied in parentheses at first mention by the semisystematic name according to the IUPAC-IUB Commission on Biochemical Nomenclature.

Modification of the acyclic skeleton by hydrogenation, dehydrogenation leading to cis, trans isomerization, cyclization, and oxygenation yields the great variety of carotenoids. Carotenoids occur as free compounds, as esters of one or more long-chain free fatty acids, and as glycosides, and at least in vertebrates they are stoichiometrically combined with proteins. Carotenoproteins are also present in photosynthetic organisms as light-harvesting pigments (secondary photoreceptors) along with chlorophylls (Song, 1980).

We will focus on the C_{40} carotenoid, specifically the carotenes, because genetic information on fungal carotenoids is available so far only for these hydrocarbons. Oxygenated carotenoids (xanthophylls) often occur in fungi as carotenoic acids.

DISTRIBUTION OF CAROTENOIDS IN FUNGI

Carotenoids in fungi are widely but not ubiquitously distributed and their occurrence in a given genus cannot be predicted as yet. Carotenoids have been detected in some 200 fungal species, but in a few species the specific carotenoids have not been identified. Approximately 100 fungal species lack carotenoids; however, their number will undoubtedly increase, as will the number of carotenoid-containing species, as more fungi are analyzed (Goodwin, 1980; Liaaen-Jensen, 1979; Valadon, 1976). Valadon(1976) advocated carotenoids as valuable taxonomic markers, particularly for differentiating between genera and reinforcing morphological and other biochemical characters. Ragan and Chapman (1978) proposed carotenic pathways for the delineation of phylogenetic relationships among fungi.

Simple carotenes, especially β-carotene and γ-carotene, are the most frequently occurring carotenoids in fungi, and β-carotene is quantitatively often the major pigment. In Chytridiomycetes and Zygomycetes, carotenes but very few xanthophylls have been found (Davies, 1961; Goodwin, 1980). Carotene accumulation is sometimes indicative of the mating type of these lower fungi (Federici and Thompson, 1979). In Ascomycetes, carotenes also predominate quantitatively, but xanthophylls occur with some frequency. Of 80 species of Ascomycetes, 39 possess xanthophylls and 12 of these belong to three families: Pezizales, Helvellaceae, and Morchellaceae (Arpin, 1968; Czeczuga, 1979; Goodwin, 1980). In Basidiomycetes and Deuteromycetes, approximately one-half of all species in which carotenoids have been found also possess xanthophylls, but carotenes are present in all of these and are usually predominant (Goodwin, 1980). In carotenoid-containing fungi, carotenes are usually present, but in certain species and strains oxidative metabolism may convert carotenes so that one or more xanthophylls become the major component of the pigmentation.

Some xanthophylls are unique to fungi, especially among the Ascomycetes: for example phillipsiaxanthin (1,1'-dihydroxy-3,4,3',4'-tetradehydro-1,2,1'2'-tetrahydro-ψ,ψ-carotene-2,2'-dione) and plectaniaxanthin (3',4'-didehydro-1',2'-dihydro-β,ψ-carotene-1'-2'diol) and their didehydro derivatives, as well as aleuriaxanthin (1',16'-didehydro-1'-2'-dihydro-β,ψ-caroten-2'-ol), along with their esters (Arpin, 1968; Korf, 1973; Valadon et al., 1980). Xanthophyll acids rare in nature, are also unique to fungi; torularhodin (3',4'-didehydro-β,ψ-caroten-16'-oic acid), neurosporoxanthin (4'-apo-β-caroten-4'-oic adic), and asperxanthin occur in Ascomycetes, Basidiomycetes, and Deuteromycetes (Simpson et al., 1964; Aasen and Liaaen-Jensen, 1965; Zajic and Kuehn, 1962). An additional distinguishing feature in fungal carotenoids is the C-3 and/or C-3, double bond found in phillipsiaxanthin, plectaniaxanthin, and in the carotenes torulene (3',4'-didehydro-β,ψ-carotene) and 3,4-didehydrolycopene (Arpin and Liaaen-Jensen, 1967a,b; Bae et al., 1971). Certain xanthophylls common in flowering plants also occur in Ascomycetes and Basidiomycetes, for example, zeaxanthin [(3R,3'R)-β,β-carotene-3,3'-diol], rubixanthin [(3R)-β,ψ-caroten-3-ol], and β-cryptoxanthin [(3R)-β,β-caroten-3-ol] (Valadon, 1976; Czeczuga, 1979). The ketoxanthophylls, typical of animal carotenes and common among algae and certain Cyanobacteria but rare in fungi, are present in certain Basidiomycetes and Deuteromycetes: canthaxanthin (β,β-carotene-4,4'-dione) astaxanthin (3,3'-dihydroxyl-β,β-carotene-4,4'-dione), and rhodoxanthin (4',5'-didehydro-4,5'-retro-β,β-carotene-3,3'-dione (Haxo, 1950; Gribanovski-Sassu and Foppen, 1967; Andrewes et al., 1976; Alasoadura and Visser, 1972). While some of the xanthophylls shared by fungi and other organisms are identical in structure, astaxanthin, the main carotenoid in *Phaffia rhodozyma,* has the opposite chirality (3R,3'R) to that of the lobster astaxanthin (Andrewes and Starr, 1976). The generalization that the absolute stereochemistry of individual carotenoids is independent of their source, therefore, is no longer valid.

EXTRACTION, ISOLATION, AND IDENTIFICATION OF CAROTENOIDS

We present here only a brief outline of general methods for the extraction, isolation, and identification of carotenoids, because detailed procedures have been published (Britton and Goodwin, 1971; Liaaen-Jensen, 1971, 1980; Davies, 1976; Moss and Weedon, 1976; Weedon, 1979; Eugster, 1979; Taylor and Ikawa, 1980).

With few exceptions, carotenoids are fat-soluble pigments and are extracted with lipid solvents. Mechanical disruption of the fungal cells with a Waring blendor, Braun (Melsungen) homogenizer, Bead Beater (Biospec Products),

French press, or pestle and mortar is usually more efficient for extraction. Quantitative recovery of astaxanthin from *P. rhodozyma* was obtained after cell-wall digestion with *Bacillus circulans*, strain *4L-12* (Johnson et al., 1978). Water-miscible solvents, such as acetone, methanol, or ethanol, are highly effective for fresh cell-homogenates, and after the water is removed from the cell extracts they are followed by extraction with water-immiscible solvents such as petroleum ether, diethyl ether, or dichloromethane. Freshly purified, deaerated solvents as well as protection from bright light, heat, oxygen, and acids are essential in all procedures because carotenoids are inherently unstable. Since chloroform usually contains HCl, the customary chloroform-methanol mixture for the extraction of neutral lipids should be avoided, unless the acid is completely removed. ζ-Carotene, lycopene, and many other carotenoids are unstable in air and should be manipulated in an inert atmosphere, for example, oxygen-free nitrogen. Pure β-carotene and phytoene, however, are reasonbaly stable in air. Butylated hydroxytoluene is useful as an antioxidant in solvents for carotenoid extraction. Most carotenoids are stable in alkali. Astaxanthin, which occurs in *Clavaria* and *Phaffia*, however, is rapidly oxidized in the presence of alkali and air. Reactions with acids or alkalis under experimental, controlled conditions, however, are highly useful for diagnostic purposes (Liaaen-Jensen, 1971; Eugster, 1979).

Even when the carotenoids are present in free form, saponification with 10-30% ethanolic KOH is advantageous because lipids are removed and do not interfere in the subsequent purification procedures. Acetone in the saponification mixture may yield artifacts, especially with carotenals. All traces of acetone must therefore be removed before alkali is added. The saponified mixture is diluted with water and partitioned against the petroleum ether or peroxide-free diethyl ether and the carotenes are recovered in the epiphase. Carotenoids with two or more hydroxyl groups, as well as carotenoic acids, remain in the hypophase and can be extracted into the ether phase after adjusting the hypophase to pH 4; carotenoids with one hydroxyl group are evenly distributed between epiphase and hypophase. Sterols in the nonsaponifiable fraction precipitate from petroleum ether in the freezer overnight or by the digitonide method. The carotenoid extract is thoroughly washed to remove all traces of alkali, dried, and the solvent evaporated under vacuum. Small volumes are evaporated with a stream of oxygen-free nitrogen. Carotenoids can be stored in inert solvents under nitrogen below $-10°C$.

Fractionation of the carotenoid extract is achieved by chromatography on a preparative column containing a single phase or a mixture of several stationary phases, such as confectionary sugar, alumina, silica gel, calcium and magnesium carbonates, or oxides or cellulose. The stationary phase, especially of thin-layer chromatograms, affects the stability of the carotenoids (Tanaka et al., 1981).

The concentrated carotenoid extract dissolved in a nonpolar solvent is applied to the top of the column and the carotenoids are eluted with solvents or solvent mixtures of increasing polarity. The eluates are conveniently monitored by analytical thin-layer chromatography (TLC) and spectrophotometry. High-performance liquid chromatography combines the high resolution power and high speed of TLC and the preparative capacity of column chromatography.

Ultraviolet (UV) and visible light (electronic) spectra indicate the type of chromophore and are a most convenient method for identifying carotenoids by comparing spectra or absorption maxima with published values (Davies, 1976). This approach, however is not without pitfalls. Several different carotenoids have the same absorption spectrum and are easily misidentified without further diagnostic tests (Valadon, 1976).

A precision mass spectrometer, conveniently interfaced with a gas chromatograph for purification of the extract and inserting the carotenoid as a volatile derivative, measures the mass of the molecular ion which may allow calculation of the molecular formula. The limitations of microanalysis can therefore be avoided. The fragmentation pattern often permits the deduction of structural characteristics that identify the carotenoids.

Infrared spectra which did not have a major role in the elucidation of carotenoid structures are nevertheless useful for indicating oxygen substitutions and allene or acetylene configurations, although the latter have not been found in fungal carotenoids. All of these methods are highly sensitive. Using modern precision instruments interfaced with computer capability, the amount of carotenoids required for analysis is usually in the 100-μg range.

The natural occurrence of cis and trans isomers also requires the determination of the stereochemistry to identify the carotenoid (Liaaen-Jensen, 1980). While several methods for producing stereomutation are available, iodine-catalyzed photoisomerization is readily achieved and monitored spectrophotometrically (Zechmeister, 1962). Cis carotenoids have less intensive absorption spectra than their all-trans isomers; the absorption bands are usually shifted to lower wavelengths and the "cis" peak appears in the UV region. Determining the absolute configuration of asymmetric units may require chiroptical methods, optical rotatory dispersion, and circular dichroism (Sturzenegger et al., 1980), and finally x-ray crystallography. The latter has been used for only very few carotenoids to date.

Quantitative determination of a carotenoid in a fraction is achieved by use of light spectrophotometry and published specific extinction coefficients for specified carotenoids. Here $E_1^{1\%}{}_{cm}$ is the extinction at a given wavelength, usually λ_{max} in a certain solvent of a 1% (wt/vol) carotenoid solution in a cuvette with a 1-cm light path. The absorption E at a given wavelength is related to x gram of carotenoid in y ml of solution by the following equation:

$$x\ g = \frac{Ey}{E_{1\ cm}^{1\%}\ 100}$$

If E is not available, a value of 2500 for a carotenoid dissolved in hexane may be used for an approximation (Davies, 1976).

BIOSYNTHESIS

Carotenoids, as all terpenoids, are synthesized from mevalonic acid (MVA) by head-to-tail condensation of the two isoprene isomers isopentenyl pyrophosphate (IPP) and dimethylallyl pyrophosphate (DMAPP). Four isoprene units linked in this manner form C_{20} geranylgeranyl pyrophosphate (GGPP). Tail-to-tail dimerization of the two allylic pyrophosphates (GGPP) results in prephytoene (octahydro-ψ,ψ-carotene) (Fig. 1) (see Davies, 1973, 1979; Davies and Taylor, 1976; Britton, 1976a,b; Goodwin, 1980, for reviews).

Evidence for the pathway from MVA to phytoene in fungi comes from incorporation studies with cell-free extracts of *Phycomyces blakesleeanus* (Yokoyama et al., 1962; Lee and Chichester, 1969; Lee et al., 1972), extracts of the *car10* mutant of *P. blakesleeanus* (Davies, 1973), and *Neurospora crassa* (Spurgeon et al., 1979; Mitzka-Schnabel and Rau, 1981; Harding and Turner, 1981). Cell-free extracts convert MVA or IPP to phytoene. The conversion of GGPP to phytoene, phytofluene, ζ-carotene, neurosporene, and lycopene and also that of prephytoene to phytoene was carried out by Lee and Chichester (1969) with partially purified extracts of the *C9* mutant of *P. blakesleeanus*. Geranylgeranylpyrophosphate was converted to prephytoenepyrophosphate by an extract from *Mycobacteria* sp. (Altman et al., 1972).

Dehydrogenation of phytoene to form conjugated double bond involves the trans β elimination of two neighboring allylic hydrogens. Four successive dehydrogenation steps yield lycopene ψ,ψ-lycopene) (Fig. 2). The dehydrogenation sequence was proposed by Porter and Lincoln (1950) before the structure of all the intermediates had been identified.

Unsaturated acyclic carotenes were obtained by incubating extracts of *P blakesleeanus* with [^{14}C] phytoene (Davies, 1973). Stepwise removal of two hydrogens alternately from either side of the central chromophore of phytoene yields phytofluene (hexahydro-ψ,ψ-carotene) and ζ-carotene (7,8,7',8'-tetrahydro-ψ,ψ-carotene) in wild-type *Phycomyces sp.* extracts. For extracts of *carR21* and *mad107* mutants of *P. blakesleeanus*, however, Davies (1973) found that the dehydrogenation sequence differs from the Porter-Lincoln pathway. Phytofluene is reduced to a mixture of two heptaenes consisting of ζ-carotene and its unsymmetrical isomer 7,8,11,12-tetrahydrolycopene (7,8,11,12-tetrahydro-ψ,ψ-carotene). The mixture of these two heptaenes constitutes the θ-carotene found in *N. crassa* (Davies, 1973). The sequence of dehydrogenation

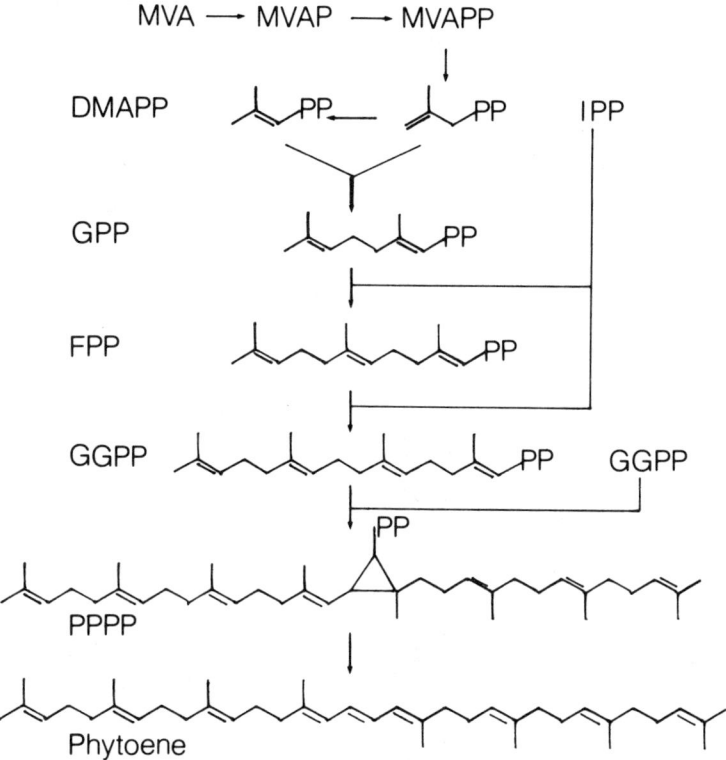

Figure 1 The biosynthetic pathway of phytoene from mevalonic acid (MVA). MVAP, mevalonic acid-5-phosphate; MVAPP, mevalonic acid-5-pyrophosphate; IPP, isopentenylpyrophosphate; DMAPP, dimethylallylpyrophosphate; GPP, geranylpyrophosphate; FPP, farnesylpyrophosphate; GGPP, geranylgeranylpyrophosphate; PPPP, prephytoenepyrophosphate.

in fungi is intermediate between that of plants and certain purple bacteria. For example, *Rhodospirillum rubrum* produces 7,8,11,12-tetrahydrolycopene (Davies, 1970) and plastids from tomato fruit 7,8,7',8'-tetrahydrolycopene (Porter and Spurgeon, 1979).

Inhibitor studies have also contributed significantly to the elucidation of the desaturation pathway. Diphenylamine (DPA) prevents dehydrogenation and causes phytoene accumulation at the expense of the unsaturated carotenes. Upon removal of DPA the unsaturated carotenes accumulate at the expense of phytoene (Olson and Knizley, 1962; Goodwin, 1952).

Phytoene from fungi is in the 15-cis configuration with 1-2% in the all-trans isomer configuration (Aung Than et al., 1972; Valadon and Mummery, 1979).

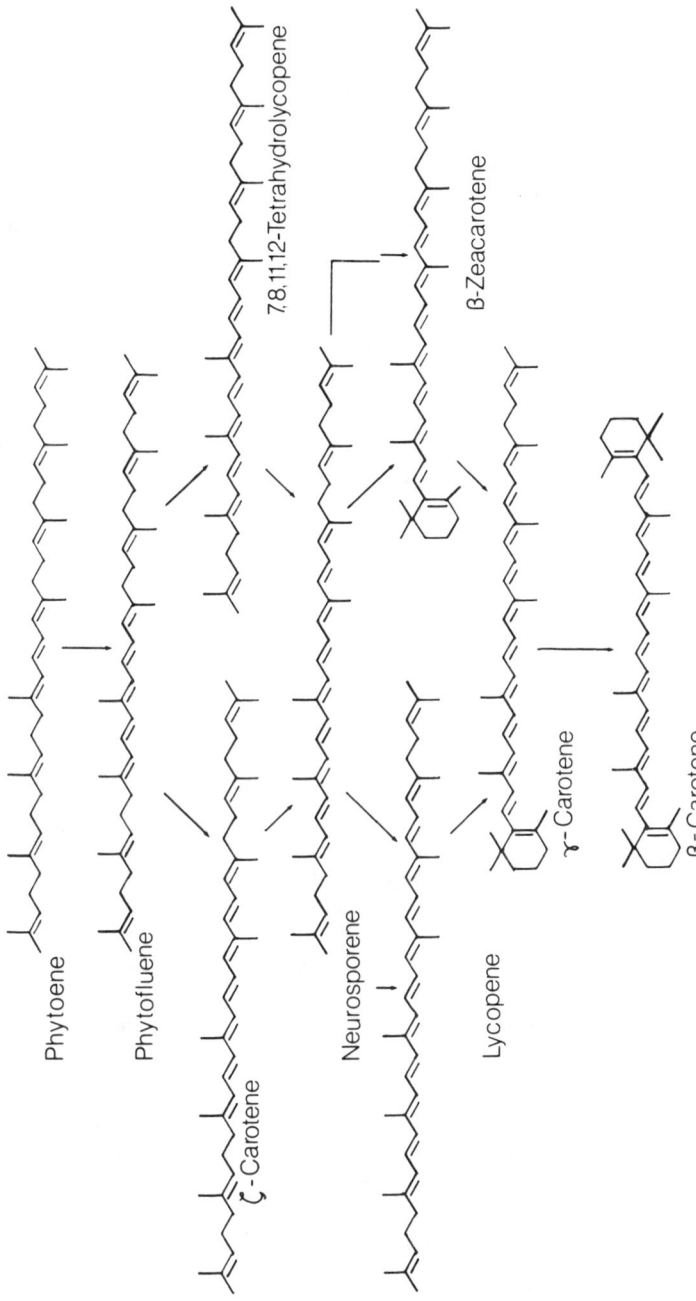

Figure 2 Dehydrogenation and cyclization sequence for the biosynthesis of β-carotene from phytoene.

The higher unsaturated carotenes, for example, ζ-carotene, 7,8,11,12-tetrahydrolycopene, and neurosporene, are in all-trans configurations, indicating that isomerization occurs at an early stage of dehydrogenation, presumably before phytofluene formation. The all-trans phytofluene pathway is more important in fungi than in plants. This difference has an important bearing on the putative dehydrogenase or dehydrogenase complex for which there is firm genetic evidence but, so far, no biochemical information. Gathering genetic evidence for the cis-trans isomerase, which may be associated with or be part of the dehydrogenase complex, will involve the screening of phytoene-accumulating mutants.

Cyclization of the unsaturated, acyclic carotenes results in the formation of the common ε or β rings. Thus lycopene yields γ-carotene (β,ψ-carotene) while neurosporene with its lycopene-type end group yields β-zeacarotene (7,8,-dihydro-β,ψ-carotene), which upon dehydrogenation results in γ-carotene. Neurosporene is, therefore, a branch point that is either dehydrogenated to lycopene or cyclized to β-zeacarotene (Fig. 2). Both pathways apparently operate in fungi. The occurrence of 7′,8′,11′,12′-tetrahydro-γ-carotene (7′,8′,11′,12′-tetrahydro-β,4-carotene), "cyclic ζ-carotene," in a DPA-treated *mad107* mutant of *P. blakesleeanus* indicates that cyclization can occur earlier than at the level of neurosporene (Davies and Rees, 1973).

Phycomyces sp. extract converts lycopene into cyclic carotenes (Davies, 1973). The *car21* mutant of *P. blakesleeanus* (Davies, 1973) and the pumpkin mutant of *Ustilago violacea* (Garber et al., 1980) accumulate lycopene rather than β-carotene, as in wild-type *P. blakesleeanus,* the yellow mutant of *U. violacea* (Garber et al., 1975), and some pink strains of *U. violacea* (O. H. Will, M. Ruddat, and E. D. Garber, unpublished work). 2-(4-Chlorophenylthio)triethylamine HCl (CPTA) applied to *Phycomyces* sp or *Blakeslea trispora* leads to lycopene accumulation (Coggins et al., 1970; Hsu et al., 1972). *Blakeslea trispora* also accumulates lycopene after treatment with a number of nitrogenous bases, such as imidazole, pyridine, and certain of their derivatives (Ninet et al., 1969).

Lycopene does not, however, appear to be the cyclization substrate in a number of fungi. *Rhizophlyctis rosea* (Davies 1961) does not produce lycopene, and the occurrence of β-zeacarotene after DPA application in *Rhodotorula* sp. (Simpson et al., 1964) and *P. blakesleeanus* (Davies et al., 1963) indicates that neurosporene is the cyclization substrate. In lycopene-accumulating mutant strains of *P. blakesleeanus* the pathways leading to β-carotene apparently shifts during development. Initially lycopene serves as the cyclization substrate; however, at a later growth stage during which β-carotene and γ-carotene accumulate, lycopene is bypassed and β-zeacarotene is the precursor for β-carotene synthesis (Hsu et al., 1974b). The presence of both cyclization pathways in the same species was determined in cell-free extracts from the mutant *C115 car42 mad107* of *P. blakesleeanus* (Bramley et al., 1977). Incorporation of labeled neurosporene, lycopene, and β-zeacarotene indicated that both cyclization pathways were quantitatively equally important. The experimental approach, however, did not establish whether the primary cyclization of neursporene or lycopene is catalyzed by separate, specific enzymes or by one nonspecific enzyme. The operation of a multienzyme aggregate or a complex of four dehydrogenases and two cyclases which catalyzes the entire sequence from phytoene to β-carotene would be consistent with the low conversion rate. Low substrate specificity of the putative enzyme aggregate could mimic two separate pathways in *Phycomyces*.

Different sensitivities to CPTA or nicotine of the primary (lycopene to γ-carotene, neursporene to β-zeacarotene) and secondary cyclization reactions (β-zeacarotene to γ-carotene, γ-carotene to β-carotene) have been interpreted as evidence for two separate cyclization enzymes (Hsu et al., 1972).

Oxygenation of the hydrocarbon carotene is the final step of xanthophyll formation. Oxygen insertion occurs at various stages of desaturation and/or cyclization and results in a large variety of xanthophyll structures which, so far, preclude the recognition of a general pathway.

Carotenogenic Enzymes

Active cell-free extracts from *P. blakesleeanus, N. crassa,* and *B. trispora* convert various precursors to carotenoids, but none of these enzyme preparations, however, has been purified to homogeneity. So far, only a carotenogenic enzyme complex from tomato fruit plastids has been characterized (Porter and Spurgeon, 1979). The enzyme complex has a molecular weight of 200,000 and requires Mn^{2+} as a cofactor, and adenosine 5′-triphosphate (ATP) possibly as an allosteric effector (Maudinas et al., 1977).

Prenyltransferase, which catalyzes the condensation between IPP and an allylic pyrophosphate to produce the next higher homolog of the allelic substrate, has been obtained as a homogeneous preparation from *Saccharomyces cerevisiae* (Eberhardt and Rilling, 1975) and *P. blakesleeanus* (Rilling, 1979). Operating in the early part of terpene synthesis, this enzyme could be as active in catalyzing steroid precursors as carotene precursors; both pathways, however, may be compartmentalized.

The enzyme system from *Phycomyces* extracts synthesizing GGPP and phytoene was in the soluble $100,000g$ fraction (Yokoyama et al., 1962). In purified extracts from *N. crassa*, the soluble $100,000g$ fraction catalyzed the synthesis of GGPP, while a particulate enzyme converted GGPP to phytoene (Spurgeon et al., 1979). Two subcellular fractions from *N. crassa*, sedimenting at $23,000g$ and $115,000g$, incorporated MVA predominantly into phytoene and, to a lesser extent, into the unsaturated carotenoids. Incorporation into unsaturated carotenoids was improved by fortifying the incubation mixture with a GGPP-synthesizing system either from *Cucurbita pepo* seeds or with the $115,000g$ soluble fraction from *N. crassa* (Mitzka-Schnabel and Rau, 1981). Approximately 80% of the carotenogenic enzymes were associated with the particulate fraction, especially in the 115,000g pellet which, based on marker enzyme activities, was enriched in endoplasmic reticulum (ER) and possibly also in plasma membrane. The floating lipid fraction which contains the bulk of the carotenoids has only trace amounts of carotenogenic activity (Mitzka-Schnabel and Rau, 1981). Cell-free extracts from an albina mutant of *N. crassa* which has 20 times more sterols than carotenoids incorporate only minimal amounts of MVA into phytoene, reflecting the competition for precursors by the sterol pathway (Bobowski et al., 1977).

The enzymes for the desaturation of phytoene to lycopene and the accompanying cis-trans isomerization which requires four dehydrogenases and one isomerase have not been characterized, nor have those for the conversion of lycopene to cyclic carotenes. Two separate enzymes are probably involved in catalyzing the primary and secondary cyclization reactions (Hsu et al., 1972).

SUBCELLULAR LOCALIZATION OF CAROTENOIDS IN FUNGI

Carotenoids occur in fungal cells predominantly in the spherosomes (lipid bodies) as lipid droplets and much smaller amounts are associated with membranes. Intracellularly separated organelles in centrifuged sporangiophores of *Phycomyces* show β-carotene to occur as the centripetal lipid droplet with a buoyant density of 1.06 g/ml. The lipid droplet from an albino mutant of *Phyomyces* sp. was colorless (Zalokar, 1969). Carotenoid synthesis in *B. trispora* appears to "fatten" the mycelium as carotenoids accumulate in increasing numbers of spherosomes (Feofilova and Pivovarova, 1976). In the *P. blakesleeanus* mutant *C115 car42 mad107*, 97% of the β-carotene was localized in oil droplets, and only 2% in a particulate fraction which sedimented with a buoyant density of 1.10 g/ml (Riley and Bramley, 1976). The organelles in this fraction were not identified. The ratio of β-carotene to protein was 0.025:1 in the membrane fraction and 0.13:1 in the lipid globules. In arthrospores of *Trichophyton mentagrophytes* carotenoids are contained in lipid granules consisting of an osmophilic matrix embedding membranous structures (Hashimoto et al., 1978).

Riley and Bramley (1976) found that differential centrifugation of fungal cell homogenates caused irreversible binding of lipids to particulate fractions. It is likely, therefore, that carotenoids reported to occur in fungal mitochondria isolated by differential centrifugation represent contaminants (Cederberg and Neujahr, 1970; Neupert and Ludwig, 1971; Keyhani et al., 1972). Carotenoids in a mitochondrial fraction isolated from *N. crassa* by differential centrifugation were considered to be contaminants (Mitzka-Schnabel and Rau, 1980). Rigorous proof for the occurrence of carotenoids in fungal mitochondria is still lacking. Ramadan-Talib and Prebble (1978) believed that carotenoids are present in mitochondrial membranes of *N. crassa* on the basis of the photoprotective action of carotenoids. In light-grown mycelium with a high carotene content (190-200 µg/g protein), quinone on the inner mitochondrion membrane is protected from photodestruction by carotenoid, which was also found in isolated inner membrane fractions. In cells with lesser amounts of carotenoid (75 µg/g protein), carotenoid is restricted to the outer membrane (Ramadan-Talib and Prebble, 1978).

In subcellular fractions from *N. crassa*, 60% of the carotenoids were detected in the floating lipids, and the remainder in a fraction with a buoyant density of 1.076 g/ml which was isolated by differential centrifugation followed by sucrose density gradient centrifugation. On the basis of nicotinamide adenine dinucleotide phosphate (NADPH) and nicotinamide adenine dinucleotide (NADH) cytochrome c reductase activities, the fraction was identified as ER enriched (Mitzka-Schnabel and Rau, 1980). In *N. crassa,* carotenoid synthesis and storage occur in different cell compartments; spherosomes function mainly for the storage of carotenoids synthesized on ER. Separation of synthesis and storage fits well with fine-structural observations of fungal cells. In *N. crassa* as well as in plants, some ER is closely associated with the developing spherosomes (Wanner et al., 1981). According to Frey-Wyssling et al. (1963), spherosomes develop from the ER. Wanner et al. (1981) presented evidence for a continuous flow of storage lipids and membrane consituents from the ER to spherosomes. Triglycerides and presumable also carotenoids accumulate within the lipophilic middle layer of one of the unit membranes along certain sites of the ER. When the spherosome reaches a certain size, the ER and the spherosome separate (Wanner, et al., 1981).

Blakeslea trispora contains carotenoids in the cell wall, which is characterized by an increased number of layers and peculiarly thickened regions (Feofilova and Pivovarova, 1976; Cederberg and Neujahr, 1970). In a mutant of *N. crassa* which does not form cell walls, however, the carotenoid composition is qualitatively the same as in the wall-containing wild type (Mitzka and Rau, 1977).

Little is known about the organization of carotenoids in membranes, even for photosynthetic membranes in which carotenoids are involved in energy transfer reactions and protection against photo-oxidation. Immunological procedures located β-carotene and several xanthophylls on the outside of thylakoids in green plants (Radunz and Schmid, 1979). Carotenoid organization has been investigated in model membranes. In dipalmitin-phosphatidylcholine liposomes, β-carotene appears to be aggregated in both the gel and liquid crystal state of the lipid bilayer, but the xanthophyll organization is phase dependent (Yamamoto and Bangham, 1978).

FUNCTION OF CAROTENOIDS

Photoprotection

Carotenoids function universally by protecting cells from the lethal combination of light and oxygen. The original observation of the photodynamic activation of a carotenoid-deficient *Rhodopseudomonas sphaeroides* (Sistrom et al., 1956) has been extended to photosynthetic and nonphotosynthetic organisms, including man (Krinsky, 1978; Burnett, 1976; Mathew-Roth, 1975). Conidia of *N. crassa* wild type containing β-carotene and γ-carotene are more resistant to visible light in the presence of photosensitizing dyes than carotenoid-deficient albino mutants (Morris and Subden, 1974; Blanc et al., 1976; Shimizu et al., 1979) or a lemon-yellow mutant containing ζ-carotene and neurosporene in a ratio of approximately 3:1 (Thomas et al., 1981). Inhibition of carotene synthesis in *N. crassa* with β-ionone rendered the wild-type cells as sensitive to photodynamic killing as the albino mutants (Morris and Subden, 1974). A pink wild-type strain of *Rhodotorula glutinis* and a yellow mutant containing mostly β-carotene and γ-carotene were also more resistant to photodynamic inactivation in red laser light than carotenoid-deficient white mutants containing only phytoene (Maxwell et al., 1966). Survival of photodynamic inactivation of *Dacryopinax spathularia* (Heterobasidiomycetes) was directly correlated with light-induced carotenoid synthesis and accumulation (Goldstrohm and Lilly, 1955). Similarly, carotenoids also protected *Fusarium aquaeductuum* in the absence of exogenous photosensitizers from photokilling by a high energy fluence rate (56 W/m^2) with near-UV light (300-400 nm) or bright sunlight (Huber and Schrott, 1980). Carotenoids, however, were not effective in protecting *N. crassa* in the presence of exogenous photosensitizing dyes, or *R. glutinis* in the presence or absence of photosensitizers from photodynamic inactivation by near-UV light, which in the experiments with *N. crassa* was shown not to involve singlet oxygen formation (Thomas et al., 1981; Chichester and Maxwell, 1969; Maxwell and Chichester, 1971).

Inhibition of the synthesis of the carotenoids which are involved in photoprotection may be the active mechanism of certain bacteriostats, fungicides, and herbicides.

The mechanisms of carotenoids protecting cells from photodynamic inactivation involve quenching of the excited triplet of the sensitizing pigment (chlorophyll in photosynthesizing cells), quenching of the singlet oxygen produced by the triplet sensitizer dye, or, finally, quenching lipid peroxidation (Krinsky, 1979; Rodgers and Bates, 1980). Only carotenoids having more than a minimum of nine conjugated double bonds, however, are effective in the exothermic energy transfer reaction with singlet oxygen in the photoreactive reaction (Mathew-Roth et al., 1974). Minimum protection from photodynamic inactivation of the lemon-yellow mutant of *N. crassa* containing more ζ-carotene (seven conjugated double bonds) than neurosporene (nine conjugated double bonds) demonstrates the importance of the chromophore lengths (Thomas et al., 1981).

More stringent structural requirements were found for carotenoids protecting the photosynthetic reaction center in *R. rubrum* (Boucher et al., 1977). Nonbacterial carotenoids failed to prevent photodynamic bleaching of the bacteriochlorophyll in reconstitution experiments, implying that carotenoid binding to specific sites of the photoreaction center, rather than their presence in a specific ration to bacteriochlorophyll, was crucial.

Protection of respiratory enzymes from photoinactivation by carotenoids has been observed in fungi and bacteria in vivo and in vitro (Prebble and Huda, 1977; Anwar and Prebble, 1977; Ramadan-Talib and Prebble, 1978). Blue light inactivation of respiration in vivo is less in carotenoid-containing wild-type strains of *N. crassa* than in the albino mutant (*FGSC 16*). In isolated mitochondria from the same strains, protection of the respiratory ubiquinone from photodestruction depends on the amount of carotenoid present in the cells (Ramadan-Talib and Prebble, 1978).

Functional properties of membrane-associated carotenoids have been investigated in model membranes as well as in vivo. Canthaxanthin was found to protect liposomes in the presence of photosensitizing dyes from singlet oxygen-induced lysis (Anderson and Krinsky, 1973). Bacterial carotenoids taken up into liposomes decrease membrane fluidity, indicating that carotenoids, at least in bacterial membranes, may function by reinforcing the lipid bilayer (Rottem and Markowitz, 1979). Based on preliminary leucine uptake experiments with *F. aquaeductuum*, irradiated with near-UV light, carotenoids may also act in vivo as membrane stabilizers (Huber and Schrott, 1980).

Photoreception

The vexing problem of the involvement of carotenoids as photoreceptors of their own biosynthesis and other light-induced processes which possess very similar action spectra appears to be resolved against carotenoids in favor of flavins (Presti and Delbrück, 1978). The evidence for the identity of the photoreceptor-mediating carotenogenesis will be discussed on p. 122; here we present those cases where carotenoids presumably function in photoreception.

Carotenoids appear to control photoinduced sporulation in *Trichoderma viride*, which is blocked by diphenylamine (DPA), a carotenoid synthesis inhibitor (Kumagai and Oda, 1969). Side effects of DPA interacting with DNA have not, however, been excluded from affecting sporulation, but DPA does not appear to inhibit carotenogenesis by binding to DNA (Valadon and Mummery, 1973).

In the mutants *carA*, *carB*, *carR*, and *carB carR* of *P. blakesleeanus* which control carotene synthesis and in which phototropism is unaffected, the threshold for photoinduced sporangiophore initiation, which in an O_2-limiting environment becomes blue light dependent, is 100-2000 times lower than in wild type (Galland and Russo, 1979a). The relationship between carotene content and sporangiophore initiation is complicated by the inhibitory effect of CO_2 accumulating in the enclosed environment (Russo et al., 1980, 1981). Nevertheless, β-carotene does not appear to be the photoreceptor for the sporangiophore initiation, because in the carotene-deficient double mutants the threshold for sporangiophore initiation does not significantly rise over mutants containing 100 times more carotene, nor does it decrease in a superproducer mutant (*B401*) synthesizing 20 times more than wild type. Galland and Russo (1979a) envisioned β-carotene as a mediator substance between the genetic products of *madA* and *madB* genes and sporangiophore initiation, while flavin may serve as a common photoreceptor for sporangiophore initiation and phototropism.

In *Leptosphaeria michotii* carotenoids are thought to stabilize the light-entrained sporulation rhythm (Jerebzoff-Quintin and Jerebzoff, 1980). Convincing evidence for the participation of carotenoids or their derivatives in a photoperiodic response comes from a carotenoid-deficient albino mite, *Amblyseius potentillae*, which fails to respond to the appropriate stimulus (Van Zon et al., 1981).

Precursor Function

Differential accumulation of carotenoids in gametangia of fungi has led to speculation about its significance without ascertaining a direct function for carotenoids in reproduction. β-Carotene serves as a precursor for trisporic acid, a sex hormone of the Mucorales (Bu'Lock et al., 1976). The conversion of β-carotene to trisporic acid probably proceeds via retinal and retinol (vitamin A) (Bu'Lock et al., 1974, 1976) (see also p. 131). Impaired sexual reproduction in a carotene-deficient mutant of *P. blakesleeanus* supports the requirement of β-carotene as a precursor (Gooday, 1974).

The accumulation of carotenoids in fungal gametangia and spores may also be related to the precursor function of carotenoids for sporopollenin, the extremely resistant material found in many fungal spores, including zygospores of the Zygomycetes, as well as in the exine of pollen. Sporopollenin is a polymer of highly cross-linked xanthophylls. β-Carotene incorporation into sporopollenin has been reported for zygospores of *Mucor mucedo* (Gooday et al., 1973) and for ascospores of *N. crassa*. Carotenoid-deficient mutants, however, also produce sporopollenin (Gooday et al., 1974). The accumulation of γ-carotene in male gametes of *Allomyces* and *Blastocladiella*, clearly not related to sporopollenin production, may be caused by an impaired tricarboxylic acid cycle making metabolites available for the isoprenoid pathway (Cantino and Hyatt, 1953) (see also p. 121).

β-Carotene may also be involved as a retinol precursor in controlling sporangiophore development in *P. blakesleeanus*. In carotene-deficient mutants of *P. blakesleeanus*, Galland and Russo (1979b) observed that in low oxygen tension which specifically inhibits sporangiophore initiation, the application of vitamin A and its derivatives will, like blue light, alleviate the inhibition. It is quite plausible that β-carotene as the retinol precursor may function as the endogenous regulator for sporangiophore differentiation in *Phycomyces*.

GENETIC ANALYSIS

The choice of a fungal species for a biochemical and genetic study of pigment biosynthesis depends on at least three criteria: (1) spores for mutagenesis, (2) a sexual or parasexual cycle for mapping mutant loci, and (3) heterokaryosis or diploidy for complementation (function) tests. Mutant strains provide blocked sites in the biosynthetic pathway so that accumulated intermediate compounds may be detected and characterized. Genetic data can furnish significant biochemical clues when the mutant loci are very closely linked. In species with no demonstrable sexual cycle, the parasexual cycle can often place mutant loci in the same or different chromosomes, a specific chromosome arm, or in sites along a chromosome arm distal from the centromere (Pontecorvo, 1958). Furthermore, the parasexual cycle yields diploid cells for complementation tests in species lacking heterokaryosis.

While color mutations furnish excellent markers for transmission genetics, they have little biochemical or enzymological value when the accumulated intermediate compound or compounds are unknown. Carotenes, particularly those in the immediate pathway from phytoene to β-carotene (Porter and Lincoln, 1950), have received considerable attention from biochemists and geneticists using plants, bacteria, and fungi as experimental material. Among the fungi, three species have been extensively investigated: *N. crassa*, *P. blakesleeanus*, and *U. violacea*. The contributions from each species will be presented separately.

The first enzyme model for carotene biosynthesis was formulated by Jensen et al. (1961) to explain the accumulation of carotenes and carotenoids in mutant strains of purple bacteria. The model was based on the composition of the accumulated intermediate compounds and the structure of the immediately "adjacent" compounds in the biosynthetic pathway. This model for prokaryotic species has been the basis for models proposed for carotenogenesis in fungal species. In the purple bacteria model, the dehydrogenation of neurosporene to lycopene is enzymatically distinct from the preceding dehydrogenations which result in the formation of neurosporene from phytoene. Furthermore, the patterns of accumulated intermediate compounds in mutant strains suggest that the three dehydrogenations in the conversion of phytoene to neurosporene may be catalyzed by a single enzyme.

Carotenogenesis in Fungi 111

Yen and Marrs (1976) mapped mutations yielding different patterns of carotene accumulation or altering the biosynthesis of bacteriochlorophyll in the purple bacterial species *Rhodopseudomonas capsulata*. The carotene mutations were assigned to five closely linked loci which were adjacent to two closely linked loci involved in bacteriochlorophyll biosynthesis. These results suggest a transcriptional basis for the coordinate response of these two types of photopigments to regulatory stimuli.

Neurospora crassa

Neurospora crassa is a saprophytic, bipolar (A,a) heterothallic species in the Ascomycetes (Pyrenomycetes), producing an octet of linearly arranged ascospores. The multinucleate homokaryotic hyphae spontaneously undergo anastomosis, yielding heterokaryotic hyphae for complementation tests when conidia of different genotypes and mating types are mixed prior to germination. The sexual cycle is readily completed on laboratory medium. The parasexual cycle has not been found in this species. Details of the life cycle have been presented by Fincham et al. (1979).

Pink wild-type strains of *N. crassa* yield a mixture of phytoene, phytofluene, neurosporene, lycopene, γ-carotene, and β-carotene (Haxo, 1949; Sheng and Sheng, 1952; Zalokar, 1954). Mutants lacking detectable carotene in the mycelium, conidiophores, and conidia are termed albino (*al*) and three loci have been identified. The *al-1* and *al-2* loci are very closely linked in chromosome IR, but are separated by the *arg-6* locus (Perkins, 1971, 1974; Perkins and Murray, 1963; Perkins et al., 1973). The *al-3* locus is in chromosome VR (Wang et al., 1971).

The *al-1* mutants lack carotene and accumulate large amounts of phytoene, suggesting a biochemical lesion for phytoene dehydrogenase (Goldie and Subden, 1973; Subden and Goldie, 1973). Certain alleles at this locus, however, yield unexpected phenotypes, such as aurescent (*35408*), with pigment in the conidiophores and conidia at the periphery of the mycelium, and mutant *ALS4*, with yellow conidia conidiophores and mycelium (Perkins and Murray, 1963; Subden and Turian, 1969). Mutant phenotypes with pigmentation may reflect "leaky" mutations for phytoene dehydrogenase. Alleles at the *al-1* locus do not complement each other (Subden and Goldie, 1973).

The *al-2* mutants also lack carotene. The genetic blocks, however, may be between geranylgeranyl pyrophosphate and phytoene or between prephytoene pyrophosphate and phytoene (Kushwaha et al., 1978). A purple mutant (MN58p) may have a "leaky" allele.

The *al-3* mutants lack carotene (Wang et al., 1971; D. D. Perkins and M. Björkman, unpublished work). The genetic block may be between isopentenyl pyrophosphate and GGPP or prephytoene pyrophosphate and phytoene (Kushawaha et al., 1978). The rosy allele yields partial pigmentation (Barratt and Ogata, 1978) and other alleles have white mycelium with a trace of pink (Wang et al., 1971). These phenotypes presumably reflect "leaky" mutations.

The yellow ($ylo\text{-}l$) mutant produces yellow carotenoids in mycelium and conidia (Garnjobst and Tatum, 1956) and the locus is in chromosome VIL (Stadler, 1956; D. D. Perkins, and M. Björkman, unpublished work). The mutation affects the synthesis of 4-β-apocarotenoic acid (Goldie and Subden, 1973) and may involve $3',4'$ double bond formation or the carboxylation of torulene or γ-carotene.

Subden and Turian (1970) used biochemical and genetic data reported by Hungate (1945), Haxo (1952), Huang (1964), and Subden and Threlkeld (1968, 1969, 1970) to construct composite genetic and complementation maps for the carotene mutations in the $al\text{-}l$ and $al\text{-}2$ loci. The $al\text{-}l$ locus is relatively long (1.5-2.0 cM) and includes two regions. Furthermore, mutations in this locus presumably yield defective phytoene dehydrogenase. Subden and Turian (1970, p. 361) proposed an enzyme model for carotenogenesis in *N. crassa*:

> Carotenoid synthesis proceeds by means of a light induced heteromultimeric protein complex that is able to perform successive desaturations and cyclizations on a given C_{40} substrate. The complex would be oriented in the cell geography such that it would associate with the primary substrate, phytoene, directly from the geranyl-geranyl pyrophosphate condensing system and then proceed with desaturation and cyclization. It is further postulated that the complex favors association with recently formed phytoene (the primary substrate) rather than endogenous phytoene and other desaturated carotenoids that constitute the "secondary substrates."

The *Neurospora* enzyme model for carotenogenesis stipulates polypeptides with dehydrogenase and cyclase activity in a heteromultimeric complex without specifying the minimal number of each polypeptide or the relationship between the dehydrogenases and cyclases to accomplish the four dehydrogenations and two cyclizations from phytoene to β-carotene. Moreover, the model does not explain the mixture of carotenes in the wild-type strains. While the $al\text{-}l$ locus with its two regions (cistrons?) is associated with phytoene dehydrogenase, the gene or genes coding for the cyclases have not yet been identified.

The site or sites of the putative heteromultimeric enzyme complex in the cell geography have not yet been related to the site or sites of carotene accumulation. The site of enzyme activity may also be the site of accumulation, but there is no compelling reason for this convenient arrangement. Carotenes are hydrophobic, a significant characteristic of these compounds in searching for the site of the proposed carotenogenic enzyme complex.

Phycomyces blakesleeanus

Phycomyces blakesleeanus is a saprophytic, bipolar (+, -) heterothallic species in the Phycomycetes (Mucorales). The fusion of two multinucleate cells eventually results in the formation of a thick-walled zygospore which, after germination, produces a germ sporangium. Prior to spore formation, the meiotic products in the germ sporangium are numerically increased by mitotic divisions (Cerdá-Olmedo, 1975; Eslava et al., 1975a,b). This species presents two significant problems for genetic studies: (1) production of heterokaryotic hyphae for complementation tests and (2) the lengthy germination of zygospores for genetic analysis. Both problems have been overcome. This species has no parasexual cycle. The details of the life cycle have been presented by Fincham et al. (1979).

Heterokaryons are produced by techniques which depend on the ability of the sporangiophores to survive and regenerate after surgical procedures. Burgeff (1914) inserted the cut end of one sporangiophore into the cut end of another with a different mating-type allele and obtained heterokaryons from sporangiophore "sprouts." Ootaki (1973) modified this technique to graft sporangiophores with different mating-type alleles to get heterokaryons. The initial cross to zygospore germination takes approximately 3 months (Torres-Martínez et al., 1980).

Wild-type strains heterokaryotic for mating-type alleles yield a mixture of phytofluene, carotene, neurosporene, lycopene, γ-carotene, and large amounts of β-carotene so that these strains are yellow (Goodwin, 1952).

Red mutants accumulating lycopene and white mutants accumulating phytoene and lacking carotene were obtained from wild-type strains (Heisenberg and Cerdá-Olmedo, 1968; Meissner and Delbrück, 1968). Heterokaryons with red and white mutant nuclei displayed different colors, ranging from almost white to orange. The quantitative analysis of lycopene, γ-carotene, and β-carotene in such heterokaryons grown for 48 hr indicated a close relationship between the relative proportions of each carotene and specific nuclear ratios (de la Guardia et al., 1971). To account for this relationship, de la Guardia et al. (1971) proposed a linearly organized aggregate of four identical dehydrogenases and two identical cyclases. Support for the enzyme aggregate model was reported by Eslava and Cerdá-Olmedo (1974) and Aragón et al. (1976).

Determination of the carotene content of the same heterokaryon and homokaryon grown for 48, 96, 168, and 192 hr showed that the enzyme aggregate hypothesis predicts the carotene content during exponential growth (48 hr), but a second mechanism for β-carotene production may function later in the growth cycle (Hsu et al., 1974b). The second mechanism may bypass lycopene, via β-zeacarotene and γ-carotene, en route to β-carotene. It may be possible to reconcile the different results by assuming a shift in the organization of the aggregated enzymes as exponential growth enters the lag phase.

In the *Phycomyces* model, the first dehydrogenase accepts newly formed phytoene, the immediate carotene precursor. After each dehydrogenation, the intermediate compound passes or is passed along the linear aggregate to the next dehydrogenase and then to each cyclase. After the second cyclization, β-carotene is released from the aggregate. Using an assumption from the *Neurospora* model, intermediate carotenes are not transferred between aggregates nor accepted from the intracellular environment. Unlike the purple bacteria model, the *Phycomyces* model calls for four identical dehydrogenases. To account for the quantitative relationships between nuclear ratios and proportions of the three carotenes in the red-white heterokaryons, the gene products (polypeptides) produced by the nuclei in each cytoplasmic domain are assumed to combine randomly to provide the dehydrogenases and cyclases in each linear enzyme aggregate. Consequently, each aggregate can include different numbers of effective and defective dehydrogenases and cyclases. When an intermediate carotene reaches a defective enzyme in any one aggregate, the carotene is released into the cellular environment, thereby explaining the mixture of carotenes in the heterokaryons. This explanation, however, cannot account for the mixture of carotenes in heterokaryons for wild-type strains.

A heterokaryon with nuclei from mutant *C5* accumulating phytoene and from mutant *C6* lacking carotene had a mixture of phytoene, phytofluene, ζ-carotene, neurosporene, and β-carotene (de la Guardia et al., 1971). The proportions of these compounds agreed with the calculated values when one assumed four identical copies of one dehydrogenase and two copies of one cyclase and random combinations of effective and defective dehydrogenases and cyclases in the enzyme aggregates.

In *F. aquaeductuum* treated with cycloheximide at different times after illumination, the carotenes appear temporally in the same sequence as in the biosynthetic pathway (Bindl et al., 1970). It appears, therefore, that the enzyme may be formed and become sequentially active, which is difficult to reconcile with the hypothesis of an enzyme aggregate that functions only after all its component enzymes are properly assembled.

Ootaki et al. (1973) obtained nine white mutants from a wild-type strain after mutagenesis. Two mutants accumulated phytoene, two mutants had a trace of phytoene, and five mutants had no detectable phytoene. Complementation tests yielded two groups of mutations in different loci: *carB* and *carA*. The mutants accumulating phytoene were assigned to locus *carB*. Complementation tests involving a *carB* mutant strain and 14 white mutants from a different wild-type strain indicated that some mutations had occurred in locus *carA* and others in locus *carB*. The major accumulated carotene in one tested heterokaryon was β-carotene.

Carotenogenesis in Fungi

Ootaki et al. (1973) isolated five red mutants after mutagenesis of a wild-type strain. Lycopene was the major accumulated carotene, but several mutants also accumulated traces of phytofluene, ζ-carotene, neurosporene, γ-carotene, and β-carotene. No strain accumulated more than 12% β-carotene (total carotene). The red mutants gave negative complementation; that is, no combination accumulated more than 12% β-carotene. The red mutations were assigned to locus *carR*, which was assumed to specify cyclase. Except for 3 mutations, complementation tests involving a representative *carR* mutation and 19 white mutations were positive. Heterokaryons involving the exceptional mutations were red, indicating no complementation. One tested heterokaryon accumulated lycopene, as did the *carR* mutant strain. These observations were explained by assuming that the exceptional mutants resulted from a single-step mutation induced by the mutagen in both the *carA* and *carR* loci.

The single-step mutation hypothesis was tested by mutagenizing a *carR* strain, isolating 10 white mutants, and including the double-step mutants in heterokaryons with a *carA*, *carB*, or *carR* strain. The results indicated that the second mutation yielding a white phenotype had occurred in the *carA* or *carB* locus. These observations supported the initial explanation of a single-step mutation for the exceptional mutations.

Torres-Martínez et al. (1980) crossed the exceptional one-step mutant and the two-step white ones with *carA*, *carR*, and single-step *carRA* mutants. Crosses between *carA* and *carR* strains did not yield either *carRA* or wild-type recombinants; crosses between *carRA* and wild-type strains did not yield either *carA* or *carR* recombinants. The potent mutagens N-methyl-N'-nitro-N-nitrosoguanidine (MNNG) and *ICR-170* were used to distinguish between a single-step mutation in a complex *carRA* locus or two simultaneous mutations in the very closely linked *carR* and *carA* loci. The wild-type strain after MNNG mutagenesis gave *carR*, *carA*, and *carRA* mutants, and after *ICR-170* mutagenesis, only *carRA* mutants. A double-step mutant was mutagenized with *ICR-170* and gave neither *carR* mutants nor wild-type revertants, whereas mutagenization with MNNG yielded *carR* mutants but no wild-type revertants. Finally, MNNG mutagenesis of a single-step *carRA* strain gave both *carR* mutants and wild-type revertants. Crosses indicated that the wild-type revertants did not result from intergenic suppression. The simultaneous loss or gain of *carR* and *carA* functions by a single-step mutation and the frequency of such mutations indicated a complex *carRA* locus rather than two closely linked *carR* and *carA* loci.

Using the pattern of one- and two-step mutations, Torres-Martínez et al. (1980) proposed a complex *carRA* locus with contiguous R and A segments to accommodate the consequences of missense, nonsense, and frameshift mutations as they produce the *carR*, *carA*, and *carRA* phenotypes. The R locus specifies a cyclase using lycopene as the first substrate and γ-carotene as the second substrate; the *A* segments code for a transfer polypeptide operating on the multienzyme aggregate. Furthermore, the *R* and *A* segments are "cotranscribed to a single mRNA and cotranslated to a single polypeptide" (Torres-Martínez et al., 1980). Cleavage of the polypeptide yields cyclase (*carR* function) and substrate transfer polypeptide (*carA* function). The bifunctional *carRA* locus model is an ingenious solution to the single- and double-step *carRA* mutations. It is not clear why the substrate transfer polypeptide is part of a complex locus including the region coding for cyclase when the transfer of substrate commences at the first or second dehydrogenation. Murillo et al. (1981), however, concluded that the *carA* region is responsible for the transfer of the carotene intermediates between the carotenogenic enzyme complexes, because *carR* mutation partially blocks the conversion of lycopene to β-carotene; the transfer of the intermediate carotene substrates between carotenoid enzyme aggregates occur in wild type, but not in heterokaryons for *carA*.

Ustilago violacea

Ustilago violacea is a phytopathogenic, bipolar (a_1, a_2) heterothallic species in the Basidiomycetes (Ustilaginales), causing anther smut in susceptible species of the Caryophyllaceae. The diploid overwintering teliospores from smutted anthers germinate on laboratory medium, undergo meiosis, and produce four haploid nuclei on a short septate promycelium. The tetrad of basidiospores produces four clones of sporidia, which multiply by budding to yield a yeastlike colony. The formation of a dikaryon by a conjugation bridge between a_1 and a_2 cells at the site of inoculation in a susceptible host initiates the formation of infectious dikaryotic hyphae. The sexual cycle is completed only in the anthers of a susceptible host species. The species has a parasexual cycle yielding recombinants by mitotic recombination or haploidization (Day and Jones, 1968, 1969). Diploid sporidia provide complementation tests (Garber et al., 1975). Tetrad analysis can be accomplished with resorting to the separation of the basidiospores in a tetrad (Garber et al., 1981). The life cycle of *Ustilago maydis* also applies to *U. violacea* (Fincham et al., 1979).

Day and Jones (1968) and Day and Day (1970) obtained yellow (*y*) and orange (*o*) colony mutants from a pink strain after UV mutagenesis. Garber et al. (1975) isolated white (*w*) mutants from colored strains after UV mutagenesis. Teliospores from smutted anthers of field-collected plants or herbarium specimens yielded *w*, *y*, pumpkin (*p*), and pink strains, indicating extensive polymorphism for sporidial colony colors in nature (Garber et al., 1978).

Pink sporidia overproduce cytochrome *c*, which is present in relatively reduced amounts in *w* strains (Will et al., 1982). Yellow sporidia accumulate β-carotene; orange sporidia, γ-carotene; and pumpkin sporidia, lycopene; and white strains lack carotene (Garber et al., 1975; Garber et al., 1980). Sporidia from different wild pink strains accumulate β-carotene, lycopene, or no carotene (O. H. Will, M. Ruddat, and E. D. Garber, unpublished data). No tested haploid wild or mutant strain has been found to accumulate phytofluene, ζ-carotene, or neurosporene. A systematic survey of wild or mutant white strains may yield such strains. The accumulation of lycopene, γ-carotene, or β-carotene in different strains of *U. violacea* indicates a Porter-Lincoln scheme for carotenogenesis in this species (Fig. 2).

The *o, p, y,* and *w* loci are very closely linked (1-2 cM); the *w* locus is in one chromsome arm and the remaining loci in the other. (Garber et al., 1975, 1980; Cattrall et al., 1978). Furthermore, the sequence of the loci could be a significant factor in formulating enzyme models for carotenogenesis in *U. violacea*: *o-p-y-*centromere-*w*.

Enzyme models for carotenogenesis in *U. violacea* accommodate either a multimeric dehydrogenase with four polypeptides for dehydrogenations combined with a dimeric cyclase with two polypeptides for cyclizations (*Neurospora* model) or four dehydrogenases and two cyclases in a linear aggregate (*Phycomyces* model). Unlike the *Phycomyces* model with two identical cyclases, the *U. violacea* models have two different polypeptides for cyclizations or cyclases. At least three questions must have answers to formulate an acceptable enzyme model for *U. violacea*. Is β-carotene the terminal product of carotenogenesis or an intermediate compound? Which locus codes for the dehydrogenation polypeptides? What is the carotene content of haploid pink strains and of diploid pink strains heterozygous for different combinations of the *o, p, y,* and *w* alleles?

At least two models can be formulated to assign a specific function to each of the four loci (*p, o, y,* and *w*) involved in carotenogenesis in *U. violacea*. These models assume that certain loci code for dehydrogenases, cyclases, enzymes in the biosynthesis of phytoene, or an enzyme using β-carotene as substrate (Fig. 3).

One model stipulates that the *p* locus codes for the dehydrogenase polypeptides, and the *o* and *y* loci for the cyclase polypeptides. The mutation from p^+ to p yields a white phenotype. The p^+o genotype is responsible for lycopene accumulation, and the p^+o^+y genotype for γ-carotene accumulation. A mutation in another, as yet unknown locus is responsible for β-carotene accumulation. It should be noted that the *p, o,* and *y* loci are very closely linked in the same chromosome arm for cotranscription and cotranslation in forming the carotenogenic complex or aggregate. This model makes three assumptions: (1) Haploid strains do not accumulate phytofluene, ζ-carotene, or neurosporene; (2) β-carotene is converted to another compound; and (3) white mutants may accumulate phytoene or prephytoene compounds.

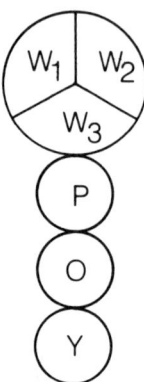

Figure 3 A model for the organization of the aggregated dehydrogenases and cyclases for carotenogenesis in *Ustilago violacea*.

A second model assumes that the w locus or a cistron in the w locus codes for the dehydrogenase polypeptides and the p and o loci for the cyclase polypeptides. The mutation from w^+ to w yields a white phenotype. The w^+p genotype is responsible for lycopene accumulation, the w^+p^+o genotype for γ-carotene accumulation, and the $w^+p^+o^+y$ genotype for β-carotene accumulation. This model makes three assumptions: (1) Haploid strains do not accumulate phytofluene, ζ-carotene, or neurosporene; (2) β-carotene is converted to another compound; and (3) the w mutants accumulate phytoene.

A survey of mutant or wild white and pink strains could reveal strains accumulating prephytoene products or phytoene (first model) or only phytoene (second model). If the survey gave two classes of white mutants, it would be important to determine the linkage relationships for the different loci.

Critical enzymological evidence is not yet available to accept or reject any of the models formulated on genetic grounds. Furthermore, it may be premature to assume that one model would apply to prokaryotic and eukaryotic species or to different eukaryotic species, particularly fungal species.

The o, p, and y loci presumably code for a single polypeptide. Crosses between wild y strains gave neither recombinant nor complementation data (Garber and Owens, 1980). The o mutation occurred once (Day and Day, 1970). The w locus, however, is complex. The origin of the w mutants merits comment and may provide clues to its role in both carotenogenesis and overproduction of cytochrome c.

The o, p, and y strains yield white mutants either spontaneously or after mutagenesis (Garber et al., 1975; M. Ruddat and E. D. Garber, unpublished data). The white mutants lack carotene and have a reduced cytochrome c

content compared with that for the pink strains (O. H. Will, M. Ruddat, and E. D. Garber, unpublished data). All of the tested white mutations map in the w locus (Garber et al., 1975; Cattrall et al., 1978; Garber, 1980). The complex w locus with two to three complementation regions may include cistrons with different functions: a structural locus for a dehydrogenation polypeptide, a regulatory locus for carotenogenesis, and a regulatory locus for cytochrome c production. One approach to at least some of these questions considers the carotene content of pink haploid and diploid strains, the cytochrome c content of strains accumulating carotene, and the presence or absence of phytoene in white strains. The pink pigmentation could mask the color associated with accumulated carotene or carotenes; the color associated with a specific carotene mutation could mask the amount of accumulated cytochrome c, which would be greater than that in w mutants.

The parasexual cycle in *U. violacea* provides diploid uninucleate sporidia so that the vagaries of nuclear ratios in heterokaryons are not a significant factor in determining the carotene content of diploids heterozygous for different combinations of carotene mutations or carotene and w mutations. Preliminary experiments indicate that diploid sporidia, unlike haploid sporidia, accumulate a mixture of carotenes. Diploid and disomic (n+1) sporidia with different combinations of carotene mutations, a carotene and a w mutation, and certain combinations of w mutations produce pink colonies (Garber et al., 1975; Cattrall et al., 1978; Garber, 1980). The carotene content of the diploid or disomic sporidia has not yet been systematically investigated. It would also be important to examine diploids heterozygous for a carotene and w mutation in trans and cis configurations.

Recombinants from crosses between o and y strains and between certain w strains are detected by their pink phenotype (Cattrall et al., 1978). These observations suggest that the o^+y^+ and $w\text{-}1^+w\text{-}2^+$ recombinants express their genotype for cytochrome c overproduction. The carotene content of these recombinants has not yet been determined and might reveal a relationship between carotenogenesis and cytochrome c overproduction.

REGULATION OF CAROTENOID PRODUCTION

The interaction of genetic and environmental factors undoubtedly regulates the course of fungal carotenogenesis. It will be difficult, however, to assess their roles and the molecular basis for the interaction as long as the dehydrogenases and cyclases, their organization into complexes or aggregates, and their subcellular localization remain enigmas. While it is customary to appeal to transcriptional, translational, and posttranscriptional control mechanisms, explanations based on such controls must be viewed with caution.

Carotenoids in Relation to Growth and Development

In several fungal species, the carotenoid content persists over a period of time without apparent quantitative or qualitative changes after a phase of rapid synthesis and accumulation and then diminishes as the culture ages (Goodwin and Willmer, 1952; Vecher and Kulikova, 1968; Hsu et al., 1974b). This pattern of carotenoid distribution in a cell over time is more obvious in static than in shake cultures (Goodwin and Willmer, 1952). During synthesis, major qualitative changes in carotenoid content alter the color of the culture. Young cells often contain large amounts of lycopene, which renders the culture red, and the culture then turns yellow-orange as γ- and β-carotene accumulate (Hsu et al., 1974b). Evidence for carotene turnover was obtained with [^{14}C]asparagine fed to *L. michotii* after the sixth day of culture (Jerebzoff-Quintin and Jerebzoff, 1980).

Carotenoids usually disappear gradually in aging fungal cultures. In *Aschersonia aleyroides* the carotene content decreased by 75% from the twentieth to the thirty-fifth day of culture (van Eijk et al., 1979). The main carotene, β-carotene, decreased proportionally to the total, while β-zeacarotene and torulene disappeared almost completely and two new cis-β-carotenes appeared (van Eijk et al., 1979).

Carotenoid accumulation occurs mainly after growth is completed (Goodwin and Willmer, 1952; Vecher and Kulikova, 1968; Hsu et al., 1974b; Hashimoto et al., 1978). Consequently, carbon depletion limits carotenoid synthesis and increased carbon supply stimulates it (Feofilova et al., 1970; Vuori and Gyllenberg, 1971). Parallel curves for growth and carotenogenesis, however, have been observed; as growth leveled off, carotene synthesis usually continued (Davies, 1973; Hsu et al., 1974b).

Carbon and Nitrogen Sources

The variety of carbon sources suitable for fungal carotenoid production is large and ranges from simple compounds, such as monosaccharides and glycerol, to complex mixtures commonly used in industrial fermentation, such as vegetable and animal oils, oil-refining by-products, kerosene, soap stock, autoclave oils, molasses, corn steep liquor, or distiller's solubles (Ninet and Renaut, 1979). Acetate as the sole C source stimulates growth and carotene production in *P. blakesleeanus* more than lactate or pyruvate (Friend et al., 1955). Acetate has a similar effect on *Rhodotorula sanniei* in the presence of a large nitrogen supply (Vuori and Gyllenberg, 1974). Yeast extract enhanced carotenogenesis in *P. blakesleeanus* grown on acetate, but not for *P. blakesleeanus* grown on glucose medium (Friend et al., 1955), while in *U. violacea* yeast extract is stimulatory in glucose medium (M. Ruddat and E. D. Garber, unpublished data). The effective component(s) of yeast extract, while still elusive, may be one or a combination of the following: thiamine, riboflavin, nicotinic acid, and vitamin A and vitamin B complex (Goodwin and Lijinsky, 1952).

β-Carotene synthesis in *P. blakesleeanus* is stimulated by leucine and valine, which function as precursors for β-hydroxy-β-methylglutaryl-coenzyme A, an important intermediate in isoprenoid synthesis (Chichester et al., 1959). Glycine is incorporated into β-carotene by *P. blakesleeanus*, presumably by the glycine-serine-pyruvate-acetyl-coenzyme A pathway (Mackinney et al., 1955). In *L. michotii*, asparagine increases carotene accumulation (Jerebzoff-Quinton and Jerebzoff, 1980).

Mutants of *Blastocladiella emersonii* and *R. rosea*, lacking α-ketoglutarate oxidase activity, produce γ-carotene. Wild-type cells synthesize γ-carotene only in the presence of high bicarbonate concentrations (Cantino, 1965). Inhibition of the tricarboxylic acid cycle by mutation or bicarbonate in wild type presumably shifts metabolites into the carotene pathway.

Temperature

In certain fungi, temperature is an important factor controlling the degree of pigmentation. *Rhodotorula gracilis* and *R. glutinis* show drastic qualitative changes with temperature (Nakayama et al., 1954; Simpson et al., 1964). *Rhodotorula rubra, Rhodotorula peneaus*, and *P. blakesleeanus* produce the same carotenoids over a range of 5-25°C (Nakayama et al., 1954; Friend and Goodwin, 1954). In the dermatophyte *T. mentagrophytes* only the arthrospores produce cartenoids. A temperature shift from 37 to 39°C inhibits carotenoid production without affecting spore formation (Hashimoto et al., 1978).

Light

Light is expected to be a crucial factor in controlling carotenogenesis, considering that carotenoids serve as photoprotection agents. Photoregulation of carotenogenesis has been demonstrated in several fungal species (see reviews by Rau, 1976; Harding and Shropshire, 1980). Most carotenogenic fungi, like flowering plants (Frosch and Mohr, 1980), produce significant amounts of carotenoid pigments in the dark, but considerably greater amounts after illumination. *Phycomyces blakesleeanus* (Garton et al., 1951; Bergman et al., 1973; Sandmann and Hilgenberg, 1978; Jayaram et al., 1979), *N. crassa* (Davies, 1973), *R. gracilis* (Simpson et al., 1964), *Penicillium slcerotium* (Mase et al., 1957), and *Sporidiobolus johnsonii* (Fiasson et al., 1972) are but a few examples. Usually photoinduction causes only quantitative changes, but in *S. johnsonii* (Fiasson et al., 1972), *R. gracilis* (Simpson et al., 1964), and *L. michotii* (Jerebzoff-Quintin and Jerebzoff, 1980) qualitative changes also occur. In 13 fungal species (Rau,

1976) which produce only traces of carotenoids in the dark, a short period of in vivo illumination drastically increases in vivo and in vitro carotenogenesis. Among those species with strict photoregulation are *N. crassa* (Went, 1901; Haxo, 1949; Rau et al., 1968), *F. aquaeductuum* (Rau, 1967a), *Verticillium agaricinum* (Valadon and Mummery, 1971), *Sphaerobolus stellatus* (Friederichsen and Engel, 1957), and *Dacryopinax spathularia* (Goldstrohm and Lilly, 1955). Critical evidence for photoregulation in *U. violacea* is not yet available.

In view of the function of carotenoids, photostimulation of carotenogenesis is intuitively acceptable, but a reduction of carotenoid accumulation by light, which has been observed in *B. emersonii* (Cantino and Horenstein, 1956), *Choanophora cucurbitarium* (Chu and Lilly, 1960), and *B. trispora* (Sutter, 1970), is surprising. It remains to be investigated whether the lower carotenoid level is the result of light-inhibited synthesis or photodestruction. In the mutant *carA5 carS42* of *P. blakesleeanus*, where the two genes are hypothesized to be in the last two positions of the sensory pathway to photocarotenogenesis, less carotene is produced in the light than in the dark (López-Días and Cerdá-Olmedo, 1980).

Photoreceptors

The identity of the photoreceptor that mediates carotene biosynthesis has not yet been elucidated despite the efforts of nearly five decades and the evidence favoring flavins. The action spectra of light-induced carotenogenesis in fungi and other organisms are all very similar, with a peak in the near UV, at or near 370 nm, and in the blue region, with a peak around 450 nm, and shoulders or smaller peaks around 420 and 480 nm (Rau, 1967a, 1975; Bergman et al., 1973; DeFabo et al., 1976). Light beyond 520 nm is not effective. The action spectrum of photocarotenogenesis resembles that of the universally occurring blue light responses (Presti and Delbrück, 1978). The identification of this pigment, the blue light receptor named cryptochrome (Gressel, 1980), currently in the forefront of photobiological interest, has inflicted biologists with the "blue light syndrome."

While carotenoids (Bünning, 1937) and flavins (Galston, 1950) have been proposed as light receptors, the time-worn attempts to match their absorption spectra with action spectra of the blue light responses have not yielded a decision between two chromophores. The paradox persists. The action spectra favor carotenoids as chromophores, while the bulk of the other evidence, although to a large extent indirect, clearly champions flavins. Carotenoids, however, cannot be completely excluded (Shropshire, 1980; DeFabo, 1980; see also p. 108).

The recently discovered biphasic fluence response of light-induced carotenogenesis in *N. crassa* (Harding, 1974; Schrott, 1980a,b, 1981), *P. blakesleeanus*

(Jayaram et al., 1979), *Fusarium oxysporum* (Schrott, 1980b), and other *Fusarium* species (quoted by Schrott, 1980b) raises the possibility of two separate photoreceptors. The large extinction coefficient of carotenoids suggested that they mediate the low-fluence response, and flavins in the high-fluence response (Jayaram et al., 1979). Sensitivity of the high-fluence component alone to the transcription and translation inhibitors cycloheximide and actinomycin D supports the hypothesis of separate photoreceptors for the low- and high-fluence components. In *N. crassa,* however, all light-induced carotenogenesis is inhibited by cycloheximide (Harding and Mitchell, 1968). *pic* mutants of *P. blakesleeanus,* which are carotene deficient only at high fluence but normal at low fluence, support the hypothesis of two photoreceptors mediating light-induced carotenogenesis (Cerdá-Olmedo, quoted in Lipson, 1980). A biphasic fluence response curve may, however, also result from rapid utilization of precursors, which would cause saturation at low fluence. More precursors may be synthesized during continuous radiation over extended time periods and produce additional carotene.

The small effect of 280-nm irradiation on carotenogenesis in *N. crassa* was interpreted as evidence for carotenoids serving as photoreceptors, because flavins possess an absorption band at 280 nm and carotenoids do not (DeFabo et al., 1976). The same conclusion was reached in a similar approach with *P. blakesleeanus.* Whitaker and Shropshire (1981) found the quantum effectiveness at 365, 405, and 445 nm much closer to the relative absorbance spectrum of β-carotene than to that of riboflavin, and they concluded, therefore, that at least for the low-fluence component controlling photocarotenogenesis in *P. blakesleeanus* β-carotene serves as the chromophore in the blue light receptor. The substantial increase in quantum effectiveness at 365 nm in the high-fluence component of the fluence response curves for photocarotenogenesis may indicate riboflavin as the photoreceptor. However, if one considers the screening by β-carotene which is synthesized during the irradiation of up to 12 hr, the increase in quantum effectiveness at 365 nm at high fluence is more likely to indicate that the same photoreceptor mediates both the high- and low-fluence components of photocarotenogenesis (Whitaker and Shropshire, 1981). Jayaram et al. (1980) found that the mutants *madA, madB,* and *madD* of *P. blakesleeanus* produce fewer carotenoids than wild type under high and low fluences, which also points to a common component for the two fluence responses without, however, identifying the photoreceptor. Identifying the blue light receptor by matching action spectra with the absorption spectra of the putative pigment is burdened with the common lack of fine structure in the action spectra, self-screening, filtering effects by overlaying pigments, differential scattering, and the possible reabsorption of the blue fluorescence excited by near-UV light, all of which distort the action spectra (DeFabo, 1980).

Carotene has a short lifetime in the excited singlet state, which was judged too short for an efficient intersystem crossing of carotenoids to the triplet state (Song and Moore, 1974). Carotenoids have therefore been ruled out as chromophores for the blue light receptor. In the proper molecular environment, however, carotenoids do transfer energy to chlorophyll with a high degree of efficiency (Song et al., 1976).

Carotenoid-deficient mutant strains of *P. blakesleeanus* (Bergman et al., 1973; Cerdá-Olmedo and Torres-Martínez, 1979; Lipson, 1980; López-Días and Cerdá-Olmedo, 1980) and the double mutants in which the leaky carotenoid production is eliminated (Presti et al., 1977) all show normal phototropism in *Phycomyces*. By implication, this evidence also rules out carotenoids as photoreceptors for all other blue light responses and favors flavins for the role. Flavin-deficient mutants of *N. crassa* were 1/4 less sensitive to light induction of carotenogenesis than the wild type and 1/80 less light sensitive for the entrainment of rhythmic conidiation, which is also a blue light-regulated process (Paietta and Sargent, 1981). This is the first time that flavin mutants have been directly linked with photocarotenogenesis, thus providing formidable support for flavin as the chromophore of the blue light receptor.

Irradiation of *F. aquaeductuum* with red light in the presence of the artificial photoreceptor methylene blue or toluidine blue induces carotenogenesis very effectively (Lang-Feulner and Rau, 1975). Flavins can carry out photosensitized oxidation, which carotenoids cannot do; and photo-oxidation of a so far unidentified substance by the concomitant reduction of the photoreceptor has been proposed as the initial light reaction (Rau, 1971) which eliminates carotenoids as light receptors in *F. aquaeductuum* carotenogenesis. The effects of H_2O_2 and dithionite on photocarotenogenesis are in accordance with this hypothesis (Theimer and Rau, 1970; Rau, 1980).

The induction of a blue light response in *P. blakesleeanus* with an action peak at 595 nm is the most convincing argument in favor of flavin, because it is the wavelength for the direct optical excitation for the transition from the ground state to the lowest triplet state of riboflavin (Delbrück et al., 1976). The value for the 595-nm peak cannot be directly measured, but is obtained by extrapolation which, while based on reasonable assumptions, is not direct and unambiguous proof (DeFabo, 1980).

Blue light responses in *Pilobolus kleinii* and coleoptiles of *Zea mays* are inhibited by substances interacting with flavins, such as sodium azide, potassium iodide, and phenylacetic acid (Page and Curry, 1966; Schmidt et al., 1977). Dithionite and hydroxylamine, which inhibit flavin-mediated photoinduction, also prevent photocarotenogenesis in *F. aquaeductuum* (Theimer and Rau, 1970). Other substances interacting with flavins, such as rotenone, dichloromethylurea, and amobarbital, did not suppress light-induced carotenogenesis in *F. aquaeductuum*. Uptake of these compounds, however, was not investigated (Rau, 1980).

Blue light-induced reduction of *b*-type cytochrome, which has been observed in *P. blakesleeanus, Dictyostelium discoideum* (Poff and Butler, 1974), and *N. crassa* (Muñoz and Butler, 1975), occurs with purified cytochrome *b* from *D. discoideum* only after the addition of flavin (Manabe and Poff, 1978). Flavin serves, therefore, as the photoreceptor, and in the photochemical reaction the excited flavin would be reduced by an unidentified donor while flavin transferred an electron to cytochrome *b*. Since *D. discoideum* lacks a blue light physiology, it is obviously necessary to link the flavin-mediated cytochrome *b* reduction with well-established blue light responses, especially because blue light-induced cytochrome *b* reduction in *P. blakesleeanus* has a low quantum efficiency and was also observed in HeLa cells, which are unlikely to possess a significant photophysiology (Lipson and Presti, 1977). It is therefore significant that plasma membrane-enriched fractions prepared from *N. crassa* (Brain et al., 1977), *P. blakesleeanus* (Schmidt et al., 1977), and coleoptiles of *Zea mays* (Leong et al., 1981) contained the photoreducible cytochrome *b* system. Even more interesting is the observation with the *poky-timex* mutant of *N. crassa*. Deficient in cytochrome *b*, it showed a lesser light-induced cytochrome *b* reduction, as well as impaired entrainment of rhythmic conidiation (discussed in Brain et al., 1977). Ninnemann and Klemm-Wolfgramm (1980), however, found that photoinduced cytochrome *b* reduction was not correlated with light-induced phase shifts of the circadian rhythm of conidiation in *N. crassa*.

Valadon and co-workers have suggested the operation of two pigments other than carotenoids or flavins controlling photocarotenogenesis in *V. agaricinum*: (1) a new pigment which has one absorption peak at 390 nm and a smaller one at 420 nm (Osman and Valadon, 1978, 1979) and (2) phytochrome (Valadon et al., 1979). Red light-induced carotenogenesis which was reversed by far-red irradiation and a far-red difference spectrum with a crude extract showing a maximum at 670 nm and a minimum at 750 nm may indeed indicate the presence of phytochrome in *V. agaricinum*. Turbidity changes in the extract, however, could have influenced the spectrometric measurements. In *F. aquaeductuum* and *N. crassa,* phytochrome did not, however, control carotenogenesis (Schrott et al., 1982). In view of the fundamental function of phytochrome in plants (Mohr, 1981), its putative presence in fungi, which were believed to lack phytochrome, requires critical reevaluation of the experimental evidence.

Equally important is the suggestion of a photochromic light receptor for phototropism in *P. blakesleeanus,* which was made on the basis of two significant observations: (1) lower saturation at 605 nm than at 450 nm in the photogeotropic equilibrium angle and (2) the partial reversal of the blue light photogeotropic response with simultaneous 450- and 605-nm irradiation (Löser and Schäfer, 1980). While the interpretation by Löser and Schäfer is in contrast to that of Delbrück et al. (1976), who attributed the action of light at 595 nm to excitation of the lowest triplet state of riboflavin, it is in agreement with Hartmann's (1977) conclusion, that the action spectra for the blue light receptor could be represented as the action spectrum of a photochrome.

Underlying the numerous attempts to identify the blue light receptor is the assumption that a single type of chromophore receives and transduces the blue light signal mediating the multitude of blue light responses. It is apparent, however, that even the light absorption system is more complex than a clean-cut "either/or" choice between flavins and carotenoids or their protein-bound forms. Considering light-controlled carotenogenesis in bacteria along with that in fungi expands the choice of photoreceptors to porphyrins (Howes and Batra, 1970) and a flavin-porphyrin complex (Weeks et al., 1973). Isolating the blue light receptor and testing it in vitro have not been achieved so far, nor have mutants been identified which are affected in the photoreceptor.

Dark Reactions

From the cycloheximide sensitivity of the temperature-dependent lag phase following irradiation and from in vitro incorporation studies of in vivo irradiated mycelium, it has been inferred that at least in fungi with an absolute light requirement, light-induced protein synthesis may involve de novo synthesis of carotenogenic enzymes (Rau, 1967b; Bindl et al., 1970; Subden and Turian, 1970; Lang and Rau, 1972; Harding, 1974; Valadon et al., 1975; Spurgeon et al., 1979; Mitzka-Schnabel and Rau, 1981). However, direct evidence for de novo synthesis of carotenogenic enzymes or their isolation from fungi and biochemical characterization is still lacking. Cycloheximide applications carried out increasingly later after illumination, when carotenoid accumulation tends to be maximum, are less and less effective, indicating that cycloheximide does not act by interfering with the catalytic activity of enzymes (Rau, 1967b).

The first light-regulated step in the carotenoid biosynthetic pathway precedes that of phytoene formation. The activity of cell-free enzyme preparations with prenyltransferase activity from *N. crassa* wild type and carotenoid-deficient mutants *al-1, al-2,* and *al-3* is stimulated by in vivo light treatment, and so is the conversion of GGPP to phytoene, which is catalyzed by a membrane-bound enzyme (Harding and Turner, 1981). Earlier investigations by Lansbergen et al. (1976) had shown that phytoene synthesis and dehydrogenation were independently regulated by light in *N. crassa* wild type and the *al-1* mutant.

While the mechanism of light-induced protein synthesis is still unknown, its regulation at the transcriptional level has been proposed (Schrott and Rau, 1975). Evidence for photoinduced de novo synthesis of mRNA with state-of-the-art methods (affinity chromatography of double labeled polysomal RNA from *F. aquaeductuum*) has been presented, but the specificity of the mRNA for the photoinduced carotenoid synthesis remains to be shown (Schrott and

Rau, 1977). As Gressel (1980) noted, it is premature to expect to find a specific mRNA for photocarotenogenesis, considering the huge eukaryotic genome and the minute change in the enzyme complement required for carotenoid formation. So far the identification of a mRNA has been achieved only for photoinduced protein synthesis coded for by the much less complex plastid genome (Bedbrook et al., 1978).

Inhibitors

Regulation of carotenogenesis by chemical inhibitors has been observed in at least four sites in the biosynthetic pathway: (1) the level of prenyltransferases where the isoprenoid pathway branches, (2) the phytoene synthetase level, (3) the dehydrogenation reactions with the resulting accumulation of phytoene, and (4) the cyclization steps with the accumulation of lycopene. In addition, cis-trans isomerization is affected. The inhibitors are reasonably specific and cause selective accumulation or even actual stimulation of the synthesis of the precursors. Some of the inhibitors, therefore, enhance the production of specific carotenoids and can be regarded as promoters or inhibitors.

Phytoene Synthetase Inhibition

Phenylpyridazinone herbicides which are effective bleaching agents in green plants, such as norflurazon (SAN 9789) and 4-chloro-5(methylamino)-2-(3-trifluoromethylphenyl)pyridazine-3(2H)one, inhibit carotene accumulation in *P. blakesleeanus,* as well as in algae and green plants. Norflurazon inhibited carotenoid synthesis by blocking the activity of phytoene synthetase in cell-free extracts of *P. blakesleeanus* (Sandmann et al., 1980).

Dehydrogenation Inhibition

Diphenylamine (DPA), the most extensively investigated inhibitor of carotenogenesis, blocks the dehydrogenation steps, resulting in the accumulation of phytoene (Goodwin, 1952; Goodwin et al., 1953; Turian and Haxo, 1954; Friend et al., 1955; Valadon and Mummery, 1966; Olson and Knizley, 1962; Davies, 1970). In the C5 mutant strain of *P. blakesleeanus* in which dehydrogenation is genetically blocked, DPA, however, stimulates phytoene synthesis (Lee et al., 1975). Sandmann and Hilgenberg (1980) claimed that there was inhibition of phytoene precursor formation by DPA along with its effect on β-carotene synthesis in *P. blakesleeanus* Stamm 1. Removal of DPA from *Verticillium alboatrum* by washing the cells allows carotene synthesis to resume at the expense of the accumulated phytoene (Valadon and Mummery, 1966). Benzophenone, especially 4-substituted benzophenone, and biphenyls as well as 9-fluorenone act similarly to DPA in *Mucor hiemalis* (Herber et al., 1972).

Treatment of *R. glutinis* and *V. albo-atrum* with β-ionone results in a decrease of unsaturated carotenoids, with a concomitant increase of the more saturated ones (Simpson et al., 1964; Valadon and Mummery, 1966), whereas in *V. agaricinum* and in the mutant strains *C5* and *C9* of *P. blakesleeanus* a general decline of the total carotenoids occurs with no significant qualitative changes (Valadon and Mummery, 1973; Lee et al., 1975). β-Ionone generally has a stimulatory effect on carotenogenesis (Reyes et al., 1964; Lee et al., 1975; see also p. 131). Methylheptenone also lowered the carotenoid content in *V. agaricinum* (Valadon and Mummery, 1973). After treatment of *V. agaricinum* with 9-fluorenone, an interesting change in the structural configuration of carotene was observed: As trans-phytoene accumulated, only cis-phytoene was present in the control, albeit in low concentration (Valadon and Mummery, 1979). Substituted dibenzylamines, which share structural similarities with fluorenes, cause accumulation of cis- and polycis-carotenoids in grapefruit (Poling et al., 1980).

The action mechanism of the dehydrogenation inhibitors is still unknown. The effect of cycloheximide applications following treatment of *V. agaricinum* with 9-fluorenone indicates an inhibition of existing dehydrogenase activity by 9-fluorenone, rather than an effect on enzyme synthesis (Valadon and Mummery, 1979), and the inhibition of carotene dehydrogenase by DPA does not occur at the level of RNA polymerase (Valadon and Mummery, 1973).

Cyclization Inhibitors

Certain N-heterocyclic compounds are effective inhibitors of the cyclization of acyclic carotenoids. Nicotine treatment of wild-type and mutant strains of *P. blakesleeanus* and *V. agaricinum* and the yellow mutant of *Ustilago violacea* results in a drastic decrease in cyclic carotenes, with a concomitant increase in lycopene (Davies, 1973; Valadon and Mummery, 1974; M. Ruddat and E. D. Garber, unpublished results). Pyridine, imidazole, and certain of their derivatives inhibit cyclization of carotenoids in *B. trispora* and *P. blakesleeanus* (Ninet et al., 1969; Feofilova et al., 1970; Davies, 1973; Elahi et al., 1973a). In the *C115* mutant strain of *P. blakesleeanus,* treatment with N-heterocyclic compounds strongly inhibited β-carotene synthesis and lycopene, γ-carotene, and phytoene accumulated. Removal of methylimidazole or 3-methylpyridine from the incubation medium resulted in resumption of carotenogenesis: The lycopene and γ-carotene content decreased, while the β-carotene content increased. It follows that the lycopene formed in the presence of these inhibitors remains available in the pool of precursors for conversion to β-carotene (Elahi et al., 1973a). In the mutant of *mad107* of *P. blakesleeanus* removal of nicotine did not, however, result in a color change from red to yellow, as lycopene did not decrease although β-carotene increased (Davies, 1973). Apparently the accumulated lycopene became metabolically inert on entering the lipid pool.

The structure-function relationship among the N-heterocyclic compounds remains to be elucidated. While 2-aminopyridine strongly inhibits cyclization, 2-formylpyridine is inactive and 3-formylpyridine allows lycopene and β-carotene to accumulate, but 4-formylpyridine in combination with β-ionone stimulates β-carotene production (Ninet et al., 1969; see also p. 131). Interspecific differences in the action of 4-hydroxypyridine, succinimide, and isonicotinoylhydrazine were found in *B. trispora,* where these compounds stimulated carotenogenesis, and in *P. blakesleeanus,* where the compounds were ineffective or acted as growth inhibitors (Ninet et al., 1969; Elahi et al., 1973a). A relationship between the pK_a of pyridine derivatives and their inhibition of cyclases has been observed in *P. blakesleeanus*; compounds with a pK_a of 6 ± 1 were most effective (Elahi et al., 1973a).

The herbicide CPTA [2-(4-chlorophenylthio)triethylamine HCl] inhibits cyclization in wild-type and mutant strains of *P. blakesleeanus,* in *B. trispora,* as well as in a wide variety of plant tissues and microorganisms (Coggins et al., 1970; Hsu et al., 1972, 1974a; Elahi et al., 1973b; Kleinig, 1974), and acyclic carotenes, especially lycopene but also γ-carotene, accumulate at the expense of β-carotene. A similar effect was observed in *R. glutinis* and *R. rubra,* where lycopene accumulated at the expense of torulene, torularhodin, and β-carotene (Hayman et al., 1974). The *carA* mutants of *P. blakesleeanus,* which produce little or no carotenes, accumulate high amounts of lycopene and some γ-carotene in response to CPTA (Murillo, 1980). In CPTA-treated mated cultures of *B. trispora,* lycopene and γ-carotene accumulated to a much larger extent than could be expected from just inhibition of β-carotene synthesis by CPTA (Hsu et al., 1972). In both organisms CPTA action is sensitive to cycloheximide. In *V. agaricinum,* the effect of CPTA on cyclization is less defined and only phytoene appears to accumulate (Valadon and Mummery, 1974). 2-(4-Chlorophenylthio)triethylamine HCl analogs, diethylethoxyamines, triethylamine, tributylamine, 2-chloroethyltrimethylammonium chloride (Cycocel—a plant growth retardant—diethylaminopropiophenone, and p-(2-dimethylaminoethyl)phenol are also effective inhibitors of cyclization in *B. trispora* and *P. blakesleeanus* (Hsu et al.,1974a; Elahi et al., 1973b, 1975). Generally, the ethoxyethers are more active than the thioethers. Certain CPTA analogs stimulate lycopene synthesis so strongly that they may become useful as lycopene inducers for the citrus industry and perhaps for fermentation. Among these lycopene inducers, aliphatic esters of 2-diethylaminoethanol, however, stimulate β-carotene production in grapefruit, presumably because of the ease with which the ester bond is hydrolyzed (Poling et al., 1977). The effects of a large number of these CPTA-analogs have been determined, but only a few have been investigated with respect to fungal carotenoids.

It is apparent that the carotenoid biosynthesis pathway can be regulated by chemicals in specific ways, although no definite conclusion on their action mechanisms can be drawn as yet. In fact, three independent action mechanisms have been suggested: (1) interference with cyclase activity (Hsu et al., 1972), (2) derepression of the gene(s) regulating the synthesis of enzymes for lycopene synthesis (Hsu et al., 1972), and (3) removal of end-product inhibition (Murillo, 1980).

Carotenoid Stimulation

The sharp increase in carotenoids in mated *B. trispora* led to the observation of the activation of carotenogenesis by trisporic acid, which functions as a sex hormone in heterothallic Mucorales (see Ciegler, 1965; Gooday, 1974; van den Ende, 1976, for reviews). Certain other members of the Mucorales, especially *M. mucedo, M. Hiemalis, P. blakesleeanus,* and *C. cucurbitarium,* also produce increased amounts of carotenoids in response to trisporic acids (Bu'Lock et al., 1976). The elevated carotenoid level observed only in the (-) mating type in *B. trispora* in response to trisporic acid (Thomas and Goodwin, 1967; van den Ende, 1968; Bekhtereva et al., 1969) contrasts with that found in both mating types of *M. mucedo* (Gooday, 1968), which may be reconciled by differences in the cellular uptake of trisporic acids (Bu'Lock et al., 1972, 1976).

Trisporic acids which are C_{18} isoprenoid carboxilic acids (Caglioti et al., 1966) stimulate carotene biosynthesis, presumably by regulating the synthesis of the limiting carotenogenic enzyme(s). Cycloheximide inhibits the effect of trisporic acid when applied simultaneously (Thomas et al., 1967). Synthesis of sterols, prenols, and ubiquinones is also enhanced along with carotenoids (Thomas et al., 1967; Bu'Lock and Osagie, 1973). Therefore the enzyme(s) regulated by trisporic acid should function at an early stage in the isoprenoid pathway, a notion which is supported by the observation of trisporic acid-stimulated phytoene synthesis in the presence of DPA (Thomas and Goodwin, 1967). Control of the enzyme(s) involved in the conversion of 5-phosphomevalonate to dimethylallylpyrophosphate was proposed as the site of action of trisporic acid (Rao and Modi, 1977). An increase in membrane-bound protease activity in mated and trisporic acid-treated *B. trispora* was interpreted as trisporic acid-mediated inactivation of the protein inhibitor of carotenogenic enzymes (Govind et al., 1981). Trisporic acid also stimulates RNA synthesis in *B. trispora,* and because of the diversity of the biochemical, physiological, and striking morphological effects of trisporic acid, Bu'Lock suggested a direct or indirect derepression of multiple genes by trisporic acid (Bu'Lock et al., 1976).

The accumulation of carotenes in sexually reproducing Mucorales may be related to their function as precursors for trisporic acids. Label from β-carotene was found in trisporic acids (Austin et al., 1970). Caroteneless mutants of *P. blakesleeanus* fail to produce trisporic acid (Sutter, 1975), and DPA, which inhibits β-carotene synthesis, also prevents trisporic acid formulation (Thomas and Goodwin, 1967). Cleavage of β-carotene into retinyl derivatives and the formation of the β-C_{18} ketone prehormones with subsequent conversion to trisporic acid was demonstrated by Bu'Lock and his co-workers (Bu'Lock et al., 1974, 1976). In organisms responsive to trisporic acids, β-carotene synthesis may therefore be controlled by a positive feedback loop.

Availability of β-carotene alone is an adequate but unlikely sufficient condition for trisporic acid synthesis, which may be controlled by an additional regulator. Cyclic adenosine 5-monophosphate (cAMP) reversed the catabolite repression of glucose on carotenogenesis in mated but not in (+) or (−) cultures of *B. trispora* (Dandekar and Modi, 1980). In *N. crassa* Kristkii et al. (1981) observed a close correlation between carotenoid formation and a light-induced decrease in cAMP levels.

Naturally occurring compounds that are structurally closely related to trisporic acids are also effective stimulators of carotene production in *B. trispora* and *P. blakesleeanus* (Dandekar et al., 1980; Murillo and Cerdá-Olmedo, 1976; Eslava et al., 1974; Bu'Lock et al., 1972; Friend et al., 1955; Reyes et al., 1964). Assessing the carotenogenic acitvity of trisporic acid, abscisic acid, α- and β-ionone, and vitamin A (ordered according to decreasing effectivity), in *B. trispora,* Dandekar et al. (1980) concluded that the length of the side chain and the presence of a keto group in the ring were critical for the biological activity of these compounds. Citrus oil increases carotene accumulation in *B. trispora,* presumably because of its ionone content (Ciegler, 1965). Reyes et al. (1964) found that β-ionone stimulated β-carotene and sterol synthesis in *P. blakesleeanus* by regulating enzyme activities involved in the conversion of 5-phosphomevalonate to dimethylallylpyrophosphate, the same site suggested for trisporic action (Rao and Modi, 1977).

The activity of trisporic acid and β-ionone is competitive, indicating that these compounds act at the same site, and their effect is sensitive to cycloheximide. It has been inferred that this group of endogenous compounds enhances carotenogenesis by gene derepression. In contrast, the synthetic inhibitors (see p. 129) regulate carotenogenesis mainly through affecting enzyme activity, although gene derepression has also been postulated for the stimulation of lycopene accumulation by tertiary amines (Hsu et al., 1972). Evidence for the action mechanism of these compounds, however, is lacking.

The pyridine derivatives isoniazid, iproniazid, and 4-formylpyridine, either alone but more effectively in combination with β-ionone or 2,6,6-trimethyl-1-

acetylcyclohexene (TAC), stimulate β-carotene production in *B. trispora* up to 3 g/liter (30 mg/g dry weight of mycelium), whereas in trisporic acid-supplemented medium only 1.2 g/liter of β-carotene were produced (Ninet and Renaut, 1979). Like dimethylformamide, α-pyrrolidan, or succinimide, the pyridine derivatives may act as permeation agents, but it cannot be excluded that they selectively stimulate enzyme activity. The position of the formyl substituent on the pyridine ring is crucial for the activity of these compounds (see also p. 129). Cyclohexane, cyclohexanone, and their trimethyl derivatives, as well as bicyclic terpenes of turpentine in combination with the pyridine derivatives, can substitute for β-ionone, but TAC was most effective (Cederberg and Neujahr, 1969; Ninet and Renaut, 1979). The discovery of this group of carotenogenesis stimulators in fungi, especially trisporic acid and the pyridine derivatives along with β-ionone, has brought the fermentation of carotenoids close to economic feasibility (Ninet and Renaut, 1979).

Certain quaternary ammonium compounds, which inhibit terpene enzymes at critical branching points of the isoprenoid pathway, enhance carotene synthesis by diverting precursors into the carotene pathway. Isopropyl-4-dimethylamino-5-methylphenyl-1-piperidinecarboxylate methyl chloride (AMO-1618) prevents cyclization of GGPP to ent-kaurene (Dennis et al., 1965) and stimulates carotenogenesis (Lee et al., 1975).

In *F. aquaeductuum* and *Cephalosporium diospyros*, which require light for carotenogenesis, p-hydroxymercuribenzoate and p-chloromercuribenozoate stimulate carotenoid formation in the dark (Rau, 1967b; Parn and Seviour, 1974). It is unlikely, however, that these compounds stimulate carotenogenesis by inhibiting the suppressor, which is normally inactivated by light, because the effect of light and p-chloromercuribenzoate are additive (Theimer and Rau, 1969). The action of these compounds is sensitive to cycloheximide (Theimer and Rau, 1972).

β-Carotene can control its own synthesis through trisporic acids by a positive feedback loop, while at high levels of carotene accumulation a negative feedback mechanism becomes operative. Addition of neurosporene, β-zeacarotene, lycopene, or γ-carotene along with Tween 80 as surfactant to enzyme extracts from the *mad-107* mutant of *P. blakesleeanus* increases the ratio of phytoene to squalene (Davies, 1973). The enzymes involved are specific, inasmuch as spirilloxanthin, a nonfungal carotenoid, is ineffective in changing this ratio. Addition of β-carotene to the same enzyme extract inhibits the cyclization reactions and neurosporene and lycopene accumulate, while addition of phytoene or squalene inhibits their synthesis by affecting the prenyltransferases through yet another feedback loop (Davies, 1973). Sensitivity of carotene synthesis to feedback control is also suggested by the observation that inhibition of β-carotene synthesis which is controlled either genetically (Ootaki et al., 1973), chemically

(Olson and Knizley, 1962; Coggins et al., 1970), or both (Murillo, 1980) generally stimulates the carotene pathway and certain intermediates of β-carotene accumulate.

With the redundancy of control mechanisms regulating the carotene pathway, one cannot help but wonder why a secondary metabolic product would be so closely guarded, rivaling the controls observed for primary metabolites.

Mutations

Mutations reducing or enhancing the production of related compounds often provide clues to the regulation of the pertinent biosynthetic pathway. In such cases, however, "leaky" mutations must be distinguished from regulatory mutations. The operon model for regulation in prokaryotic species introduced the now familiar terms "repression" and "derepression" to accompany "feedback inhibition." Repression and derepression are applied to altered phenotypes due to mutations.

Mutations for high β-carotene or lycopene accumulation have been found in *P. blakesleeanus* (Murillo and Cerdá-Olmedo, 1976; Murillo et al., 1978). Three recessive noncomplementing mutations (*carS*) produced intensely yellow colonies after mutagenization of wild-type spores (Murillo and Cerdá-Olmedo, 1976). One *carS* strain produced two-step white mutants. One *carS carB* strain accumulated high levels of phytoene; five *carS carA* strains accumulated β-carotene in medium supplemented with vitamin A; and one *carS carRA* mutant had a trace of lycopene and did not respond to vitamin A. Finally, one intensely yellow mutant did not respond to vitamin A and accumulated more β-carotene than the parental strain. The *carS* mutations do not appear to influence phytoene dehydrogenation or lycopene cyclization. The β-carotene content of heterokaryons with *carA* and *carS* nuclei indicated different functions for each locus. Consequently, the *carS* gene is distinct from the other genes (*carA, carB*, and *carR*) involved in the biosynthesis of carotenes in *P. blakesleeanus*. Murillo and Cerdá-Olmedo (1976) speculated that the *carS* gene product may be a diffusible cytoplasmic regulator of carotenogenesis.

Murillo et al. (1978) obtained carotene-superproducing mutants of *P. blakeleeanus* which simultaneously and constitutively displayed several of the stimulator effects previously noted for photoinduction, certain chemicals, regulatory mutations, the interaction of hyphae with different mating-type alleles, and heterokaryons with nuclei carrying different mating-type alleles. For example, mutations in the *carS* locus resulted in high β-carotene yields in the dark, and mutations in the *carR* locus high lycopene yields. These observations support the hypothesis of independence for the stimulatory effects of photoinduction, vitamin A, *carS* mutations, and sexual interaction (Murillo and Cerdá-Olmedo, 1976).

A model of the sequence and interaction of the sensory transduction pathways under genetic control which regulate photocarotenogenesis in *P. blakesleeanus* have been proposed by López-Días and Cerdá-Olmedo (1980). The pertinent genes are *madA* and *madB, picA, picB* and *carA*, and *carS*. They do not yet provide insight into their relationships at the molecular level. These attempts to dissect the regulatory machinery for carotenogenesis in *P. blakesleeanus*, however, provide a model for comparable efforts in other fungal species, particularly those in the Phycomycetes.

Mutations in photoreceptor-regulating carotenoid biosynthesis may yield albino phenotypes. Such mutations should be distinguishable from those responsible for a defective dehydrogenase, which would also yield albino phenotypes. Although crosses between different albino mutants may indicate different loci, very closely linked loci could be interpreted as a complex locus with complementation regions (cistrons?). Both *N. crassa* and *U. violacea* have complex *al* or *w* loci, respectively, with two to three complementation regions.

Harding and Shropshire (1980) viewed *N. crassa* as the fungal species most likely to yield regulatory mutations of carotene biosynthesis by identifying regulatory albino mutants. In this species, conidia, unlike mycelium, synthesize carotenoid pigments in the dark. Consequently, mutants with pigmented conidia and albino mycelium should retain the structural genes for carotenogenesis but have regulatory mutations. The white-collar mutation of *N. crassa* (*wc*, chromosome VII) is responsible for pigmented conidia produced by albino mycelium. Harding and Shropshire (1980) presented unpublished data from R. E. Subden on the incorporation of 1-$[^{14}C]$ isopentenyl pyrophosphate (IPP) into phytoene by the wild-type, *al-1, al-2, al-3,* and *wc* strains cultured in the light and in the dark. The wild-type, *al-1*, and *wc* strains had low activity levels for converting IPP into phytoene in the dark and, except for the *wc* strain, high activity levels for light conversion. No detectable conversion was noted for *al-2* and *al-3* strains in the light or the dark. These observations suggested a reasonable scheme for the photoregulation of carotenogenesis in *N. crassa* (Harding and Shropshire, 1980). The *wc*$^+$ product is required for the photoinduction process; the *al-2* and *al-3* loci code for structural genes converting IPP to phytoene. The *al-1* locus codes for an enzyme (dehydrogenase?) for the conversion of phytoene to the carotene pigments. The scheme calls for the regulation of the *al-1, al-2,* and *al-3* structural genes by the light induction process. Finally, the scheme predicts mutations with the *wc* phenotype at different loci, one group of such mutations lacking the photoreceptor for photoinduction. The mutant *al-2* of *N. crassa* is defective in the particulate enzyme, while *al-3* is defective in the soluble enzymes which convert IPP to GGPP. The *wc-1* mutant has the membrane-bound enzyme activity in the dark-grown mycelium, but it is not light regulated since in vivo light treatment does not increase activity (Harding and Turner, 1981).

The three basic issues concerning carotenogenesis in fungal species continue to resist solution in molecular terms: *organization* and *site* of the carotenogenic enzymes and their *regulation* by photoinduction, chemicals, interaction of mating-type alleles, and mutations. Recombinant DNA and nucleotide sequencing technology may provide answers not yet attainable by methods which were successful for other biosynthetic pathways.

ACKNOWLEDGMENTS

We wish to acknowledge support from the Charles L. and Frances K. Hutchinson Fund and the Dr. Wallace C. and Clara A. Abbott Memorial Fund, The University of Chicago.

REFERENCES

Aasen, A. J., and Liaaen-Jensen, S. (1965). Fungal carotenoids. II. The structure of the carotenoic acid neurosporoxanthin. Acta Chem. Scand. *19*:1843-1853.

Alasoadura, S. O., and Visser, S. A. (1972). Pigment study of *Sphaerobolus stellatus*. Mycopathol. Mycol. Appl. *47*:295-300.

Altman, L. J., Ash, L., Kowerski, R. C., Epstein, W. W., Larson, B. R., Rilling, H. C., Muscio, F., and Gregonis, D. G. (1972). Prephytoene pyrophosphate. A new intermediate in the biosynthesis of carotenoids. J. Am. Chem. Soc. *94*:3257-3259.

Anderson, S. M., and Krinsky, N. I. (1973). Protective action of carotenoid pigments against photodynamic change to liposomes. Photochem. Photobiol. *18*:403-408.

Andrewes, A. G., and Starr, M. P. (1976). (3R,3'R)-astaxanthin from the yeast *Phaffia rhodozyma*. Phytochemistry *15*:1009-1011.

Andrewes, A. G., Phaff, H. J., and Starr, M. P. (1976). Cartenoids of *Phaffia rhodozyma*, a red-pigmented fermenting yeast. Phytochemistry *15*:1003-1007.

Anwar, M., and Prebble, J. (1977). The photoinactivation of the respiratory chain in *Sarcina lutea* (*Micrococcus luteus*) and protection by endogenous carotenoid. Photochem. Photobiol. *26*:475-481.

Aragón, C. M. G., Murillo, F. J., Guardia, de la, M. D., and Cerdá-Olmedo, E. (1976). An enzyme complex for the dehydrogenation of phytoene in *Phycomyces*. Eur. J. Biochem. *63*:71-75.

Arpin, N. (1968). Les carotenoides des discomycètes: Essai chimiotaxinomique. Doctoral thesis, Université de Lyon, Lyon.

Arpin, N., and Liaaen-Jensen, S. (1967a). Recherches chimiotaxinomiques sur les champions. Fungal carotenoids. IV. Les caroténoides de *Phillipsia carminea* (Pat.) le Gal, isolement et identificatin d'une xanthophylle naturelle nouvelle. Bull. Soc. Chim. Biol. *49*:527-536.

Arpin, N., and Liaaen-Jensen, S. (1967b). Recherches chimiotaxonomiques sur les champignons. Fungal carotenoids. III. Nouveaux cartenoides, notamment sous forme d'esters tertiaires, isolés de *Plectaria coccinea* (Scop. ex Fr.) Fuck. Phytochemistry 6:995-1005.
Aung Than, Bramley, P. M., Davies, B. H., and Rees, A. F. (1972). The stereochemistry of phytoene. Phytochemistry 11:3187-3192.
Austin, D. J., Bu'Lock, J. D., and Drake, D. (1970). The biosynthesis of trisporic acids from β-carotene via retinal and trisporol. Experientia 26:348-349.
Bae, M., Lee, T. H., Yokoyama, H., Boettger, H. G., and Chichester, C. O. (1971). The occurrence of plectaniaxanthin in *Cryptococcus laurentii*. Phytochemistry 10:625-629.
Barratt, R. W., and Ogata, W. N. (1978). *Neurospora* stock list: Ninth revision (June 1978). Neurospora Newsl. 25:29.
Bedbrook, J. R., Link, G., Coen, D. M., Bogorad, L., and Rich, A. (1978). Maize plastid gene expressed during photoregulated development. Proc. Nat. Acad. Sci. USA 75:3060-3064.
Bekhtereva, M. N., Feofilova, E. P., Sergeeva, L. N., Boltyanskaya, E. B., Kuvshinova, V. I., Murygina, V. P., and Berezinkov, V. M. (1969). Influence of the beta factor (trisporic acids) on the formation of carotenoids by minus strains of *Bl. trispora*. Mikrobiologiya 38:328-337.
Bergman, K., Eslava, A. P., and Cerdá-Olmedo, E. (1973). Mutants of *Phycomyces* with abnormal phototropism. Mol. Gen. Genet. 123:1-16.
Bindl, E., Lang, W., and Rau, W. (1970). Untersuchungen über die lichtabhängige Carotinoidsynthese VI. Zeitlicher Verlauf der Synthese der einzelnen Carotinoide bei *Fusarium aquaeductuum* unter verschiedenen Induktionsbedingungen. Planta 94:156-174.
Blanc, P. L., Tuveson, R. W., and Sargent, M. L. (1976). Inactivation of carotenoid-producing and albino strains of *Neurospora crassa* by visible light, black light, and ultraviolet radiation. J. Bacteriol. 125:616-625.
Bobowski, G. C., Barker, W. G., and Subden, R. E. (1977). The conversion of [2-^{14}C] mevalonic acid into triterpenes and tetraterpenes by cell-free extracts of a *Neurospora crassa* albino mutant. Can. J. Bot. 55:2137-2141.
Boucher, F., Van der Rest, M., and Gingras, G. (1977). Structure and function of carotenoids in the photoreaction center from *Rhodospirillum rubrum*. Biochim. Biophys. Acta 461:339-357.
Brain, R. D., Freeberg, J. A., Weiss, C. V., and Briggs, W. R. (1977). Blue light-induced absorbance changes in the membrane fractions from corn and *Neurospora*. Plant Physiol. 59:948-952.
Bramley, P. M., Aung Than, and Davies, B. H. (1977). Alternative pathways of carotene cyclization in *Phycomyces blakesleeanus*. Phytochemistry 16:235-238.
Britton, G. (1976a). Later reactions of carotenoid biosynthesis. Pure Appl. Chem. 47:223-236.

Britton, G. (1976b). Biosynthesis of carotenoids. In *Chemistry and Biochemistry of Plant Pigments*, vol. 1, 2nd ed., T. W. Goodwin (Ed.). Academic, London, pp. 262-327.
Britton, G., and Goodwin, T. W. (1971). Biosynthesis of carotenoids. Methods Enzymol. *18C*:654-701.
Bu'Lock, J. D., and Osagie, A. U. (1973). Prenols and ubiquinones in single-strain and mated cultures of *Blakeslea trispora*. J. Gen. Microbiol. *76*:77-83.
Bu'Lock, J. D., Drake, D., and Winstanley, D. J. (1972). Specificity and transformation of the trisporic acid series of fungal sex hormones. Phytochemistry *11*:2011-2018.
Bu'Lock, J. D., Jones, B. E., Taylor, D., Winskill, N., and Quarrie, S. A. (1974). Sex hormones in mucorales. The incorporation of C_{20} and C_{18} precursors into trisporic acids. J. Gen. Microbiol. *80*:301-306.
Bu'Lock, J. D., Jones, B. E., and Winskill, N. (1976). The apocarotenoid system of sex hormones and prohormones in mucorales. Pure Appl. Chem. *47*: 191-202.
Bünning, E. (1937). Phototropismus und Carotinoide. III. Weitere Untersuchungen an Pilzen und höheren Pflanzen. Planta *27*:583-610.
Burgeff, H. (1914). Untersuchungen über Variabiltät, Sexualität and Erblichkeit bei *Phycomyces nittens* Kunze I. Flora *107*:259-316.
Burnett, J. H. (1976). Functions of carotenoids other than in photosynthesis. In *Chemistry and Biochemistry of Plant Pigments*, Vol. 1, 2nd ed., T. W. Goodwin (Ed.). Academic, London, pp. 655-679.
Caglioti, L., Cainelli, T., Camerino, B., Mondelli, R., Prieto, R., Quilico, A., Salvatori, T., and Selva, A. (1966). The structure of trisporic-C acid. Tetrahedron Suppl. *7*:175-187.
Cantino, E. C. (1965). Relations of metabolism to cell development in plants. In *Encyclopedia of Plant Physiology*, Vol. XV/1, A. Lang (ed.). Springer-Verlag, Heidelberg, pp. 213-233.
Cantino, E. C., and Hyatt, M. T. (1953). Carotenoids and oxidative enzymes in the aquatic Phycomycetes *Blastocladiella* and *Rhizophlyctis*. Am. J. Bot. *40*:688-694.
Cantino, E. C., and Horenstein, G. A. (1956). The stimulatory effect of light upon growth and CO_2 fixation in *Blastocladiella*. I. The S. K. I. Cycle. Mycologia *48*:777-799.
Cattrall, M. E., Baird, M. L., and Garber, E. D. (1978). Genetics of *Ustilago violacea*. III. Crossing over and nondisjunction. Bot. Gaz. Chicago *139*:266-270.
Cederberg, E., and Neujahr, H. Y. (1969). Activation of β-carotene synthesis in *Blakeslea trispora* by certain terpenes. Acta Chem. Scand. *23*:957-961.
Cederberg, E., and Neujahr, H. Y. (1970). Distribution of β-carotene in subcellular fractions of *Blakeslea trispora*. Experientia *26*:366-367.

Cerdá-Olmedo, E. (1975). The genetics of *Phycomyces blakesleeanus*. Genet. Res. *25*:285-296.
Cerdá-Olmedo, E., and Torres-Martínez, S. (1979). Genetics and regulation of carotene biosynthesis. Pure Appl. Chem. *51*:631-637.
Chichester, C. O., and Maxwell, W. A. (1969). The effects of high intensity visible and ultraviolet light on the death of microorganism. Life Sci. Space Res. *7*:11-18.
Chichester, C. O., Yokoyama, H., Nakayama, T. O. M., Lutkon, A., and Mackinney, G. (1959). Leucine metabolism and carotene bisoynthesis. J. Biol. Chem. *234*:598-602.
Chu, I. S., and Lilly, V. G. (1960). Factors affecting the production of carotene by *Choanephora cucurbitar*. Mycologia *52*:80-96.
Ciegler, A. (1965). Microbial carotenogenesis. Adv. Appl. Microbiol. *7*:1-34.
Coggins, C. W., Jr., Henning, G. L., and Yokoyama, H. (1970). Lycopene accumulation induced by 2-(4-chlorophenylthio)-triethylamine hydrochloride. Science *168*:1589-1590.
Czeczuga, B. (1979). Investigations on carotenoids in fungi. VI. Representatives of the Helvellaceae and Morchellaceae. Phyton Horn Austria *19*:225-232.
Dandekar, S., and Modi, V. V. (1980). Involvement of cyclic AMP in carotenogenesis and cell differentiation in *Blakeslea trispora*. Biochim. Biophys. Acta *628*:398-406.
Dandekar, S., Modi, V. V., and Jani, U. K. (1980). Chemical regulators of carotenogenesis by *Blakeslea trispora*. Phytochemistry *19*:795-798.
Davies, B. H. (1961). The carotenoids of *Rhizophlyctis rosea*. Phytochemistry *1*:25-29.
Davies, B. H. (1970). A novel sequence for phytoene dehydrogenation in *Rhodospirillum rubrum*. Biochem. J. *116*:93-99.
Davies, B. H. (1973). Carotene biosynthesis in fungi. Pure Appl. Chem. *35*:1-28.
Davies, B. H. (1976). Carotenoids. In *Chemistry and Biochemistry of Plant Pigments*, Vol. 2, 2nd ed., T. W. Goodwin (Ed.). Academic, London, pp. 38-165.
Davies, B. H. (1979). Solved and unsolved problems of carotenoid formation. Pure Appl. Chem. *51*:623-630.
Davies, B. H., and Rees, A. F. (1973). $7', 8', 11', 12'$-Tetrahydro-γ-carotene: A novel carotene from *Phycomyces blakesleeanus*. Phytochemistry *12*:2745-2750.
Davies, B. H., and Taylor, R. F. (1976). Carotenoid biosynthesis—The early steps. Pure Appl. Chem. *47*:211-221.
Davies, B. H., Villoutreix, J., Williams, R. J. H., and Goodwin, T. W. (1963). The possible role of β-zeacarotene in carotenoid cyclization. Biochem. J. *89*:96P.
Day, A. W., and Day, L. L. (1970). Ultraviolet light sensitive mutants of *Ustilago violacea*. Can. J. Genet. Cytol. *12*:891-904.
Day, A. W., and Jones, J. K. (1968). The production and characteristics of diploids in *Ustilago violacea*. Genet. Res. *11*:63-81.

Day, A. W., and Jones, J. K. (1969). Sexual and parasexual analysis in *Ustilago violacea*. Genet. Res. *14*:195-221.
DeFabo, E. (1980). On the nature of the blue light photoreceptor: Still an open question. In *The Blue Light Syndrome*, H. Senger (Ed.). Springer-Verlag, Berlin, pp. 187-197.
DeFabo, E. C., Harding, R. W., and Shropshire, W., Jr., (1976). Action spectrum between 260 and 800 nanometers for the photoinduction of carotenoid biosynthesis in *Neurospora crassa*. Plant Physiol. *57*:440-445.
Delbrück, M., Katzir, M., and Presti, D. (1976). Responses of *Phycomyces* indicating optical excitation of the lowest triplet state of riboflavin. Proc. Nat. Acad. Sci. USA *73*:1969-1973.
Dennis, D. T., Upper, C. D., and West, C. A. (1965). An enzymic site of inhibition of gibberellin biosynthesis by AMO 1618 and other plant growth retardants. Plant Physiol. *40*:948-952.
Eberhardt, N. L., and Rilling, H. C. (1975). Prenyltransferase from *Saccharomyces cerevisiae*. Purification to homogeneity and molecular properties. J. Biol. Chem. *250*:863-866.
Eijk van, G. W., Mummery, R. S., Roeymans, H. J., and Valadon, L. R. G. (1979). A comparative study of carotenoids of *Aschersonia aleyroides* and *Aspergillus giganteus*. Antonie van Leeuwenhoek *45*:417-422.
Elahi, M., Chichester, C. O., and Simpson, K. L. (1973a). Biosynthesis of carotenoids by *Phycomyces blakesleeanus* mutants in the presence of nitrogenous heterocyclic compounds. Phytochemistry *12*:1627-1632.
Elahi, M., Lee, T. H., Simpson, K. L., and Chichester, C. O. (1973b). Effect of CPTA (2-(4-chlorophenylthio)-triethylamino hydrochloride) and Cycocel (2-chloroethyl)-trimethyl ammonium chloride) on the biosynthesis of carotenoids by *Phycomyces blakesleeanus* mutants. Phytochemistry *12*:1633-1639.
Elahi, M., Glass, R. W., Lee, T. -C., Chichester, C. O., and Simpson, K. L. (1975). The effect of CPTA analogs and other nitrogenous compounds on the biosynthesis of carotenoids in *Phycomyces blakesleeanus* mutants. Phytochemistry *14*:133-138.
Ende van den, H. (1968). Relationship between sexuality and carotene synthesis in *Blakeslea trispora*. J. Bacteriol. *96*:1298-1303.
Ende van den, H. (1976). *Sexual Interactions in Plants*, Academic, London.
Eslava, A. P., and Cerdá-Olmedo, E. (1974). Genetic control of phytoene dehydrogenation in *Phycomyces*. Plant Sci. Lett. *2*:9-14.
Eslava, A. P., Alvarez, M. I., and Cerdá-Olmedo, E. (1974). Regulation of carotene biosynthesis in *Phycomyces* by vitamin A and β-ionone. Eur. J. Biochem. *48*:617-623.
Eslava, A. P., Alvarez, M. I., Burke, P. V., and Delbrück, M. (1975a). Genetic recombination in sexual crosses of *Phycomyces*. Genetics *80*:445-462.
Eslava, A. P., Alvarez, M. I., and Delbrück, M. (1975b). Meiosis in *Phycomyces*. Proc. Nat. Acad. Sci. USA *72*:4076-4080.

Eugster, C. H. (1979). Characterization, chemistry and stereochemistry of carotenoids. Pure Appl. Chem. *51*:463-506.
Federici, B. A., and Thompson, S. N. (1979). Beta-carotene in the gametophytic phase of *Coelomomyces dodgei*. Exp. Mycol. *3*:281-284.
Feofilova, E. P., and Pivovarova, T. M. (1976). Differentiation of *Blakeslea trispora* mycelium in relation to carotene production. Mikrobiologiya *45*:854-860.
Feofilova, E. P., Bekhtereva, M. N., and Kozlova, Y. I. (1970). Influence of certain stimulators and pyridine derivatives on carotenoid biosynthesis by the (−) strain of *Blakeslea trispora*. Mikrobiologiya *39*:389-395.
Fiasson, J. L., Trouilloud, M., and Grange, A. (1972). Variation du contenu pigmentaire de *Sporidiobolus johnsonii* (Sporobolomycétacées) sous l'influence de divers facteurs de milieu. Rev. Mycol. *37*:48-59.
Fincham, J. R. S., Day, R. R., and Radford, A. (1979). *Fungal Genetics* 4th ed. University of California, Berkeley.
Frey-Wyssling, A., Grieshaber, E., and Mühlethaler, K. (1963). Origin of spherosomes in plant cells. J. Ultrastruct. Res. *8*:506-516.
Friederichsen, J., and Engel, H. (1957). Beiträge zur Kenntnis des Abschussrhythmus und des Farbstoffs von *Sphaerobolus stellatus* (Thode) Pers. Planta *49*:578-587.
Friend, J., and Goodwin, T. W. (1954). Studies in carotenogenesis. 12. The effect of temperature and thiamine concentrations on carotenogenesis by *Phycomyces blakesleeanus*. Biochem. J. *57*:434-437.
Friend, J., Goodwin, T. W., and Griffiths, L. A. (1955). Studies in carotenogenesis. 15. The role of carboxylic acids in the biosynthesis of β-carotene by *Phycomyces blakesleeanus*. Biochem. J. *60*:649-655.
Frosch, S., and Mohr, H. (1980). Analysis of light-controlled accumulation of carotenoids in mustard (*Sinapsis alba* L.) seedlings. Planta *148*:279-286.
Galland, P., and Russo, V. E. A. (1979a). Photoinitiation of sporangiophores in *Phycomyces* mutants deficient in phototropism and in mutants lacking β-carotene. Phytochem. Phytobiol. *29*:1009-1014.
Galland, P., and Russo, V. E. A. (1979b). The role of retinol in the initiation of sporangiophores of *Phycomyces blakesleeanus*. Planta *146*:257-262.
Galston, A. W. (1950). Riboflavin, light, and the growth of plants. Science *111*:619-624.
Garber, E. D. (1980). Genetics of *Ustilago violacea*. VIII. Fine structure of the white locus. Bot. Gaz. Chicago *141*:479-482.
Garber, E. D., and Owens, A. E. (1980). Genetics of *Ustilago violacea*. VI. Characterization of white and yellow strains by recombination and complementation. Bot. Gaz. Chicago *141*:79-84.
Garber, E. D., Baird, M. L., and Chapman, D. J. (1975). Genetics of *Ustilago violacea*. I. Carotenoid mutants and carotenogenesis. Bot. Gaz. Chicago *136*:341-346.
Garber, E. D., Baird, M. L., and Weiss, L. M. (1978). Genetics of *Ustilago violacea*. II. Polymorphism of color and mutational requirements of sporidia from natural populations. Bot. Gaz. Chicago *139*:261-265.

Garber, E. D., Ruddat, M., and Merza, A. P. (1980). Genetics of *Ustilago violacea*. VII. The pumpkin locus. Bot. Gaz. Chicago *141*:210-212.
Garber, E. D., Will, O. H., III, and Kokontis, J. M. (1981). Genetics of *Ustilago violacea*. XI. The pseudohyphal mutation and tetrad analysis. Bot. Gaz. Chicago *142*:589-591.
Garnjobst, L., and Tatum, E. L. (1956). A temperature-independent riboflavin-requiring mutant of *Neurospora crassa*. Am. J. Bot. *43*:149-157.
Garton, G. A., Goodwin, T. W., and Lijinsky, W. (1951). Studies in carotenogenesis. I. General conditions governing β-carotene synthesis by the fungus *Phycomyces blakesleeanus* Burgeff. Biochem. J. *48*:154-163.
Goldie, A. H., and Subden, R. E. (1973). The neutral carotenoids of wild type and mutant strains of *Neurospora crassa*. Biochem. Genet. *10*:275-284.
Goldstrohm, D. D., and Lilly, V. G. (1955). The effect of light on the survival of pigmented and nonpigmented cells of *Dacryopinax spathularia*. Mycologia *57*:612-623.
Gooday, G. W. (1968). Hormonal control of sexual reproduction in *Mucor mucedo*. New Phtyol. *67*:815-821.
Gooday, G. W. (1974). Fungal sex hormones. Annu. Rev. Biochem. *43*:35-49.
Gooday, G. W., Fawcett, P., Green, D., and Shaw, G. (1973). The formation of fungal sporopollenin in the zygospore wall of *Mucor mucedo*: A role for the sexual carotenogenesis in the mucorales. J. Gen. Microbiol. *74*:233-239.
Gooday, G. W., Green, D., Fawcett, P., and Shaw, G. (1974). Sporopollenin formation in the ascospore wall of *Neurospora crassa*. Arch. Microbiol. *101*: 145-151.
Goodwin, T. W. (1952). Studies in carotenogenesis. 3. Identification of the minor polyene components of the fungus *Phycomyces blakesleeanus* and a study of their synthesis under various cultural conditions. Biochem. J. *50*: 550-558.
Goodwin, T. W. (1980). *The Biochemistry of Carotenoids*, Plants, Vol. 1, Chapman and Hall, London.
Goodwin, T. W., and Lijinsky, W. (1952). Studies in carotenogenesis. 2. Carotene production by *Phycomyces blakesleeanus*: The effect of different amino acids when used in media containing low concentrations of glucose. Biochem. J. *50*:268-273.
Goodwin, T. W., and Willmer, J. S. (1952). Studies in carotogenesis. 4. Nitrogen metabolism and carotene synthesis in *Phycomyces blakesleeanus*. Biochem. J. *51*:213-217.
Goodwin, T. W., Jamikorn, M., and Willmer, J. S. (1953). Studies on carotenogenesis. 7. Further observations concerning the action of diphenylamine in inhibiting the synthesis of β-carotene in *Phycomyces blakesleeanus*. Biochem. J. *53*:531-538.
Govind, N. S., Metha, B., Sharma, M., and Modi, V. V. (1981). Protease and carotenogenesis in *Blakeslea trispora*. Phytochemistry *20*:2483-2485.
Gressel, J. (1980). Blue light and transcription. In *The Blue Light Syndrome*, H. Senger (Ed.). Springer-Verlag, Berlin, pp. 133-153.

Gribanovski-Sassu, O., and Foppen, F. H. (1967). The carotenoids of the fungus *Epicoccum nigrum* Link. Phytochemistry 6:907-909.

Guardia de la, M. D., Aragón, C. M. G., Murillo, F. J., and Cerdá-Olmedo, E. (1971). A carotenogenic enzyme aggregate in *Phycomyces*: Evidence from quantitative complementation. Proc. Nat. Acad. Sci. USA 68:2012-2015.

Harding, R. W. (1974). The effect of temperature on photoinduced carotenoid biosynthesis in *Neurospora crassa*. Plant Physiol. 54:142-147.

Harding, R. W., and Mitchell, H. K. (1968). The effect of cycloheximide on carotenoid biosynthesis in *Neurospora crassa*. Arch. Biochem. Biophys. 128:814-818.

Harding, R. W., and Shropshire, W., Jr. (1980). Photocontrol of carotenoid biosynthesis. Annu. Rev. Physiol. 31:217-238.

Harding, R. W., and Turner, R. V. (1981). Photoregulation of the carotenoid biosynthetic pathway in albino and white collar mutants of *Neurospora crassa*. Plant Physiol. 68:745-749.

Hartmann, K. M. (1977). Aktionsspektrometrie. In *Biophysik*, W. Hoppe, W. Lohmann, H. Markl, and H. Ziegler (Eds.). Springer-Verlag, Berlin, pp. 197-222.

Hashimoto, T., Pollack, J. H., and Blumenthal, H. J. (1978). Carogenogenesis associated with arthrosporulation of *Trichophyton mentagrophytes*. J. Bacteriol. 136:1120-1126.

Haxo, F. (1949). Studies on the carotenoid pigments of *Neurospora*. I. Composition of the pigment. Arch. Biochem. 20:400-421.

Haxo, F. (1950). Carotenoids of the mushroom *Cantharellus cinnobarinus*. Bot. Gaz. Chicago 112:228-231.

Haxo, F. (1952). Carotenoid formation by mutant strains of *Neurospora crassa*. Biol. Bull. 103:286.

Hayman, E. P., Chichester, C. O., and Simpson, K. L. (1974). Effects of CPTA upon carotenogenesis and lipoidal constituents in *Rhodotorula* species. Phytochemistry 13:1123-1128.

Heisenberg, M., and Cerdá-Olmedo, E. (1968). Segregation of heterokaryons in the asexual cycle of *Phycomyces*. Mol. Gen. Genet. 102:187-195.

Herber, R., Maudinas, B., and Villoutreix, J. (1972). Influence de différents composés chimniques sur la caroténogenèse de *Mucor hiemalis*. Phytochemistry 11:3461-3464.

Howes, C. D., and Batra, P. P. (1970). Mechanism of photoinduced carotenoid synthesis: Further studies on the action spectrum and other aspects of carotenogenesis. Arch. Biochem. Biophys. 137:175-180.

Hsu, W. J., Yokoyama, H., and Coggins, C. W. (1972). Carotenoid biosynthesis in *Blakeslea trispora*. Phytochemistry 11:2985-2990.

Hsu, W. J., Poling, S. M., and Yokoyama, H. (1974a). Effect of amines on carotenogenesis of *Blakeslea trispora*. Phytochemistry 13:415-419.

Hsu, W. J., Ailion, D. C., and Delbrück, M. (1974b). Carotenogenesis in *Phycomyces*. Phytochemistry 13:1463-1468.

Huang, P. C. (1964). Recombination and complementation of albino mutants in *Neurospora*. Genetics 49:453-469.

Huber, A., and Schrott, E. L. (1980). Photokilling and protective mechanisms in *Fusarium aquaeductuum*. In *The Blue Light Syndrome*, H. Senger (Ed.), Springer-Verlag, Berlin, pp. 299-308.

Hungate, M. V. G. (1945). A genetic study of albino mutants of *Neurospora crassa*. M. A. thesis, Stanford University, Stanford, Calif.

Jayaram, M., Presti, D., and Delbrück, M. (1979). Light-induced carotene synthesis in *Phycomyces*. Exp. Mycol. *3*:42-52.

Jayaram, M., Leutwiler, L., and Delbrück, M. (1980). Light-induced carotene synthesis in mutants of *Phycomyces* with abnormal phototropism. Photochem. Photobiol. *32*:241-245.

Jensen, S. L., Cohen-Bazire, G., and Stanier, R. Y. (1961). Biosynthesis of carotenoids in purple bacteria. A re-evaluation based on consideration of chemical structure. Nature *192*:1168-1172.

Jerebzoff-Quintin, S., and Jerebzoff, S. (1980). Carotenogenesis and asparagine in *Leptosphaeria michotii* (West) Sacc. Protoplasma *104*:43-54.

Johnson, E. A., Villa, T. G., Lewis, M. J., and Phaff, H. J. (1978). Simple method for the isolation of astaxanthin from the basidiomycetous yeast. *Phaffia rhodozyma*. Appl. Environ. Microbiol. *35*:1155-1159.

Keyhani, J., Keyhani, E., and Goodgal, S. H. (1972). Studies on the cytochrome content of *Phycomyces* spores during germination. Eur. J. Biochem. *27*: 527-534.

Kleinig, H. (1974). Inhibition of carotenoid synthesis in *Myxococcus fulvus* (Myxobacterales). Arch. Microbiol. *97*:217-226.

Korf, R. P. (1973). Discomycetes and Tuberales. In *The Fungi*, Vol. 4A, G. C. Ainsworth, F. K. Sparrow, and A. S. Sussman (Eds.). Academic, London, pp. 249-319.

Krinsky, N. I. (1978). Non-photosynthetic functions of carotenoids. Philos. Trans. R. Soc. London B. *284*:581-590.

Krinsky, N. I. (1979). Carotenoid protection against oxidation. Pure Appl. Chem. *51*:649-660.

Kristkii, M. S., Sokolovskii, V. Y., Belozerskeya, T. A., and Chernysheva, E. K. (1981). Participation of cyclic AMP in light-dependent regulation of carotenoid synthesis in *Neurospora crassa*. Dokl. Akad. Nauk SSR *258*:759-762.

Kumagai, T., and Oda, Y. (1969). An action spectrum for photoinduced sporulation in the fungus *Trichoderma viride*. Plant Cell Physiol. *10*:387-392.

Kushwaha, S. C., Kates, M., Renaud, R. L., and Subden, R. E. (1978). The terpenyl pyrophosphates of wild type and tetraterpene mutants of *Neurospora crassa*. Lipids *13*:352-355.

Lang, W., and Rau, W. (1972). Untersuchungen über die lichtabhängige Carotinoidsynthese. IX. Zum Induktionsmechanismus der carotinoidbildenden Enzyme bei *Fusarium aquaeductuum*. Planta *106*:345-354.

Lang-Feulner, J., and Rau, U. (1975). Redox dyes as artificial photoreceptors in light-dependent carotenoid synthesis. Photochem. Photobiol. *21*:179-183.

Lansbergen, J. C., Renaud, R. L., and Subden, R. E. (1976). Phytoene photoinduction in a carotenoid mutation of *Neurospora crassa*. Can. J. Bot. *54*: 2445-2448.
Lee, T. -C., and Chichester, C. O. (1969). Geranylgeranyl pyrophosphate as the condensing unit for enzymatic synthesis of carotenes. Phytochemistry *8*: 603-609.
Lee, T. -C., Lee, T. H., and Chichester, C. O. (1972). Phytoene biosynthesis: Possible mechanisms for the coupling of geranylgeranyl pyrophosphate. Phytochemistry *11*:681-687.
Lee, T. -C., Rodriguez, D. B., Karasawa, K., Lee, T. H., Simpson, K. L., and Chichester, C. O. (1975). Chemical alteration of carotene biosynthesis in *Phycomyces blakesleeanus* and mutants. Appl. Microbiol. *30*:988-993.
Leong, T. -Y., Vierstra, R. D., and Briggs, W. R. (1981). A blue light-sensitive cytochrome-flavin complex from corn coleoptiles. Further characterization. Photochem. Photobiol. *34*:697-703.
Liaaen-Jensen, G. (1971). Isolation Reactions. In *Carotenoids*. O. Isler (Ed.). Birkäuser-Verlag, Basel, pp. 61-188.
Liaaen-Jensen, S. (1979). Carotenoid: A chemosystematic approach. Pure Appl. Chem. *51*:661-675.
Liaaen-Jensen, S. (1980). Stereochemistry of natural occurring carotenoids. Progr. Chem. Org. Nat. Prod. *39*:123-172.
Lipson, E. D. (1980). Sensory transduction in *Phycomyces* photoresponses. In *The Blue Light Syndrome*, H. Senger (Ed.). Springer-Verlag, Berlin, pp. 110-118.
Lipson, E. D., and Presti, D. (1977). Light-induced absorbance changes in *Phycomyces* photomutants. Photochem. Photobiol. *25*:203-208.
López-Días, I., and Cerdá-Olmedo, E. (1980). Relationship of photocarotenogenesis to other behavioural and regulatory responses in *Phycomyces*. Planta *150*:134-139.
Löser, G., and Schäfer, E. (1980). Phototropism in *Phycomyces*: A photochromic sensor pigment. In *The Blue Light Syndrome*, H. Senger (Ed.). Springer-Verlag, Berlin, pp. 244-250.
Mackinney, G., Chichester, C. O., and Nakayama, T. (1955). The incorporation of glycine carbon into β-carotene in *Phycomyces blakesleeanus*. Biochem. J. *60*:xxxvii-xxxviii.
Manabe, K., and Poff, K. L. (1978). Purification and characterization of the photoreducible b-type cytochrome from *Dictyostelium discoideum*. Plant Physiol. *61*:961-966.
Mase, Y., Rabourn, W. J., and Quackenbush, F. W. (1957). Carotene production by *Penicillium sclerotiorum*. Arch. Biochem. Biophys. *68*:150-156.
Mathew-Roth, M. M. (1975). Therapy of human photosensitivity. Photochem. Photobiol. *22*:302-303.
Mathew-Roth, M. M., Wilson, T., Fujimori, E., and Krinsky, N. I. (1974). Carotenoid chromophore length and protection against photosensitization. Photochem. Photobiol. *19*:217-222.

Maudinas, B., Bucholtz, M. L., Papastephanou, C., Katayar, S. S., Briedis, A. V., and Porter, J. W. (1977). The partial purification and properties of a phytoene synthesizing enzyme system. Arch. Biochem. Biophys. *130*:354-362.

Maxwell, W., and Chichester, C. O. (1971). Photodynamic responses in *Rhotorula glutinis* in the absence of added sensitizers. Photochem. Photobiol. *13*: 259-273.

Maxwell, W. A., MacMillan, J. D., and Chichester, C. O. (1966). Function of carotenoids in protection of *Rhodotorula glutinis* against irradiation from a gas laser. Photochem. Photobiol. *5*:567-577.

Meissner, G., and Delbrück, M. (1968). Carotenes and retinal in *Phycomyces* mutants. Plant Physiol. *43*:1279-1283.

Mitzka, U., and Rau, W. (1977). Composition and photoinduced biosynthesis of the carotenoids of a protoplast-like *Neurospora crassa* "slime" mutant. Arch. Microbiol. *111*:261-263.

Mitzka-Schnabel, U., and Rau, W. (1980). The subcellular distribution of carotenoids in *Neurospora crassa*. Phytochemistry *19*:1409-1413.

Mitzka-Schnabel, U., and Rau, W. (1981). Subcellular site of carotenoid biosynthesis in *Neurospora crassa*. Phytochemistry *20*:63-69.

Mohr, H. (1981). Licht und Entwicklung—das Phytochromsystem der Pflanzen. Naturwissenschaften *68*:193-200.

Morris, S. A. C., and Subden, R. E. (1974). Effects of ultraviolet radiation on carotenoid containing and albino strains of *Neurospora crassa*. Mutat. Res. *22*:105-109.

Moss, G. P., and Weedon, B. C. L. (1976). Chemistry of the carotenoids. In *Chemistry and Biochemistry of Plant Pigments*, Vol. 1, 2nd ed., T. W. Goodwin (Ed.). Academic, London, p. 148.

Muñoz, V., and Butler, W. L. (1975). Photoreceptor pigment for blue light in *Neurospora crassa*. Plant Physiol. *55*:421-426.

Murillo, F. J. (1980). Effect of CPTA on carotenogenesis by *Phycomyces carA* mutants. Plant Sci. Lett. *17*:201-205.

Murillo, F. J., and Cerdá-Olmedo, E. (1976). Regulation of carotene synthesis in *Phycomyces*. Mol. Gen. Genet. *148*:19-24.

Murillo, F.J., Calderoń, I. L., López-Días, L., and Cerdá-Olmedo, E. (1978). Carotene-superproducing strains of *Phycomyces*. Appl. Environ. Microbiol. *36*:639-642.

Murillo, F. J., Torres-Martínez, S., Aragón, C. M. G., and Cerdá-Olmedo, E. (1981). Substrate transfer in carotene biosynthesis in *Phycomyces*. Eur. J. Biochem. *119*:511-516.

Nakayama, T. O. M., Mackinney, G., and Phaff, H. J. (1954). Carotenoids in asporogenous yeasts. Antonie van Leeuwenhoek. J. Microbiol. Serol. *20*: 217-228.

Neupert, W., and Ludwig, G. D. (1971). Sites of biosynthesis of outer and inner membrane proteins of *Neurospora crassa* mitochondria. Eur. J. Biochem. *19*:523-532.

Ninet, L., and Renaut, J. (1979). Carotenoids. In *Microbial Technology*, Vol. 1, 2nd ed., H. J. Pippler and D. Perlman (Eds.). Academic, New York, pp. 529-544.

Ninet, L., Renaut, J., and Tissier, R. (1969). Activation of the biosynthesis of carotenoids by Blakeslea trispora. Biotechnol. Bioeng. *11*:1195-1210.

Ninnemann, H., and Klemm-Wolfgramm, E. (1980). Blue-light-controlled conidiation and absorbance change in *Neurospora* are mediated by nitrate reductase. In *The Blue Light Syndrome*, H. Senger (Ed.). Springer-Verlag, Berlin, pp. 238-243.

Olson, J. A., and Knizley, H. (1962). The effect of diphenylamine on carotenoid, sterol and fatty acid synthesis in *Phycomyces blakesleeanus*. Arch. Biochem. Biophys. *97*:138-145.

Ootaki, T. (1973). A new method for heterokaryon formation in *Phycomyces*. Mol. Gen. Genet. *121*:49-56.

Ootaki, T., Lightly, A. C., Delbrück, M., and Hsu, W. -J. (1973). Complementation between mutants of *Phycomyces* deficient with respect to carotenogenesis. Mol. Gen. Genet. *121*:57-70.

Osman, M., and Valadon, L. R. G. (1978). Effect of light quality on the photoinduction of carotenoid synthesis in *Verticillium agaricinum*. Microbios *18*:229-234.

Osman, M., and Valadon, L. R. G. (1979). Studies on the near-UV effect on carotenogenesis in *Verticillium agaricinum* Microbios *23*:53-64.

Page, R. M., and Curry, G. M. (1966). Studies on phototropism of young sporangiophores of *Pilobolus kleinii*. Photochem. Photobiol. *5*:31-40.

Paietta, J., and Sargent, M. L. (1981). Photoreception in *Neurospora crassa*: Correlation of reduced light sensitivity with flavin deficiency. Proc. Natl. Acad. Sci. USA *78*:5573-5577.

Parn, P., and Seviour, R. J. (1974). Pigments induced by organomercurial compounds in *Cephalosporium diospyros*. J. Gen. Microbiol. *85*:228-236.

Perkins, D. D. (1971). Gene order in the albino region of linkage group I. Neurospora Newsl. *18*:14.

Perkins, D. D. (1974). Osmotic mutants. Neurospora Newsl. *21*:25.

Perkins, D. D., and Murray, N. N. (1963). New markers and linkage data. Neurospora Newsl. *4*:26.

Perkins, D. D., Smith, M. R., and Galeazzi, D. R. (1973). New markers and linkage data. Neurospora Newsl. *20*:45.

Poff, K. L., and Butler, W. L. (1974). Absorbance changes induced by blue light in *Phycomyces blakesleeanus* and *Dictyostelium discoideum*. Nature *248*: 799-801.

Poling, S. M., Hsu, W. -J., Koehrn, F. J., and Yokoyama, H. (1977). Chemical induction of β-carotene biosynthesis. Phytochemistry *16*:551-555.

Poling, S. M., Hsu, W. J., and Yokoyama, H. (1980). Chemical induction of poly-cis carotenoid biosynthesis. Phytochemistry *19*:1677-1680.

Pontecorvo, G. (1958). *Trends in Genetic Analysis*, Columbia University, New York.

Porter, J. W., and Lincoln, R. E. (1950). *Lycopersicon* selections containing a high content of carotenes and colourless polytenes. II. The mechanism of carotene biosynthesis. Arch. Biochem. *27*:390-403.

Porter, J. W., and Spurgeon, S. L. (1979). Enzymatic synthesis of carotenes. Pure Appl. Chem. *51*:609-622.

Prebble, J., and Huda, S. (1977). The photosensitivity of malate oxidase system of a pigmented strain and a carotenoidless mutant of *Sarcina lutea* (*Micrococcus luteus*). Arch. Microbiol. *113*:39-42.

Presti, D., and Delbrück, M. (1978). Photoreceptors for biosynthesis, energy storage and vision. Plant Cell Environ. *1*:81-100.

Presti, D., Hsu, W. J., and Delbrück, M. (1977). Phototropism in *Phycomyces* mutants lacking β-carotene. Photochem. Photobiol. *26*:403-405.

Radunz, A., and Schmid, G. H. (1979). On the localization and function of the xanthophylls is in the thylakoid membrane. Ber. Dtsch. Bot. Ges. *92*:437-443.

Ragan, M. A., and Chapman, D. J. (1978). *A Biochemical Phylogeny of the Protists*, Academic, New York.

Ramadan-Talib, Z., and Prebble, J. (1978). Photosensitivity of respiration in *Neurospora* mitochondrion. Biochem. J. *176*:767-775.

Rao, S., and Modi, V. V. (1977). Carotenogenesis: Possible mechanism of action of trisporic acid in *Blakeslea trispora*. Experientia *33*:31-33.

Rau, W., (1967a). Untersuchungen über die lichtabhängige Carotinsynthese. I. Das Wirkungsspektrum von *Fusarium aquaeductuum*. Planta *72*:14-28.

Rau, W. (1967b). Untersuchungen über die lichtabhängige Carotinoidsynthese. II. Ersatz der Lichtinduktion durch Mercuribenzoat. Planta *74*:263-277.

Rau, W. (1971). Untersuchungen über die lichtabhängige Carotinoidsynthese. VII. Reversible Unterbrechung der Reaktionskette durch Cycloheximide und anaerobe Bedingungen. Planta *101*:251-264.

Rau, W. (1975). Zum Mechanismus der Photoregulation von Morphosen am Beispiel der Carotinoidsynthese. Ber. Dtsch. Bot. Ges. *88*:45-60.

Rau, W. (1976). Photoregulation of carotenoid biosynthesis in plants. Pure Appl. Chem. *47*:237-243.

Rau, W. (1980). Blue light-induced carotenoid biosynthesis in microorganisms. In *The Blue Light Syndrome*, H. Senger (Ed.). Springer-Verlag, Berlin, pp. 283-298.

Rau, W., Lindemann, I., and Rau-Hund, S. (1968). Untersuchungen über die lichtabhängige Carotinoidsynthese. III. Die Farbstoffbildung von *Neurospora crassa* in Submerskultur. Planta *80*:309-316.

Reyes, P., Chichester, C. O., and Nakayama, T. O. M. (1964). The mechanism of β-ionone stimulation of carotenoid and ergosterol biosynthesis in *Phymyces blakesleeanus*. Biochim. Biophys. Acta *90*:578-592.

Riley, G. J. P., and Bramley, P. M. (1976). The subcellular distribution of carotenoids in *Phycomyces blakesleeanus* C115 car42 mad107(-). Biochim. Biophys. Acta *450*:429-440.

Rilling, H. C. (1979). Prenyltransferase. Pure Appl. Chem. *51*:597-608.

Rodgers, M. A., and Bates, A. L. (1980). Kinetic and spectroscopic features of some carotenoid triplet states sensitized by singlet oxygen. Photochem. Photobiol. *31*:533-537.

Rottem, S., and Markowitz, O. (1979). Carotenoids act as reinforcers of the *Acholeplasma laidlawii* lipid bilayer. J. Bacteriol. *140*:944-948.

Russo, V. E. A., Galland, P., Toselli, M., and Volpi, L. (1980). Blue light induced differentiation in *Phycomyces blakesleeanus*. In *The Blue Light Syndrome*, H. Senger (Ed.). Springer-Verlag, Berlin, pp. 563-569.

Russo, V. E. A., Phol, U., and Volpi, L. (1981). Carbon dioxide inhibits phorogenesis in *Phycomyces* and blue light overcomes this inhibition. Photochem. Photobiol. *34*:233-236.

Sandmann, G., and Hilgenberg, W. (1978). Förderung der β-Carotinsynthese durch Licht bei *Phycomyces blakesleeanus* Bgff. Biochem. Physiol. Pflanz. *172*:401-407.

Sandmann, G., and Hilgenberg, W. (1980). Carotinbildung und Vorstufenakkumulation unter dem Einfluss von Diphenylamin in Hell- und Dunkelkulturen von *Phycomyces blakesleeanus*. Biochem. Physiol. Pflanz. *175*:237-242.

Sandmann, G., Bramley, P. M., and Böger, P. (1980). The inhibitory method of action of the pyridazinone herbicide norflurazon on a cell-free carotenogenic enzyme system. Pestic. Biochem. Physiol. *14*:185-191.

Schmidt, W., Hart, J., Filner, P., and Poff, K. L. (1977). Specific inhibition of phototropism in corn seedlings. Plant Physiol. *60*:736-738.

Schrott, E. L. (1980a). Fluence response relationship of carotenogenesis in *Neurospora crassa*. Planta *150*:174-179.

Schrott, E. L. (1980b). Dose response and related aspects of carotenogenesis in *Neurospora crassa*. In *The Blue Light Syndrome*, H. Senger (Ed.). Springer-Verlag, Berlin, pp. 309-318.

Schrott, E. L. (1981). The biphasic fluence response of carotenogenesis in *Neurospora crassa*: Temporary insensitivity of the photoreceptor system. Planta *151*:371-374.

Schrott, E. L., and Rau, W. (1975). Versuche zum direkten Nachweis einer Beteiligung von m-RNS an der Photoinduktion der Carotinoidsynthese. Ber. Dtsch. Bot. Ges. *88*:233-243.

Schrott, E. L., and Rau, W. (1977). Evidence for a photoinduced synthesis of poly(A) containing mRNA in *Fusarium aquaeductuum*. Planta *136*:45-48.

Schrott, E. L., Huber-Willer, A., and Rau, W. (1982). Is phytochrome involved in the light-mediated carotenogenesis in *Fusarium aquaeductuum* and *Neurospora crassa*? Photochem. Photobiol. *35*:213-216.

Sheng, T. C., and Sheng, G. (1952). Genetic and non-genetic factors in pigmentation of *Neurospora crassa*. Genetics *37*:264-269.

Shimizu, M., Egashira, T., and Takahama, U. (1979). Inactivation of *Neurospora crassa* by singlet molecular oxygen generated by photosensitized reaction. J. Bacteriol. *138*:293-296.

Shropshire, W., Jr. (1980). Carotenoids as primary photoreceptors in blue-light responses. In *The Blue Light Syndrome*, H. Senger (Ed.). Springer-Verlag, Berlin, pp. 172-186.

Simpson, K. L., Nakayama, T. O. M., and Chichester, C. O. (1964). Biosynthesis of yeast carotenoids. J. Bacteriol. *88*:1688-1694.
Sistrom, W. R., Griffiths, M., and Stanier, R. Y. (1956). The biology of a photosynthetic bacterium which lacks colored carotenoids. J. Cell. Comp. Physiol. *48*:473-515.
Song, P. S. (1980). Spectroscopic and photochemical characterization of flavoproteins and carotenoproteins as blue light photoreceptors. In *The Blue Light Syndrome*, H. Senger (Ed.). Springer-Verlag, Berlin, pp. 157-171.
Song, P. S., and Moore, T. A. (1974). On the phototropism and phototaxis: Is a carotenoid the most likely candidate? Photochem. Photobiol. *19*:435-441.
Song, P. S., Prasad, K., Prézelin, B. B., and Haxo, F. T. (1976). Molecular topology of the photosynthetic light-harvesting pigment complex, peridinin-chlorophyll a-protein, from marine dinoflagellates. Biochemistry *15*:4422-4427.
Spurgeon, S. L., Turner, R. V., and Harding, R. W. (1979). Biosynthesis of phytoene from isopentenyl pyrophosphate by a *Neurospora* enzyme system. Arch. Biochem. Biophys. *195*:23-29.
Stadler, D. R. (1956). A map of linkage group VI of *Neurospora crassa*. Genetics *41*:528-543.
Sturzenegger, V., Buchecker, R., and Wagnière, G. (1980). Classification of the CD spectra of carotenoids. Helv. Chim. Acta *63*:1074-1092.
Subden, R. E., and Goldie, A. H. (1973). Biochemical analyses of isoallelic series at the *al-1* locus of *Neurospora crassa*. Genetica *44*:615-620.
Subden, R. E., and Threlkeld, S. F. H. (1968). Genetic and complementation studies of a new carotenoid mutant in *Neurospora crassa*. Can. J. Genet. Cytol. *10*:351-356.
Subden, R. E., and Threlkeld, S. F. H. (1969). Some aspects of complemention with carotenogenic *al* loci in *Neurospora crassa*. Experientia *25*:1106-1107.
Subden, R. E., and Threlkeld, S. F. H. (1970). Genetic fine structure of albino (*al*) region of *Neurospora crassa*. Genet. Res. *15*:139-146.
Subden, R. E., and Turian, G. (1969). Improved techniques for study of carotenoid intermediates in *Neurospora*. Neurospora Newsl. *15*:8
Subden, R. E., and Turian, G. (1970). A mechanism for carotenoid synthesis in fungi involving a multifunctional enzyme complex. Mol. Gen. Genet. *108*:358-364.
Sutter, R. P. (1970). Effect of light on β-carotene accumulation in *Blakeslea trispora*. J. Gen. Microbiol. *64*:215-221.
Sutter, R. P. (1975). Mutations affecting sexual development in *Phycomyces blakesleeanus*. Proc. Nat. Acad. Sci. USA *72*:127-130.
Tanaka, Y., Katayama, T., Simpson, K. L., and Chichester, C. O. (1981). Stability of carotenoids on silica gel and other absorbents. Nippon Suisan Gakkaishi *47*:799-811.
Taylor, R. F., and Ikawa, M. (1980). Gas chromatography, gas chromatography-mass spectrometry, and high pressure liquid chromatography of carotenoids and retenoids. Methods Enzymol. *67*:233-261.
Theimer, R. R., and Rau, W. (1969). Mutants of *Fusarium aquaeductuum* lacking

photoregulation of carotenoid synthesis. Biochim. Biophys. Acta *177*:180-181.
Theimer, R. R., and Rau, W. (1970). Untersuchungen über die lichtabhängige Carotinoidsynthese. V. Aufhebung der Lichtinduktion durch Reduktionsmittel und Ersatz des Lichts durch Wasserstoffperoxid. Planta *92*:129-137.
Theimer, R. R., and Rau, W. (1972). Untersuchungen über die lichtabhängige Carotinoidsynthese. VIII. Die unterschiedlichen Wirkungsmechanismen von Licht und Mercuribenzoat. Planta *106*:331-343.
Thomas, D. M., and Goodwin, T. W. (1967). Studies on carotenogenesis in *Blakeslea trispora*. I. General observations on synthesis in mated and unmated strains. Phytochemistry *6*:355-360.
Thomas, D. M., Harris, R. C., Kirk, J. T. O., and Goodwin, T. W. (1967). Studies on carotenogenesis in *Blakeslea trispora*. Phytochemistry *6*:361-366.
Thomas, S. A., Sargent, M. L., and Tuveson, R. W. (1981). Inactivation of normal and mutant *Neurospora crassa* conidia with visible light and near UV: Role of $^1O_2$, carotenoid composition and sensitizer location. Phytochem. Phytobiol. *33*:349-354.
Torres-Martinez, S., Murillo, F. J., and Cerdá-Olmedo, E. (1980). Genetics of lycopene cyclization and substrate transfer in β-carotene biosynthesis in *Phycomyces*. Genet. Res. *36*:299-309.
Turian, G., and Haxo, F. T. (1954). Minor polyene components in the sexual phase of *Allomyces javanicus*. Bot. Gaz. Chicago *115*:254-260.
Valadon, L. R. G. (1976). Carotenoids as additional taxonomic characters in fungi: A review. Trans. Br. Mycol. Soc. *67*:1-15.
Valadon, L. R. G., and Mummery, R. S. (1966). Inhibition of carotenoid synthesis in a mutant of *Verticillium albo-atrum*. J. Gen. Microbiol. *45*:531-540.
Valadon, L. R. G., and Mummery, R. S. (1971). Effect of light on nucleic acids, protein and carotenoids of *Verticillium agaricinum*. Microbios. *4*:227-240.
Valadon, L. R. G., and Mummery, R. S. (1973). Effect of certain inhibitors of carotenogenesis in *Verticillium agaricinum*. Microbios *7*:173-180.
Valadon, L. R. G., and Mummery, R. S. (1974). Carotenogenesis in *Verticillium agaricinum* in response to nicotine and to CPTA. Microbios *10A*:97-104.
Valadon, L. R. G., and Mummery, R. S. (1979). The effect of 9-fluorenone on protein synthesis and carotenogenesis in *Verticillium agaricinum*. Microbios Lett. *6*:129-135.
Valadon, L. R. G., Travis, R. L., and Key, J. L. (1975). Light-induced activation of cytoplasmic protein synthesis in *Verticillium agaricinum*. Physiol. Plant *34*:196-200.
Valadon, L. R. G., Osman, M., and Mummery, R. S. (1979). Phytochrome mediated carotenoid synthesis in the fungus *Verticillium agaricinum*. Photochem. Photobiol. *29*:605-607.
Valadon, L. R. G., Mummery, R. S., van Eijk, G. W., Roeymans, H. J., and Britton, G. (1980). Taxonomic implications of the carotenoids of *Iodophanus carneus*. Trans. Br. Mycol. Soc. *74*:187-190.
Van Zon, A. Q., Overmeer, W. P. J., and Veerman, A. (1981). Carotenoids func-

tion in photoperiodic induction of diapause in a predacious mite. Science 213:1131-1133.
Vecher, A. S., and Kulikova, A. (1968). Changes in polyene compounds at various stages of development of Rhodotorula gracilis. Mikrobiologiya 37: 558-560.
Vuori, A., and Gyllenberg, H. E. (1971). Effect of regulation of nitrogen metabolism on the carotenoid synthesis in Rhodotorula sanniei. Zentralbl. Bakteriol. Parasitenkd. Infectionskr. Hyg. Abt. 2 126:552-559.
Vuori, A. T., and Gyllenberg, H. E. (1974). Nitrogen metabolism of the acetate-stimulated carotenoid synthesis in Rhodotorula sanniei. Zentralbl. Bakteriol. Parasitenkd. Infektionskr. Hyg. Abt. 2 129:68-71.
Wang, S. S., Magill, J. M., and Phillips, R. L. (1971). Auxotrophic and visible mutations in white-spore (ws-1). Neurospora Newsl. 18:16.
Wanner, G., Formanek, H., and Theimer, R. R. (1981). The ontogeny of lipid bodies (spherosomes) in plant cells. Planta 151:109-123.
Weedon, B. C. L. (1979). Carotenoid research—Past, present and future. Pure Appl. Chem. 51:435-445.
Weeks, O. B., Saleh, F. K., Wirahadikusumah, M., and Berry, R. A. (1973). Photoregulated carotenoid biosynthesis in non-photosynthetic microorganisms. Pure Appl. Chem. 35:63-80.
Went, F. A. F. C. (1901). *Monilia sitophila* (Mont) Sacc. ein technischer Pilz Javas. Zentralbl Bakteriol. Parasitenkd. Infektionskr. Hyg. Abt. 2 7:544-550, 591-598.
Whitaker, B. D., and Shropshire, W., Jr. (1981). Spectral sensitivity in the blue and near ultraviolet for light-induced carotene synthesis in *Phycomyces* mycelia. Exp. Mycol. 5:243-252.
Will, O. H., III, Ruddat, M., and Garber, E. D. (1982). Characterization of the pigment in pink sporidial colonies of *Ustilago violacea* as cytochrome c. Exp. Mycol. 6:253-258.
Yamamoto, H. Y., and Bangham, A. D. (1978). Carotenoid organization in membranes. Thermal transition and spectral properties of carotenoid-containing liposomes. Biochim. Biophys. Acta 507:119-127.
Yen, H. -C. and Marrs, B. (1976). Map of genes for carotenoid and bacteriochlorophyll biosynthesis in *Rhodopseudomonas capsulata*. J. Bacteriol. 126:619-629.
Yokoyama, H., Nakayama, T. O. M., and Chichester, C. O. (1962). Biosynthesis of β-carotene by cell-free extracts of *Phycomyces blakesleeanus*. J. Biol. Chem. 237:681-686.
Zajic, J. E., and Kuehn, H. H. (1962). Biosynthesis of yellow pigments by *Aspergillus niger*. Mycopathol. Mycol. Appl. 17:149-158.
Zalokar, M. (1954). Studies on the biosynthesis of carotenoids in *Neurospora crassa*. Arch. Biochem. Biophys. 70:561-567.
Zalokar, M. (1969). Intracellular centrifugal separation of organelles in *Phycomyces*. J. Cell Biol. 41:494-509.
Zechmeister, L. (1962). *Cis-Trans Isomeric Carotenoids, Vitamins A and Arylpolyenes*. Springer-Verlag, Vienna.

6
Evolution and Secondary Pathways

Hans Zähner and Heidrun Anke / Universität Tübingen, Tübingen, Federal Republic of Germany

Timm Anke / Universität Kaiserslautern, Kaiserslautern, Federal Republic of Germany

INTRODUCTION

Definition

Many prokaryotic and eukaryotic organisms, bacteria, blue-green algae, fungi, plants, and lower animals produce metabolites of great value for human use but with no apparent function for the producing organism. These "secondary" metabolites were often produced during a well-defined time span at a certain stage of the producer's life cycle. In many cases the appearance of secondary metabolites was observed after cessation of vegetative growth and during sexual reproduction or the formation of spores or cysts. Some of these observations, especially in bacteria and fungi, led to a definition of secondary metabolites as products not essential for growth and of no apparent function for the producing cell (Aharonowitz and Demain, 1980).

Occurrence of Secondary Metabolites Among Different Organisms

The occurrence of secondary metabolism is a feature not common to all living organisms, but restricted to certain taxa. Among animals, coral reef coelenterates (e.g., sponges and corals) and other marine invertebrates are the most prolific producers of secondary metabolites (Faulkner, 1977; von Berlepsch, 1980). Within the plant kingdom, from which a great abundance of secondary metabolites has been isolated, only few groups (e.g., Poales) are poor producers. Few metabolites have been reported from blue-green algae, whereas several thousand secondary metabolites have been reported from bacteria and fungi (Bérdy et al., 1980).

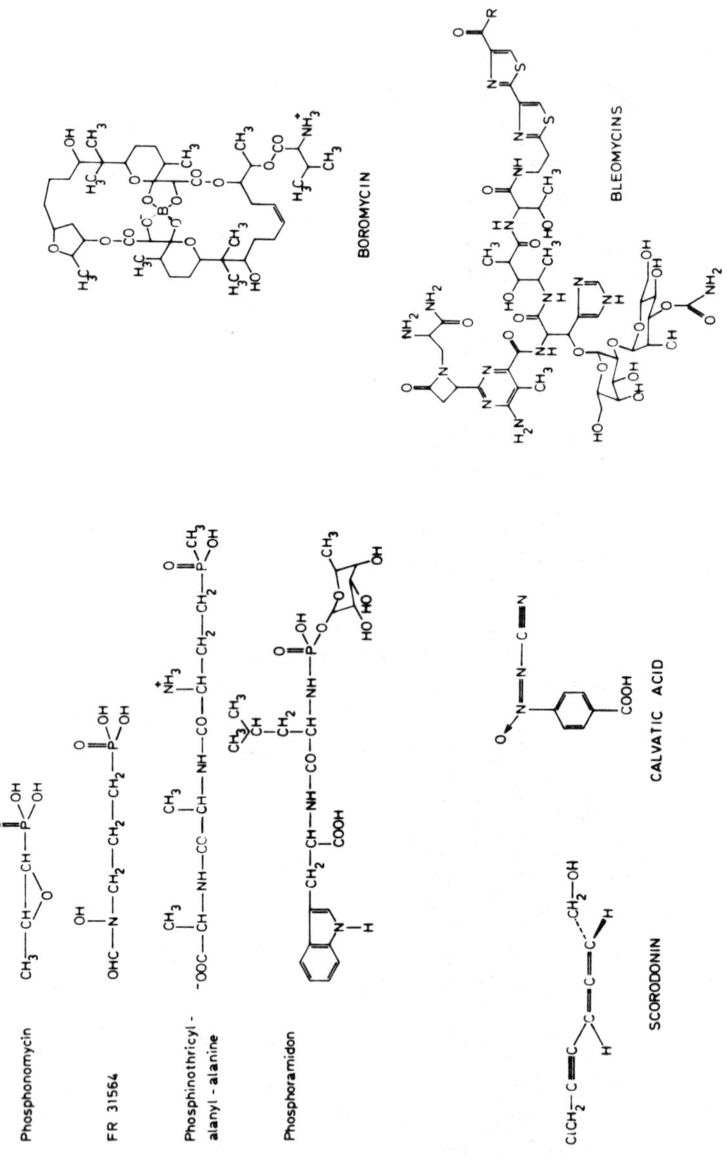

Figure 1 Secondary metabolites with unusual structures.

Comparison of Primary and Secondary Metabolism

Intermediary (primary) metabolism and its products are reasonably well known, but our knowledge of secondary pathways and metabolites is still very limited.

Primary metabolism consists of a limited number of well-known intermediates and products, the production of which is usually well balanced and essentially identical for all organisms. Therefore primary metabolites have also been termed "general" metabolites (Martin and Demain, 1980). Secondary metabolism is genotypically and phenotypically specific (Bu'Lock, 1975). In artificial cultures, the production of these "special" metabolites is sensitive to culture conditions (Weinberg, 1974). Overproduction is frequently observed. Secondary metabolites are usually produced as groups of closely related chemical families, sometimes with bizarre structures containing unusual chemical linkages. Some examples are shown in Fig. 1. While in primary metabolism only derivatives of phosphorous acid, for example, esters and anhydrides, are found, among secondary metabolites amides of phosphorous acid and derivatives of phosphonic and hypophosphorous acid are also present.

The occurrence of certain types of secondary metabolites may be a characteristic feature of an order, family, genus, species, or even a single strain. In some cases, for example, the plant families *Solanaceae* (Hegnauer, 1962, 1963, 1964, 1966, 1969, 1973) and *Compositae* (Bohlmann, 1980), where the chemical constituents, alkaloids, terpenoids, and polyins, have been extensively investigated, the pattern of distribution of secondary metabolites is of great taxonomic value. Of the secondary metabolites produced by fungi, some pigments (Eugster, 1973; Steglich, 1980), toxins like the aflatoxins, amanitins, or gyromitrin (Wyllie and Morehouse, 1977), or terpenoids (Anke and Steglich, 1981) seem to be restricted to certain genera and species. Within the bacteria, the *Actinomycetales* exhibit the most prolific secondary metabolism. Over 3000 different antibiotics have been isolated from the genus *Streptomyces* alone. In this genus some compounds seem to be produced by a single strain of one species, whereas others, like leupeptin and oxytetracycline, are found in the cultures of many different streptomycetes (Kurylowicz, 1976; Umezawa, 1979).

Variation of Secondary Metabolites

The phenotypically specific features of secondary metabolism are illustrated by the variation of the composition of anthraquinonic mixtures produced by *Aspergillus cristatus*. Depending on the medium, the mixture is composed of either only two anthraquinones (erythroglaucin and physcion) or at least 15 different compounds, as shown in Fig. 2 (Anke et al., 1980). Anthraquinones are not only produced by ascomycetes; they also are found in basidiomycetes (Steglich, 1980), and many more are known to occur in plants (Thomson, 1971). This

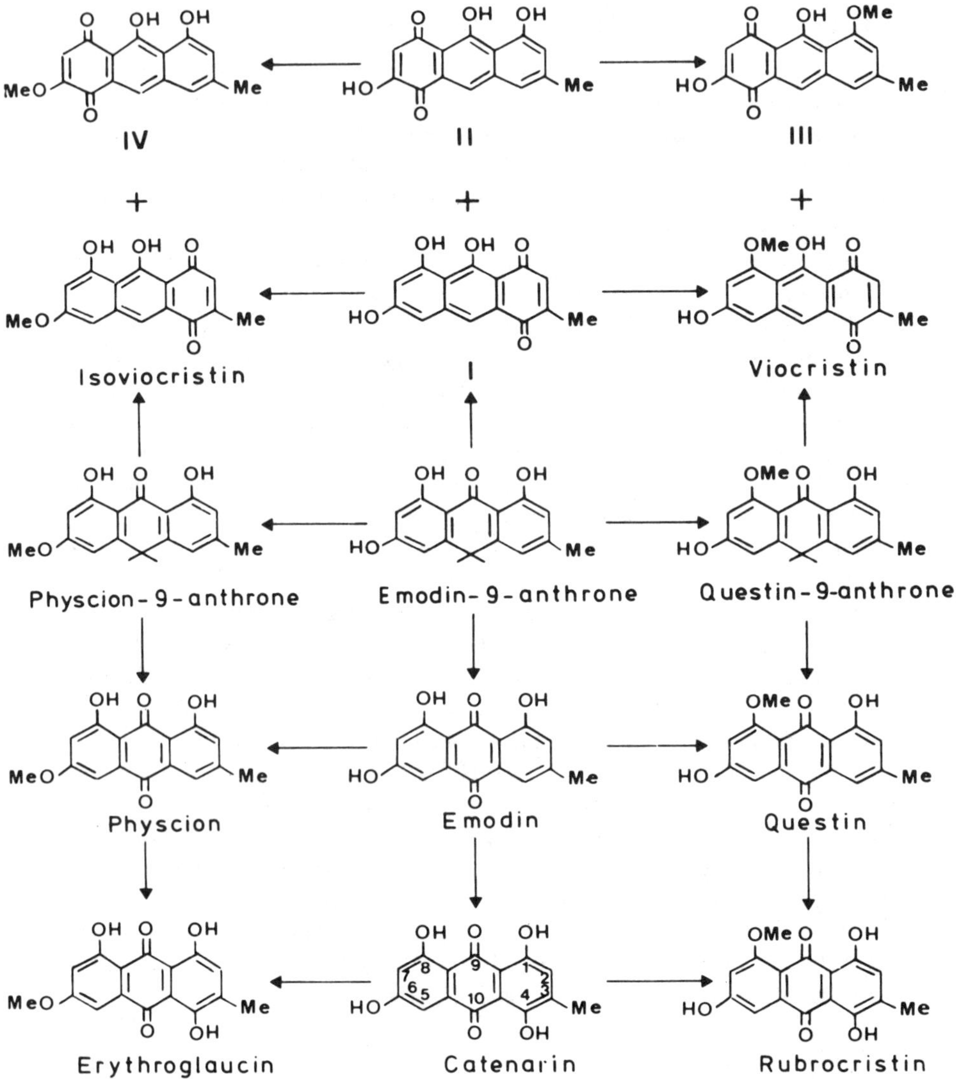

Figure 2 Anthraquinones from *Aspergillus cristatus*.

example illustrates that secondary metabolites are formed in great variety. Their number seems to be "unlimited," and probably only a small percentage has been isolated so far.

However, the spectrum of variation of secondary metabolism cannot be evaluated by looking only at one chemical group but, rather, by observing an

Evolution and Secondary Pathways

entire biosynthetically related family, for example, the polyketides, to which the anthraquinones belong. The variation in the case of the polyketides consists of the number of acetate or malonyl units; the usage of propionyl-coenzyme A or methyl malonyl as starter or extender units; different branching and cyclization types; different reduction and hydroxylation patterns; introduction of chlorine, bromine, or "extra" oxygens; introduction of methyl and alkyl groups, including the formation of methoxy and alkoxy derivatives; combinations with other fatty acids, amino acids, sugars, aromatic acids, and isoprenoid units; the condensation of two chains; and dimerization reactions. After combining all the different possibilities (shown here only for the polyketides), it appears that the many hundreds of polyketides isolated so far constitute only a tiny fraction of the millions of theoretically possible compounds. This may reflect the fact that during evolution new metabolites continuously appeared, many of which were subsequently discarded. It may also reflect the fact that our knowledge covers only a small percentage of the compounds in existence today.

Interrelationship Between Primary and Secondary Metabolism

Despite the great diversity of chemical structures, all secondary metabolites are derived from simple precursors drawn from intermediary metabolism. As shown in Fig. 3, the main branching points leading to secondary products are the following: *acetyl-coenzyme A,* leading to polyketides, polyins, terpenoids, steroids, or carotenoids; *shikimate,* from which aromatic compounds can be derived; *amino acids,* which are the precursors of peptides and alkaloids, and *glucose* for the biosynthesis of glycosides and aminoglycosides.

Regulation of Secondary Metabolism

In complex and nutritionally rich media, production of secondary metabolites usually, but not always, starts after the cellular growth rate has slowed down. By using media which allow only low growth rates, production of secondary metabolites can be initiated at an earlier fermentation stage, with trophophase and idiophase occurring simultaneously or overlapping each other.

In fungi, as well as in bacilli and streptomycetes, the production of secondary metabolites coincides with differentiation: for example, endospore formation and production of peptide antibiotics in bacilli (Lee et al., 1975), production of alkaloids and formation of penicilli and conidiospores in *Penicillium cyclopium* (Luckner, 1980), production of cephalosporin and arthrospore formation in *Cephalosporium acremonium* (Martin and Demain, 1978), and ergot alkaloid production and formation of chlamydospores in *Claviceps purpurea* (Spalla and Marnati, 1978). In this fungus alkaloid production is restricted to the heterokaryotic stage and the appearance of conidia is associated with the loss

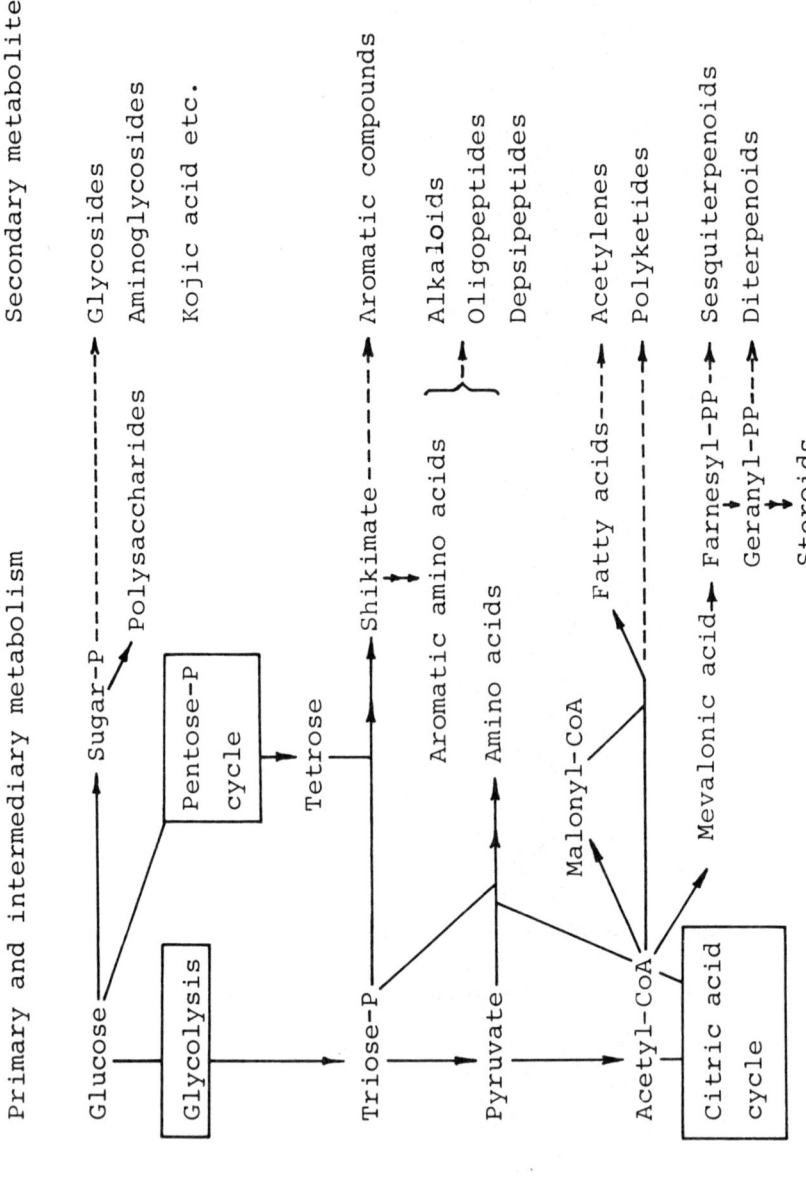

Figure 3 Interrelationship between primary and secondary metabolism.

Evolution and Secondary Pathways

of ergot-producing ability. These observations have often led to the conclusion that some of the secondary metabolites may act as differentiation effectors. However, in many cases mutants have been obtained in which differentiation occurs without detectable production of secondary metabolites (Luckner, 1980; Katz and Demain, 1977; Vandamme, 1981).

There exist several mechanisms which regulate the production of secondary metabolites: firstly, overall or *general* regulatory controls which couple the production of these metabolites with low growth rates and, secondly, *specific* regulation, either strain specific or product specific.

The general mechanisms include carbon catabolite, nitrogen metabolite, and phosphate regulation (for details see Martin and Demain, 1980). The exact mechanisms involved in these cases are still unknown.

Specific regulation includes, among other phenomena, stimulation of streptomycin biosynthesis by the A factor (Khokhlov and Tovarova, 1979), the increase of alkaloid synthesis by a hormone-like compound in *Penicillium cyclopium* (Luckner, 1980), the effects of methionine and norleucine on the production of cephalosporin in *Cephalosporium acremonium* (Drew and Demain, 1977), and feedback inhibition of penicillin biosynthesis by penicillin (Gordee and Day, 1972). This end-product regulation is observed in the biosynthesis of many antibiotics (Martin, 1978).

Another factor that influences secondary metabolism is the availability of primary metabolites and intermediates serving as precursors in the biosynthesis of secondary metabolites. Since enzymes involved in secondary metabolism often exhibit high K_m values and little specificity for their substrates, the amino acids in gramicidins can be exchanged, for example. The availability of precursors is an important factor in the regulation of secondary metabolism. Thus any factor changing the pool size of such a precursor will influence the production of secondary metabolites derived from this precursor.

A THEORETICAL APPROACH TO THE PHENOMENA OF SECONDARY METABOLISM

Secondary Metabolism as a Biochemical Playground

The large number of secondary metabolites already known has led to many speculations concerning a reason for the existence of secondary metabolism. It has been proposed that secondary metabolism is an outgrowth of unbalanced primary metabolism, that secondary metabolites may be the products of detoxification of toxic intermediates, and that they may be helpful in suppressing competitors in the natural environment or may play a role in cellular differentiation, especially during sporulation. Certainly these assumptions are of value only in

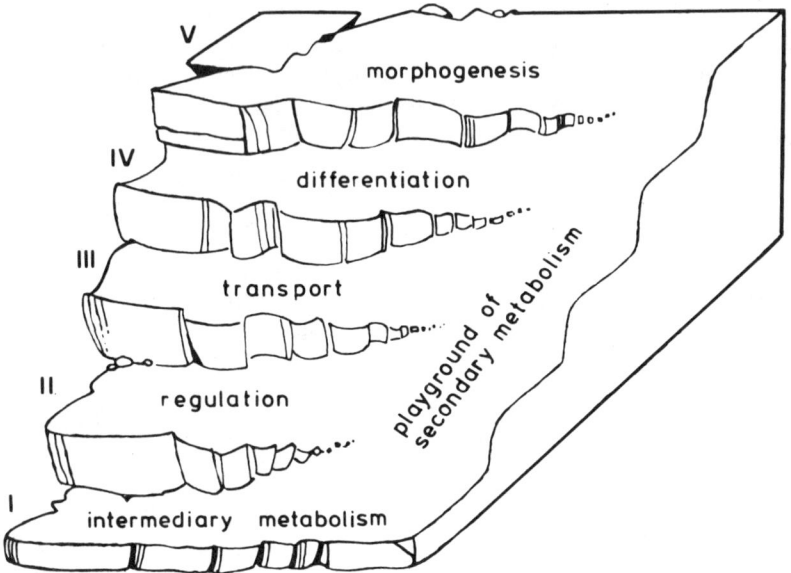

Figure 4 Connection of secondary metabolism with the five primary cellular events.

a few cases. A more generalized view on secondary metabolism has been recently offered by Zähner (1978, 1979). Besides and closely connected to the five distinct levels—intermediary metabolism, regulation, transport, differentiation, and morphogenesis, as shown in Fig. 4—we have the diffuse field of secondary metabolism, an area where biochemical evolution is taking place continuously. The existence of this "playground" is dependent mainly on the supply of surplus precursors and energy from intermediary metabolism. Starting from given precursors from intermediary metabolism, in secondary metabolism the evolution of biochemical pathways can proceed in all directions provided that the metabolites are not toxic to the producing organism, at least not at the time of their appearance.

Conditions for the Presence of a Sizable Secondary Metabolism

The main selection principle during the evolution of organisms which can afford the "luxury" of a biochemical playground cannot be fast growth and short generation times. These selection principles are most important for organisms grow-

Evolution and Secondary Pathways 161

ing in substrates rich in easily accessible carbon and nitrogen sources, for example, *Enterobacteriaceae, Lactobacillaceae,* and yeasts. *Short generation times* require an extremely well-regulated intermediary metabolism; there is no place or time for metabolic distractions. Microorganisms growing only in *extreme environments* like hot sulfur springs (*Sulfolobus* spp.) and burning coal piles (*Thermoplasma* spp.) and those able to use substrates which cannot be metabolized by others (methanotrophic bacteria) do not seem to have an important secondary metabolism. Evolution has led them to perfectly adapt to their environment, which is concomitant with a loss of adaptability.

A characteristic feature of the more prolific producers of secondary metabolites among microorganisms (the streptomycetes, bacilli, pseudomonads, ascomycetes, basidiomycetes, and Fungi Imperfecti) is their great *adaptability* to changing environmental conditions such as substrate, pH, temperature, and oxygen and water availability. This adaptability is essential for soil inhabitants, especially for those with long generation times like the streptomycetes and especially the fungi. The ability to use a wide variety of substrates of different composition which enter primary metabolism at different points is often correlated with a less stringent regulation. This can lead to an enlargement of the pool sizes for the metabolites at the branching points of primary and secondary metabolism.

A View of the Genetics of Secondary Metabolism

Biosynthesis

A biochemical playground requires great genetic flexibility. Some genes coding for biosynthetic enzymes of secondary metabolism are located on plasmids. Among them are enzymes involved in the synthesis of methylenomycin (Wright and Hopwood,1976; Hopwood, 1979), leupeptin, kanamycin (Umezawa, 1979), and chloramphenicol (Akagawa et al., 1975).

Mobility of Genetic Information

A glance at the occurrence of identical secondary metabolites in organisms separated very early in evolution—for example, the maytansines in *Maytenus* and *Nocardia* (Brufani, 1977; Sneden and Beemsterboer, 1980), β-lactams in ascomycetes and actinomycetes (Cooper, 1980), lysergic acid in *Claviceps* and *Convolvolus* (Hoffmann, 1961; Gröger, 1978), and trichothecenes in *Fusarium* and *Baccharis megapotamica* (Kupchan et al., 1976)—may lead to the speculation that the biosynthetic pathways leading to these products have been acquired by direct transfer of the corresponding genes. It is conceivable that this event might have taken place during a symbiont-host or parasite-host and interrelationship or simply by close contact and uptake of genetic material during

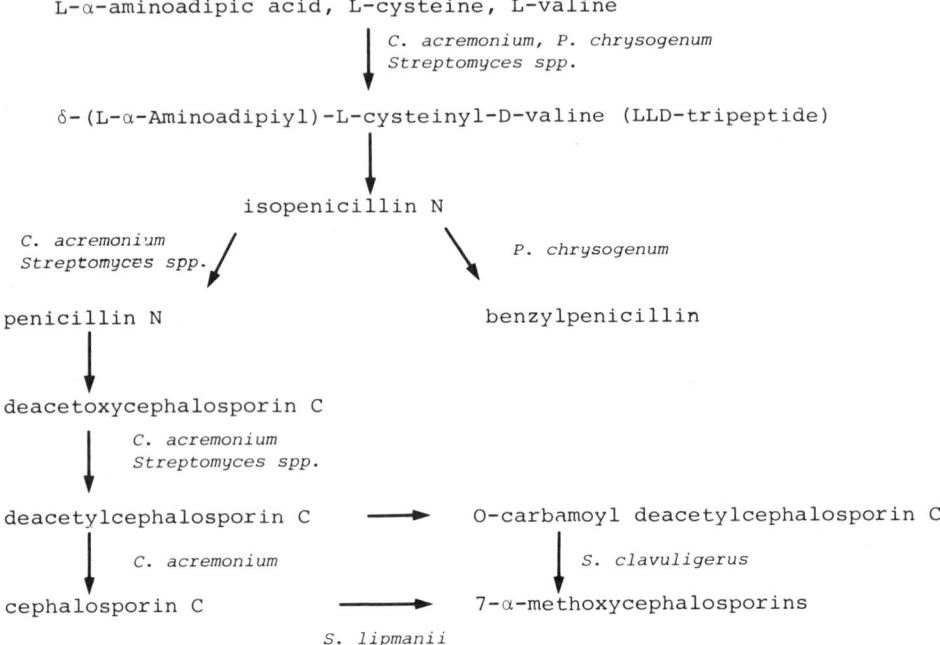

Figure 5 Biosynthesis of penicillins and cephalosporins in *Cephalosporium acremonium*, *Penicillium chrysogenum*, and *Streptomyces* spp.

proglonged growth in a same habitat, for example, soil. It is also conceivable that lytic enzymes produced by soil organisms, such as lysozyme or chitinases, may lead to protoplast formation and fusion, provided that the osmotic pressure is suitable. Genetic recombination by protoplast fusion and outgrowth to vegetative cells has been demonstrated for streptomycetes (Hopwood and Merrick, 1977) and higher fungi (Peberdy, 1979). Even recombinants from different genera can be obtained.

The transfer of genetic material is a well-established fact for the transfer of plasmid-mediated antibiotic resistance among enteric bacteria or the transfer of the Ti plasmid from *Agrobacterium tumefaciens* to its plant host (Chilton, 1980; Thomashow et al., 1980). If the assumption of genetic transfer were true for the maytansine, lysergic acid, or β-lactam genes, one would expect identical biosynthetic pathways with identical or very similar enzymes. In the case of the β-lactams, identical biosynthetic pathways are assumed, as shown in Fig. 5 (O' Sullivan et al., 1980; Baldwin et al., 1981); however, a comparison of the isolated enzymes has not yet been undertaken.

Evolution and Secondary Pathways 163

On the other hand, the occurrence of very similar metabolites among distant taxa could also result from the accumulation of the same intermediary precursors. This, we think, is most likely for the terpenoids predominantly found in basidiomycetes, plants, and marine animals. Only a few enzymes are necessary for the biosynthesis of sesquiterpenes from farnesyl pyrophosphate, an intermediate of steroid and sesquiterpene biosynthesis. The broad distribution of the terpenes also seems to point to this latter hypothesis.

Transitions from Secondary to Primary Metabolism. Functional Equivalence of Compounds Derived from Different Biosynthetic Pathways

A continuous evolution of biochemical pathways in secondary metabolism must occasionally lead to compounds which offer an advantage to the producing organism. It would make sense that once such a useful pathway has been developed, it can be made part of primary metabolism. The sideramines and other compounds involved in iron transport offer a good example of such a transition. Though very few iron-free sideramines, for example, nocardamin from a species of *Nocardia* (Keller-Schierlein and Prelog, 1961) and desferritriacetylfusigen (Anke, 1977) from fungi, exhibit antimicrobial activities, it was because of this property that they were originally detected. Today it has been clearly established that the sideramines and related compounds play an essential role in the solubilization and uptake of iron. Their biosynthesis is regulated by the concentration of soluble iron in the substrate. As has been shown for rhodotorulic acid synthetase (Anke and Diekmann, 1972) and fusigen synthetase (Anke et al., 1973), these key enzymes could not be detected as long as a sufficient amount of iron was present in the culture media; these enzymes are detectable only in iron-deficient cells. Because of their features in function and regulation, the sideramines are primary metabolites. There are important arguments, however, that the sideramines and other iron transport compounds are derived from secondary metabolism. In streptomycetes and fungi there is a great variety of compounds, all containing hydroxamate groups which are involved in the complexing of iron. Usually one strain produces several of these compounds (Diekmann, 1973). *Enterobacteriaceae* have solved the problem of iron transport by "inventing" enterochelin-type compounds which contain dihydroxybenzoic acid (Rosenberg and Young, 1974). With exception of some strains of *Microbacterium lacticum* and the sideramine or hemin auxotrop fungus *Pilobulus kleinii*, a loss of the ability to produce these chelators results only in slower growth. In this case iron can be taken up as complexes of organic acids such as citrate. Another feature of secondary metabolism, the occurrence of different compounds in different taxa, is clearly shown for the iron transport compounds. The size and conditions for the playground of secondary metabolism result in

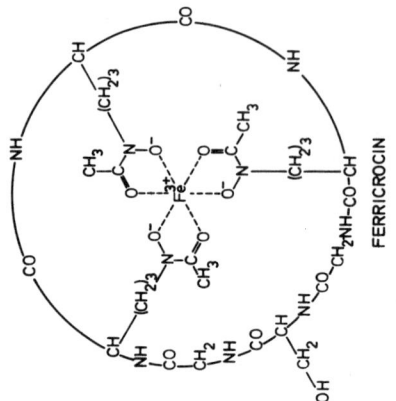
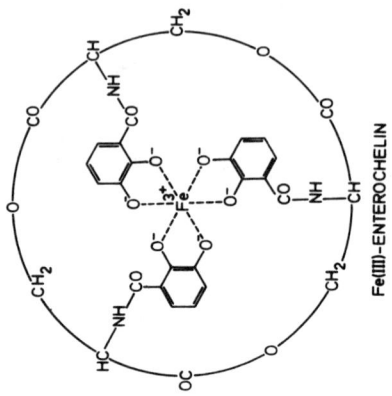
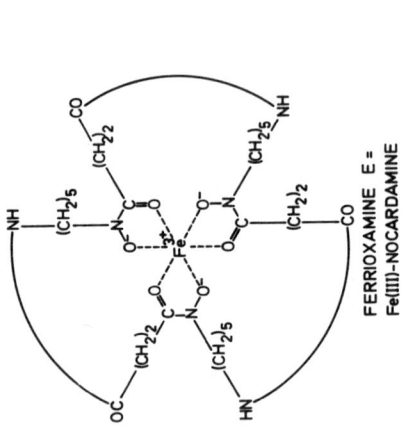
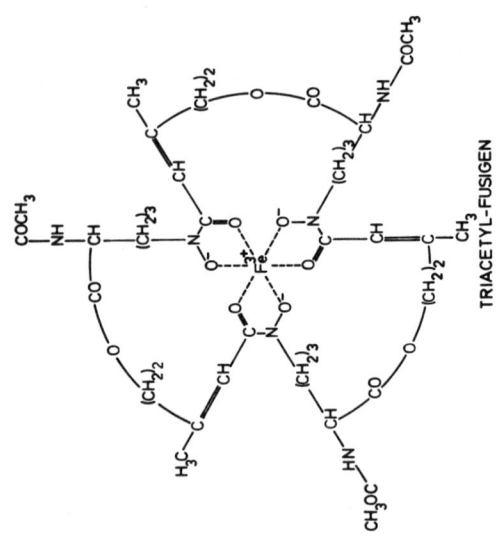

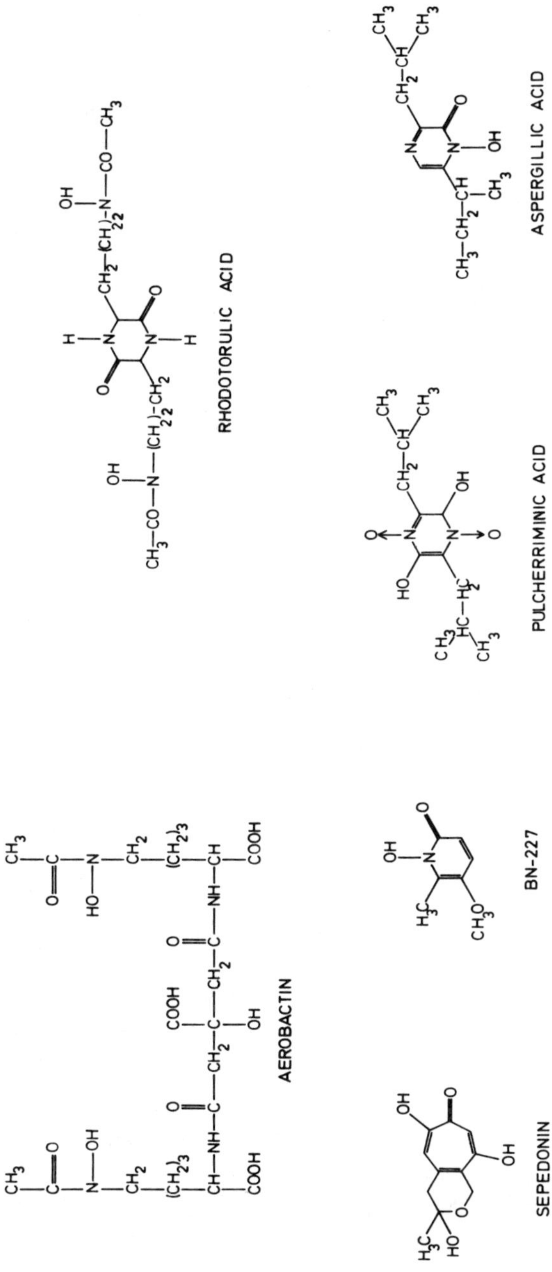

Evolution and Secondary Pathways

Figure 6 Iron-chelating compounds from bacteria and fungi.

metabolites that are different from organism to organism. Therefore for a given problem, for example, the transport of iron into cells where many solutions are possible, a variety of chemically different molecules have been derived, all serving the same purpose. In Fig. 6 the structures of different iron chelators are given, illustrating this chemical diversity. The functional equivalence of structurally completely different metabolites is, in our opinion, an important feature of the postulated "playground."

While the sideramines and enterochelin, the iron transport compounds with the highest efficiency, have already become part of primary metabolism, there are other metabolites which might be considered at the border line between primary and secondary metabolism. Among these are the tropolones, pulcherriminic acid, and aspergillic acid, all forming ion complexes which can be used in iron transport (Akers et al., 1980). Although these compounds have acquired a function, their biosynthesis is not yet subject to regulation by the iron content of the medium. The antibiotically active ionophores can be considered as another example: The potassium-specific compound valinomycin might be involved in the potassium uptake of the producing streptomycetes; its biosynthesis, however, is not dependent on the potassium content of the media and its distribution seems to be limited to very few strains.

Inhibitors of protein, RNA, and DNA synthesis, like the aminoglycosides, kirromycin, amanitins, actinomycins, and coumermycins, could possibly play a role in various steps of differentiation. A role for the amanitins in the formation of fruit bodies of *Amanita* species has been proposed (Johnson and Preston, 1979).

Metabolites like gramicidin S in *Bacillus brevis* (Nandi et al., 1981) which retard germination and simultaneous outgrowth of spores or 2'-hydroxy-4-imino-2,5-cyclohexadienone in *Agaricus bisporus* (Mize et al., 1980) confer obvious advantages to the producing organism; hence the chances that some of the spores will germinate when the conditions are favorable for a sufficiently long period of time to allow reproduction are greatly enhanced.

Genes for the biosynthesis of secondary metabolites which have assumed an important function in primary metabolism have to be withdrawn from the "playground" and stabilized and kept, for example, in the chromosome. They also have to become subject to regulation.

Transfer of Information from Organisms with Secondary Metabolism to Others Having None

According to the symbiont theory, during evolution whole organisms, bacteria, or blue-green algae with their complete genetic information and metabolism were incorporated into early eukaryotic cells and developed into mitochondria and chloroplasts (for a review see Fredrick, 1981). The rationale for this process was the usefulness of part of the prokaryotic metabolism to the host and the huge amount of time saved by making use of preexisting genetic information instead of developing it de novo. During this process genes for the enzymatic sequences essential for the eukaryote were kept, partly in the organelle and partly integrated into the chromosomes of the nucleus. Information that conferred no advantage to the eukaryotic host and which was unimportant for the maintenance of the organelle was eliminated.

We now propose that once a useful biochemical pathway has been invented in secondary metabolism, it can be utilized by other organisms, even those which cannot afford secondary metabolism, for example, *Enterobacteriaceae*, especially if this confers an essential advantage for survival.

The genes responsible for resistance to aminoglycoside antibiotics might serve as an example. Aminoglycosides are produced by streptomycetes, bacilli, and pseudomonads. Except for adenylation, all enzymes for inactivation of these antibiotics can be detected in the producers, phosphorylation being part of the biosynthetic pathway (Courvalin et al., 1981). Resistance of pathogenic strains of *Escherichia coli, Salmonella,* and *Shigella* to these antibiotics results from adenylation, acetylation, or phosphorylation. The comparatively rapid appearance of resistant strains makes a new development of the inactivating enzymes unlikely. According to the hypothesis of Davies and Benveniste (1974), the genes (plasmids) conferring resistance were transferred from the producers, for example, from *Pseudomonas* to the sensitive *Enterobacteriaceae*, where they are maintained and expressed as long as the cells are exposed to aminoglycosides.

In some pathogenic strains of *E. coli,* genes for biosynthesis of the iron transport compound aerobactin are located on a plasmid (Stuart et al., 1980; Braun, 1981). Since the virulence of pathogenic *Enterobacteriaceae* is strongly correlated to the ability to mobilize and take up iron from the infected host (Williams, 1979), possession of this plasmid confers an important advantage. In the context of our proposition, it would be very interesting to know the plasmid's origin and distribution.

CONCLUDING REMARKS

The playground of secondary metabolism is probably a very early relic of biochemical evolution. In lower organisms it is turned on in times of stress, for example, when there is a scarce supply of nutrients, much like the memory and thoughts of a human being trying to improve his or her condition of life. Once a useful invention is made, it can be fixed on mobile genes and then communicated to others by genetic transfer, a process for which mankind has invented the pencil and paper and books and journals. Even organisms which have lost secondary metabolism and adaptability resulting from selection for fast growth and short generation times in nutritionally rich media can in some cases participate in the development of secondary metabolism by genetic transfers.

REFERENCES

Aharonowitz, Y., and Demain, A. L. (1980). Thoughts on secondary metabolism. Biotechnol. Bioeng. *22* (Suppl. 1):5-9.

Akagawa, H., Okanishi, M., and Umezawa, H. (1975). A plasmid involved in chloramphenicol producing *Streptomyces venezuelae*: Evidence from genetic mapping. J. Gen. Microbiol. *90*:336-346.

Akers, H. A., Abrego, V. A., and Garland, E. (1980). Thujaplicins from *Thuja plicata* as iron transport agents for *Salmonella typhimurium*. J. Bacteriol. *141*:164-168.

Anke, H. (1977). Metabolic products of microorganisms. 163. Desferritriacetylfusigen, an antibiotic from *Aspergillus deflectus*. J. Antibiot. *30*:125-128.

Anke, H., Anke, T., and Diekmann, H. (1973). Biosynthesis of sideramines in fungi. Fusigen synthetase from extracts of *Fusarium cubense*. FEBS Lett. *36*:323-325.

Anke, H., Kolthoum, I., and Laatsch, H. (1980). Metabolic products of microorganisms. 192. The anthraquinones of the *Aspergillus glaucus* group. II. Biological activity. Arch. Microbiol. *126*:231-236.

Anke, T., and Diekmann, H. (1972). Metabolic products of microorganisms. 112. Biosynthesis of sideramines in fungi. Rhodotorulic acid synthetase from extracts of *Rhodotorula glutinis*. FEBS Lett. *27*:259-262.

Anke, T., and Steglich, W. (1981). Screening of basidiomycetes for the production of new antibiotics. In *Advances in Biotechnology*, vol. 1, M. Moo-Young, C. W. Robinson, and C. Vezina (Eds.). Pergamon, Toronto, pp. 35-40.

Baldwin, J. E., Keeping, J. W., Singh, P. D., and Vallejo, C. A. (1981). Cell-free conversion of isopenicillin N into deacetoxycephalosporin C by *Cephalosporium acremonium* mutant M-0198. Biochem. J. *194*:649-651.

Bérdy, J., Aszalos, A., Bostian, M., and McNitt, K. L. (Eds.) (1980). In *Handbook of Antibiotic Compounds, Macrocyclic Lactone (Lactam) Antibiotics*, Vol. 2. CRC Press, Boca Raton, pp. 1-2.

Berlepsch, K. von (1980). Drugs from marine organisms. Naturwissenschaften *67*:338-342.

Bohlmann, F. (1980). Neues über die Chemie der Compositen. Naturwissenschaften *67*:588-594.

Braun, V. (1981). *Escherichia coli* containing the Col V plasmid produce the iron ionophore aerobactin. FEMS Lett. *11*:225-228.

Brufani, M. (1977). The ansamycins. In *Topics in Antibiotic Chemistry, Aminoglycosides and Ansamycins*, vol. 1, P. G. Sammes (Ed.). Wiley, New York, pp. 96-213.

Bu'Lock, J. D. (1975). Secondary metabolism in fungi and its relationship to growth and development. In *The Filamentous Fungi, Industrial Mycology,* Vol. 1, J. E. Smith and D. R. Berry (Eds.). Arnold, London, pp. 35-58.

Chilton, M.-D. (1980). *Agrobacterium* Ti plasmids as a tool for genetic engineering in plants. In *Genetic Engineering of Osmoregulation,* D. W. Rains, R. C. Valentine, and A. Hollaender (Eds.). Plenum, New York, pp. 23-31.

Cooper, R. D. G. (1980). New β-lactam antibiotics. In *Topics in Antibiotic Chemistry, Mechanisms of Action of Nalidixic Acid and Its Congeners, New β-Lactam Antibiotics,* Vol. 3, P. G. Sammes (Ed.). Ellis Horwood, Chichester, pp. 39-199.

Courvalin, P., Carlier, C., and Collatz, E. (1981). Evolutionary relationships between plasmid-mediated aminoglycoside-modifying enzymes from gram-positive and gram-negative bacteria. In *New Trends in Antibiotics: Research and Therapy,* G. Gialdroni Grassi and L. D. Sabath (Eds.). Elsevier/North-Holland, New York, pp. 95-109.

Davies, J. E., and Benveniste, R. E. (1974). Enzymes that inactivate antibiotics in transit to their targets. In *Mode of Action of Antibiotics on Microbial Walls and Membranes,* M. R. J. Salton and A. Thomasz (Eds.). New York Academy of Sciences, New York, pp. 130-136.

Diekmann, H. (1973). Siderochromes (Iron (III)-trihydroxamates). In *Handbook of Microbiology, Microbial Products,* Vol. 3, A. I. Laskin and H. U. Lechevalier (Eds.). CRC Pres, Cleveland, pp. 449-457.

Drew, S. W., and Demain, A. L. (1977). Effect of primary metabolites on secondary metabolism. Annu. Rev. Microbiol. *31*:342-356.

Eugster, C. H. (1973). Pilzfarbstoffe, ein Überblick aus chemischer Sicht mit besonderer Berücksichtigung der *Russulae.* Z. Pilzk. *39*:45-96.

Faulkner, D. J. (1977). Interesting aspects of marine natural products chemistry. Tetrahedron *33*:1421-1443.

Frederick, J. F. (Ed.) (1981). *Origins and Evolution of Eukaryotic Intracellular Organelles.* New York Academy of Sciences, New York.

Gordee, E. Z., and Day, L. E. (1972). Effect of exogenous penicillin on penicillin biosynthesis. Antimicrob. Agents Chemother. *1*:315-322.

Gröger, D. (1978). Ergot alkaloids—Recent advances in chemistry and biochemistry. In *Antibiotics and Other Secondary Metabolites,* R. Hütter, T. Leisinger, J. Nüesch, and W. Wehrli (Eds.). Academic, London, pp. 201-217.

Hegnauer, R. (1962, 1963, 1964, 1966, 1969, 1973). *Chemotaxonomie der Pflanzen,* Vols. 1-6. Birkhäuser Verlag, Basel.

Hoffmann, A. (1961). Die Wirkstoffe der mexikanischen Zauberdroge Ololiuqui. Planta Med. *9*:354-367.

Hopwood, D. A. (1979). Genetics of antibiotic production by actinomycetes. J. Nat. Prod. *42*:596-602.

Hopwood, D. A., and Merrick, M. J. (1977). Genetics of antibiotic production. Bacteriol. Rev. *41*:595-635.

Johnson, B. C., and Preston, J. F. (1979). Unique amanitin resistance of RNA synthesis in isolated nuclei from *Amanita* species accumulating amanitins. Arch. Microbiol. *122*:161-167.

Katz, E., and Demain, A. L. (1977). The peptide antibiotics of *Bacillus*: Chemistry, biogenesis, and possible functions. Bacteriol. Rev. *41*:449-474.
Keller-Schierlein, W., and Prelog, V. (1961). Stoffwechselprodukte von Actinomyceten. 30. Mitt. Über das Ferrioxamin E, ein Beitrag zur Konstitution des Nocardamins. Helv. Chim. Acta *44*:1981-1985.
Khokhlov, A. A., and Tovarova, I. I. (1979). Autoregulator from *Streptomyces griseus*. In *Regulation of Secondary Product and Plant Hormone Metabolism*, M. Luckner and K. Schreiber (Eds.). Pergamon, New York, pp. 133-145.
Kupchan, S. M., Jarvis, B. B., Dailey, R. G., Bright, W., Bryan, R. F., and Shizuri, Y. (1976). Baccharin, a novel potent antileukemic trichothecene triepoxide from *Baccharis megapotamica*. J. Am. Chem. Soc. *98*:7092-7093.
Kurylowicz, W. (1976). *Antibiotics, a Critical Review*. Polish Medical Publishers, Warsaw.
Lee, S. G., Littau, V., and Lipmann, F. (1975). The relation between sporulation and the induction of antibiotic synthesis and of amino acid uptake in *Bacillus brevis*. J. Cell. Biol. *66*:233-242.
Luckner, M. (1980). Alkaloid biosynthesis in *Penicillium cyclopium*—Does it reflect general features of secondary metabolism? J. Nat. Prod. *43*:21-40.
Martin, J. F. (1978). Manipulation of gene expression in the development of antibiotic production. In *Antibiotics and Other Secondary Metabolites. Biosynthesis and Production*, R. Hütter, T. Leisinger, J. Nüesch, and W. Wehrli (Eds.). Academic, New York, pp. 19-37.
Martin, J. F., and Demain, A. L. (1978). Fungal development and metabolite formation. In *The Filamentous Fungi, Developmental Biology*, Vol. 3, J. E. Smith and D. R. Berry (Eds.). Arnold, London, pp. 426-450.
Martin, J. F., and Demain, A. L. (1980). Control of antibiotic biosynthesis. Microbiol. Rev. *44*:230-251.
Mize, P. D., Jeffs, P. W., and Boekelheide, K. (1980). Structure determination of the active sulfhydryl reagent in gill tissue of the mushroom *Agaricus bisporus*. J. Org. Chem. *45*:3540-3543.
Nandi, S., Lazaridis, I., and Seddon, B. (1981). Gramicidin S and respiratory activity during the developmental cycle of the producer organism *Bacillus brevis* Nagano. FEMS Lett. *10*:71-75.
O'Sullivan, J., Huddleston, J. A., and Abraham, F. R. S. (1980). Biosynthesis of penicillins and cephalosporins in cell-free systems. Philos. Trans. R. Soc. London B *289*:363-365.
Peberdy, J. F. (1979). Fungal protoplasts: Isolation, reversion, and fusion. Annu. Rev. Microbiol. *33*:21-39.
Rosenberg, H., and Young, I. G. (1974). Iron transport in enteric bacteria. In *Microbial Iron Metabolism*, J. B. Neilands (Ed.). Academic, New York, pp. 67-82.
Sneden, A. T., and Beemsterboer, G. (1980). Normaytansine, a new antileukemic ansa macrolide from *Maytenus buchananii*. J. Nat. Prod. *43*:637-640.

Spalla, C., and Marnati, M. P. (1978). Genetic aspects of the formation of ergot alkaloids. In *Antibiotics and Other Secondary Metabolites. Biosynthesis and Production,* R. Hütter, T. Leisinger, J. Nüesch, and W. Wehrli (Eds.). Academic, London, pp. 219-232.

Steglich, W. (1980). Pigments of higher fungi (macromycetes). In *Pigments in Plants,* F.-C. Czygan (Ed.). Gustav Fischer Verlag, Stuttgart, pp. 393-412.

Stuart, S. J., Greenwood, K. T., and Luke, R. K. J. (1980). Hydroxamate-mediated transport of iron controlled by Col V plasmids. J. Bacteriol. *143*: 35-42.

Thomashow, M. F., Nutter, R., Postle, K., Chilton, M. D., Blattner, E. R., Powell, A., Gordon, M. P., and Nester, E. W. (1980). Recombination between higher plant DNA and the Ti plasmid of *Agrobacterium tumefaciens.* Proc. Nat. Acad. Sci. USA 77:6448-6452.

Thomson, R. H. (1971). Anthraquinones. In *Naturally Occurring Quinones,* Academic, London, pp. 367-535.

Umezawa, H. (1979). Recent chemical studies of bioactive microbial products: Genetics, active structures, development of effective agents with potential usefulness. Heterocycles *13*:23-47.

Vandamme, E. J. (1981). Properties, biogenesis and fermentation of the cyclic decapeptide antibiotic, gramicidin S. In *Topics in Enzyme and Fermentation Biotechnology,* Vol. 5, A. Wiseman (Ed.). Ellis Horwood, Chichester, pp. 187-261.

Weinberg, E. D. (1974). Secondary metabolism: Control by temperature and inorganic phosphate. *Dev. Ind. Microbiol.* 15:70-81.

Williams, P. H. (1979). Novel iron uptake system specified by Col V plasmids: An important component in the virulence of invasive strains of *Escherichia coli.* Infect. Immun. 26:925-932.

Wright, L. F., and Hopwood, D. A. (1976). Identification of the antibiotic determined by the SCP 1 plasmid of *Streptomyces coelicolor* A3 (2). J. Gen. Microbiol. *95*:96-106.

Wyllie, T. D., and Morehouse, L. G. (Eds.) (1977). *Mycotoxic Fungi, Mycotoxins, Mycotoxicoses, Mycotoxic Fungi and Chemistry of Mycotoxins,* Vol. 1, Marcel Dekker, New York.

Zähner, H. (1978). The search for new secondary metabolites. In *Antibiotics and Other Secondary Metabolites. Biosynthesis and Production.* R. Hütter, T. Leisinger, J. Nüesch, and W. Wehrli (Eds.). Academic, London, pp. 1-17.

Zähner, H. (1979). What are secondary metabolites? Folia Microbiol. *24*:435-443.

Molecular Aspects of Morphogenesis

7
Fungal Nucleic Acids

Shelby N. Freer / Agricultural Research Service, United States Department of Agriculture, Peoria, Illinois

INTRODUCTION

Developmental biologists assume that the major control of cellular differentiation results from the differential expression of selected genes. The production of new cell types or different metabolic processes (secondary metabolism) presumably requires the expression of formerly quiescent genes. Since these developmental processes are so complex, ideally one would study the regulation and function of a single gene. However, prior to 1970, it was virtually impossible to isolate DNA molecules corresponding to a single gene in sufficient quantity for analysis at the molecular level. Recent developments in recombinant DNA research make the isolation and identification of a specific DNA fragment from any organism possible. Specific genes can be identified, amplified in vivo, and reisolated in sufficient quantity for study in vitro. The opportunity now exists to analyze the structure of specific genes and relate this to gene expression in vivo.

Several authors have recently reviewed the subject of nucleic acids in filamentous fungi (Brambl et al., 1978; Haber et al., 1977; Lovett, 1975, 1976; Lovett et al., 1977; Schmit and Brody, 1976; Van Etten et al., 1976, 1977, 1981). These reviews describe the general properties of DNA and RNA in fungi and the molecular events associated with fungal differentiation (sporulation and germination). Most of this material will not be covered in this chapter. Instead, this chapter will describe some of the techniques used in recombinant DNA technology, briefly consider the organization and transcription of the fungal genome, and describe in moderate detail some of the results that the application of cloning techniques to filamentous fungi have generated. Lastly, I will suggest some fundamental problems that might be attacked and possible practical benefits that might be reaped by applying recombinant DNA techniques to fungal systems.

CONSTRUCTION AND IDENTIFICATION OF RECOMBINANT DNA MOLECULES

The development of recombinant DNA technology has transpired over the last 10 years because of several key discoveries. The isolation of mutant *Escherichia coli* strains unable to restrict foreign DNA (cleave it specifically and degrade it) laid part of the foundation. Another major development was the isolation and characterization of bacterial restriction endonucleases. These enzymes cleave double-stranded (ds) DNA at specific sites within the DNA, thus generating DNA fragments of defined sizes. The majority of these DNA fragments will neither transform nor replicate efficiently in host cells. DNA vehicles have been constructed to overcome this problem. These vehicles (vectors) are most commonly either phage λ, yeast 2-μm DNA, bacterial plasmids that carry antibiotic resistance genes, or derivatives of the above. After insertion of the DNA fragments into the vectors, covalent bonds are formed with phage T4 ligase. The chimeric DNA molecules are inserted into appropriate host organisms (usually *Saccharomyces cerevisiae* strains or *E. coli* strains unable to restrict foreign DNA) via transformation. As the host organism grows, it replicates (amplifies) the chimeric vehicles and thus the foreign DNA. By creating a large collection, or library, of independent recombinant DNA molecules, it is possible to clone the entire genome of an organism. For example, the *S. cerevisiae* genome contains about 1.4×10^4 kilobase pairs (kbp) (Lauer et al., 1977) of DNA. If 4600 independent recombinant clones are obtained with an average yeast DNA insert size of 15 kbp, the probability of having any 15 kbp sequence present in the collection would be 99% (Clarke and Carbon, 1976).

The genomic library would not be very useful if there was no way to screen the recombinant DNA collections for specific recombinant molecules. Various methods have been developed to identify clones that contain desired DNA sequences or genes.

One of the most common methods of analyzing a collection of recombinant DNA molecules is by colony or plaque hybridization (Grunstein and Hogness, 1975; Benton and Davis, 1977). The DNA from colonies containing recombinant DNA or recombinant λ phage DNA is first transferred to a nitrocellulose filter; next, purified complementary radioactive nucleic acids are hybridized to the nitrocellulose-bound DNA. If a colony or plaque contains a DNA sequence complementary to the probe, these DNAs will hybridize and can be detected by autoradiography. The colonies or plaques that contain the desired sequences can then be isolated. With this technique, the ribosomal (r) RNA genes from several organisms have been isolated, including *S. cerevisiae* (Bell et al., 1977), *Neurospora crassa* (Free et al., 1979), and *Mucor racemosus* (Cihlar and Sypherd, 1980). Similarly, the genes encoding yeast transfer (t) RNA have been cloned (Olson et al., 1979).

Another hybridization probe method to screen libraries involves the synthesis of complementary (c) DNA. Either purified messenger (m) RNA of RNA fractions enriched for a specific mRNA are used to direct the synthesis of cDNA. The cDNA can then be used as a radioactive hybridization probe, as discussed above. In addition, it is possible (with DNA polymerase) to form double-stranded (ds) DNA from the original single-stranded cDNA. This dsDNA can be inserted into an appropriate vector, amplified, and used to screen the library for the native gene. Since cDNA is complementary to the mRNA, comparative analysis of a gene and its transcript can be performed.

For many of the genes one wishes to clone, a convenient hybridization probe does not exist. An alternate procedure, which has proven useful in *S. cerevisiae*, is to complement mutants of *E. coli* with recombinant plasmids containing yeast DNA inserts. The desired clone would produce a functional yeast gene product that complements the mutation in *E. coli*. This strategy has been surprisingly successful with yeast, as Clarke and Carbon (1978) estimated that about 25% of the complementation experiments have worked. As an example, the yeast *ura3* gene (encoding orotidine 5′-phosphate decarboxylase) was selected for its ability to relieve the *pyrF* mutation in *E. coli* (Bach et al., 1979). Recently, it has become possible to perform similar complementation experiments using yeast as the host organism. The major discovery that advanced this system was the development of a transformation system in yeast (Hinnen et al., 1978; Struhl et al., 1979). Theoretically, it is now possible to isolate any yeast gene for which there exists a mutant yeast. This development also is important because now the expression of eukaryotic genes can be studied in a eukaryotic organism.

Rapid advances in techniques to analyze and modify genes have paralleled the advances in the isolation of genes. It is now possible not only to synthesize complete genes but also to sequence relatively large DNA molecules (~ 250 bases) in a few hours. Two recent volumes of *Methods in Enzymology* edited by Wu (1978) and Grossman and Moldave (1980) have described various recombinant DNA techniques in detail.

PROPERTIES OF THE FUNGAL GENOME

To better understand how nucleic acids might regulate differentiation, it is imperative to characterize the genomic DNA and to investigate how DNA is organized and transcribed in the fungal genome. The general properties of the fungal genome have been determined from vegetative cells; however, the general properties of DNA from vegetative cells and other developmental stages are probably identical.

The size of the haploid fungal genome ranges from about 0.5×10^{10} daltons in *Mucor rouxii* to 3×10^{10} daltons in *Coprinus lagopus* (Stork, 1974).

Some recent papers have reported values of 1.7×10^{10} daltons for *Aspergillus nidulans* (Timberlake, 1978) and *N. crassa* (Krumlauf and Marzluf, 1979), 2.3×10^{10} daltons for *Schizophyllum commune* (Dons et al., 1979), 2.7×10^{10} daltons for *Achlya bisexualis* (Hudspeth et al., 1977), and 0.9×10^{10} daltons for *S. cerevisiae* (Lauer et al., 1977). In yeast, the nuclear DNA is distributed among about 17 nuclear chromosomes. From sucrose density gradient analysis, the average mass of a chromosome is estimated to be 6×10^8 daltons (Petes and Fangman, 1972). Since this value is close to that predicted from the amount of nuclear DNA in yeast ($9.0 \times 10^9/17 = 5.3 \times 10^8$), it is assumed that each chromosome is composed of a single DNA molecule.

Chromatin Structure

The chromosomal DNA of higher plants and animals is associated with a class of proteins rich in basic amino acids called histones. In general, the histone group in most eukaryotic cells is comprised of five main proteins, termed H1, H2A, H2B, H3, and H4. The amount of histone protein present in the nucleus of higher eukaryotic cells is approximately equal to the weight of DNA. The proteins, except for H1, are present in approximately equimolar amounts.

For years, a controversy existed as to whether histones were present in fungal chromatin (Horgen and Silver, 1978; Morris, 1980); however, with the development of improved extraction procedures, it is clear that fungi contain histonelike proteins. The histones extracted from *A. nidulans* (Feldon et al., 1976) and *N. crassa* (Goff, 1976) are similar but probably not identical to those of higher eukaryotes.

Histones and DNA interact to form nucleoprotein particles that have repeating units 70-80 Å in diameter (~ 200 bp of DNA per unit) when viewed with the electron microscope. At low ionic strength, the nucleoprotein particles, or nucleosomes, appear as "beads on a string." The nucleosomes are composed of two molecules each of H2A, H2B, H3, and H4 and a "string" of 40-60 bp of DNA complexed with one molecule of H1 (Kornberg, 1977). This basic chromatin structure exists in fungi with minor variations. The basic repeat unit in fungi is 150-160 bp long, instead of 200 bp. However, the length of DNA associated with the nucleosome particles is 140 bp. Thus the major differences between fungi and higher eukaryotes must lie in the length of the spacer unit between particles (10-20 bp versus 40-60 bp). Nonhistone acidic proteins are also associated with eukaryotic chromatin; however, their function is not known.

Chromosomal Organization

The chromosomal DNA of several eukaryotes has been analyzed by DNA-DNA and DNA-RNA reassociation techniques. These experiments have shown that the genomes of higher eukaryotic organisms are comprised of both unique (single-copy) and reiterated DNA components. The unique DNA contains the majority of the sequences coding for mRNA, whereas the reiterated component contains some structural genes (histones) as well as genes for rRNA and tRNA (Lewin, 1980). The amount of reiterated DNA in higher eukaryotic organisms is 50% or more, whereas in fungi the amount is much less. For example, 16% of the DNA in *A. bisexualis* (Hudspeth et al., 1977), 8% in *N. crassa* (Krumlauf and Marzluf, 1979), 7% in *S. commune* (Dons et al., 1979; Ullrich et al., 1980), 5% in *S. cerevisiae* (Lauer et al., 1977), and 2-3% in *A. nidulans* (Timberlake, 1978) are reiterated.

The organization of the unique and repetitive DNA in higher eukaryotes appears to follow two general patterns. The predominant pattern, found in all major animal phyla and flowering plants, is characterized by the interspersion of the repetitive sequences throughout the genome. The relatively short repetitive sequences (200-400 bp in length) are linked to single-copy DNA at intervals averaging 1000-2000 bp (Davidson et al., 1975). This arrangement is called the "short-period interspersion pattern." Since the repetitive DNA exists in a relatively highly ordered pattern, it has been posutlated that this DNA may play a central role in the coordinate regulation of sets of unlinked genes (Davidson and Britton, 1979). However, not all organisms display the short-period interspersion pattern. Crain et al. (1976) found that the repetitive DNA in *Drosophila* is organized in long stretches (4000-10,000 bp in length) that are interspersed with single-copy DNA at intervals of at least 10,000 bp. This pattern of DNA organization is termed the "long-period interspersion pattern."

Fungal DNAs are not organized in either of the above patterns. The repetitive and single-copy DNA sequences in *A. bisexualis* contain 2.7×10^4 and 1.35×10^5 bp, respectively (Hudspeth et al., 1977). These values are approximately five times longer than those estimated for *Drosophila* (Manning et al., 1975; Crain et al., 1976). The values for the repetitive and single-copy DNA sequence lengths are estimated at 1.1×10^4 and 2.6×10^7 bp, respectively, in *A. nidulans* (Timberlake, 1978), and at $4.7-5.0 \times 10^4$ (Dons et al., 1979; Ullrich et al., 1980) and 2.9×10^7 bp (Dons et al., 1979), respectively, in *S. commune*. Similarly, the repetitive DNA sequence length in *N. crassa* has been estimated to be 7.3×10^4 bp long (Krumlauf and Marzluf, 1979). Since only 8% of the *N. crassa* genome (2.7×10^7 bp) is repetitive DNA, there can be at most 30 such sequences in the entire genome. Likewise, the maximum number of repetitive sequences in *Achyla*, *A. nidulans,* and *S. commune* is 250, 60, and 50, respectively.

Experiments designed to determine the location of the repetitive sequences indicate that these sequences are not distributed randomly throughout the genome. For example, only 3.6% of the *A. bisexualis* single-copy DNA is contiguous to repetitive sequences at a 5000-bp fragment length, and only 4% of the *N. crassa* single-copy DNA is linked to repetitive DNA at fragment lengths up to 10,200 bp. Therefore the repetitive DNA must be organized in very long clusters and is not interspersed throughout the genome. Other reassociation experiments using either rRNA or cloned rDNA show that possibly all of the reitereated DNA in *A. nidulans* (Timberlake, 1978) and about 90% of the reiterated DNA in *N. crassa* (Krumlauf and Marzluf, 1980) code for rRNA. Thus fungi, which exhibit coordinate control of unlinked genes, do so in the absence of extensive amounts of reiterated DNA. The current models for gene regulation (Davidson and Britton, 1979) may not apply to fungi. It should be noted that the limits of detection of reiterated DNA in the DNA-DNA reassociation experiments are about 18-25 bp; that is, if fungi contain reiterated DNA of less than 18 bp in length interspersed with single-copy DNA, it would not be detected.

The number of potential genes in fungi can be estimated from the amount of single-copy DNA present, assuming that the average gene is about 1200 bp long. The *Achyla* genome has the potential to encode about 30,000 structural genes (Hudspeth et al., 1977), *A. nidulans* about 22,000 (Timberlake, 1978), *N. crassa* about 18,000 (Krumlauf and Marzluf, 1979), and *S. commune* about 25,000 (Dons et al., 1979; Ullrich et al., 1980). During vegetative growth, not all of these potential genes are expressed. *Achyla* contains about 1900 different mRNAs (Timberlake et al., 1977), *N. crassa* about 2000 (Wong and Marzluf, 1980), and *S. commune* about 13,500 (Zantinge et al., 1981). Of the different genes expressed in fungi, not all are expressed to the same extent. In general, a few genes (20-30) are expressed 200-1000 times per cell, 200-400 genes are expressed 20-150 times per cell, and the remaining genes are expressed 1-11 times per cell (Hereford and Rosbash, 1977). This general pattern of gene expression is found in both *Achyla* and *N. crassa*. For example, *Achyla* contains 29 different genes expressed 1000 times per cell, 220 genes expressed 140 times per cell, and 3000 genes expressed 11 times per cell (Rozek et al., 1978). Wong and Marzluf (1980) determined that *N. crassa* mRNA is also distributed into three abundance classes: a high-abundance, low-complexity class containing 10 different mRNA species; a moderate-abundance class containing 150 different mRNAs; and a low-abundance, high-complexity class of 1800 mRNAs. The average number of copies per cell of the high-, medium-, and low-abundance classes of mRNA is estimated to be 75, 12, and 1, respectively. The pattern of mRNA complexity is slightly different in *S. commune*. Zantinge et al. (1981) found only medium- and low-abundance classes, that is, 600 different mRNA sequences expressed 100 times per monokaryon and 13,250 different sequences expressed about 5 times per cell.

Transcription

Most types of RNA in eukaryotic organisms undergo some type of posttranscriptional modification. For example, the initial rRNA transcript in yeast is a large RNA molecule ($\sim$ 35 S), which contains the 5.8 S, 18 S, and 26 S rRNA sequences. This large precursor rRNA is processed (cleaved) to yield the mature rRNAs (Petes, 1980; Long and Dawid, 1980). The initial transcripts of many of the yeast tRNAs are about 100 nucleotides long. These precursor tRNAs are cleaved to about 75 nucleotides in length, and some bases are modified to yield the mature tRNAs (Johnson et al., 1980). Similarly, the yeast 5 S RNA is derived from a precursor RNA that is about 10 nucleotides larger (Tekamp et al., 1980).

The mRNA of higher eukaryotes also undergoes extensive posttranscriptional modifications. The primary transcripts of many eukaryotic genes (termed heterogeneous nuclear RNA or HnRNA) are large RNA molecules that may be several times the size of the mature mRNA. The HnRNAs contain sequences (introns) that do not appear in the mature mRNA, interspersed with sequences (exons) that form the functional mRNA. Thus the HnRNA must be processed to remove the introns and rejoin the exons to form a functional mRNA (Crick, 1979). The sequences in the genes that encode the introns are commonly called intervening sequences.

Fungal mRNAs differ from those of higher eukaryotic organisms in that they do not appear to be derived from large HnRNA precursors (Timberlake et al., 1977). If the average fungal mRNA is derived from a larger precursor molecule, this primary transcript is at most 10-20% larger than the mature mRNA (Firtel and Lodish, 1973; Freer et al., 1977). Recently, however, Gallowitz and Sures (1980) have sequenced the *S. cerevisiae* actin gene, which was previously cloned with pBr322 (Gallowitz and Seidel, 1980). The DNA sequence data show that the actin gene contains a 304-bp intervening sequence. From this one example, it is too early to determine whether intervening sequences will be a common occurrence in the nuclear genes of fungi. It should be noted that intervening sequences have been found in yeast mitochondrial genes (Perlman et al., 1980).

Two other posttranscriptional processing events occur in the formation of functional mRNA. Adenylate residues are added to the 3'-hydroxyl end of the RNA. As many as 100 adenylate residues are added to the RNAs, giving rise to the polyadenylate (A) segment. The polyA segment is rapidly processed to 30-50 nucleotides in the mature mRNA (Freer et al., 1977). The second modification that mRNAs undergo is the addition of a 7-methylguanosine triphosphate to the 5' end of the mRNA (cap structure). The cap structure has been identified in mRNAs from *N. crassa* and *S. cerevisiae* (Banerjee, 1980) and presumably is present in all fungi.

APPLICATION OF RECOMBINANT DNA TECHNOLOGY TO FUNGAL SYSTEMS

To date, there has been limited application of recombinant DNA technologies to filamentous fungi. The rRNA genes from various fungi have been cloned and/or analyzed by restriction endonuclease mapping-hybridization probe techniques. I will not discuss the results from these experiments, since they have recently been reviewed by Russell and Wilkerson (1980). Instead, I will discuss in moderate detail other results obtained from using recombinant DNA techniques with filamentous fungi.

Fungal Differentiation

Fungal sporulation and germination have been studied extensively because these processes are of fundamental interest to mycologists and represent model systems for studying differentiation. As mentioned previously, developmental biologists hypothesize that the major control of differentiation is due to differential expression of genes. It follows, therefore, that if mRNA or protein populations are compared between different developmental stages, differences will be observed. The experimental approach most commonly employed in the past was to pulse-label the protein population of the various cell types with radioactive amino acids, separate the in vivo labeled proteins by either one- or two-dimensional polyacrylamide gel electrophoresis, and visualize the peptides by autoradiography. This approach has had limited success. Hopper et al. (1974) and Trew et al. (1979) were unable to detect proteins specific to yeast sporulation. This same strategy applied to fungal spore germination has been more successful, since changes in protein populations during germination were detected (Freer and Van Etten, 1978; Van Etten et al., 1981).

It has been shown that the differentiated state of animals and higher plant cells is characterized by distinct mRNA sequence sets (Galau et al., 1976; Kamalay and Goldberg, 1979). Recently, this has also been shown in *A. nidulans* (Timberlake, 1980). Through RNA-cDNA reassociation analysis, it was determined that vegetative cells of *A. nidulans* contain 5600-6000 diverse mRNAs. Furthermore, approximately 1300 diverse mRNAs not present in vegetative cells were detected in conidia or sporulating cultures. Of these, about 300 were present only in conidia. The remainder were present specifically during sporulation, but were absent from spores. Timberlake (1980) also showed that the majority of the mRNAs present in vegetative cells were also present in conidiating cultures. Thus cell differentiation in *A. nidulans* involves the synthesis of new mRNAs and not the loss of preexisting ones.

In further work, Zimmerman et al. (1980) successfully isolated developmentally regulated *A. nidulans* sporulation genes. To achieve this, *A. nidulans* nuclear DNA was treated with various amounts of the restriction endonuclease EcoR1 to generate random complete and partial digestion products. The DNA fragments greater than 12 kbp were ligated into λ phage, and the recombinant λ phage DNA was packaged in vitro and amplified by infection and growth in *E. coli.* Approximately 95% of the phages were recombinants, and greater than 95% of the *A. nidulans* genome was randomly represented in the library.

To screen the library for developmental specific genes, a cDNA probe was prepared via the "cascade hybridization" technique (Timberlake, 1980). Poly-A(+) RNA (mRNA) was isolated from vegetative cells and condiating cultures. The conidiating cultures contain vegetative cells as well as conidiophores and mature conidia; thus the mRNA population from these cultures contains essentially all of the diverse mRNAs produced by *A. nidulans*. Next, [^{32}P] cDNA was prepared from the conidiating culture mRNA. To enrich for the developmental sequences in the cDNA preparation, the vegetative sequences were selectively removed by repeated hybridization of the cDNA to mass excesses of vegetative mRNA. After each hybridization, the unreacted cDNA was purified by chromatography on hydroxylapatite. The cDNA remaining unreacted after three challenges with vegetative mRNA was then hybridized with mRNA from sporulating cultures. The cDNA that bound to hydroxylapatite was purified and used to screen the λ *A. nidulans* library for developmental sequences by the plaque hybridization technique (Benton and Davis, 1977).

A total of 37 randomly selected λ clones that reacted with the [^{32}P] cDNA probe were further screened with cDNA prepared from the mRNA fraction from either vegetative cells, conidiating cultures, or mature conidia. Only one clone gave a strong signal with vegetative cell cDNA, whereas about 5% of the clones gave weak signals with all three probes. Approximately 95% of the clones gave moderate to strong signals with developmental cDNA and weak or undetectable signals with vegetative cell cDNA. Further analysis enabled the identification of clones that encode mRNAs which preferentially accumulate in spores, whereas others encode mRNAs that are regulated in development but absent from or present in low levels in mature conidia. Thus conidiation in *A. nidulans* is most certainly controlled by the differential expression of selected genes.

Several of the clones containing *A. nidulans* developmental DNA inserts were analyzed further to determine the number of coding regions present and to estimate the extent to which the regulated mRNAs change cellular concentration during conidiation. PolyA(+) RNA from vegetative and conidiating cultures was electrophoretically fractionated and transferred to diazophenylthioether paper (Alwine et al., 1977). Radioactive DNA probes prepared from the individual recombinant λ phage were purified and used to probe the fractionated mRNA. It was shown that one of the clones that contained only spore-specific sequences encoded five mRNAs; that is, the probe hybridized to five mRNAs of different molecular weights. Furthermore, the authors were able to estimate that the developmentally regulated mRNAs were 10-100 times more prevalent in differentiating cells than in vegetative cells.

Assuming that the developmentally regulated genes are randomly dispersed within the genome, only 1 in 300 clones containing at least one spore sequence would be expected to contain two additional ones by chance association. The probability for clones containing more than three sequences is much lower. Thus the regulated coding sequence in *A. nidulans* appear to be arranged in linked sets (Zimmerman et al., 1980).

Neurospora crassa Cloning System

From the above example, one can see the elegant experiments and exciting results obtained from using the λ phage cloning system in conjunction with hybridization probe techniques. The other major cloning system, that is, the plasmid-vector system, has also been used with filamentous fungi with some equally interesting results.

The inducible quinic acid (*qa*) catabolic pathway of *N. crassa* is composed of a four-gene cluster. Three of the loci, *qa-2, qa-3,* and *qa-4*, encode the structural genes for catabolic dehydroquinase, quinate dehydrogenase, and dehydroshikimate dehydrase, respectively. The fourth gene, *qa-1*, encodes a regulatory protein which, in conjunction with the inducer, quinic acid, controls the expression of the three structural genes. The genetic order has been established as *qa-1, qa-3, qa-4, qa-2*, and the *qa* cluster is tightly linked to the *me-7* gene (Case et al., 1979). The catabolic dehydroquinase, *qa-2*, gene has been shown to be able to complement amino acid auxotrophs deficient in the biosynthetic dehydroquinase isoenzyme (*arom-9* mutants) (Giles and Case, 1975). Thus Vapnek et al. (1977) postulated that the *qa-2* gene should also complement an *aroD*⁻ mutant in *E. coli*. The *E. coli aroD* gene product catalyzes the same biosynthetic reaction as the *N. crassa arom-9* gene product.

To test the hypothesis, *N. crassa* and *E. coli* plasmid *pBr322* DNAs were purified, mixed together, and fragmented with the Hind III restriction endonuclease. The endonuclease was inactivated by heating the reaction mixture to 65°C, and the DNA fragments were randomly ligated together with phage T4 ligase. This DNA preparation, which contained *N. crassa* sequences inserted into pBr322, was used to transform *aroD*⁻ *E. coli* mutants. The complementation of the *aroD*⁻ mutation was screened for by growth of *E. coli* on minimal media. Vapnek et al. (1977) were able to isolate an *E. coli* transformant which had the properties expected of a strain carrying a *pBr322* recombinant plasmid that complemented the chromosomal *aroD* mutation.

Further analysis of this strain showed that the complementation of the *aroD* mutation was indeed plasmid carried. The enzyme extracted from strains containing the recombinant plasmid was identical to the *N. crassa qa-2* genetic product by the criteria of the heat stability, purification characteristics, and immunological reactivity. The molecular weight of the cloned enzyme was 220,000 daltons, identical to *N. crassa* catabolic dehydroquinase; the *E. coli* biosynthetic enzyme has a molecular weight of approximately 40,000 daltons. Furthermore, since the *N. crassa* enzyme is composed of about 20 subunits of 10,000 daltons, assembly of eukaryotic subunits must have taken place in *E. coli*.

A transformation system utilizing *N. crassa* as the host has recently been developed (Case et al., 1979). The *N. crassa* strain used as a recipient in these experiments contained both a *qa-2* and an *arom-9* mutation. This double mutant lacks both the catabolic and biosynthetic dehydroquinase activities and is thus unable to grow on minimal media without aromatic amino acid supplementation. The transforming DNA used was the above *pBr322* recombinant plasmid that contains the *N. crassa qa-2* gene (*pVK88*). Expression of the *pVK88 qa-2* gene will enable the *N. crassa* double mutant to grow on minimal media. Spheroplasts of the *N. crassa* mutant were prepared with the β-glucuronidase enzyme, *pVK88* DNA was transformed into the cells with the aid of polyethylene glycol 4000, and the cells were plated onto a minimal regeneration media. As many as 30 transformants per microgram of input DNA were isolated. Enzyme assays showed the presence of catabolic dehydroquinase.

Tetrad analysis of transformed isolates indicated that the inheritance of the *qa-2*$^+$ phenotype is Mendelian, not cytoplasmic. Three types of transformants—replacement, linked insertion, and unlinked duplication—were observed from a sample of 14 independently occurring transformants. In replacement types, the *qa-2*$^+$ phenotype is closely linked to the *me-7* gene, and the activities of the three *qa* enzymes are present at normal levels. No bacterial sequences were detectable; thus the *qa-2*$^-$ gene in the recipient has apparently been replaced by the *qa-2*$^+$ gene from *pVK88*. The linked insertion-type transformants also show normal activity of the *qa-2* enzyme and close linkage of the *qa-2*$^+$ phenotype and the *me-7* gene; however, the *qa-4* enzyme is absent. Thus the *pVK88 qa-2*$^+$ gene appears to have recombined into the *N. crassa qa-2* locus in an abnormal manner such that the *qa-2* locus is active but the *qa-4* gene is inactivated. The duplication-type transformants have normal enzyme activities for the three *qa* enzymes, but the *qa-2*$^+$ gene is unlinked to the *me-7* gene. This type of transformant suggests that there are *N. crassa* DNA sequences located in the *qa* region carried on *pVK88* that are reiterated in the genome and permit insertion at more than one site.

POTENTIAL AREAS OF INVESTIGATION

Researchers have begun to apply some of the recombinant DNA techniques to the study of fungal development, yet our understanding of the processes that control differentiation in filamentous fungi is still meager. With the cloning of developmental genes, our knowledge in this area should increase dramatically. By comparing the DNA sequences of genes encoding developmental and constitutive enzymes, it may be possible to deduce that portion of the gene which controls expression. Besides the sporulation and germination systems, sexual compatibility in filamentous fungi should be an excellent system for studying genetic regulation. In these systems, either two or four genes control differentiation. These systems have the added benefit that within the fungi the basic requirement of having sexual genes prevails; however, in certain fungi the sexual compatibility genes must be different (heterothallic), whereas others require identical mating-type loci to sexually reproduce (homothallic). The various mechanisms by which these genes control sexual development in fungi will prove to be fascinating.

Fungi are noted for their variability. In filamentous fungi it is common for such characteristics as antibiotic production, mycotoxin production, and virulence to become attenuated after several transfers of a culture in the laboratory. Such variations are strain specific and usually occur without any other observable alterations in the fungus. The cause of this variability, to the best of my knowledge, is unknown. A factor that has hindered the solution to these problems is that many of the organisms do not have a sexual stage. If, however, either all or part of the genes responsible for these characteristics in the wild-type strain could be cloned, they could be used as hybridization probes to determine the cause of the attenuation. If the variation is the result of either the loss of a gene or the movement of a gene within a chromosome or between chromosomes, by treating the nuclear DNA with various restriction enzymes and following a specific gene with the hybridization probes the cause of the variations might be determined.

The unique properties that enable certain fungi to invade and colonize specific plants are largely unknown. The newer molecular biological techniques might lend themselves well to the study of these phenomena. If the "pathogenicity" genes from a particular organism could be cloned, the functions of the genetic products encoded by these genes might be determined. Also, if the genes responsible for virulence were cloned, one could possibly discover what determines the virulence of an organism. The studies would have much value in and by themselves; however, they could also lead to effective biological control of plant diseases. For example, if organisms cloned in the laboratory were identical to the parents except that attenuated virulence genes were introduced, these organism might be capable of competing with the more virulent parent organism in the field and thus be able to reduce disease severity.

The applications of recombinant DNA technology that could be of both scientific and economic value are limitless. One commercial application might be to increase the yield of a particular genetic product. Chimeric plasmids consisting of the yeast *ura3* gene and part or all of the yeast 2-μm plasmid linked to a bacterial plasmid were constructed. When this chimeric plasmid was used to transform *ura3⁻* yeast, the specific activity of orotidine 5′-monophosphate decarboxylase (coded by *ura3*) was 10- to 30-fold higher than in the wild type (Gerbaud et al., 1979). This significant increase in enzyme activity may be partly due to a genetic dosage effect, since there are about 100 copies of the plasmid per cell.

Another application would be to try to improve the rate of synthesis of a particular end product. If the pathway for a particular product and the rate-limiting enzymatic step are known, by cloning this enzyme and thus increasing the internal concentration of the enzyme the pathway should function at a more rapid rate. The same strategy should also work if the substrate concentration is the rate-limiting factor, that is, clone the gene responsible for the rate-limiting enzyme in the substrate biosynthetic pathway and thus increase the substrate concentration.

An alternate approach to increasing the rate of product formation or the amount of product formed is to alter the enzyme. If a cloned gene can be specifically altered, a mutant enzyme with a lower K_m, higher V_{max}, altered regulatory sequence, or broader substrate range could be tailored. Thus the same final results as mentioned above could be achieved without putting additional demands upon an organism's capacity to synthesize protein. The techniques involved in "site-directed mutation" are currently being developed in various laboratories.

Another potential use for recombinant DNA technology is the specific "tailoring" of an organism by transferring a particular activity to a more desirable host. Assume that a microbe produces a desirable enzyme but has a slow growth rate. If the gene for this enzyme could be cloned and expressed in a host with a faster growth rate, the commercial production of the enzyme might become more favorable economically. In addition, assuming that all things are equal, it is less expensive to harvest a mycelial fungus via filtration than to harvest a bacterium via centrifugation. Thus if the gene for a product could be introduced and expressed in a filamentous fungus rather than in a bacterium, the final cost of the product should be reduced owing to reduced recovery/removal costs.

An alternate means of reducing product cost is to use inexpensive substrates as carbon/energy sources. If the amylase, "lignase," and/or cellulase complex genes were cloned into an industrial microbe, the range of growth substrates might be expanded, thus allowing the utilization of renewable substrates (wood chips, newpaper, corn stover, etc.). Several groups are now attempting to clone the cellulase complex into *S. cerevisiae* to produce ethanol from cellulosic wastes. If this cloning is achieved, it should eventually reduce the cost of ethanol and allow utilization of an underutilized resource. Once cloned, these genes might be transferable to any industrial microbe, thus reducing the substrate costs of many fermentations.

Secondary metabolites are noted for their unique structures and chemical diversity. Lactones, quinones, coumarins, and naphthalenes are a few of the classes of compounds that have been found in secondary metabolites. Secondary metabolites typically exist as members of closely related chemical families which differ slightly, such as in the degree of methylation, hydroxylation, and so on. The application of recombinant DNA technology to secondary metabolic systems might be extremely rewarding. For example, the basic structures of many secondary metabolites are believed to be synthesized via head-to-tail condensations of acetyl and/or malonyl groups. If the genes encoding the condensation enzyme(s) could be cloned and expressed in an organism that lacks the modifying enzymes, the potential exists to produce simple, unmodified ring compounds that may be useful as chemical feedstock. Alternately, the modification enzymes are of potential value. Assume that an antibiotic is produced via fermentation and that for maximum activity the antibiotic must first be purified and then chemically methylated. If a "methylase gene" from a different organism were cloned into the antibiotic producer, the potential exists for the host organisms to produce a methylated antibiotic. Because of the myriad of diverse chemical structures that exist among secondary metabolites, I am sure that fungi contain many unique and potentially useful enzymes. If the genes encoding these enzymes can be isolated and controlled in defined systems, the potential exists to generate many new and useful chemicals, in addition to gaining insight into the processes regulating secondary metabolism.

CONCLUSION

Scientists are just starting to apply recombinant DNA techniques to filamentous fungal systems. In the next few years, more single genes will be cloned and analyzed in detail. With this, our understanding of the basic processes involved in differentiation and gene expression will increase dramatically. In addition, because of the unique enzyme activities that fungi process, many new and novel compounds and processes may be developed.

ACKNOWLEDGMENTS

I thank James Van Etten and his colleagues for providing a reprint of their review article which will appear in *The Fungal Spore: Morphogenetic Controls* (Academic, New York).

REFERENCES

Alwine, J. L., Kemp, D. J., and Stark, G. R. (1977). Method for detection of specific RNAs in agarose gels by transfer to diazobenzyloxymethyl paper and hybridization with DNA probes. Proc. Nat. Acad. Sci. USA *74*: 5350-5354.

Bach, M. -L., Lacroute, F., and Botstein, D. (1979). Evidence for transcriptional regulation of orotidine-5'-phosphate decarboxylase in yeast by hybridization of mRNA to the yeast structural gene cloned in *Escherichia coli*. Proc. Nat. Acad. Sci. USA *76*:386-390.

Banerjee, A. K. (1980). 5'-Terminal cap structure in eukaryotic messenger ribonucleic acids. Microbiol. Rev. *44*:175-205.

Bell, G. I., DeGennaro, L. J., Gelfand, D. H., Bishop, R. J., Valenzuela, P., and Rutter, W. J. (1977). Ribosomal RNA genes of *Saccharomyces cerevisiae*. J. Biol. Chem. *252*:8118-8125.

Benton, W. D., and Davis, R. W. (1977). Screening λgt recombinant clones by hybridization to single plaques in situ. Science *196*:180-182.

Brambl, R., Dunkle, L. D., and Van Etten, J. L. (1978). Nucleic acids and protein synthesis during fungal spore germination. In *The Filamentous Fungi*, Vol. 3, J. E. Smith and D. R. Berry (Eds.). Edward Arnold, London, pp. 94-118.

Case, M. E., Schweizer, M., Kushner, S. R., and Giles, N. H. (1979). Efficient transformation of *Neurospora crassa* by utilizing hybrid plasmid DNA. Proc. Nat. Acad. Sci. USA *76*:5259-5263.

Cihlar, R. L., and Sypherd, P. S. (1980). The organization of the ribosomal RNA genes in the fungus *Mucor racemosus*. Nucleic Acids Res. *8*:793-804.

Clarke, L., and Carbon, J. (1976). A colony bank containing synthetic Col E1 hybrid plasmids representeative of the entire *E. coli* genome. Cell *9*:91-99.

Clarke, L., and Carbon, J. (1978). Functional expression of cloned yeast DNA in *Escherichia coli*: Specific complementation of argininosuccinate lyase (arg H) mutations. J. Mol. Biol. *120*:517-532.

Crain, W. R., Eden, F. C., Pearson, W. R., Davidson, E. H., and Britton, R. J. (1976). Absence of short period interspersion of repetitive and nonrepetitive sequences in the DNA of *Drosophila melanogaster*. Chromosoma *56*: 309-326.

Crick, F. (1979). Split genes and RNA splicing. Science *204*:264-271.

Davidson, E. H., and Britton, R. J. (1979). Regulation of gene expression: Possible role of repetitive sequences. Science *204*:1052-1059.

Davidson, E. H., Galau, G. A., and Angerer, R. C., and Britton, R. J. (1975). Comparative aspects of DNA organization in metazoa. Chromosoma *51*: 253-260.

Dons, J. J. M., De Vries, O. M. H., and Wessels, J. G. H. (1979). Characterization of the genome of the basidiomycete *Schizophyllum commune*. Biochim. Biophys. Acta *563*:100-112.

Feldon, R. A., Sanders, M. M., and Morris, R. N. (1976). Presence of histones in *Aspergillus nidulans*. J. Cell Biol. *68*:430-439.

Firtel, R. A., and Lodish, H. F. (1973). A small nuclear precursor of messenger RNA in the cellular slime mold *Dictyostelium discoideum*. J. Mol. Biol. *79*: 295-314.
Free, S. J., Rice, P. W., and Metzenberg, R. L. (1979). Arrangement of the genes coding for ribosomal ribonucleic acids in *Neurospora crassa*. J. Bacteriol. *137*:1219-1226.
Freer, S. N., and Van Etten, J. L. (1978). Changes in messenger RNAs and protein synthesis during germination of *Rhizopus stolonifer* sporangiospores. Exp. Mycol. *2*:313-325.
Freer, S. N., Mayama, M., and Van Etten, J. L. (1977). Synthesis of polyadenylate-containing RNA during germination of *Rhizopus stolonifer* sporangiospores. Exp. Mycol. *1*:116-127.
Galau, G. A., Klein, W. H., Davis, M. M., Wold, B. J., Britton, R. J., and Davidson, E. H. (1976). Structural gene sets active in embryos and adult tissues of the sea urchin. Cell *7*:487-505.
Gallowitz, D., and Seidel, R. (1980). Molecular cloning of the actin gene from yeast *Saccharomyces cerevisiae*. Nucleic Acids Res. *8*:1043-1059.
Gallowitz, D., and Sures, I. (1980). Structure of a split yeast gene: Complete nucleotide sequence of the actin gene in *Saccharomyces cerevisiae*. Proc. Nat. Acad. Sci. USA *77*:2546-2550.
Gerbaud, C., Fournier, P., Blanc, H., Aigle, M., Heslot, A., and Guerineau, M. (1979). High frequency of yeast transformation by plasmids carrying part or entire 2-μm yeast plasmid. Gene *5*:233-253.
Giles, N. H., and Case, M. E. (1975). Two pairs of isozymes in the aromatic biosynthetic and catabolic pathways in *Neurospora crassa*. In *Isozymes*, Vol. 2, C. L. Market (Ed.). Academic, New York, pp. 865-876.
Goff, C. G. (1976). Histones of *Neurospora crassa*. J. Biol. Chem. *251*:4131-4138.
Grossman, L., and Moldave, K. (Eds.) (1980). *Nucleic Acids, Part I, Methods in Enzymology*, Vol. 65. Academic Press, New York, 968 pp.
Grunstein, M., and Hogness, D. S. (1975). Colony hybridization: A method for the isolation of cloned DNAs that contain a specific gene. Proc. Nat. Acad. Sci. USA *72*:3961-3965.
Haber, J. E., Wejksnora, P. J., Wygal, D. D., and Lai, E. Y. (1977). Controls of sporulation in *Saccharomyces cerevisiae*. In *Eukaryotic Microbes as Model Developmental Systems*, D. H. O'Day and P. A. Horgen (Eds.). Marcel Dekker, New York, pp. 129-154.
Hereford, L. M., and Rosbash, M. (1977). Number and distribution of polyadenylated RNA sequences in yeast. Cell *10*:453-462.
Hinnen, A., Hicks, J. B., and Fink, G. R. (1978). Transformation of yeast. Proc. Nat. Acad. Sci. USA *75*:1929-1933.
Hopper, A. K., Magee, P. T., Welch, S. K., Friedman, M., and Hall, B. D. (1974). Macromolecular synthesis and breakdown in relation to sporulation and meiosis in yeast. J. Bacteriol. *119*:619-628.
Horgen, P. A., and Silver, J. C. (1978). Chromatin in eukaryotic microbes. Annu. Rev. Microbiol. *32*:249-284.

Hudspeth, M. E. S., Timberlake, W. E., and Goldberg, R. B. (1977). DNA sequence organization in the water mold *Achlya*. Proc. Nat. Acad. Sci. USA *74*:4332-4336.
Johnson, J. D., Ogden, R., Johnson, P., Abelson, J., Dembeck, P., and Itakura, K. (1980). Transcription and processing of a yeast tRNA gene containing a modified intervening sequence. Proc. Nat. Acad. Sci. USA *77*:2564-2568.
Kamalay, J. C., and Goldberg, R. B. (1979). Regulation of structural gene expression in tobacco. Cell *19*:935-946.
Kornberg, R. D. (1977). Structure of chromatin. Annu. Rev. Biochem. *46*:931-954.
Krumlauf, R., and Marzluf, G. A. (1979). Characterization of the sequence complexity and organization of the *Neurospora crassa* genome. Biochemistry *18*:3705-3713.
Krumlauf, R., and Marzluf, G. A. (1980). Genome organization and characterization of the repetitive and inverted repeat DNA sequences in *Neurospora crassa*. J. Biol. Chem. *255*:1138-1145.
Lauer, G. D., Roberts, T. M., and Klotz, L. C. (1977). Determination of the nuclear content of *Saccharomyces cerevisiae* and implications for the organization of DNA in yeast chromosomes. J. Mol. Biol. *114*:507-526.
Lewin, B. (1980). *Gene Expression,* Vol. 2, Wiley, New York, pp. 488-492.
Long, E. O., and Dawid, I. B. (1980). Repeated genes in eukaryotes. Annu. Rev. Biochem. *49*:727-764.
Lovett, J. S. (1975). Growth and differentiation of the water mold *Blastocladiella emersonii*: Cytodifferentiation and the role of ribonucleic acid and protein synthesis. Bacteriol. Rev. *39*:345-404.
Lovett, J. S. (1976). Regulation of protein metabolism during spore germination. In *The Fungal Spore: Form and Function,* D. J. Weber and W. M. Hess (Eds.). Wiley, New York, pp. 189-242.
Lovett, J. S., Gong, C. -S., and Johnson, S. A. (1977). Zoospore germination and early development of *Blastocladiella emersonii.* In *Eukaryotic Microbes as Model Developmental Systems,* D. H. O'Day and P. A. Horgen (Eds.). Marcel Dekker, New York, pp. 402-424.
Manning, J. E., Schmidt, C. W., and Davidson, N. (1975). Interspersion of repetitive and nonrepetitive DNA sequences in the *Drosophila melanogaster* genome. Cell *4*:141-156.
Morris, R. N. (1980). Chromosome structure and the molecular biology of mitosis in eukaryotic micro-organisms. In *The Eukaryotic Microbial Cell,* G. W. Gooday, D. Lloyd, and A. P. J. Trinci (Eds.). Cambridge University, Cambridge, pp. 41-76.
Olson, M. V., Hall, B. D., Cameron, J. R., and Davis, R. W. (1979). Cloning of the yeast tyrosine transfer RNA genes in bacteriophage lambda. J. Mol. Biol. *127*:285-296.
Perlman, P. S., Alexander, N. J., Hanson, D. K., and Mahlar, H. R. (1980). Mosaic genes in yeast mitochondria. In *Gene Structure and Expression,* D. H. Dean, L. F. Johnson, P. C. Kimball, and P. S. Perlman (Eds.). Ohio State University, Columbus, pp. 211-253.

Petes, T. D. (1980). Molecular genetics of yeast. Annu. Rev. Biochem. *49*:845-876.
Petes, T. D., and Fangman, W. L. (1972). Sedimentation properties of yeast chromosomal DNA. Proc. Nat. Acad. Sci. USA *69*:1188-1191.
Rozek, C. E., Orr, W. C., and Timberlake, W. E. (1978). Diversity and abundance of polyadenylated RNA from *Achlya ambisexualis*. Biochemistry *17*: 716-722.
Russell, P. J., and Wilkerson, W. M. (1980). The structure and biosynthesis of fungal cytoplasmic ribosomes. Exp. Mycol. *4*:281-337.
Schmit, J. C., and Brody, S. (1976). Biochemical genetics of *Neurospora crassa* conidial germination. Bacteriol. Rev. *40*:1-41.
Stork, R. (1974). Molecular mycology. In *Molecular Microbiology*, J. B. G. Kwasinski (Ed.). Wiley, New York, pp. 423-477.
Struhl, K., Stinchcomb, D. T., Scherer, S., and Davis, R. W. (1979). High-frequency transformation of yeast: Autonomous replication of hybrid DNA molecules. Proc. Nat. Acad. Sci. USA *76*:1035-1039.
Tekamp, P. A., Garcea, R. L., and Rutter, W. J. (1980). Transcription and *in vitro* processing of yeasts 5S rRNA. J. Biol. Chem. *255*:9501-9506.
Timberlake, W. E. (1978). Low repetitive DNA content in *Aspergillus nidulans*. Science *202*:973-975.
Timberlake, W. E. (1980). Developmental gene regulation in *Aspergillus nidulans*. Dev. Biol. *78*:497-510.
Timberlake, W. E., Shumard, D. S., and Goldberg, R. B. (1977). Relationship between nuclear and polysomal RNA populations of *Achlya*: A simple eukaryotic system. Cell *10*:623-632.
Trew, B. J., Friesen, J. D., and Moens, P. B. (1979). Two-dimensional protein patterns during growth and sporulation in *Saccharomyces cerevisiae*. J. Bacteriol. *138*:60-69.
Ullrich, R. C., Droms, K. A., Doyon, J. D., and Specht, C. A. (1980). Characterization of DNA from the basidiomycete *Schizophillum commune*. Exp. Mycol. *4*:123-134.
Van Etten, J. L., Dunkle, L. D., and Knight, R. H. (1976). Nucleic acids and fungal spore germination. In *The Fungal Spore: Form and Function*, D. J. Weber and W. M. Hess (Eds.). Wiley, New York, pp. 243-300.
Van Etten, J. L., Dunkle, L. D., and Freer, S. N. (1977). Germination of *Rhizopus stolonifer* sporangiospores. In *Eukaryotic Microbes as Model Developmental Systems*, D. H. O'Day and P. A. Horgen (Eds.). Marcel Dekker, New York, pp. 372-401.
Van Etten, J. L., Dahlberg, K. R., and Russo, G. M. (1981). Nucleic acids. In *The Fungal Spore: Morphogenetic Controls*, G. Turian and H. R. Hohl (Eds.). Academic, New York, pp. 277-299.
Vapnek, D., Hautala, J. A., Jacobson, J. W., Giles, N. H., and Kushner, S. R. (1977). Expression in *Escherichia coli* K-12 of the structural gene for catabolic dehydroquinase of *Neurospora crassa*. Proc. Nat. Acad. Sci. USA *74*: 3508-3512.

Wong, L. -J. C., and Marzluf, G. A. (1980). Sequence complexity and abundance classes of nuclear and polysomal polyadenylated RNA in *Neurospora crassa*. Biochim. Biophys. Acta *607*:122-135.

Wu, R. (Ed.) (1979). *Recombinant DNA. Methods in Enzymology,* Vol. 68. Academic Press, New York, 555 pp.

Zantinge, B., Hoge, J. H. C., and Wessels, J. G. H. (1981). Frequency and diversity of RNA sequences in different cell types of the fungus *Schizophyllum commune*. Eur. J. Biochem. *113*:381-389.

Zimmerman, C. R., Orr, W. C., Leclerc, R. F., Barnard, E. C., and Timberlake, W. E. (1980). Molecular cloning and selectin of genes regulated in *Aspergillus* development. Cell *21*:709-715.

8
Controls for Development and Differentiation of the Dikaryon in Basidiomycetes

Carlene Allen Raper / Wellesley College, Wellesley, Massachusetts

INTRODUCTION

As a developmental system in eukaryotes the differentiation of the specialized heterokaryon known as the dikaryon in higher fungi has been considered a relatively simple and, in some ways, ideal object for the study of factors controlling development. The dikaryon, capable of independent and indefinite vegetative propagation, yet also capable of fruiting, is unique to this group of organisms. In fact, the predominant life-style of basidiomycetes in nature appears to be the dikaryon, in which haploid nuclei of two compatible types associate and divide conjugately in pairs, generally one pair per cell, over long periods of time. Days, months, even years may intervene between fertilization via hyphal fusion (plasmogamy) and nuclear fusion (karyogamy), which is immediately followed by meiosis in the basidium of the fruiting body and the production of sexual spores. It is this unique feature of basidiomycetes that permits a study of the effects of two different genomes, as carried by separate nuclei, upon one another within a common cytoplasm. The two can be put together in various paired combinations and subsequently isolated from one another, thus allowing comparison of their expressions together and separately, before, during, and after their association with one another. This adds an intriguing dimension to the investigation of dikaryosis as a developmental system.

In this chapter, I focus upon those factors that are known to be associated with the development and maintenance of the dikaryon while placing particular emphasis upon the controlling elements involved. Only those studies relevant to this subject will be considered. In view of the fact that most of the information on the development of the dikaryon (sometimes called sexual morphogenesis) has been obtained through intensive studies of only a few representative basidiomycetes—principally the homobasidiomycete *Schizophyllum commune*, but also certain species of *Coprinus*—I will concentrate on these fungi almost exclusively, with only occasional reference to the results of relevant research on other forms.

Since the discovery of the dikaryon by Bensaude (1918) and Kniep (1920) the many factors involved in its development have been studied intensively during the past three decades by a variety of investigators. The studies have produced information at several levels: morphological, cytological, genetic, physiological, biochemical, and molecular. In so far as possible, I will attempt to integrate this information with the aim of trying to understand the principal elements of control in the process. Material treated comprehensively in previously published review articles will be described only briefly, with references to those articles. Other, more recent information, concerning, for example, molecular aspects of the system and the genetic control of nuclear behavior in heterokaryons, will be considered in greater detail here. Some speculation as to the meaning of the knowledge now available will be attempted, with a strong appreciation of the numerous gaps that remain in our understanding of this subject.

FACTORS INVOLVED IN DIFFERENTIATION OF THE DIKARYON

Morphogenesis

Within the Life Cycle

Beyond the three basic types of life cycle known for other organisms, haploid, diploid, and haplodiploid, the basidiomycetes have, characteristically and uniquely, a fourth type known as a haplodikaryotic cycle. The dikaryophase, characterized by the stable association of paired haploid nuclei, is the counterpart of a vegetative diplophase and, by and large, the genetic and physiological equivalent of a diploid. This fourth type of life cycle has haploid and dikaryotic phases of varying proportions preceding a single cell generation stage in which the nucleus is diploid just prior to meiosis.

Dikaryosis is a specialiced form of heterokaryosis. Heterokaryosis, the association of distinct genomes carried by separate nuclei in a common cytoplasm, is known to occur regularly only in the fungi. The indefinite form, in which the ratio of component nuclear types is indefinite and changeable, is found in most fungi; the specialized form, dikaryosis, is genetically balanced, in that each cell of the mycelium contains two different haploid nuclei, and the nuclear types throughout the mycelium are maintained in a strict ratio of 1:1. Although the dikaryon occurs in ascomycetes, within the tissue of the ascus, just prior to karyogamy and meiosis, it is incapable of indefinite vegetative propagation in this class of fungi—the dikaryotic cells of ascomycetes are obligate parasites on the homokaryotic mycelium of the maternal parent. Only in basidiomycetes is the dikaryon capable of indefinite propagation, during which two haploid genomes can functionally interact over extended periods of time while maintaining their identities in separate nuclei.

The Dikaryon in Basidiomycetes

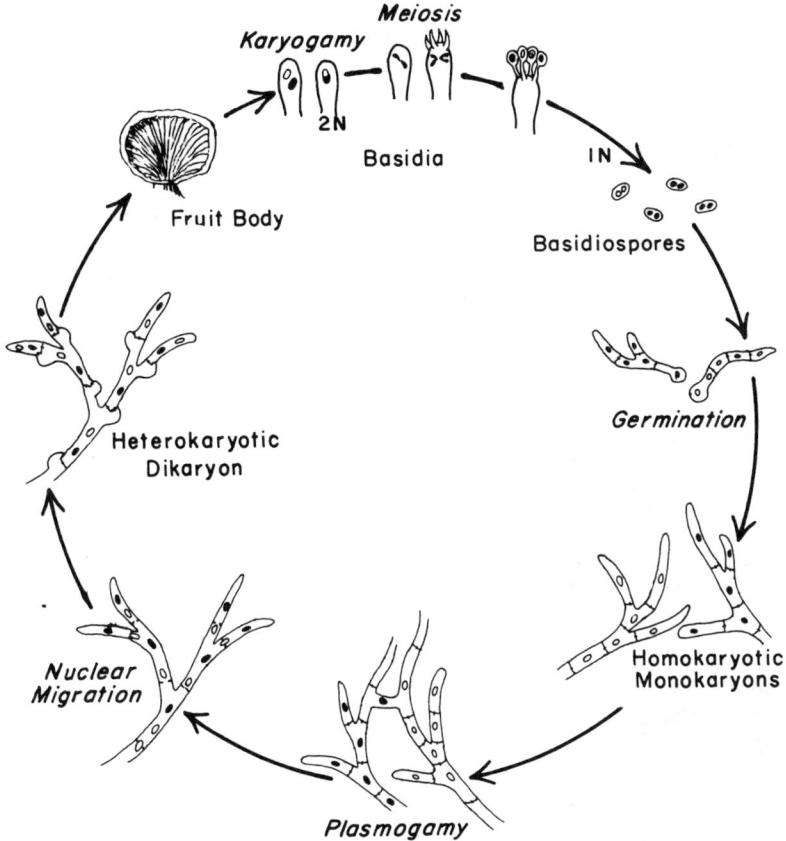

Figure 1   Life cycle of a typical basidiomycete such as *S. commune*.

The position of the dikaryon in the life cycle of the basidiomycete is indicated in Fig. 1. Figure 1 illustrates the cycle of a typical member of the subclass Hymenomycetes. In the majority of higher basidiomycetes examined, haploid basidiospores germinate to produce haploid homokaryotic mycelia with uninucleate cells known as monokaryons. The nuclei in a given germling are genetically identical, having descended, through mitosis, from a single postmeiotic nucleus. The homokaryotic monokaryon, capable of indefinite vegetative propagation, mates by hyphal fusion (plasmogamy) with another compatible monokaryon to establish the dikaryon, which is also capable of indefinite vegetative propagation. The transition from mated monokaryons to the dikaryon normally occurs through the reciprocal migration of nuclei from each mate into and throughout the hyphae of the other. Under appropriate environmental conditions, the mature dikaryon is induced to produce fruiting bodies containing basidial cells in which karyogamy, meiosis, and spore formation occur in immediate succession.

The series of events as described here is not the same for all species. Variations, including asexual spore formation, homokaryons with multinucleate cells, an imprecisely structured dikaryon, and heterokaryosis in basidiospore formation, are found elsewhere (for a review, see Raper, 1966a; Raper, 1978a,b; Casselton, 1978; Arita, 1979). Nevertheless, for our purposes, this definition of life cycle for a typical basidiomycete will suffice as an appropriate point of reference.

As is apparent from Fig. 1, changes in morphology occur throughout the life cycle. We will concentrate on those associated with the development of the dikaryon from the homokaryotic state.

General Scheme of Morphogenesis

Morphogenesis of the typical dikaryon of basidiomycetes has been described numerous times (see Raper, 1966b; Raper, J. R., and Raper, C. A., 1973; and Raper, 1978a, for reviews). Essentially, the change in morphology is from a homokaryotic mycelium composed of uninucleate cells and unadorned septa to a dikaryotic mycelium composed of binucleate cells with a fused hook cell (clamp connection) at each septum. The intervening steps in the establishment of the dikaryon from its initiation via hyphal fusion between two compatible monokaryons are listed and illustrated graphically in Fig. 2 and are briefly described as follows:

1. Nuclear migration. Nuclei of each mate penetrate and migrate throughout the preestablished mycelium of the other.
2. Nuclear pairing. When the invading nuclei reach apical cells, a migrant nucleus ultimately pairs with a resident nucleus, usually centrally, in each apical cell.
3. Hook cell formation. Upon cell enlargement, a basipetally directed outgrowth occurs adjacent to the paired nuclei.
4. Conjugate division. The two nuclei divide synchronously, one along the hyphal axis and the other into the lateral hook.
5. Hook cell septation. Two septa, each separating daughter nuclei, are laid down simultaneously, one perpendicular to the hyphal axis and one across the base of the hook, thus creating three new cells: the new binucleate apical cell, the uninucleate hook cell, and a uninucleate subapical cell.
6. Hook cell fusion. The tip of the hook cell then fuses with the subapical cell and the temporarily entrapped nucleus quickly moves to reestablish the dikaryon in the subapical cell, thus restoring the exact 1:1 ratio of parental types in each new hyphal cell.

The genetic control of this sequence of events will be considered later.

The Dikaryon in Basidiomycetes

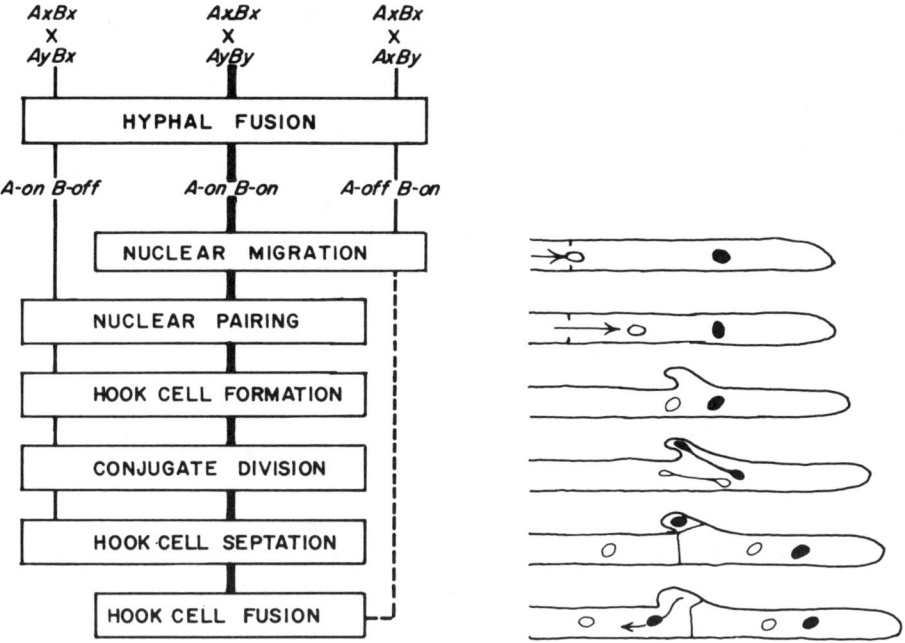

Figure 2 General scheme of morphogenesis in the establishment of the dikaryon and its regulation by the A and B incompatibility factors in matings between monokaryons of a typical basidiomycete such as *S. commune*.

The Cytology of Morphogenesis

The cytological details of the developing dikaryon have not resolved the mystery of nuclear migration, but they do provide some clues as to intracellular and intercellular changes that accompany this process.

The septum of basidiomycetes, with its central dolipore complex, serves as a barrier to the passage of nuclei. Normally, in the monokaryon, a septum separating daughter nuclei forms immediately after mitosis; hence each cell, following cytokinesis, contains one nucleus only. Nevertheless, cytoplasmic continuity between cells is maintained by small perforations in the parenthosomes surrounding the central swelling in the dolipore (Moore and McAlear, 1961; Bracker and Butler, 1963; Girbardt, 1965). An early event in the process of nuclear migration, following hyphal anastamosis between compatible strains, is the dissolution of this septal barrier, usually starting with the dolipore and its encompassing parenthosomes (Giesy and Day, 1965; Jersild et al., 1967; Koltin and Flexer, 1969; Marchant and Wessels, 1973, 1974; Mayfield, 1974).

Observations of migrating nuclei by phase microscopy of living cells in the vicinity of early fusions between compatible hyphae have been difficult to make. This, perhaps, results from the problem of identifying rare early anastomoses and/or the lack of much nuclear movement for long periods (several hours) following the initial fusion. Those few such cytological studies that have been made present a confusing picture that is difficult to interpret. Lange (1966) observed disrupted septa, nonsynchronous nuclear division, and irregular, oscillatory movement of nuclei within several cells adjacent to anastamosed, compatible hyphae over a period of 12 hr in *Polystictus versicolor*. Bistis (1970) followed nuclear behavior in hyphal cells receiving nuclei from germinated oidia of *Clitocybe truncicola*. His observations indicate both cytoplasmic and septal disruption as well as nuclear division in the many cells along the pathway of invading nuclei as they proceed toward the hyphal apex. Niederpruem (1980a) could discern no nuclear movement in the vicinity of newly formed hyphal fusions within 3 hr after such anastamoses in *S. commune*. He did, however, observe nuclear movements in so-called "migration hyphae" in regions 20-40 cells behind their hyphal apices, yet some distance away from initial hyphal anastamosis between mates. The cells of these regions, many of which were anucleated, contained many vacuoles and were bounded by incomplete septa. About 10-20 anucleated cells were traced toward the hyphal apex, where aggregates of 3-12 nuclei per cell were seen. These nuclear aggregates divided synchronously and, subsequently, nuclear divisions were preceded by hook cell formation. Ultimately, dikaryotic cells developed from the division of multikaryotic cells through nuclar reduction by various means involving (1) permanent entrapment of nuclei in the hook cells, (2) the movement of nuclei into branches emanating from hook cells, or (3) the unequal distribution of nuclear progeny in cytokinesis. Bistis (1970) also noted an irregular distribution of nuclei and abnormal hook cell formation prior to the establishment, at the hyphal apex, of regularly reproducing dikaryotic cells with normal fused hook cells otherwise known as clamp connections.

The analysis by electron microscopy of fixed hyphal cells in which nuclei were migrating, that is, "migration hyphae," revealed the presence of fibrillar bundles approximately 300 nm in diameter, each composed of many fibrils 10 nm in diameter (Raudaskoski, 1972, 1973; Raudaskoski and Koltin, 1973). The bundles extended as much as 270 nm throughout and/or across cells. The fibrils appeared to terminate at a point of electron-dense material. Microtubules, relatively frequent in migration hyphae, were seen closely associated with the membranes of interphase nuclei. Both microfibrils and microtubules were seen to extend across degraded intercellular septa. Microtubules have also been observed in association with intracellular nuclear movement following conjugate division in the dikaryon (Girbardt, 1968).

The involvement of microtubules in nuclear migration is also implicated indirectly by results from studies on the effects of inhibitors of microtubule protein assembly. The microtubule inhibitors podophyllotoxin and two of its derivatives, in concentrations that did not inhibit growth, delayed the establishment of the dikaryon (as signaled by the appearance of hook cells) by up to 8 hr, or by about 20%, in matings between compatible strains of *Schizophyllum* (Omerod et al., 1976). Omerod and co-workers suggested that this delay reflects a preferential inhibitory effect on microtubule assembly in the process of nuclear migration prior to the establishment of the dikaryon, as distinct from an effect on microtubule assembly in the process of nuclear division. The microtubule inhibitor griseofulvin was also shown by these authors to cause some delay in the establishment of the dikaryon, although the effects were erratic. Raudaskoski and Huttunen (1977) demonstrated that griseofulvin caused alterations of nuclear positioning, with a consequent disruption of nuclear distribution in the dikaryon of *Schizophyllum.* They attributed the nuclear displacement to a disruption of nuclear migration associated with the maintenance of the dikaryon before, during, and after nuclear division and suggested that this is due to abnormal function of microtubules.

These cytological studies on living and fixed material clearly document the dissolution of septa as a requisite to nuclear migration. They implicate nuclear division, vacuolization, and the formation of microfibrils and microtubules as concomitants of the process. It is speculated that the latter may serve to orient and move the migrating nuclei toward the hyphal apices. The primary motive force of nuclear migration, however, remains a mystery.

Time Course

A consideration of the time course of nuclear migration leads to the conclusion that this process must involve an oriented, independent movement of invading nuclei. The rate of nuclear migration, as measured principally by the sampling of mated mycelia for the expression of genetically tagged invading nuclei (Snider and Raper, 1958; Prevost, 1962; Ross, 1976), but also by direct observations through phase microscopy (Niederpruem, 1980a), far exceeds that of hyphal growth. In *Schizophyllum,* rates range from 500 to 3000 μm/hr, exceeding the calculated rate for hyphal growth, about 100 μm/hr, by 5-30 times. The independence of nuclear migration from hyphal growth is further indicated by the difference in their respective temperature coefficients, which for growth is 2, as compared to 6 for nuclear migration (Snider and Raper, 1958). The all-time record of 40,000 μm/hr for the rate of nuclear migration was found by Ross (1976) in *Coprinus congregatus.*

Niederpruem (1980a) observed two types of nuclear migration in live preparations of *Schizophyllum*: one associated with pulsating cytoplasm and directed primarily toward the hyphal apex, with an occasional backward movement; and another, more commonly observed process in which nuclei move in the apical direction not only independently of cytoplasmic flow, but also independently of the movement of other visible organelles. The speed of migration ranged from 400 to 2600 μm/hr, and migrating nuclei in interphase were seen to traverse several cells with disrupted septa. The velocity of migration associated with cytoplasmic flow exceeded the growth rate 10- to 20-fold. The velocity of migration of the independent type varied from as fast as that associated with cytoplasmic flow to a much slower rate approaching that of hyphal growth. Both types, involving a range of speeds, occurred successively for the same nucleus, over an extended period and distance, in a given hypha.

The lack of any very extensive exchange of cytoplasm following mating also indicates that cytoplasmic flow cannot account for all or even a major part of nuclear migration. Casselton and Condit (1972) demonstrated this with genetically marked mitochondria in *Coprinus cinereus*. Watrud and Ellingboe (1973) found evidence for the transfer of cobalt-stained mitochondria from donor to recipient cells as far as 1.2 cm from the position of the donor mycelium in compatible and some incompatible matings of *Schizophyllum*. The transfer of these cytoplasmic organelles, however, was limited and appeared to be independent of nuclear migration, thus suggesting that the two events are unrelated.

The overall process of nuclear migration, from the time of hyphal anastamosis to the conversion of both mates to stable dikaryons, may take from 2 to 5 days, depending upon the specific strains involved and their cultural conditions.

The role of nuclear division in this process is not known, although nuclear division has been observed in so-called migration hyphae. Nuclear movement immediately following mitosis appears to be associated with microtubules and its speed is comparable to that of migrating nuclei (Girbardt, 1968). Normally, nuclear division requires 6-12 min and occurs in actively growing hyphal tip cells at intervals of about every 1-2 hr (Girbardt, 1960; Niederpruem et al., 1971). Whether or not the forces orienting and moving daughter nuclei immediately following mitosis are comparable to those associated with migrating nuclei, nuclear division per se cannot account for nuclear migration.

The time course for the appearance and development of hook cells in the developing dikaryon of *S. commune* has been examined recently by Niederpruem (1980b). No hook cells were seen until about 40 hr after mating. Dikaryotic cells with clamp connections (fused hook cells) had developed on most of the mycelial periphery at about 2 days. A minority of peripheral hyphae were seen at this time to contain multinucleate cells bounded by septa with fused hook cells. This condition persisted through several rounds of synchronous nuclear division, accompanied by hook cell formation and fusion, until the number of nuclei was, by the processes of exclusion mentioned earlier, reduced to two per cell. The dikaryotic cells reproduced regularly as dikaryons thereafter. In all cases observed, the nuclei maintained a near-central position in the cell just prior to hook cell formation and division. The duration of the process from the beginning of hook cell development, through mitosis, to the fusion of the hook cell with the subapical cell was approximately 20-40 min. Girbardt (1968), with the aid of time-lapse cinemicrography, had previously timed the entire process of hook cell formation, nuclear division, and movement of postdivision nuclei at 20-30 min. in the dikaryon of *P. versicolor*. Correlated ultrastructural analysis revealed, in association with moving nuclei, abundant microtubules and a bilobed granular structure called the "kinetochore equivalent," in or near the nuclear envelope. Parag (1970), in observations of *S. commune*, timed the period from the beginning of hook cell formation to the beginning of nuclear division at 4 min, conjugate (synchronous) nuclear division and subsequent rapid movement of daughter nuclei away from each other at 6-7 min. and the formation of septa at about 7 min. The movement of daughter nuclei was clocked at 80 μm/min.

In conclusion, the development of the hook cell first occurs when invading nuclei approach apical hyphal cells, but may begin, in multikaryotic cells, several cell divisions before the establishment of the strictly structured dikaryon. In the strictly structured dikaryon, the beginning of hook cell development precedes mitosis, and mitosis is followed in succession by septal formation and hook cell fusion.

Genetic Controls

The "Master Switch Genes"

Nature and function: The genes responsible for initiating and regulating dikaryosis in basidiomycetes are termed incompatibility genes. Depending upon the species, there may be anywhere from one to four such genes, each capable of existing in any one of a number of alternate states. In the minority of basidiomycetes studied, the multiple alleles of one gene, located in the A incompatibility factor, control dikaryosis; in the majority of known basidiomycetes, the multiple alleles of four genes, located in pairs within each of two incompatibility factors, A and B, regulate dikaryosis. The A and B factors are unlinked, but the two genes within each factor, $A\alpha$ and $A\beta$, $B\alpha$ and $B\beta$ are linked. The functions and fine structure of the incompatibility genes have been studied most extensively in the bifactorial species *S. commune*; consequently, this form is examined in depth as an example of the other related forms.

Schizophyllum commune is typical of other basidiomycetes in the extensiveness of genetic polymorphism that exists at each of its incompatibility loci. In a worldwide sample of natural populations, the number of alleles that have been estimated to exist at the $A\alpha$ and $A\beta$ loci are 9 and 32, respectively; for $B\alpha$ and $B\beta$, the number of alleles are 9 and 9 (Raper et al., 1960; Koltin et al., 1967; Stamberg and Koltin, 1972). Matings in which the two partners are homoallelic for all four loci are incompatible ($A=B=$). Complete compatability leading to dikaryosis occurs only when the two mates differ in at least one of the genes of the A factor, $A\alpha$ or $A\beta$, and at least one of the genes of the B factor, $B\alpha$ or $B\beta$ ($A\neq B\neq$). Heteroallelism for both genes of both factors also results in complete compatibility. If the alleles are the same for each mate at both genes of one factor but different at one or both genes of the other factor, that is, $A=B\neq$ or $A\neq B=$, distinct hemicompatible interactions occur. These have served to identify those parts of the sequence of events leading to dikaryosis that are regulated separately by the A and B factors.

In general, genes of the A factor regulate pairing of invading with resident nuclei, hook cell formation, conjugate nuclear division, and hook cell septation following nuclear division. Genes of the B factor regulate nuclear migration and, in the presence of hook cells determined by the genes of the A factor, hook cell fusion (see Fig. 2). Neither the so-called A sequence nor B sequence alone is sufficient to establish a true dikaryon; both must operate to complete the sexual cycle through dikaryosis and formation of fruiting bodies and sexual spores (Raper, 1966b).

Fine structure: The complexity of the incompatibility genes is indicated not only by their pleiotropic functions and the multiplicity of their alleles, but also by the variety of discernable alterations achieved by mutation in at least one such gene (see below). Nevertheless, major attempts to resolve the incompatibility locus into physically separable functional parts through intragenic recombination have succeeded only in identifying that part of the B locus which is responsible for regulating migration of invading nuclei (Raper, C. A., and Raper, J. R., 1973; see also next section). Despite extensive attempts to obtain new alleles through intragenic recombination, none were found within a frequency of 10^{-6} (Raper, 1966b; Y. Koltin, personal communication).

Curiously enough, wild-type alleles of the $B\alpha$ and $B\beta$ series appear to be associated with deletions of various sizes. This was indicated by a recombinational analysis of $B\alpha$ and $B\beta$ alleles in all possible paired combinations (Stamberg and Koltin, 1971, 1973). Recombination in samples up to 5000 ranged from 8.9 to 0%. It was concluded from the pattern of this analysis that certain specific combinations of $B\alpha$ and $B\beta$ alleles are incapable of recombining with one another as a result of overlapping deletions. Stamberg and Koltin proposed as the simplest explanation of the results that the various wild-type alleles of both $B\alpha$ and $B\beta$ are intimately associated with deletions of various sizes which approach and possibly even overlap each other in the region between the two genes. The authors further postulated that such deletions may account for the origin of specific incompatibility alleles. Extensive attempts, however, to generate new alleles, comparable to wild type, through mutagenesis, have consistently failed, despite the fact that x-rays, which are known to cause chromosomal aberrations, including deletions, were used in most of these attempts (see next section). Nevertheless, recombinational analysis of wild-type alleles with a series of mutant alleles deficient in function and suspected of representing chromosomal damage in various degrees of severity has supported the idea that at least some of the wild-type alleles as well as some mutant alleles are associated with chromosomal aberrations, deletions, or possible inversions which interfere with recombination between $B\alpha$ and $B\beta$ loci (Stamberg et al., 1977; see also next section).

Mutational analysis: Mutations of the "master switch genes" have produced a variety of phenotypes in which regulation of dikaryosis is altered in numerous ways. In a comprehensive mutational analysis of the two genes of the *B* incompatibility factor in *Schizophyllum*, a selection process designed to detect mutation from a wild-type allele of one specificity to that of another has yielded only mutations constitutive for *B* factor function. The B sequence alone operates continually in such mutants in the absence of mating, and they are called primary or *A-off B-on* mutants, where *off* and *on* mean that the relevant sequence of morphogenesis is inoperative and operative, respectively (Parag, 1962; Koltin, 1968; Raudaskoski et al., 1976; Koltin et al., 1979). Such *A-off B-on* mutants resemble the *A=B≠* heterokaryon in morphology.

In the total of 34 constitutive mutations generated and located in the genes of the *B* factor, five of the nine known alleles in the $B\alpha$ series and six of the nine known alleles in the $B\beta$ series were mutated. A mutation in one or the other of the two loci is sufficient to "turn on" the events of the B sequence. The phenotypes of these constitutive mutations are essentially of two types: those that have lost their allelic specificities and those that have retained them. The former recognize their parental types as being different from themselves and are consequently capable of "turning on" the B sequence in matings with the parent progenitor; the latter recognize their parental types as being the same as themselves and cannot "turn on" the B sequence to initiate nuclear migration in the parent progenitor. The two types have been recovered in approximately equal frequencies, and the average frequency of induced mutation for both types does not exceed 4.3×10^{-8}. Most of the mutations were induced by 70-kV x-rays; a few were obtained with nitrogen mustard or acridine and one occurred spontaneously.

Similar mutations have been obtained in an *Aβ* allele of *Schizophyllum* (Raper et al., 1965) and in the *A* factor of *C. cinereus* (Day, 1960).

The failure to demonstrate the generation of new alleles comparable to those found in nature, either by intragenic recombination or mutagenesis, prompted an intensive effort to generate secondary mutations in primary mutant strains, with the prospect in mind that subsequent mutation might be the key to the derivation of new, fully functional alleles. The secondaries, the majority of which were induced in a single *Bβ* allele of *Schizophyllum* (Raper and Raudaskoski, 1968; Raudaskoski, 1970; C. A. Raper and J. R. Raper, 1973), occurred at an average frequency of 1.2×10^{-5}, a factor of about 300 greater than for primary mutations. The phenotypes ranged from a reversion to parental type, through a number of intermediate stages of functional impairment, to a complete absence of function for the whole *B* factor. (All of the secondary mutants were selected for a change from the *A-off B-on* to *A-off B-off* phenotype.) The majority expressed varying degrees of deficiency in regulating the B sequence when combined with wild-type alleles in matings. Among the deficiencies, the most common was the inability to accept migrating nuclei, but in some, the failure to donate nuclei was also evident. The deficiency in those mutants incapable of performing any B function at all extends through both loci, *Bα* as well as *Bβ*. On the basis of recombinational analysis, the latter is thought to represent a deletion of the entire *B* factor (C. A. Raper and J. R. Raper, 1973; Stamberg et al., 1977). In fact, recombinational analysis of several of the secondary B mutants implicate deletions of and between *Bα* and *Bβ*, with a positive correlation between the size of the deletion and the degree or impairment. Nevertheless, all mutants appear completely normal in their growth and morphology as isolated, unmated mycelia. This suggests that the interval between *Bα* and *Bβ* carries no genes that are indispensible to the vegetative survival of the organism. In no case was the equivalent of a new or different allele generated by secondary mutation. The mechanism of origin of the numerous alleles found in nature remains elusive.

An interpretation of the various mutations obtained in the *B* factor of *Schizophyllum* alone suggests a complex function for incompatibility genes in general. For a single locus of the factor, two major functions are apparent: one for the recognition of self versus nonself in allelic interactions and one for the initiation and regulation of the B sequence of morphogenesis. Those primary mutations that have constitutive function, that is, express the B sequence continually in the absence of mating, but retain specificity of the parental allele are interpreted as alterations in the regulatory region of the gene. Those primary mutations that not only have constitutive function but fail to recognize the parental allele as identical to themselves are thought to represent alterations affecting both regions of the gene. The fact that one of the latter, upon subsequent mutagenesis, reverted to parental type with respect to both nonconstitutive function and allelic specificity suggests that the primary mutation was at a point in the gene possibly affecting the secondary or tertiary structure of its product. Another of the secondary mutations derived from a constitutive mutant of this latter type reverted to nonconstitutive function, but remained altered in its specificity. It represents the closest approach yet to the generation of a new allele equivalent to those found in nature. Its expression differs from that of a natural allele in that the mutant is capable of donating, but not receiving, nuclei when mated with its grand progenitor, that is, the wild-type progenitor from which it was derived by a two-step process of mutagenesis. Therefore the specificity of this mutant is not entirely different from that of the original grand progenitor. The majority of secondary mutants, characterized by various qualities and degrees of regulatory disfunction in their interactions with other wild-type alleles, appear to represent changes in different parts of the regulatory portion of the incompatibility gene. On the basis of their phenotypes, one can postulate a subdivision of this regulatory region into three separately mutable parts: one for the acceptance of migrant nuclei, one for the donation of migrant nuclei, and one for the fusion of hook cells in established dikaryons. One of these functions, the acceptance of nuclei is apparently epistatic to that of hook cell fusion. Furthermore, the site specific for this particular function is separable by crossover from the other parts of the gene.

In view of these considerations, the following tentative structure for the $B\beta$ locus has been proposed: The gene may consist of two major parts—one variable region coding for allele specificity (Sp) and involved in the recognition phenomenon, and one constant region coding for function in the initiation and regulation of the B sequence of morphogenesis. The latter region is composed of at least three parts: *NA* for acceptance of migrant nuclei, *ND* for donation of migrant nuclei, and *HF* for hook cell fusion. The proposed arrangement is illustrated in Fig. 3.

This array of *B* mutations provides some clues about the overall action of the gene. The fact that the constitutive mutation (with *A-off B-on* phenotype) is dominant to all other *B* alleles, wild type as well as mutant, and the likelihood that the *B-always off* mutation is a deletion of the entire *B* factor suggest that the *B* gene operates via a positive control mechanism. In other words, the product of the gene normally functions in conjunction with the product of another gene of the same series, but of different specificity, to activate morphogenesis. The alternative possibility, that the gene by itself functions to repress an otherwise constitutive process of morphogenesis and that the interaction between alleles results in derepression, seems less likely. This interpretation, however, is only tentative. Our understanding of the mode of operation of the incompatibility genes remains highly speculative.

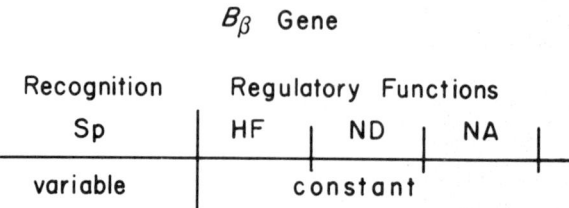

Figure 3 Proposed structure of the $B\beta$ locus in *S. commune* as derived by mutational analysis. The notation Sp stands for allele specificity; NA, nuclear accepatance in mating; ND, nuclear donation in mating; and HF, hook cell fusion. The latter three regions control specific parts of the B sequence of morphogenesis and are thought to be constant among $B\beta$ alleles. The first, Sp, is involved in a recognition phenomenon which activates the regulatory region in matings between two monokaryons carrying different allelic specificities. This region is thought to be variable among $B\beta$ alleles.

Genes Specifying Details of Differentiation

The epithet "master switch genes" for the genetic components of the incompatibility factors seems all the more appropriate in consideration of those other genes, scattered throughout the genome, that come to expression only when sexual morphogenesis is "switched on" by the incompatibility genes. The activities of these scattered loci, studied in their mutant state as modifiers of the normal course of morphogenesis, appear to be regulated by the activity of the incompatibility genes themselves. In the context of prokaryotic genetics, they are considered comparable to structural genes subject to regulation by regulatory elements situated elsewhere in the genome. Consequently they are viewed as secondary controls of the specific aspects of differentiation relevant to the development of the dikaryon. Of this complex, over 80 so-called "modifier mutations," representing 12 phenotypes with respect to specific effects on morphogenesis, have been analyzed. Some cause blocks or alterations to specific steps in the A sequence, for example, nuclear pairing or hook cell formation; others alter the B sequence by, for example, blocking septal breakdown and nuclear migration or inducing the abnormal development of hook cells; still others affect steps common to both sequences, such as hook cell fusion (C. A. Raper and J. R. Raper, 1964, 1966, and unpublished work).

Most of these loci specifying the details of morphogenesis are not linked to their regulating elements. An exception is a cluster of nine genes, expressed in mutated form as blocks to nuclear migration, which are linked by 10-20% recombination with the regulating genes of the B factor (Dubovoy, 1975). In addition to this cluster, three other genes affecting the B sequence, as well as several loci affecting the degree of recombination between the $B\alpha$ and $B\beta$ genes, are all linked to the B factor. The entire complex might be thought of as a kind of "super gene" with linked elements that are functionally related to one another (Raper, J. A., and Raper, C. A. 1973; Koltin and Stamberg, 1972).

Physiological and Biochemical Concomitants of Morphogenesis

Growth

A comparison of growth curves for unmated wild-type monokaryons, dikaryons, and the hemicompatible $A=B\neq$ heterokaryon with the indefinite ratio of component nuclear types reveals similar rates of growth, by weight, for monokaryons and dikaryons, but a relatively lower growth rate for the latter heterokaryon, in which only the B sequence is expressed (Rich and Deppe, 1976). The growth curves in all cases are typical of those for biological populations in general, in that they consist of three phases: an initial lag phase, a log phase, and a postexponential phase. Although the rate of growth during the log phase is comparable for the monokaryon and dikaryon, the timing of this phase is different, commencing and finishing somewhat earlier for the dikaryon than for the monokaryon. The log phase for the $A=B\neq$ heterokaryon starts earlier and ends earlier, with a rate about half that of the other two types. Growth rates for the homokaryotic mutant mimics of these heterokaryons and, in addition, for the mutant mimic of the $A\neq B=$ heterokaryon were also studied. These homokaryotic mutants, phenotypically comparable to the heterokaryons they mimic, carry constitutive mutations in the incompatibility genes of the A factor or B factor or both, resulting, accordingly in an A-on B-off phenotype, as in the $A\neq B=$ heterokaryon; A-off B-on, as in the $A=B\neq$ heterokaryon; and A-on B-on, as in the $A\neq B\neq$ dikaryon.

In the growth studies, each homokaryotic mutant mimic produced results comparable to those obtained with the relevant heterokaryon, A-off B-on or A-on B-on. In addition, the growth pattern of the A-on B-off mutant (a mimic of the $A\neq B=$ heterokaryon, which was not tested) was similar to that of the wild-type monokaryon and the dikaryon.

In summary, the timing of growth phase, the growth rate, and the resulting accumulated weight of material are comparable for all phenotypes except the A-off B-on type. For a given colony with an unlimited food supply, the rate of growth and the total accumulation of material possible for the A-off B-on phenotype is about half that of the other three types.

Enzymes Associated with Development

A physiological basis for the lower growth capacity of the A-off B-on phenotype has been sought in biochemical studies. Efficiency of growth for the A-off

B-on phenotype is far less than that for the other phenotypes (Hoffman and Raper, 1971). The molar growth of the *A-off B-on* mutant is about 10 times less than that of the *A-off B-off* wild-type strain, coisogenic for all but the *B* gene. The adenosine triphosphate (ATP) yield for the wild-type *A-off B-off* strain is close to the theoretical maximum of 36 mol ATP per mol glucose, as compared to about 3 mol ATP per mol glucose for the *A-off B-on* mutant. An analysis of the respiratory response of isolated mitochondria to ADP demonstrated a 50% reduction in the increase of respiratory rate in the *A-off B-on* phenotype, thus indicating at least a partial uncoupling of energy metabolism and a mitochondrial malfunction (Hoffman and Raper, 1972). Adenosine triphosphate production and adenosine triphosphatase *(ATPase)* activity in isolated mitochondria from the *A-off B-on* phenotype, however, showed the same normal values as for mitochondria from the *A-off B-off* phenotype (Hoffman and Raper, 1974). Furthermore, mycelial growth of both types was sensitive to N,N-dicyclohexylcarbodiimide (DCCD), which is thought to inhibit phosphorylation; this indicated the presence of an active coupling system in the *A-off B-on* phenotype. Nevertheless, the greater inhibition of growth by Krebs cycle intermediates of the *A-off B-on* mutant as compared to the wild type indicated an abnormal dependency of the mutant on glycolysis. A general uncoupling of energy efficiency was also implicated by the fact that the mutant had a growth yield on ethanol seemingly as good as that on glucose. From these mixed data, Hoffman and Raper suggested that the less efficient metabolism of the *A-on B-off* phenotype reflects a general, albeit partial, uncoupling of energy production from energy conservation that is possibly due to a cytoplasmic, extramitochondrial ATPase that does not release Pi. Alternatively, a malfunction localized in the mitochondria could account, at least partially, for this inefficiency.

Wessels (1978), on the basis of studies comparing accumulated cell constituents in different cell types, offered another possible explanation for the low growth efficiency of the *A-off B-on* phenotype. A comparison of cell constituents in the *A-off B-on* mutant with those of a coisogenic *A-off B-off* strain (a modified *A-off B-on* mutant containing another mutation that blocks the B sequence) showed that accumulation of glycogen, a specific glucan, and triglycerides is lower in the *A-off B-on* phenotype, whereas accumulation of protein, chitin, and polar lipids is about the same in both phenotypes. Furthermore, extracts of mycelia with the *A-off B-on* phenotype showed increased specific activities of a glucanase, amylase, and lipase, but not of chitinase or protease (Wessels, 1969; J. G. Wessels and D. J. Niederpruem, 1967, unpublished work). Biochemical and ultrastructural studies showed elevation in the activity of another lytic enzyme, acid phosphatase, in the *A-off B-on* phenotype (Raudaskoski, 1976; I. Charvat, personal communication). Wessels (1969) therefore suggested that normal or above-normal synthetic processes are accompanied by increased degradation of certain polymers synthesized in the *A-on B-off* phenotype, and that it is this relatively high rate of degradation by a number of elevated hydrolytic enzymes that accounts for the waste of energy and low molar growth yield characteristic of this phenotype. Perhaps the energy "wasted" with respect to growth is utilized instead for the process of nuclear migration.

The continual synthesis and dissolution of septa in the *A-off B-on* phenotype is a case in point. The septum in *Schizophyllum* differs from the hyphal wall in that it lacks the component known as S-glucan, a glucan with α-1,3 linkages that is soluble in alkali. Its components, chitin and R-glucan, a glucan with β-1,3 and β-1,6 linkages, insoluble in alkali, are components also of the hyphal wall (Sietsma et al., 1977). The R-glucan is susceptible to digestion by an enzyme complex known as R-glucanase, which is found in elevated amounts in the *A-off B-on* phenotype (Wessels, 1969; Wessels and Niederpruem, 1967; Wessels and Koltin, 1972). This same glucan was shown to be significantly more vulnerable to enzymatic degradation as a component of the cross-wall than as a component of the lateral wall (Janszen and Wessels, 1970; Wessels and Marchant, 1974). The digestion of R-glucan in the cross-wall results in sufficient dissolution of the septal apparatus to allow free passage of nuclei (see Wessels, 1978, for a review). Septal synthesis occurs concomitantly with septal breakdown in the *A-off B-on* phenotype (Niederpruem, 1971), and this, together with the phenomenon of continual nuclear migration, can account for the irregular distribution of component nuclei in this phenotype, as described by Raper (1966b).

The cross-walls of the *A-off B-off* and *A-off B-on* phenotypes were more resistant to dissolution by R-glucanase (and also chitinase) than were those of the *A-on B-on* phenotype (Wessels and Marchant, 1974). This suggests that somehow the A sequence, when *on*, alters the structure of the cross-walls in such a way as to protect them from enzymatic digestion. Even though activity of R-glucanase is very low in newly formed, actively growing dikaryons, it rises sharply when glucose is depleted in the culture medium (Wessels, 1966; Wessels and Niederpruem, 1967). The high level of R-glucanase, after catabolite repression ends, starts to degrade R-glucan in the walls of preformed mycelium while leaving the septa intact. The R-glucan in this context is thought to serve as a major reserve polysaccharide for the enlargement of the pilei in fruit body formation (Niederpruem and Wessels, 1969; Sietsma et al., 1977). The fact that the *A-on B-on* dikaryon can act as a donor but not as a receptor of compatible migrating nuclei in matings with *A-off B-off* monokaryons [in so-called "dimon" matings (Raper, 1966b)] could be due to the invulnerability of its cross-walls to enzymatic degradation by R-glucanase.

The whole process of establishing a dikaryon from the mating of two compatible monokaryons can be thought of as a dynamic one in which hyphal fusion is followed, within a few hours, by the recognition of alternate alleles between genes of the *B* factor to initiate the B sequence. The B sequence activates the elevation of specific hydrolytic enzymes, including R-glucanase, which is responsible for the dissolution of septa. It also "turns on" those processes associated with the migration of invading nuclei toward hyphal apices—among these are microtubule and microfibril assembly. This entire sequence is accompanied by low growth efficiency. Only after the invading nuclei reach the hyphal apices to establish a nearly equal ratio of nuclear types, some 40 hr later, is the A sequence initiated by allelic interaction of the *A* incompatibility genes. Among the first processes of the A sequence is catabolite repression of R-glucanase and the modification of newly synthesized septa so that they are no longer subject to degradation by R-glucanase. Continued operation of the A sequence reverses the high energy-consuming processes characteristic of the B sequence to restore a normal, energy-efficient growth pattern. This may be accomplished by repression of other hydrolytic enzymes such as amylase, lipase, and acid phosphatase. Also included in the operation of the A sequence is the pairing of nuclei of opposite types, their synchronous division, and hook cell formation and septation. A second, or late, part of the B sequence then comes into play in the fusion of hook cells. Henceforth the two sequences operate jointly to maintain the energy-efficient, highly structured heterokaryon known as the dikaryon.

Involvement of cAMP

Catabolite (glucose) repression of R-glucanase in the establishment of the dikaryon implicates cyclic nucleotides as mediators in this morphogenetic process. Attempts to demonstrate this have been somewhat frustrated by the failure of cells to take up cyclic nucleotides readily. Studies in which a combination of adenosine 3′,5′-cyclic monophosphate (cAMP) and theophylline (an inhibitor of phosphodiesterase, resulting in higher endogenous levels of cAMP) was applied exogenously to cells in various stages of development demonstrated an effect on development (P. Gladstone, unpublished results). Glucose repression of hook cell formation had been noted earlier in two mutant strains of *Schizophyllum*: one, a diploid carrying $A\neq B\neq$, and the other an *A-on B-off* strain in which the $A\beta$ allele was mutated to constitutive function for A sequence (P. Gladstone, 1972, 1973, unpublished results). Hook cell formation, a normal expression of *A-on* phenotypes, was expressed at irregular intervals in actively growing colonies of both these mutants, but this expression increased significantly upon depletion or removal of glucose in the medium. Gladstone demonstrated a greater and more consistent effect (up to a 50-fold increase in hook cell formation on glucose-fed colonies) by the exogenous application of cAMP (0.5-1 mM) in the presence of theophylline (1-10 mM), thus indicating that endogenous elevation of cAMP is required for the development of this particular event in morphogenesis. The effect of cAMP in overcoming glucose repression of hook cell formation was reversible. Induction of hook cells by cAMP required a period of 3 days, with maximum response at 9-13 days; reversion to septa without hook cells required longer, more irregular periods. Cyclic guanosine 5′-monophosphate (GMP) could not substitute for cAMP in this function, nor did it inhibit the cAMP effect in comparable concentrations. Cyclic AMP was not effective in inducing hook cell formation in the wild-type *A-off B-off* monokaryon, nor did Gladstone note any effect of cAMP on the *A-off B-on* or *A-on B-on* phenotypes. Endogenous cAMP was detected in all phenotypes in average amounts of 50 pmol/mg protein, but there is no certain information as to relative levels. Measurements, however, of intracellular cAMP levels [by a Gilman isotope dilution assay (Gilman, 1970)] in relation to the percentage of hook cells expressed in the *A-on B-off* mutant indicated a direct correlation between endogenous cAMP and hook cell formation.

Another, related study showed a similar effect of the methylxanthines theophylline and caffeine on hook cell formation in *A-on B-off* mutants and also an adverse effect on hook cell fusion in *A-on B-on* dikaryons (Dubovoy and Munoz, 1979). Effects of caffeine or theophylline were tested in concentrations of 1 mg/ml incorporated into a minimal medium containing glucose. In the presence of these inhibitors of phosphodiesterase, hook cell fusion to basipetal cells occurred in the incidence of only about 5%, as compared to virtually 100% in the absence of these compounds. The effect was also seen to be reversible. These methylxanthines appeared neither to affect the early part of the B sequence, nuclear migration, nor to trigger any developmental effect in wild-type *A-off B-off* monokaryons.

Both studies suggest a role for cAMP in sexual development as initiated by the genes of the *A* and *B* factors. Cyclic AMP alone, in the absence of an *A-on* and/or *B-on* trigger, cannot initiate any phase of development, but it appears to play some regulatory role within the A sequence (in hook cell formation) and within the late part of the B sequence (hook cell fusion) once these sequences are turned on. The specific effects of exogenously applied cAMP and methylxanthines indicate that the early part of the B sequence, nuclear migration, is unaffected by cAMP. Endogenous levels may play a role, however, in view of the fact that R-glucanase, which is necessary for nuclear migration, is glucose repressible. The normal operation of the A sequence and late B sequence does seem to require appropriate levels of cAMP; relatively high levels for hook cell formation and possibly lower levels for hook cell fusion. It is of some interest to note that cAMP was shown to induce fruiting in *Coprinus macrorhizus* (Uno and Ishikawa, 1973), but this has not been demonstrated in *Schizophyllum*, despite attempts to do so.

It is tempting to speculate that certain of the mutations that modify secondary controls of sexual morphogenesis may represent alterations in the regulation of or by cAMP. For example, those that block hook cell fusion in the *A-on*

B-off or *A-on B-on* genotypes may be relatively low in cAMP or cAMP receptor protein (CRP) or possibly insensitive to a cAMP-CRP complex; those that induce hook cell formation in *A-off B-on* and simultaneously block hook cell fusion in the *A-on B-on* genotypes may have a relatively high level of cAMP or CRP; and that common class of modifiers that block nuclear migration and which have been shown to be low in R-glucanase in the *A-off B-on* genotype (Wessels and Koltin, 1972) may be low in cAMP, and so on. Such speculations remain to be tested.

Molecular Corollaries

Total Nuclear Genome

The genome size of *S. commune*, as determined by reassociation analysis of sheared nuclear DNA, is 22.8×10^9 daltons (d), representing approximately 3.7×10^7 nucleotide pairs (NTP) which is about 8.8 times the size of the genome of *Escherichia coli* (Dons et al., 1979; Ullrich et al., 1980a). This size is well within the range of 10^7-10^8 NTP for other fungi analyzed (Ohja and Dutta, 1978; Dusenbery, 1975; Hudspeth et al., 1977; Ohja et al., 1977). The repetitive fraction of DNA in *Schizophyllum* represents only about 7-10% of the genome, with a reptition frequency of 49-50. This falls within the lower end of the range found for other fungi, which is 2-20% of the genome (Dutta, 1974; Timberlake, 1978). The repetitive sequences in *Schizophyllum* are not interspersed among the single-copy sequences (Dons et al., 1979; Ullrich et al., 1980b) as is the case in many animals and higher plants (Davidson, 1977). Most of the relatively small amount of repetitive DNA found in *Schizophyllum* could be accounted for by those sequences coding for ribosomal RNA which were shown to require about 6% of the total nuclear genome (Dons and Wessels, 1980). The remainder might consist of a small number of reiterated genes such as the histone gene.

The G+C content of nuclear DNA, calculated from its buoyant density and melting profile, is 57%, as contrasted to 22-25% for mitochondrial DNA (Dons et al., 1979; Ullrich et al., 1980a). On the basis of cesium chloride profiles and melting patterns, mitochondrial DNA appears to contain interspersed (A+T)-rich sequences. The DNA in the two genomes, nuclear and mitochondrial, therefore, are distinct and separable. Total DNA from the dikaryon was found to be indistinguishable from that of its component monokaryons on the basis of reassociation kinetics (Ullrich et al., 1980a).

The nuclear DNA of *Schizophyllum* is packaged into the haploid number of 11 chromosomes (Carmi et al., 1978). This was determined by three-dimensional reconstructions of synaptonemal complexes as seen by electron microscopy (EM) in the pachytene stage of meiosis. Earlier studies, principally by light microscopy, variously reported four and eight chromosomes (Haapala and Nien-

The Dikaryon in Basidiomycetes

stedt, 1976; Radu et al., 1974), and seven linkage groups have been defined (Frankel and Ellingboe, 1977). In view of the small size of this fungal nucleus, however, the latest, most comprehensive EM analysis is taken as the most reliable.

Protein Profiles of Differentiation

Synthesis of distinct proteins during sexual morphogenesis from the monokaryon to the dikaryon is implied by a two-dimensional gel-electrophoretic study of the pattern of proteins associated with distinct stages of this differentiative process (de Vries et al., 1980). A comparison of protein profiles in coisogenic strains of *A-off B-off, A-on B-off, A-off B-on,* and *A-on B-on* phenotypes revealed differences of up to 7% among the 700 or so proteins detectable after pulse labeling with [^{35}S] methionine. Coisogenic monokaryons (*A-off B-off*) differed from each other by only 2%, whereas these same two monokaryons and the dikaryon (*A-on B-on*) derived from them differed by 6.6-7.7%. Most of these differences fell into two categories of nearly equal size: proteins specifically absent ("switched off") and proteins specifically present ("switched on") in the *A-on B-on* as compared to the *A-off B-off* state of morphogenesis. The 22 "switched-on" proteins unique to the *A-on B-on* phenotype were, on the average, slightly larger and more acidic than the 20 "switched-off" proteins found only in the *A-off B-off* monokaryons. Surprisingly, an extensive overlap in protein profiles was found in the *A-on B-off* and *A-off B-on* phenotypes (mutant mimics of the $A{\neq}B{=}$ and $A{=}B{\neq}$ heterokaryons, respectively). These two phenotypes, although morphologically and physiologically very distinct, appear to share all but about 1% of the proteins analyzed, and all but two of these were found in either the monokaryon or the dikaryon. Furthermore, the two *A-on B-on* phenotypes, one as represented by the $A{\neq}B{\neq}$ heterokaryotic dikaryon and one as represented by the homokaryotic dikaryon in which both *A* and *B* had been mutated to constitutive function, had a 98% similarity instead of the expected 100%.

In interpreting these results, one must take into consideration the fact that quantitative differences were not noted with certainty, and that only rapidly synthesized sulfur-containing polypeptides with an apparent molecular weight between 17,000 and 130,000 and an isoelectric point between pH 5 and 7 were seen. The actual number of polypeptides present in *Schizophyllum* must be much higher than the approximate 700 examined. Analysis of the frequency and diversity of messenger RNA sequences (see below) indicates that this organism is capable of coding for about 10,000 different proteins. The number resolved in this study of protein profiles therefore would represent only about 7% of its protein coding capacity. Perhaps greater differences in protein profiles for the different developmental phenotypes would be seen in the analysis of other types of proteins.

Nevertheless, the relatively minor protein differences detected by de Vries et al. (1980) seem somewhat difficult to reconcile with the other morphological, physiological, and genetic parameters that have been defined for morphogenesis. Furthermore, earlier studies, with less revolving power, demonstrated apparent protein differences of considerable magnitude. Antigenic differences between the dikaryon and its component homokaryotic monokaryons were seen (Raper and Esser, 1961), and a comparison of protein and isozyme patterns for different phenotypes by a one-dimensional electrophoretic analysis indicated large and significant differences (Wang and Raper, 1969, 1970). These isozyme differences, however, could not be confirmed in a subsequent study (Ullrich, 1977); they could have resulted from artifacts of proteolytic digestion in the dikaryon as a result of the longer growth period used in the Wang-Raper study.

Relevant RNA Complexities

Contrary to expectations, a comparison of messenger RNA complexities in *Schizophyllum* revealed little difference between phenotypes (Zantinge et al., 1979). In general, isolated messenger, identified as polyadenylated RNA (polyA + RNA), constituted 2.5% of the total RNA. These polyA + RNAs have an average length of 1100 nucleotides and a polyA tract approximately 33 nucleotides long. The populations of polyA + RNA from the two phenotypes *A-off B-off* and *A-on B-on* were compared by hybridizing each population separately, to saturation, to gap-translated single-copy DNA. In each case, 12-14% of the single-copy DNA was hybridized, as indicated by resistance to S1 nuclease digestion. No significant qualitative differences in polyA + RNA populations were detected in competition experiments where RNAs from the different phenotypes were mixed in the hybridization reaction. Furthermore, the saturation level of total RNA from each type was shown to be comparable and also qualitatively similar. This comparable level of hybridization to DNA for polyA + RNA and total RNA suggests that there is no significant coding class of RNA without polyA tracts. There is, therefore, no evidence for the presence of a large complex class of nonpolyadenylated, so-called heterogeneous nuclear RNA (hnRNA) as the primary transcript which must be processed before serving as messenger in translation to protein product.

On the basis of the percentage of hybridization of single-copy DNA to polyA + RNA, assuming asymmetric transcription, and accounting for the 80% ability of the gap-translated DNA to react, it was calculated that about 32.5% of single-copy DNA sequences are transcribed in all phenotypes, thus indicating the presence of 10,000-13,500 average-size mRNAs, a value of RNA complexity similar to the lowest values reported for higher eukaryotes (Davidson, 1977). It appears, therefore, that in *Schizophyllum* less than 45% of the total DNA (about 34% single-copy plus 7-10% repetitive) is transcribed into stable RNA.

The lack of any evidence for qualitative differences in mRNA populations from the different phenotypes is in striking contrast to changes in transcription observed with similar methods during development in other fungi such as *Dictyostelium discoideum* (Firtel, 1972) and *Neurospora crassa* (Dutta and Chaudhuri, 1975), where qualitative changes involving as much as 38% of the genome were observed.

The possibility that differeing phenotypes may differ significantly in the 1000 or less rare RNA sequences, not detectable by the first methods used for *Schizophyllum*, gained little support from hybridization studies with copy DNA (cDNA) which were designed to isolate rarer RNA sequences correlated with phenotype (Zantinge et al., 1981). By making two kinds of cDNA from polyA + RNAs, one from the *A-off B-off* phenotype and one from the *A-on B-on* phenotypes, and hybridizing the cDNA of one phenotype with the polyA + RNA from the other, the unhybridized portion of cDNA in each case could be detected and isolated. Using such enriched cDNA preparations, the presence of 300-400 rare polyA + RNAs (mRNAs) specific to cell type was revealed. This represented a difference of 2-3% between the two phenotypes. On the other hand, no specific differences beyond 1% were detectable with the same techniques when total RNAs were analyzed. The comparative kinetics of RNA-DNA hybridization for polyA + RNA versus total RNA indicated that only 25% of the complex, polydispersed RNA sequences in total RNA are polyadenylated (Zantinge et al., 1979). Zantinge and co-workers therefore suggested the possibility that differences in polyadenylation of individual sequences in total RNA might account for the discrepancies observed in the RNA-cDNA hybridization studies. Although messenger RNA, at least for the most part, has been found to contain polyA tracts, the function of these polyA tracts in translation is not understood. Furthermore, the procedure used for the recovery of polyA + RNA may eliminate some proportion of the polyA + fraction of RNA. The tentative conclusion here is that the complex nonpolyadenylated RNAs do not represent a very different population from the polyadenylated RNAs in *Schizophyllum*.

The authors concluded from all these results that no appreciable transcriptional control is involved in the regulatory activities of the incompatibility genes. If control of genetic expression at the transcriptional level does occur, they estimated that fewer than about 100 rare RNA sequences can be involved. This was substantiated further by the near identity of protein patterns observed after cell-free translation of total RNA from the different phenotypes (Zantinge et al., 1981). The pattern of proteins synthesized in vitro differed markedly from those produced in vivo (de Vries et al., 1980). The approximate 6% difference in proteins of the *A-off B-off* versus *A-on B-on* phenotypes observed in vivo as contrasted to the less than 1% (average 0.4%) difference found in vitro may be due to protein modification after translation, at the level beyond transcription and processing of RNA sequences. Possibly those 22 heavier, more acidic proteins unique to the *A-on B-on* phenotype in the living cell represent a class of post-translationally modified polypeptides that are specifically relevant to expression of the incompatibility genes in the dikaryon.

Such speculations, however, must be tempered by several considerations: for example, the possibility that the methods of analyses used are not sensitive enough to detect the sought for differences, even though such differences have been detected by similar methods in other organisms, or the possibility that quantitative differences in the availability of specific mRNAs at different stages of development may play a role. Even so, the molecular corollaries of sexual morphogenesis in *Schizophyllum* appear, at the moment, to stand in contrast to those found for other developing systems.

Regulation of Nuclear Behavior

In the Dikaryon

A key event in the development of the dikaryon from two compatible parent monokaryons is the establishment of synchronous division in the nuclei of the two parental types. The nuclei of each type, while in the parent cytoplasm, divide according to an independently established mitotic cycle. When the two types come together in mixed cytoplasm at the point of hyphal fusion, they are most likely in different phases of that cycle. In order to establish synchrony between the two types, some adjustments must be made. In the case of many basidiomycetes, such as *Schizophyllum,* this whole process is further complicated by the phenomenon of nuclear migration, which requires a period of at least 48 hr in matings between two colonies established simultaneously by subculture. During this period a foreign nucleus from one parent is constantly invading the cells of the other and is thus being continually subjected to those factors in the foreign cytoplasm that regulate division of the resident nucleus. It is perhaps logical to speculate that after a certain lag period the invading nucleus becomes entrained to the mitotic cycle of the resident nucleus and the synchrony of the ultimate dikaryotic pair is thus established, but there is no evidence for this.

Recent research in our laboratory has suggested instead that mitosis in the dikaryon is regulated through interaction between the two genomes and that the genes primarily responsible for this are those of the *B* incompatibility factor. Evidence for regulation of nuclear division by the interaction of alleles of the *B* incompatibility genes has come from studies of nuclear behavior following separation of component nuclei from the dikaryon, "dedikaryotization."

Depsite the fact that the dikaryon maintains a strict 1:1 ratio of the two component nuclear types, dedikaryotization usually results in recovery of the two nuclear types in asymmetric ratios as judged by the survival of the resulting monokaryons (C. A. Raper, 1978a, unpublished results). Such nuclear selection has been noted by others in *Lenzites trabea* (Kerruish and Da Costa, 1963), *P. versicolor* (Lange, 1966), *Pholiota nemeko* (Arita, 1979), and *Schizophyllum* (Leonard et al., 1978). We have shown by genetic analysis that specific alleles of the *B* incompatibility genes in *Schizophyllum* play a key role in determining the ratio of nuclear recovery after dedikaryotization (C. A. Raper, unpublished results). First we demonstrated, in five separate experiments on one particular dikaryon containing *A51 B51:A41 B41*, that one nuclear type, *A51 B51*, was always recovered in the majority by 68-100%. In a total sample of 326, 303 were of the majority type. The method of separation, recovery, and identification of nuclear types was by protoplast formation, regeneration, [according to the methods of de Vries and Wessels (1972)], and mating-type tests. In these experiments, several variables suspected of having a possible influence on this result were tested. None of these, for example, the age of the dikaryon, maternal versus nonmaternal origin of the protoplasts, the age of the protoplasts at the time of regeneration, the age of regenerates at the time of isolation for growth and testing, and the method of selecting regenerates, affected the results. Protoplasts from each of the two unmated monokaryotic strains, *A51 B51* and *A41 B41* separately, regenerated in approximately equal frequency, thus indicating an equal capability in recovery of the two nuclear types before dikaryotic association. A comparison of these results with those of Leonard et al. (1978), in which a chemical means of separating nuclear types from comparable dikaryons was used, suggested that the method of dedikaryotization had no influence on the result. Leonard and co-workers isolated 154 monokaryotic hyphal tips from different *A51 B51:A41 B41* dikaryons treated with sodium cholate (which inhib-

Table 1 Preferential Recovery of *B51* over *B41* from Dikaryons of Progeny from *A51 B51* × *A41 B41*

	Percentage of *A51*	Percentage of *B51*
A51 B51 × *A41 B41*	100	100
	85	85
	94	94
A41 B51 × *A51 B41*	0	100
	21	79
	35	65
	2	98
	29	71

Table 2 Hierarchy of B Factors from Dikaryons

	B14	B51	B35	B41	B42	B4
Percentage of						
B14	–	77	67	67		67
B51		–		89	97	87
B35			–	69		
B41				–	70	
B42					–	62
B14 > B51 > B35 > B41 > B42 > B4						

its hook cell fusion and allows proliferation of the hook cell into monokaryotic hyphae). A total of 94% of the surviving isolated tips were of the $A51\ B51$ type.

In order to determine whether or not the preferential recovery of $A51\ B51$ over $A41\ B41$ was correlated with the specificities of either the A or B incompatibility factors, we obtained and identified several meiotic progeny from the $A51\ B51:A41\ B41$ dikaryon (i.e., $A51\ B51$, $A41\ B41$, $A51\ B41$, and $A41\ B51$) and mated these in compatible pairs ($A51\ B51 \times A41\ B41$ and $A51\ B41 \times A41\ B51$) to establish several new dikaryons. The component nuclei of each dikaryon were then separated out by the protoplast method and tested for A and B types. In all eight dikaryons so analyzed, the ratio of nuclear recovery was significantly asymmetric and the majority of the types of nucleus recovered in each case always carried $B51$; in five of the eight cases the major nucleus carried $A41$, in the other three, $A51$ (see Table 1). Thus B factor, and not A factor, specificity, was implicated in the regulation of this asymmetric recovery of nuclear types.

A subsequent comprehensive analysis of the relationship of many different B factor specificities, in paired dikaryotic combinations, of strains collected from all over the world (original collection of J. R. Raper) not only confirmed the role of the B factor in the phenomenon, but revealed a hierarchical order of B factor specificies in the determination of which nucleus of a pair will most likely survive dedikaryotization. For example, in all paired combinations of $B14$ with $B51$, $B35$, $B41$, and $B4$, the nucleus carrying $B14$ is the major type recovered; $B51$ predominates over $B35$, $B41$, and $B4$; $B35$ predominates over $B41$, and so on. Consequently, the hierarchical order of these B specificities is $B14 > B51 > B35 > B41 > B4$ (Table 1). Bearing in mind that each B specificity is determined by a particular combination of $B\alpha$ and $B\beta$ alleles, we next examined a series of paired B factors in which different $B\alpha$ alleles were opposed in contrast, while the $B\beta$ alleles were held in common ($B\alpha \neq$ and $B\beta \neq$), and vice versa ($B\alpha = B\beta \neq$). Such an analysis showed that the hierarchical order of B factor specificity

Table 3 Hierarchy of $B\beta$'s

	B14 $\alpha 3\text{-}\beta 6$	B6 $\alpha 3\text{-}\beta 7$	B26 $\alpha 3\text{-}\beta 1$	B35 $\alpha 3\text{-}\beta 3$	B41 $\alpha 3\text{-}\beta 2$
Percentage of					
$\alpha 3\text{-}\beta 6$	–	54	80	67	67
$\alpha 3\text{-}\beta 7$		–	89	84	48
$\alpha 3\text{-}\beta 1$			–	71	51
$\alpha 3\text{-}\beta 3$				–	69
$\beta 6 = \beta 7 > \beta 1 > \beta 3 > \beta 2$					

is a consequence of the hierarchical order of the $B\alpha$ and $B\beta$ specificities combined. It was found for example, that when $B\beta$ specificity was held in common, $B\alpha 3 > B\alpha 2 > B\alpha 2'$, and when $B\alpha$ specificity was held in common $B\beta 6 = B\beta 7 > B\beta 1 > B\beta 3 > B\beta 2$ (Table 3). As would be expected from this, $B\alpha 3\text{-}B2$, which is $B41$, predominates over $B\alpha 2\text{-}\beta 2$, which is $B42$, and $B\alpha 3\text{-}\beta 3$, which is $B35$, predominates over $B\alpha 3\text{-}\beta 2$, and so on. Variations in the percentage of the major nucleus recovered ranged from 100% down to about 50%, or not significantly different from 1:1 in a few cases, but the overall pattern strongly suggests an ordered, functional difference in B gene specificities (C. A. Raper and T. Hardenberg, unpublished results).

A study by C. A. Raper and S. DiBartolomeis, currently underway, has taken this analysis to a comparison of strains carrying mutations of the $B\alpha 3\text{-}\beta 2$ alleles in an effort to determine whether or not specific alterations or impairment of B function by mutation effects changes in the position of these alleles in the established hierarchical order for wild-type B alleles. The results so far are startling: It appears that the degree of impairment of B function, from slight to complete is correlated with a progressive movement toward the upper end of the hierarchy! The constitutive mutant of $B\beta 2$ (*mut 1*) in which B is always "turned on," but otherwise functioning normally in the regulation of the B sequence shifted to the lower end of the hierarchy; the secondary mutant of this constitutive primary mutant, which was a revertant to wild type (*mut 2*), had shifted position back to that of the wild-type progenitor, $B\alpha 3\text{-}\beta 2$; whereas those secondary mutants representing progressive degrees of functional impairment were shifted to a higher position in the hierarchy, superior not only to the immediate progenitor (*mut 1*) and grand progenitor $B\alpha 3\text{-}\beta 2$, but higher with respect to another wild-type B, $B\alpha 3\text{-}\beta 3$, which is superior to $B\alpha 3\text{-}\beta 2$. In fact, one of these secondary mutants (*mut 5*), which not only had impaired function but carried partial deletions within the B factor, displayed this latter shift. Furthermore, a mutant in which the entire B factor was functionless, probably because of

Table 4 Hierarchy of *B41* Mutants

Mutations	B35 α3-β3	B41 α3-β2	B42 α2-β2	mut 1
Percentage of				
mut 6 (α and β deleted)	—[a]	—	—	97
mut 5 (β partially deleted)	85	—		93
mut 4 (β function impaired)	75	73		100
mut 3 (β slightly impaired)	74	89		99
mut 2 (β of *1* reverted to β2)	21	—		77
mut 1 (β constitutive function)	0	25	12	—
B41 (Wild-type progenitor)	31	—	70	78
6, 5, 4, 3 > 2 = wild-type progenitor > 1				

[a]Cannot be tested.

deletion (*mut 6*), predominated over its immediate progenitor in which B function is *always on* (see Table 4). In summary, B deleted > B wild type > B *always on*. (See C. A. Raper and J. R. Raper, 1973, and Stamberg et al., 1977, for descriptions of these mutants. Note: They are designated with different symbols in these texts.)

These results suggested that a function of the *B* genes, previously undetected, is the inhibition in self of nuclear survival immediately following dedikaryotization. What we are detecting here may be a lag in the "turning off" of a nuclear function that normally prevails in the dikaryon. The hierarchy of wild-type *B*'s suggests that Bα and Bβ alleles differ in their ability to perform this function.

A basis for this B function in the selective discrimination against self after dedikaryotization has been sought in cytological studies. We suspect that B function has an inhibitory effect on nuclear division, perhaps involving a relative sensitivity in response to mitotic inhibitor(s)—a process of possible importance to the maintenance of synchronous division in the dikaryon. Results from preliminary experiments by C. A. Raper, A. Mayerhoff, and S. Nelson (unpublished results), in which the rate of nuclear division in dedikaryotized monokaryotic protoplasts was monitored over a period of time, have supported this idea. Over a 48-hr period after protoplast formation, about 30% of monokaryotic protoplasts from the dikaryon *A41 B41:A51 B51* remained monokaryotic. Whereas virtually all of the monokaryotic protoplasts isolated from each of the component, unmated monokaryotic mycelia, *A41 B41* and *A51 B51*, had become multikaryotic (Fig. 4). The genotypes of the dedikaryotized monokaryons which remain monokaryotic are not known. They may be constituted primarily of *A41 B41* type. This seems logical, since protoplasts containing the *A41 B41* nucleus are greatly discriminated against in recovery via protoplast regeneration.

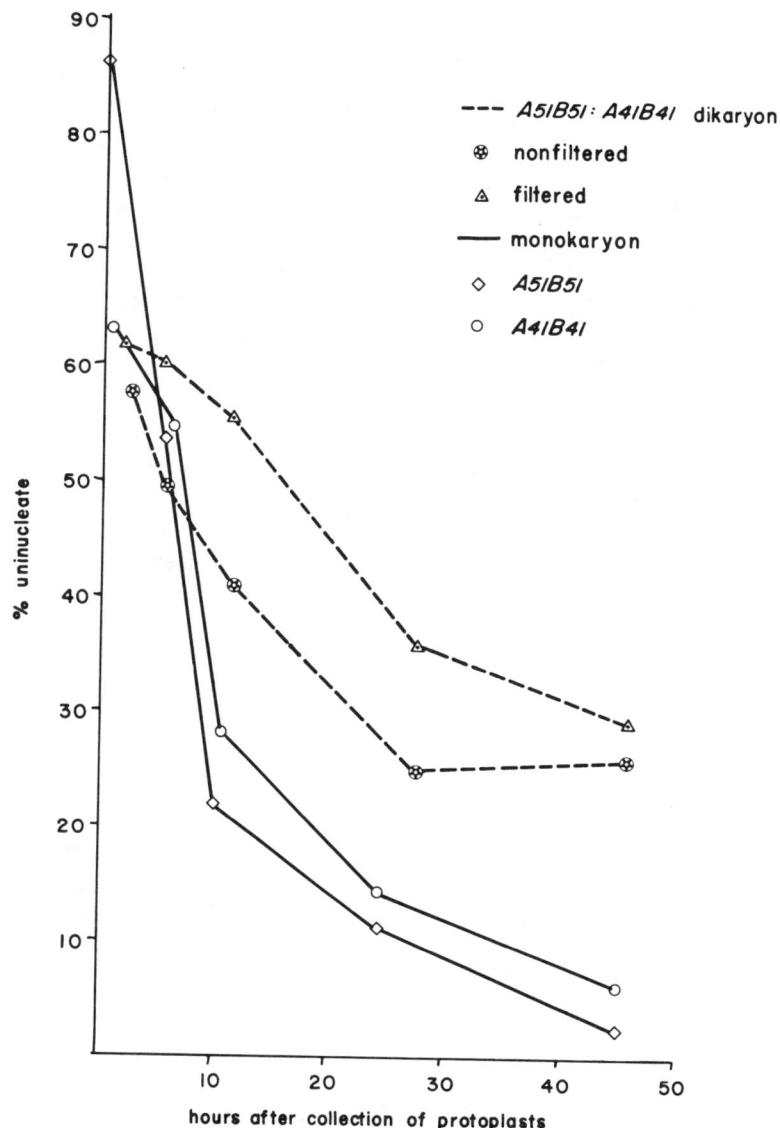

Figure 4 Comparison of fractions of protoplasts in which no nuclear division has occurred over a period of time after protoplast formation in protoplasts derived from a dikaryon and its two component, unmated monokaryons. Filtered protoplasts from the dikaryon are enriched for uninucleate protoplasts at time 0 by size separation in filtration through a 5-μm Nucleopore filter.

Tests involving nutritional mutations are anticipated in an attempt to identify that class of dedikaryotized nuclei that fail to divide.

In the Hemicompatible $A=B\neq$ Heterokaryon

Ratios of nuclear types in the $A=B\neq$ heterokaryon are most often highly disparate (Snider and Raper, 1965). A sampling of nuclear ratios in such heterokaryons containing B factors of known position in the hierarchy established by the criterion of nuclear selectivity from the dikaryon has indicated (C. A. Raper, A. Mayerhoff, and F. Dooneief, unpublished results) a comparable hierarchical order with respect to effect on nuclear ratios in the $A=B\neq$ heterokaryon, but in reverse order! For example, in the $A=B\neq$ heterokaryons $A41\ B41$: $A41\ B51$ or $A51\ B41:A51\ B51$, the nucleus carrying $B41$ is always in the majority. Similarly, in $A=B\neq$ heterokaryons containing $B41$ and $B42$, the $B42$ nucleus is in the majority. Hence the hierarchy is $B42 > B41 > B51$, an order opposite to that seen for nuclear recovery from the dikaryon. The results of this study have come from analyses of about two dozen heterokaryons to date, but are as yet incomplete. They also include a few discrepancies in other B factor combinations. Assuming verification of this phenomenon, particularly as it pertains to the mutated B factors,* yet another dimension to the effect of the B genes on nuclear survival (via division?) may be operating within this context. Here perhaps a function of the B genes is the stimulation in self of nuclear division or, to put it another way, the determination of a relative sensitivity to mitotic stimulator(s).

As previously stated, whether or not the genes of the B factor affect mitotic cycles in the process of nuclear migration is not yet known. If so, the effect is not apparent from observations of dividing nuclei immediately following hyphal fusion between compatible types. Lange (1966) observed nonsynchronous division in nuclei of both types within 12 hr after fusion in *P. versicolor,* but neither type appeared to divide more frequently than the other. Logically, one might expect a stimulation of mitosis in the invading nucleus. Perhaps this does occur, but at a later time in the process. Furthermore, the genes of the A factor may well have some regulating effect on mitosis in the dikaryon that is as yet undetected.

In any case, these studies on regulation of nuclear behavior by alleles of the B factor genes suggest the existence of yet another part of a B gene locus, for example, in the $B\beta$ locus (see Fig. 3) a region for mitotic regulation (MR) might

---

*A recent analysis of nuclear ratios in $A=B\neq$ heterokaryons, each containing muts 3 and 4 (see Table 4), versus the grand progenitor $B\alpha 3$-$\beta 2$ revealed in each case that the major nucleus carried the wild-type progenitor alleles. This is supporting evidence for the reverse order of B factor alleles in the $A=B\neq$ heterokaryons as compared to the dikaryon.

be added to the other four: specificity (*sp*), hook cell fusion (*HF*), nuclear donation (*ND*), and nuclear acceptance (*NA*). Although the function of mitotic regulation appears to be correlated with specificity, it is difficult to speculate on the placement of this region with respect to the others.

SPECULATIONS

The entire process of dikaryotization obviously involves a myriad of factors, many of which operate sequentially: the expression of numerous genes that determine the specific details of differentiation, morphological changes, changes in the appearance of cellular organelles (microtubules, vacuoles), alterations in septal structure, differing growth characteristics and metabolic functions, the presence of specific enzymes (R-glucanase), involvement of the so-called "second messenger," cAMP, differing patterns of protein synthesis, and alterations in nuclear division and movement. The uniquely intriguing aspect of this developmental pattern is its triggering and regulation by the interaction of differing alleles in the four genetic loci of the incompatibility factors. The following questions remain: What is the nature of these "master switch genes"? What is the nature of their products? How do they work?

Any attempt to speculate on this must take into consideration a number of known characteristics of these genes in addition to their pleiotropic effects:

1. Each has an extensive series of multiple alleles.
2. Each has the ability to recognize self versus nonself via internuclear communication within the heterokaryotic cells.
3. The consequence of interallelic recognition is not only pleiomorphic, but nearly identical for all paired combinations of alleles except possibly for the function of mitotic regulation by the *B* genes.
4. Initiation of the morphogenetic process, or the "turning on," by nonself recognition requires a significant lag period; the "turning off" process, for example, after dedikaryotization, also requires a lag period.
5. New alleles, equivalent to wild type, cannot be derived by intra-allelic recombination in a frequency greater than 10^{-6}, nor can they be generated by mutagenesis, either as a primary or secondary event, in frequencies greater than 4×10^{-8}.
6. Some natural alleles of $B\alpha$ and $B\beta$ appear to be associated with deletions, as indicated by their inability to recombine with one another.
7. Wild-type alleles, when exposed to mutagens, mutate to constitutive function in a frequency of 4.3×10^{-8}, and such constitutive mutations, dominant in function to all other mutations and wild-type alleles, can revert to parental type.

8. Other mutations resulting in various degrees of functional impairment, all the way to deletion of the entire incompatibility factor, can be induced.
9. The expression of many other genes, scattered throughout the genome, is regulated by the incompatibility genes.
10. Cyclic AMP plays a regulatory role in the process of morphogenesis as "turned on" by the incompatibility genes.
11. The appearance of unique proteins, possibly derived by posttranslational modification rather than by significant changes in transcription of the genome, is correlated with incompatibility gene expression.

On the basis of all these considerations, the following speculations can be made about the structure and function of these genes.

1. The two genes of a single factor are similar to one another, the pair probably having originated from a single genetic locus by duplication. Each gene has evolved a distinct series of alternate states and the expression of each gene is independently controlled. Furthermore, both pairs of incompatibility genes $A\alpha$-β and $B\alpha$-β, probably originated earlier from a single genetic locus. One pair of genes was translocated to another chromosome and then the two factors subsequently diverged in function.

2. The derivation of the specific alleles of the genes is complex, probably involving the accumulation of a multiple series of specific mutations.

3. Each gene is complex in structure, with a variable region for specificity and a constant region for function. The constant region is divided into several parts.

4. Each gene complex codes for several products, one of which is involved in the recognition of a comparable but different product from another gene of the same allelic series. The other products of the complex serve to initiate and regulate sexual morphogenesis. The product active in the recognition phenomenon may be a protein, as postulated by Kuhn and Parag (1972), or a nucleic acid, as suggested by Ullrich (1978). The products which function to initiate and regulate sexual morphogenesis are most likely proteins—possibly enzymes —that play a role in specific modifications (via phosphorylation, acetylation, or the like) of existing proteins that result in new functions for these proteins. Or, perhaps, some of these gene products are specific receptor proteins for cAMP or are hormone-like stimulators of cAMP. The possibility that incompatibility genes code for activator molecules that "turn on" a battery of other genes bearing the appropriate receptor sites in their regulatory regions (e.g., the genes specifying the details of sexual morphogenesis), as might be suggested by the Britten-Davidson model for gene regulation (Britten and Davidson, 1969), seems reasonable, but the bulk of current information on the molecular genetics of this system does not provide evidence for this.

5. Whatever the products may be, they appear to operate intracellularly, that is, within the cells of a single hyphal system, and outside or between nuclei.

Several models have been proposed to explain the mode of operation of the incompatibility genes in basidiomycetes (Prevost, 1962; Raper, 1966b; Kuhn and Parag, 1972; Pandey, 1977; Ullrich, 1978). None have been proven or disproven, although in the light of current knowledge, some seem more plausible than others (see Raper, 1966b, and Raper, 1978a, for a discussion).

The most direct approach to a characterization of an incompatibility gene would be to isolate, clone, and sequence it. A characterization of the product(s) of such a gene, assuming that its transcript is translatable, might be achieved, after identifying the clone in an appropriate biological test, by using that amplified DNA segment as a probe for the isolation of mRNA, which could then be translated in vitro to yield identifiable product(s).

A library of the entire genome of *S. commune* is currently being cloned (R. C. Ullrich and C. A. Specht, personal communication), as is the genome of *C. cinereus* (P. Pukkila, personal communication). The problem of identifying those clones which contain the genes of interest remains unsolved. Nutritional markers closely linked to the $A\alpha$ and $A\beta$ genes in both *S. commune* and *C. cinereus* might be used to identify the DNA fragments containing the *A* genes, but the available markers may be too distant. A biological test for the incompatibility genes might be carried out in which the test cells transformed by the DNA clone of interest would be recognized by a phenotypical change to the relevant stage of differentiation in the sexual cycle, for example, from a mycelium that does not form hook cells to one that does.

Despite numerous attempts to locate nutritional markers close to the *B* factor genes, none have been found. The *B* factor genes might be isolated, however, through the use of a mutant strain containing a deletion for the entire *B* factor (e.g., *mut 6*, Table 4). DNA enriched for the *B* genes could be isolated by driving hybridization (double strandedness) of total, denatured DNA from a wild-type (*B* containing) strain with denatured DNA from the mutant (*B* deleted) strain and collecting the unhybridized, single-stranded DNA that remained. The single-stranded DNA should contain the $B\alpha$ and $B\beta$ alleles of the wild-type strain that had no complement in the genome of the *B*-deleted strain. This single-stranded DNA could be made double stranded in vitro and ligated into a suitable vector for cloning. The resulting library of DNA clones could be used as probes against total DNA from each of the two parent strains, *B* containing and *B* deleted. Any cloned DNA which hybridized with DNA from the *B*-containing strain but failed to hybridize with DNA from the *B*-deleted strain would be tentatively designated as a clone containing a sequence within the deleted region, possibly a sequence for a *B* factor gene. The enrichment process could be repeated for the isolation of several wild-type B factors and mutated B genes, including constitutively functioning mutant B genes (e.g., *mut 1*, Table 4). The several emerging clones thus identified could then be tested with one another for cross-hybridization. Cross-hybridization would indicate complementary DNA sequences in each pair tested and the degree of homology could be analyzed by electron microscopy. The cloned DNAs might then be used as probes in "Northern blot" hybridization experiments with RNA from the parent B-containing strain versus RNA from the B-deleted strain. The detection of transcript from the B-containing strain and the failure to detect transcript from the B-deleted strain would be a further indication of a clone containing B gene. Subsequently, one could attempt to isolate sufficient transcript for in vitro translation. The use of DNA from designated clones as biological probes would, at present, have to depend upon transformation experiments in which, for example, test cells would contain a mutation constitutive for A factor function and a transformant would be identified by its ability to form a dikaryon mimic and fruiting bodies.

Whether or not these methods of identification can be accomplished remains to be seen. The problem of analyzing the biological effects of these genes remains unsolved. A relatively simple bioassay for an incompatibility gene would therefore be highly desirable. This might be developed by one of several means.

A bioassay for alleles of the B factor might be developed by exploiting their apparent effects on mitosis. As is the case with the two alternate alleles of the mating-type locus in the hemiascomycete *Saccharomyces cerevisiae*, there may be a signal-response mechanism responsible for cell cycle regulation (Bucking-Throm et al., 1973; Wilkinson and Pringle, 1974). Possibly the alleles of $B\alpha$ and $B\beta$ in *Schizophyllum* code for molecules specifically involved in regulation of the mitotic cycle. The observation of specific effects on mitosis or DNA synthesis in test cells of *Schizophyllum*, as has been done in *S. cerevisiae*, might serve as an appropriate bioassay for B factor genes. The usefulness of such a test would depend upon whether or not the effective molecules are diffusable and can be shown to affect test cells when applied from outside the cells.

Another possibility worth exploring is that specific proteins associated with the nuclear envelope or plasmalemma are involved in regulation of the developing dikaryon. Since nuclear behavior is central to the process of dikaryosis, alleles of the incompatibility genes may code for specific proteins in the nuclear envelope which play a role in nuclear migration and/or mitosis or in recognition of other nuclei carrying incompatibility alleles of compatible type. There is no known precedent, but there are cases of possible relevance in, for example, higher plants: in the mobility and recognition phenomena apparent in the behavior of pollen tube nuclei during fertilization, and in the role of the host nucleus leading the development of the infection thread during the invasion of the plant root system by nitrogen-fixing bacteria, for example, in the infection of legumes by *Rhizobium*. In both cases, as in the basidiomycetes, a signal-response phenomenon involving the nucleus is implicated, and the site for this could be in the nuclear envelope. Proteins associated with the nuclear envelope have not been characterized in any of these systems. However, the protein profile of the nuclear membranes of erythrocytes have been compared with that of liver cells in the chick (Jackson, 1976). Jackson identified four and seven major polypeptides in each, respectively. Three of these were common to the nuclear envelopes of the two types of cells; the remainder were specific for cell type. This suggests that differentiation of cell types is correlated with differentiation of proteins in the nuclear envelope. This interpretation is tenuous, however, because the outer membrane of the erythrocyte nucleus was not intact in these experiments.

The genetics of the incompatibility system in basidiomycetes is somewhat analogous to that of the histocompatibility system in higher animals. In the latter system, proteins associated with the plasmalemma are specific for the multiple alleles of the histocompatibility-gene complex (see Ploegh et al., 1981, for a review.) The histocompatibility system operates via cell-cell recognition. In the incompatibility system, there is no evidence for cell-cell recognition; the recognition phenomenon appears to occur entirely within the cell, but between nuclei. Therefore the characterization of nuclear membranes from cells of different known genotypes and phenotypes with respect to the incompatibility alleles might reasonably be attempted first. For example, profiles of nuclear membrane proteins might be compared for two different monokaryons coisogenic for all but the incompatibility alleles, for the dikaryon containing the genomes of these same two coisogenic compatible strains, and for coisogenic strains carrying constitutively functioning mutations of the *A* and *B* incompatibility genes. If consistent differences are found, then this study could be expanded to include other alleles and combinations of the incompatibility genes. Some correlation between allelic specificity and specificity of nuclear membrane proteins and/or between phenotype and nuclear membrane proteins might be detected. Such proteins might subsequently be identified by the use of monoclonal antibody techniques.

Identification and characterization of proteins specific for alleles of the incompatibility genes could open the way to identification of cloned DNA fragments containing such genes. The clones might be screened for the in vitro translation of the relevant proteins.

A third possibility for the development of a bioassay for incompatibility genes is a search for enzymes known to be involved in specific types of protein modification, for example, phosphorylation, acetylation, methylation, and so on, and the correlation of such enzymes with the function of the incompatibility genes. A first step might be an attempt to identify some common type of modification in the "switched on" versus "switched off" proteins of the dikaryon, detected by de Vries et al. (1980). If a particular type or types of modification can be identified, then enzymes known to be required for such modification(s) might be detected and correlated with, for example, dikaryosis versus monokaryosis. The current evidence suggests that it would be very difficult, if not impossible, to detect and identify proteins specific for incompatibility gene function by methods of in vitro translation of the entire genome. This may be due to the fact that incompatibility genes act in vivo to elaborate enzymes required for specific modification of proteins relevant to dikaryosis.

A fourth approach might be an exploration of the possibility that extranuclear molecules responsible for morphogenesis could be detected by membrane fusion experiments in which the test cells are protoplasts. The cell-free fractions to be tested would be packaged in membranes to form artificially made vesicles. These vesicles would be fused with the test protoplasts. Such a procedure, described previously (Raper, 1978a), would allow the direct incorporation of active cell fractions in the absence of a membrane barrier. The effects would be detected in the recipient protoplasts after regeneration to the hyphal mode. A requisite to this would be the demonstration that molecules effective in morphogenesis can be transferred via cytoplasmic transplant experiments. In such experiments anucleate protoplasts made from a sexually differentiated mycelium, for example, *A-on B-on,* would be fused with nucleate protoplasts from an *A-off B-off* mycelium, and an appropriate morphogenetic expression, for example, hook cell formation, would be detected in the regenerated fusion products. If this can be accomplished, the existence of a relatively long-lasting cytoplasmic element in morphogenesis would be implicated. Through the use of modifier mutations and mutations in the incompatibility loci as a means of dissecting the effect genetically, it might then be possible to trace the origin of effect to the product(s) of the regulatory part of an incompatibility gene. A bioassay for the effective molecular agents might then be devised by fractionating the extract from donor cells into various classes of proteins, nucleic acids, and so on, and packaging the fractions in membrane-bound vesicles, either according to defined methods with erythrocyte ghosts or lipid vesicles in animal systems or by methods employing membranes derived from the fungus itself. The effectiveness of a particular fraction would be measured by morphogenetic effects in the regenerating products of fusion between the loaded membrane-bound vesicles and the nucleate test protoplasts. Ultimately, any effective fraction might be further characterized and identified as a product of an incompatibility gene. Such methods might also be employed to enhance the chances for the transformation process with cloned fragments of DNA, as mentioned earlier.

Regenerating protoplasts have a "memory" for the specific phenotype from which they are derived: Protoplasts from *A-off B-off* mycelium regenerate as monokaryons; protoplasts from the *A-on B-off* mutant regenerate to exhibit the typical morphology of that phenotype, that is, formation of nonfused hook cells, within a few cell generations; similarly, regenerating protoplasts from the *A-off B-on* mutant display the morphology of the typical, so-called "flat" mycelium, at least within 100 cell generations; and the *A-on B-on* mutant protoplasts also regenerate to form hook cells, mostly fused, within a few cell generations (C. A. Raper, unpublished results). Protoplasts derived from the *A-on B-on* heterokaryotic dikaryon, however, are varied in phenotype. Regenerates from dikaryotic protoplasts produce unfused hook cells at first and then fused hook cells within about 15 cell generations, whereas regenerates of the dedikaryotized monokaryotic protoplasts produce unfused hook cells for periods up to hundreds of cell generations before expressing a regular monokaryotic phenotype with unadorned septa (Wessels et al., 1976; C. A. Raper, unpublished results). This suggests either the existence of some cytoplasmic determinant for hook cell development that persists over long periods of time, or that there is a considerable lag period in the time required for "turning off" the expression of those genes that determine hook cell formation.

Protoplast fusion between nutritionally complementary strains has been accomplished with the use of polyethylene glycol, although the incidence of fusion is highly variable. Regeneration of fusion products of compatible monokaryons, which are recognized by their ability to grow on minimal medium, results in a monokaryotic phenotype for hundreds of cell generations before any evidence of sexual morphogenesis and dikaryosis becomes apparent. Evidently, a long period of time is required for the "turning on" of sexual morphogenesis. In contrast, regenerating fusion products of protoplasts from the *A-on B-on* mutant fused with protoplasts from the *A-off B-off* wild type develop hook cells regularly within the first 15 or so cell generations and fused hook cell formation persists indefinitely thereafter (P. Gryczka and C. A. Raper, unpublished results). This suggests the possibility of detecting effects of regulatory molecules in regenerating fusion products of anucleate protoplasts from the *A-on B-on* mutant × nucleate protoplasts from the *A-off B-off* wild type. The success of such an experiment will depend upon the ability to achieve a fairly high incidence of fusion. Then the fusion products must be recognized by their morphology or by some selective means involving, for example, the use of a cytoplasmic, drug-resistant mutant. A cytoplasmically determined mutant for resistance to chloramphenicol has been generated and identified in an *A-on B-on* mutant strain, but it has not yet been shown to operate effectively as a selective device in such experiments (P. Gryczka and C. A. Raper, unpublished results). Anucleate protoplasts for use in this kind of experiment may be selected when a combination of appropriate osmoticum and filters is used. The prospect of devising a biological test for products of the incompatibility genes by this approach seems difficult but not altogether discouraging.

The incompatibility genes of basidiomycetes each appear to have multiple products, for recognition and for function in regulating dikaryosis. The detection and identification of these products might reasonably employ any one or more of several approaches. A few of these have been suggested here. Other possibilities are left to the imagination of the reader. The system remains fascinating as a challenge to those who hope to understand more than is presently known about controls for development in eukaryotes.

ACKNOWLEDGMENTS

The work on regulation of nuclear behavior described here was supported in part by the Netherlands Organization for the Advancement of Pure Research and by the William and Flora Hewlett Foundation Grant of Research Corporation of America. Assistance in this study by Geraldine Kaye, Tamarah Hardenberg, Andrea Mayerhoff, Susan Nelson, Susan DiBartolomeis, Patricia Gryczka, and Felice Dooneief is greatfully acknowledged.

REFERENCES

Arita, I. (1979). Cytological studies on *Pholiota*. Rep. Tottori Mycol. Inst. *17*: 1-67.
Bensaude, M. (1918). Sur la sexualité chez les champignons basidiomycètes. C. R. Acad. Sci. *165*:286-289.
Bistis, G. N. (1970). Dikaryotization in *Clitocybe truncicola*. Mycologia *62*:911-924.
Bracker, C. E., and Butler, E. E. (1963). The ultrastructure and development of septa in hyphae of *Rhizoctonia solani*. Mycologia *55*:35-58.
Britten, R. J., and Davidson, E. H. (1969). Gene regulation for higher cells: A theory. Science *165*:349-357.
Bucking-Throm, E., Duntze, W., Hartwell, L. H., and Manney, T. R. (1973). Reversible arrest of haploid yeast cells at the initiation of DNA synthesis by a diffusible sex factor. Exp. Cell Res. *76*:99-110.
Carmi, P., Holm, P. B., Koltin, Y., Rasmussen, S. W., Sage, J., and Zickler, D. (1978). The pachytene karyotype of *Schizophyllum commune* analyzed by three dimenstional reconstruction of synaptonemal complexes. Carlsberg Res. Commun. *43*:117-132.
Casselton, L. A. (1978). Dikaryon formation in higher basidiomycetes. In *The Filamentous Fungi*, Vol. 3, J. E. Smith and D. R. Berry (Eds.). Edward Arnold, London, pp. 275-297.
Casselton, L. A., and Condit, A. (1972). A mitochondrial mutant of *Coprinus lagopus*. J. Gen. Microbiol. *72*:521-527.

Davidson, E. H. (1977). *Gene Activity in Early Development*, 2nd ed. Academic, New York.
Day, P. R. (1960). Mutations affecting the *A* mating-type locus of *Corpinus lagopus*. Heredity *15*:23.
Dons, J. J. M., and Wessels, J. G. H. (1980). Sequence organization of the nuclear DNA of *Schizophyllum commune*. Biochim Biophys. Acta *607*:385-396.
Dons, J. J. M., de Vries, O. M. H., and Wessels, J. G. H. (1979). Characterization of the genome of the basidiomycete *Schizophyllum commune*. Biochim. Biophys. Acta *563*:100-112.
Dubovoy, C. (1975). A class of genes affecting *B* factor-regulated development in *Schizophyllum commune*. Genetics *82*:423-428.
Dubovoy, C., and Munoz, A. (1979). Effect of methylxanthenes in sexuality of *Schizophyllum commune*. Mycologia *71*:855-861.
Dusenbery, R. L. (1975). Characterization of the genome of *Phycomyces blakesleeanus*. Biochim. Biophys. Acta *378*:363-377.
Dutta, S. K. (1974). Repeated DNA sequences in fungi. Nucleic Acids Res. *1*: 1411-1419.
Dutta, S. K., and Chaudhuri, R. K. (1975). Differential transcription of non-repeated DNA during development of *Neurospora crassa*. Dev. Biol. *43*:35-41.
Firtel, R. A. (1972). Changes in the expression of single-copy DNA during development of the cellular slime mold *Dictyostelium discoideum*. J. Mol. Biol. *66*:363-377.
Frankel, C., and Ellingboe, A. H. (1977). New mutations and a 7 chromosome linkage map of *Schizophyllum commune*. Genetics *85*:417-427.
Giesy, R. M., and Day, P. R. (1965). The septal pores of *Coprinus lagopus* (Fr.) *sensu Buller* in relation to nuclear migration. Am. J. Bot. *52*:287-294.
Gilman, A. G. (1970). A protein binding assay for adenosine $3',5'$ cyclic monophosphate. Proc. Nat. Acad. Sci. USA *67*:305-312.
Girbardt, M. (1960). Licht und elektronenoptische Untersuchungen an *Polystictus versicolor* (L.) VI. Der Nucleolus-formenwandel. Ein Beitrag zur Frage der Kernbewegung. Planta *55*:365-380.
Girbardt, M. (1965). Perspecktiven elektronischer Zellforschung. Naturwiss. Rundsch. *18*:345-349.
Girbardt, M. (1968). Ultrastructure and dynamics of the moving nucleus. In *Aspects of Cell Motility, XXI Symposium of the Society of Experimental Biology*. Cambridge University, New York, pp. 249-259.
Gladstone, P. (1972). Genetics of heritable diploidy in *Schizophyllum*. Ph.D. thesis, Harvard University, Cambridge.
Gladstone, P. (1973). Glucose repression of clamp production in diploids of *Schizophyllum commune*. Genetics *74*:595.
Haapala, O. K., and Nienstedt, I. (1976). Chromosome ultrastructure in the basidiomycete fungus *Schizophyllum commune*. Heriditas *84*:49-60.

Hoffman, R. M., and Raper, J. R. (1971). Genetic restriction of energy conservation in *Schizophyllum*. Science *171*:418-419.
Hoffman, R. M., and Raper, J. R. (1972). Lowered respiratory response to adenosine diphosphate of mitochondria isolated from a mutant *B* strain of *Schizophyllum commune*. J. Bacteriol. *110*:780-781.
Hoffman, R. M., and Raper, J. R. (1974). Genetic impairment of energy conservation in development of *Schizophyllum*: Efficient mitochondria in energy-starved cells. J. Gen. Microbiol. *82*:65-75.
Hudspeth, M. E. S., Timberlake, W. E., and Goldberg, R. B. (1977). DNA sequence organization in the water mold *Achlya*. Proc. Nat. Acad. Sci. USA *74*:4332-4336.
Jackson, R. L. (1976). Polypeptides of the nuclear envelope. Ph.D. thesis, Harvard University, Cambridge.
Janszen, F. H. A., and Wessels, J. G. H. (1970). Enzymatic dissolution of hyphal septa in a basidiomycete. Antonie van Leeuwenhoek J. Microbiol. Serol. *36*: 255-257.
Jersild, R., Mishkin, S., and Niederpruem, D. J. (1967). Origin and ultrastructure of complex septa in *Schizophyllum commune* development. Arch. Mikrobiol. *57*:20-32.
Kerruish, R. M., and Da Costa, E. W. (1963). Monokaryotization of cultures of *Lenzites trabea* (Pers.) Fr. and other wood-destroying basidiomycetes by chemical agents. Ann. Bot. *27*:653-669.
Kniep, H. (1920). Über morphologische und physiologische Geschlechtsdifferenzierung (Untersuchungen an Basidiomyzeeten). Verh. Phys. Med. Ges. Wurzburg *46*:1-18.
Koltin, Y. (1968). The genetic structure of the incompatibility factors of *Schizophyllum commune*: Comparative studies of primary mutations in the *B* factor. Mol. Gen. Genet. *102*:196-203.
Koltin, Y., and Flexer, A. S. (1969). Alteration of nuclear distribution in *B*-mutant strains of *Schizophyllum commune*. J. Cell Sci. *4*:739-749.
Koltin, Y., and Stamberg, J. (1972). Suppression of a mutation disruptive to nuclear migration in *Schizophyllum commune* by a gene linked to the *B* factor. J. Bacteriol. *109*:594-598.
Koltin, Y., Raper, J. R., and Simchen, G. (1967). Genetic structure of the incompatibility factors of *Schizophyllum commune*: The *B* factor. Proc. Nat. Acad. Sci. USA *57*:55-63.
Koltin, Y., Stamberg, J., Bawnik, N., Tamarkin, R., and Werczberger, R. (1979). Mutational analysis of natural alleles in and affecting the *B* incompatibility factor of *Schizophyllum*. Genetics *93*:383-391.
Kuhn, J., and Parag, Y. (1972). Protein subunit aggregation model for self-incompatibility in higher fungi. J. Theor. Biol. *35*:77-91.
Lange, V. I. (1966). Das Bewegungsverhalten der Kerne in fusionierten Zellen von *Polystictus versicolor* (L.). Flora *158*:487-497.
Leonard, T. J., Gaber, R. F., and Dick, S. (1978). Internuclear genetic transfer in dikaryons of *Schizophyllum commune*. II. Direct recovery and analyses of recombinant nuclei. Genetics *89*:685-693.

Marchant, R., and Wessels, J. G. H. (1973). Septal structures in normal and modified strains affecting septal dissolution in *Schizophyllum commune*. Arch. Microbiol. *96*:175-182.
Marchant, R., and Wessels, J. G. H. (1974). An ultrastrucutral study of septal dissolution in *Schizophyllum commune*. Arch. Microbiol. *96*:175-182.
Mayfield, J. E. (1974). Septal involvement in nuclear migration in *Schizophyllum commune*. Arch. Microbiol. *95*:115-174.
Moore, R. T., and McAlear, J. H. (1961). Fine structure of Mycota. 8. On the aecidial stage of *Uromyces caladii*. Phytopathol. Z. *42*:297-304.
Niederpruem, D. J. (1971). Kinetic studies of septum synthesis, erosion and nuclear migration in a growing B-mutant of *Schizophyllum commune*. Arch. Mikrobiol. *75*:189-196.
Niederpruem, D. J. (1980a). Direct studies of dikaryotization in *Schizophyllum commune*: I. Live intercellular nuclear migration patterns. Arch. Microbiol. *128*:162-171.
Niederpruem, D. J. (1980b). Direct studies of dikaryotization in *Schizophyllum commune*: II. Behavior and fate of multikaryotic hyphae. Arch. Microbiol. *128*:172-178.
Niederpruem, D. J., and Wessels, J. G. H. (1969). Cytodifferentiation and morphogenesis in *Schizophyllum commune*. Bacteriol. Rev. *33*:505-535.
Niederpruem, D. J., Jersild, R. A., and Lane, P. L. (1971). Direct microscopic studies of clamp connection formation in growing hyphae of *Schizophyllum commune*. I. The dikaryon. Arch. Mikrobiol. *78*:268-280.
Ohja, M., and Dutta, S. K. (1978). Nuclear control of differentiation. In *The Filamentous Fungi*, Vol. 3, J. E. Smith and D. R. Berry (Eds.). Edward Arnold, London, pp. 8-27.
Ohja, M., Turler, H., and Turian, G. (1977). Characterization of *Allomyces* genome. Biochim. Biophys. Acta *478*:377-391.
Omerod, W., Francis, S., and Margulis, L. (1976). Delay in the appearance of clamp connections in *Schizophyllum commune* by inhibitors of microtubule protein assembly. Microbios *17*:189-205.
Pandey, K. K. (1977). Generation of multiple genetic specificities: Origin of genetic polymorphism through gene regulation. Theor. Appl. Genet. *49*:85-93.
Parag, Y. (1962). Mutations in the B incompatibility factor of *Schizophyllum commune*. Proc. Nat. Acad. Sci. USA *48*:743-750.
Parag, Y. (1970). Genetics of tetrapolar sexuality in higher fungi: The B factor, common-B heterokaryosis and parasexuality. In *USDA Final Report 1965-1970. Project No. FG-IS-228*, 159 pp.
Ploegh, H. L., Orr, H. T., and Strominger, J. L. (1981). Major histocompatibility antigens: The human (HLA-A, -B, -C) and the murine (H-2K, H-2D) class I molecules. Cell *24*:287-299.
Prevost, G. (1962). Etude génétique d'un basidiomycète: *Coprinus radiatus* Fr. ex Bolt. Thesis, Université de Paris, Paris.

Radu, M., Steinlauf, R., and Koltin, Y. (1974). Meiosis in *Schizophyllum commune*: Chromosomal behavior and the synaptonemal complex. Arch. Microbiol. *98*:301-310.

Raper, C. A. (1978a). Control of development by the incompatibility system in basidiomycetes. In *Genetics and Morphogenesis in the Basidiomycetes*, M. N. Schwalb and P. G. Miles (Eds.). Academic, New York, pp. 3-29.

Raper, C. A. (1978b). Sexuality and breeding. In *The Biology and Cultivation of Edible Mushrooms*. S. T. Chang and W. Hays (Eds.). Academic, New York, pp. 83-113.

Raper, C. A., and Raper, J. R. (1964). Mutations affecting heterokaryosis in *Schizophyllum commune*. Am. J. Bot. *51*:503-513.

Raper, C. A., and Raper, J. R. (1966). Mutations modifying sexual morphogenesis in *Schizophyllum*. Genetica *54*:1151-1168.

Raper, C. A., and Raper, J. R. (1973). Mutational analysis of a regulatory gene for morphogenesis in *Schizophyllum*. Proc. Nat. Acad. Sci. USA *70*:1427-1431.

Raper, J. R. (1966a). Life cycles, basic patterns of sexuality, and sexual mechanisms. In *The Fungi*, Vol. 2, G. C. Ainsworth and A. Sussman (Eds.). Academic, New York, pp. 473-511.

Raper, J. R. (1966b). *Genetics of Sexuality in Higher Fungi*. Ronald Press, New York, 283 pp.

Raper, J. R., and Esser, K. (1961). Antigenic differences due to the incompatibility factors in *Schizophyllum commune*. Z. Verebungsl. *92*:439-444.

Raper, J. R., and Raper, C. A. (1973). Incompatibility factors: Regulatory genes for sexual morphogenesis in higher fungi. Brookhaven Symp. Biol. *25*:19-39.

Raper, J. R., and Raudaskoski, M. (1968). Secondary mutations at the $B\beta$ incompatibility locus of *Schizophyllum*. Heredity *23*:109-117.

Raper, J. R., Baxter, M. G., and Ellingboe, A. H. (1960). The genetic structure of the incompatibility factors of *Schizophyllum commune*: the A factor. Proc. Nat. Acad. Sci. USA *44*:889-900.

Raper, J. R., Boyd, D. H., and Raper, C. A. (1965). Primary and secondary mutations at the incompatibility loci in *Schizophyllum*. Proc. Nat. Acad. Sci. USA *53*:1324-1332.

Raudaskoski, M. (1970). A new secondary $B\beta$ mutation in *Schizophyllum* revealing functional differences in wild $B\beta$ alleles. Hereditas *64*:259-266.

Raudaskoski, M. (1972). Occurrence of microtubules and microfilaments, and origin of septa in dikaryotic hyphae of *Schizophyllum commune* during intercellular nuclear migration. Arch. Mikrobiol. *86*:91-100.

Raudaskoski, M. (1973). Light and electron microscopy study of unilateral mating between a secondary and a wild-type strain of *Schizophyllum commune*. Protoplasma *76*:35-48.

Raudaskoski, M. (1976). Acid phosphatase activity in the wild-type and B-mutant hyphae of *Schizophyllum commune*. J. Gen Microbiol. *94*:373-379.

Raudaskoski, M., and Huttunen, E. (1977). The effect of griseofulvin on nuclear distribution in a dikaryon of *Schizophyllum commune.* Mykosen *20*:339-348.

Raudaskoski, M., and Koltin, Y. (1973). Ultrastructural aspects of a mutant of *Schizophyllum commune* with continuous nuclear migration. J. Bacteriol. *116*:981-988.

Raudaskoski, M., Stamberg, J., Bawnik, N., and Koltin, Y. (1976). Mutational analysis of natural alleles at the *B* incompatibility factor of *Schizophyllum commune*: α2 and β6. Genetics *83*:507-516.

Rich, M. A., and Deppe, C. S. (1976). Control of fungal development. I. The effects of two regulatory genes on growth in *Schizophyllum commune.* Dev. Biol. *53*:21-29.

Ross, I. K. (1976). Nuclear migration rates in *Coprinus congregatus*: A new record? Mycologia *68*:418-422.

Sietsma, J. H., Rast, D., and Wessels, J. G. H. (1977). The effect of carbon dioxide on fruiting and on degradation of a cell wall glucan in *Schizophyllum commune.* J. Gen. Microbiol. *102*:385-389.

Snider, P. J., and Raper, J. R. (1958). Nuclear migration in the basidiomycete *Schizophyllum commune.* Am. J. Bot. *45*:538-546.

Snider, P. J., and Raper, J. R. (1965). Nuclear ratios and complementations in common-A heterokaryons of *Schizophyllum commune.* Am. J. Bot. *52*: 547-552.

Stamberg, J., and Koltin, Y. (1971). Selectively recombining *B* incompatibility factors of *Schizophyllum commune.* Mol. Gen. Genetics *113*:157-165.

Stamberg, J., and Koltin, Y. (1972). The organization of the incompatibility factors in higher fungi: The effects of structure and symmetry on breeding. Heredity *30*:15-26.

Stamberg, J., and Koltin, Y. (1973). The origin of specific incompatibility alleles: A deletion hypothesis. Am. Nat. *107*:35-45.

Stamberg, J., and Koltin, Y. (1974). Recombinational analysis at an incompatibility locus of *Schizophyllum.* Mol. Gen. Genet. *135*:45-50.

Stamberg, J., Koltin, Y., and Tamarkin, A. (1977). Deletion mapping of wild-type and mutant alleles at the *B* incompatibility factor of *Schizophyllum.* Mol. Gen. Genet. *157*:183-187.

Timberlake, W. E. (1978). Low repetitive DNA content in *Aspergillus nidulans.* Science *202*:973-975.

Ullrich, R. C. (1977). Isozyme patterns and cellular differentiation in *Schizophyllum commune.* Mol. Gen. Genet. *156*:157-161.

Ullrich, R. C. (1978). On the regulation of gene expression: Incompatibility in *Schizophyllum.* Genetics *88*:709-722.

Ullrich, R. C., Droms, K. A., Doyon, J. D., and Specht, C. A. (1980a). Characterization of DNA from the basidiomycete *Schizophyllum commune.* Exp. Mycol. *4*:123-124.

Ullrich, R. C., Kohorn, B. D., and Specht, C. A. (1980b). Absence of short-period repetitive-sequence interspersion in the basidiomycete *Schizophyllum commune.* Chromosoma *81*:371-378.

Uno, I., and Ishikawa, T. (1973). Purification and identification of the fruiting inducing substances in *Coprinus macrorhizus*. J. Bacteriol. *113*:1240-1248.

Vries de, O. M. H., Hoge, J. H. C., and Wessels, J. G. H. (1980). Translation of RNA from *Schizophyllum commune* in a wheat germ and rabbit reticulocyte cell-free system. Biochim. Biophys. Acta *607*:373-378.

Vries de, O. M. H., and Wessels, J. G. H. (1972). Release of protoplasts from *Schizophyllum commune* by a lytic enzyme preparation from *Trichoderma viride*. J. Gen. Microbiol. *73*:13-22.

Wang, C. S., and Raper, J. R. (1969). Protein specificity and sexual morphogenesis in *Schizophyllum commune*. J. Bacteriol. *99*:291-297.

Wang, C. S., and Raper, J. R. (1970). Isozyme patterns and sexual morphogenesis in *Schizophyllum commune*. Proc. Nat. Acad. Sci. USA *66*:882-884.

Watrud, L. S., and Ellingboe, A. H. (1973). Use of cobalt as a mitochondrial vital stain to study cytoplasmic exchange in matings of the basidiomycete *Schizophyllum commune*. J. Bacteriol. *115*:1151-1158.

Wessels, J. G. H. (1966). Control of cell-wall glucan degradation during development in *Schizophyllum commune*. Antonie van Leeuwenhoek J. Microbiol. Serol. *32*:341-355.

Wessels, J. G. H. (1969). Biochemistry of sexual morphogenesis in *Schizophyllum commune*: Effect of mutations affecting the incompatibility system on cell-wall metabolism. J. Bacteriol. *98*:697-704.

Wessels, J. G. H. (1978). Incompatibility factors and the control of biochemical processes. In *Genetics and Morphogenesis in the Basidiomycetes*, M. N. Schwalb and P. G. Miles (Eds.). Academic, New York, pp. 81-104.

Wessels, J. G. H., and Koltin, Y. (1972). R-Glucanase activity and susceptibility of hyphal walls to degradation in mutants of *Schizophyllum* with disrupted nuclear migration. J. Gen. Microbiol. *71*:471-475.

Wessels, J. G. H., and Marchant, R. (1974). Enzymic degradation of septa in wall preparations from a monokaryon and a dikaryon of *Schizophyllum commune*. J. Gen. Microbiol. *83*:359-368.

Wessels, J. G. H., and Niederpruem, D. J. (1967). Role of a cell-wall glucan-degrading enzyme in mating of *Schizophyllum commune*. J. Bacteriol. *94*:1594-1602.

Wessels, J. G. H., Hoeksema, H. L., and Stemerding, D. (1976). Reversion of protoplasts from dikaryotic mycelium of *Schizophyllum commune*. Protoplasma *89*:317-321.

Wilkinson, L. E., and Pringle, J. R. (1974). Transient G1 arrest of *S. cerevisiae* cells of mating type by a factor produced by cells of mating type *a*. Exp. Cell Res. *89*:175-187.

Zantinge, B., Dons, H., and Wessels, J. G. H. (1979). Comparison of poly (A)-containing RNA's in different cell types of the lower eukaryote *Schizophyllum commune*. Eur. J. Biochem. *101*:251-260.

Zantinge, B., Hoge, J. H., and Wessles, J. G. H. (1981). Frequency and diversity of RNA sequences in different cell types of the fungus *Schizophyllum commune*. Eur. J. Biochem. *113*:381-389.

9
Hormones and Sexuality in Fungi

Graham W. Gooday / University of Aberdeen, Aberdeen, Scotland

INTRODUCTION

As is abundantly clear from other chapters in this book, fungi produces a vast and bewildering array of metabolites. Nearly all of these have no obvious function for their producing organism. This chapter, however, deals with metabolites that have a very profound significance for their producing organisms—the fungal sex hormones. These all have very potent biological activities. Indeed, without these activities none of the metabolites would have been characterized, except in the case of trisporic acid production by *Blakeslea trispora*, which is out of all proportion for its hormonal activity, and has probably resulted from a loss of regulatory control. With this exception, the fungal sex hormones are all produced in very small amounts. This has hindered their characterization as well as that of the many other specific hormone systems that must exist to regulate sexual differentiation in fungi (Raper, 1973). The hormone systems chemically identified so far cover a wide taxonomic spectrum: one from the Oomycetes, one from the Chytridiomycetes, one (probably the only one) from the Zygomycotina, one from the Hemiascomycetes, and two from the Heterobasidiomycetes. The exceptions to this list are the slime molds, the euascomycetes, and the homobasidiomycetes, the first two of which now present very good biological evidence for sex hormones; however, we still await chemical characterizations of the molecules themselves.

These metabolites all cause a total qualitative switch in differentiation in the recipient cell, from asexual reproduction or normal vegetative growth to sexual morphogenesis. Thus their action is quite distinct from that of plant growth substances, which Trewavas (1981) argued strongly should not be regarded as hormones but, rather, as integrating agents to which different tissues have different quantitative sensitivities. Rather, the term "hormone" is clearly fully justified for these fungal metabolites (Gooday, 1974). O'Day (1981) argued for the use of the term "pheromone," i.e., a hormone acting at a distance on another individual. This distinction seems unnecessary in the microbes on tautological grounds, as there is no need to distinguish between inter- and intraorganismic hormones as there is, for example, in the insects.

Unfortunately we have too few examples to make wide-ranging generalizations. The hormones whose structures we know fall into two classes: terpenoids and hydrophobic peptides.

TERPENOIDS

Sirenin

Species of the chytrid genus *Allomyces* are water molds, living saprophytically on submerged vegetable or animal matter. The haploid vegetative cells will differentiate to give male and female gametangia, often in pairs on the same hypha (Turian, 1969; Pommerville, 1981). Thus typically, *Allomyces macrogynus* in culture forms a pair of gametangia, the male being terminal and the female subterminal. Each is multinucleate and subtended by a complete septum separating it from the rest of the hypha. When in a nutrient-poor medium, gametogenesis occurs, with synchronous cleavage giving rise to small (about 8 μm in diameter) male gametes, bright orange with γ-carotene, and larger (about 11 μm) colorless female gametes, both of which swim free through exit pores that result from localized lysis of the gametangial walls (Hatch, 1938; Pommerville, 1981). Apart from size and color, the two gametes are structurally similar, being uniflagellate, approximately spherical cells. Their behaviors are different, however, with the female cells swimming sluggishly, with frequent changes in direction, so that they tend to stay in one place, and the male cells swimming faster, with fewer changes in direction, so that they tend to cover much larger distances (Pommerville, 1978). When they come into contact, plasmogamy can occur, resulting in the formation of a biflagellate zygote (Hatch, 1938; Pommerville and Fuller, 1976). This swimming cell eventually settles to germinate as the diploid vegetative thallus, which produces either diploid zoospores as propagules for further diploid thalli or, via meiosis, haploid zoospores as propagules for the haploid vegetative thalli.

Machlis (1958) showed that the meeting process of male and female gametes was not random; rather, the female gametangium and the swimming female gametes produce a potent attractant for the male gametes. Machlis called this attractant sirenin, after the Sirens who attempted to lure Odysseus to his death. Machlis devised increasingly sensitive bioassays that involved counting the number of male gametes settling on a membrane the other side of which was the test solution (Machlis 1973a,b), and found that they were sensitive to sirenin over a very wide concentration range, 10^{-10} to 10^{-5} M. Of the five types of swimming cells produced by *Allomyces* species—male and female gametes, zygotes, and

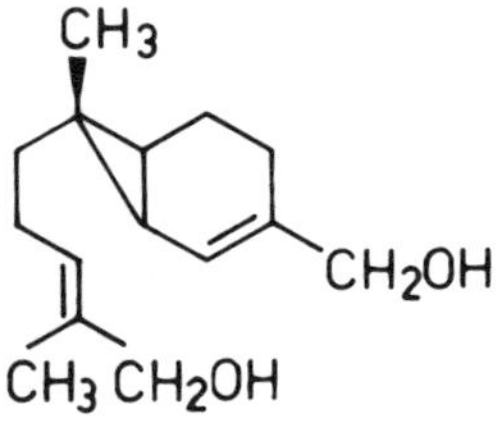

Figure 1 The structure of sirenin.

diploid and haploid zoospores—only the male gametes responded to sirenin, and only the female gametes produced it (Carlile and Machlis, 1965).

In order to characterize sirenin, Machlis et al. (1966, 1968) grew large-scale cultures of a predominantly female hybrid, originally produced by Emerson and Wilson (1954) in a study of speciation in these fungi. Sirenin proved to be a remarkable bicyclic sesquiterpene (Fig. 1) with a cyclopropane ring (Nutting et al., 1968). Sirenin and a range of isomers and derivatives were soon synthesized and the specificity of the chemotactic response was studied (Plattner et al., 1969; Plattner and Rapoport, 1971). Only the synthetic l-sirenin, identical to the natural product, was active (Machlis, 1973b).

Sirenin is produced only in small quantities (about 1 μM in the culture filtrate, even by the female hybrid), and no details of its biosynthesis have yet been reported. Presumably its precursor is farnesyl pyrophosphate. Related natural products include the corresponding hydrocarbon sesquicarene, the essential oil of the fruit of the plant *Schisandra chinensis,* and the insect juvenile hormones from the balsam fir, juvabione and dehydrojuvabione.

The secretion of sirenin does not occur until gametogenesis is completed, but actinomycin D and cycloheximide will inhibit subsequent formation of sirenin if added at any time during the first 10 and 20 min of gametogenesis, respectively (Pommerville, 1981).

The behavioral response of male gametes to sirenin has been analyzed elegantly by Pommerville (1977, 1978, 1981), who showed that a pulse of sirenin added to a suspension of cells changes their smooth swimming pattern with few changes in direction to one of many rapid changes in direction. This can clearly be interpreted as a "hunting" behavior in the apparent vicinity of a female cell, and is indeed the swimming pattern observed as a male cell approaches a female cell prior to fusion. The sirenin produced by the female cell brings about a very precise orientation and attraction of the male gametes, favoring efficient mating. The mechanisms of chemoreception and transduction of the signals necessarily involved in these responses have yet to be elucidated, but by analogy with other

Table 1 A Comparison of Male and Female Gametes of *Allomyces*

Property	Male	Female
Position of gametangium		
Allomyces macrogynus	Apical	Subapical
Allomyces arbusculus	Subapical	Apical
Pigmentation	γ-Carotene	None
Size (μm in diameter)	8	11
Swimming characteristics		
Velocity (μm sec^{-1})	100	60
Smooth-swimming distance (μm)	50	65
Turn angle	60°	80°
No of turns per minute	70	100
Production of sirenin	No	Yes
Response to sirenin		
Uptake of sirenin	Yes	No
Inactivation of sirenin	Yes	No
Change in swimming	Yes	No
Specificities for plasmogamy and karyogamy	To female	To male

Sources: Emerson and Wilson (1954), Hatch (1938), Turian and Ojha (1969), Machlis (1973b), Pommerville (1977, 1978, 1981), and Carlile and Gooday (1978).

systems, one suspects the presence of membrane-bound receptors and a subsequent temporal assessment of changes in concentration of sirenin. Calcium ions are required for the chemotactic response to sirenin (Machlis, 1973a). Localized transmembrane movements of this cation are implicated in many bioelectrically regulated responses of cells (Harold, 1982), but unfortunately these *Allomyces* gametes are too small for investigation by current techniques that might elucidate whether such a mechanism was operating here. Sirenin, taken up by male gametes following first-order kinetics from 5-400 nM solutions, is rapidly inactivated and hence cannot be reextracted (Machlis, 1973b). Presumably inactivation occurs by further metabolism. In addition, Klapper and Klapper (1977) described a low molecular weight inhibitor of the male chemotactic response.

    Male and female gametes originally had identical genomes, since they originate from a cenocytic thallus grown from a uninucleate haploid zoospore. As gametes they are now very different (Table 1). A plausible explanation of this phenomenon is that sexual phenotype in *Allomyces* species is regulated by a gene "cassette" mechanism directly analogous to that elucidated for the homothallic mating-type switches in *Saccharomyces cerevisiae* (Klar et al., 1981).

Thus all of the properties listed in Table 1 would be regulated by the allelic mating-type genes *mt male* or *mt female*, but the *mt* gene would be interchangeable from stored silent copies of *male* and *female* in the genome. Differentiation to give a pair of gametangia would be from a pair of nuclei, one *mt male* and one *mt female*. Turian and Ojha (1969) have described the transformation of one of the properties in Table 1, namely, an inversion of polarity of gametangia of *A. macrogynus*, from epigynous to hypogynous, by DNA isolated from *Allomyces arbuscula*. In the above model, this would represent the transformation of one or more of the many genes regulated by the *mt* locus.

Antheridiol and Oogoniol

Species of the Oomycete genus *Achlya* are water molds, living saprophytically on submerged vegetable and animal matter. In a series of elegant experiments with heterothallic species, Raper demonstrated that sexual differentiation in these fungi is regulated by complementary hormones, diffusing from hypha to hypha (Raper, 1951). The female cells release a hormone that switches the male cells from vegetative growth to the production of many short antheridial branches. The male cells then release a second hormone which diffuses back to the female cells, switching them in turn from vegetative growth to the production of oogonial initials.

In 1942, Raper and Haagen-Smit obtained 2 mg of active compound (termed hormone A at that time) from 1440 liters of culture filtrate of female mycelium. This was far too little for characterization by contemporary chemical techniques. This original sample was kept and later identified as substantially pure female hormone, antheridiol. Antheridiol was finally characterized as a sterol (Fig. 2, II) by McMorris and Barksdale (1967) and Arsenault et al. (1968). The yield is very low, at about 6×10^{-9} M in the culture filtrate. Antheridiol and a range of isomers and derivatives were synthesized and their biological activities assessed by Barksdale et al. (1974). The stereochemistry of the side chain proved to be very important, so that the natural product (antheridiol, 22*S*, 23*R*) was active at 6 pg/ml, whereas its isomers erythro (22*R*, 23*S*), threo (22*S*, 23*S*), and threo (22*R*, 23*R*) required 20, 20, and 400 ng/ml, respectively, to elicit branching responses in male cells, and even these responses may have been the result of slight contamination with the 22*S*, 23*R* isomer.

The complementary hormone produced by the male cells, termed oogoniol, proved much more difficult to characterize. In most situations it was only produced in the presence of antheridiol, which complicated attempts at purification. It was finally purified from culture filtrates of the homothallic *Achlya heterosexualis*, which produces it constitutively (Barksdale and Lasure, 1974). It is also a sterol, esterified at C-3 with isobutyrate, propionate, and acetate in oogoniols 1, 2, and 3, respectively (Fig. 2, III; McMorris et al., 1975; McMorris,

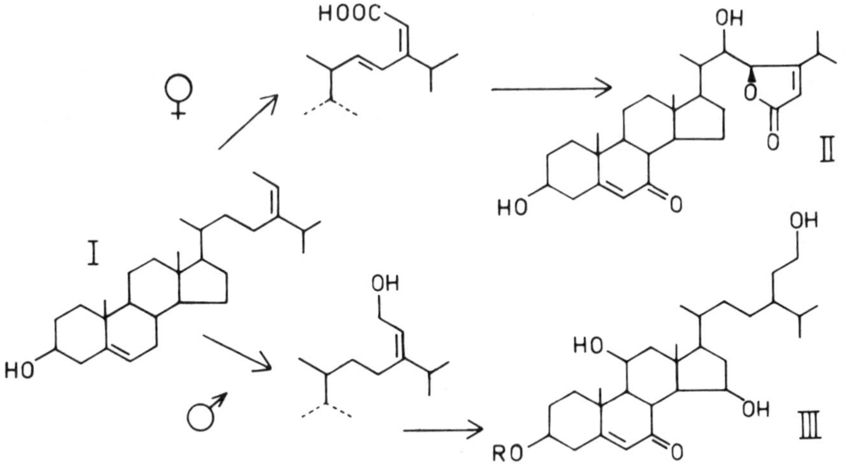

Figure 2 Suggested biosyntheses of antheridiol and oogoniols-1, -2, and -3: (I) Fucosterol, (II) antheridiol, and (III) oogoniol [R = isobutyryl (-1), propanyl (-2), acetyl (-3)].

1978). A minor metabolite, 24(28)-dehydro-oogoniol 1, proved more active in the bioassay than oogoniol 1 by a factor of about 100 (McMorris, 1978).

Fucosterol (Fig. 2.I) is the major sterol in *Achlya* species (Popplestone and Unrau, 1973). Feeding experiments with tritiated fucosterol indicate that it is the biosynthetic precursor of antheridiol, and with tritiated fucosterol and deuterated methionine that it is the biosynthetic precursor of the oogoniols (Popplestone and Unrau, 1974; McMorris, 1978; McMorris and White, 1977). The respective pathways suggested by these authors are shown in Fig. 2.

The addition of antheridiol to male mycelium of *Achlya ambisexualis* elicits a series of morphological responses, with increasingly higher concentrations (Horgen, 1981):

1. Antheridiol branching. This is the basis of the bioassay, which consists of counting the number of branches formed. It is very sensitive, detecting at least a 10^{-11} M solution.
2. Chemotropism of the antheridial branches. This was shown by adsorbing antheridiol onto particles of plastic, which then caused the branches to grow toward and wrap around them.
3. Formation of a septum delimiting the antheridium.
4. Meiosis in the antheridium.

The addition of oogoniol to female mycelium elicits the differentiation and delimitation of the oogonia, with accompanying meiosis to give rise to the oospores.

Accompanying these morphological responses to antheridiol is a series of biochemical responses:

1. The synthesis and release of oogoniol. At one time it seemed an attractive idea that oogoniol might be a bioconversion product from antheridiol, as it is produced in much smaller quantities as a result of the addition of antheridiol, but there is no evidence for this (see the discussion following McMorris, 1978). Oogoniol synthesis is inhibited by cycloheximide (McMorris and White, 1977).
2. An induced metabolism of antheridiol to inactive metabolites (Musgrave and Nieuwenhuis, 1975).
3. An increase in the activity and release of cellulase (Mullins, 1973). Mullins correlated this with the concomitant antheridial branching, suggesting that the cellulase activity is involved in the localized wall softening required for the formation of branches.
4. A marked enhancement of synthesis of protein, rRNA, and mRNA, of histone acetylation, and of transcription of *Achlya* chromatin in the presence of *Achlya* cytosol (Silver and Horgen, 1974; Horgen and Ball, 1974; Horgen, 1981; Timberlake, 1976).

Despite the marked and apparently fundamental quantitative changes in macromolecular synthesis described in item (4), Rozek and Timberlake (1980) and Gwynne and Brandhorst (1980) have failed to find any qualitative changes in messenger RNA or protein populations synthesized following treatment with antheridiol. They concluded that antheridiol-induced differentiation is accompanied by no major qualitative transcriptional and translation changes. Gwynne and Brandhorst (1980) detected a specific quantitative change, a marked increase in labeling, 30 min after antheridiol addition, of a protein of molecular weight about 60,000 (not attributable to cellulase). Since no equivalent increase in labeling of a mRNA species was detected by Rozek and Timberlake (1980) and the increase in protein labeling was only partly inhibited by actinomycin D, they suggested that this is the result of a posttranscriptional modification.

Trisporic Acid

Trisporic acid is the sex hormone of the Mucorales. Its discovery was serendipitous, as it was first identified not as a hormone but as a secondary metabolite produced in large amounts during fermentations by mated cultures of *B. trispora* (Caglioti et al., 1964, 1967). As it was produced in such large amounts, its action

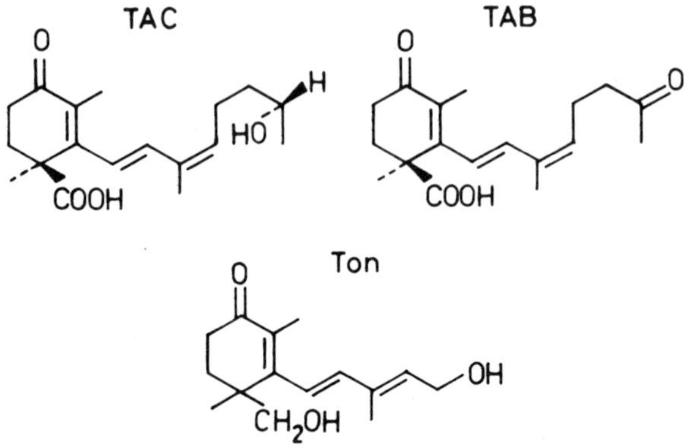

Figure 3 Structures of trisporic acid C (*TAC*), trisporic acid B (*TAB*), and the C_{15} degradation product trisporone (*Ton*).

as a hormone was at first unsuspected, until it was directly shown to be identical to the sex hormone produced in tiny amounts by *Mucor mucedo* (see reviews by Gooday, 1973; van den Ende, 1976).

Sexuality in the Mucorales, an unpredictable mystery to nineteenth century mycologists, was elucidated by Blakeslee (1904). In the heterothallic species the two mating types, designated (+) and (-), are indistinguishable when grown apart. They each grow vegetatively and reproduce asexually, with the production of sporangiospores. When they come into contact, however, their differentiation is totally switched, so that they form the characteristic sexual hyphae, zygophores. These grow chemotropically into contact, and plasmogamy occurs, leading to the formation of the characteristically large, thick-walled resting spore, the zygospore, in which karyogamy and eventually meiosis will occur. That the switch to sexual differentiation is controlled by diffusible hormones is readily shown by demonstrating that zygophore formation occurs when the two mating types are physically separated by a membrane or even by an air gap (Gooday, 1978; Mesland et al., 1974). That this hormone system is not species specific, but universal among the Mucorales, is shown by mutual zygophore formation (but not, of course, zygospore formation) when (+) and (-) strains of different species are grown together (Blakeslee and Cartledge, 1927). This phenomenon allowed Blakeslee and all subsequent workers to assign the mating type to any new isolate by reference back to the original two strains of *Rhizopus stolonifer* that Blakeslee arbitrarily designated (+) and (-). Some strains are self-fertile, that is, homothallic, forming zygospores by fusion of adjacent zygophores formed on the same mycelium. Homothallic strains also typically show interspecific sexual reactions as described above for heterothallic strains.

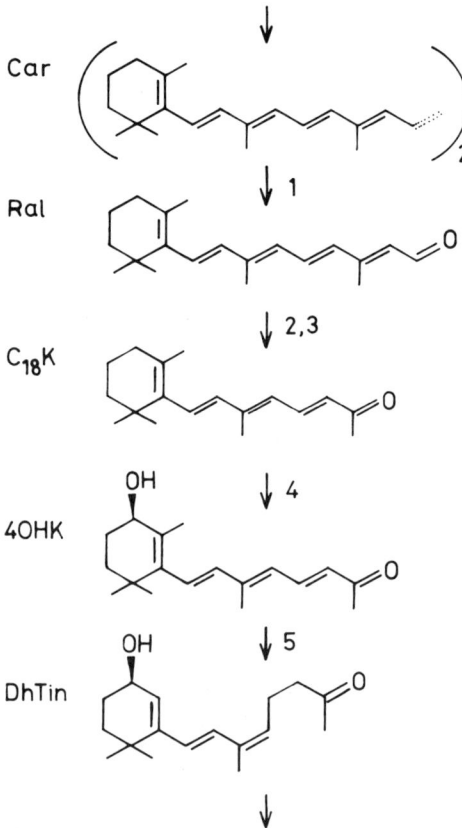

Figure 4 Suggested biosynthetic steps of trisporic acid that are common to both mating types: β-Carotene (*Car*), retinal (*Ral*), β-C_{18}-ketone ($C_{18}K$), 4-hydroxy-β-C_{18}-ketone (*4OHK*), and 4-dihydrotrisporin B (*DhTin*). The numbers refer to enzyme activities in Table 2.

Three species have been used for nearly all research on trisporic acid:

1. *Blakeslea trispora,* which produces large amounts in mated fermentations ($\sim 10^{-3}$ M) (Bu'Lock and Osagie, 1973).
2. *Mucor mucedo,* which produces much smaller amounts ($\sim 10^{-6}$ M, i.e., more in keeping with its hormonal role, but which provides an excellent and sensitive bioassay, detecting 10 nM (Gooday, 1978).

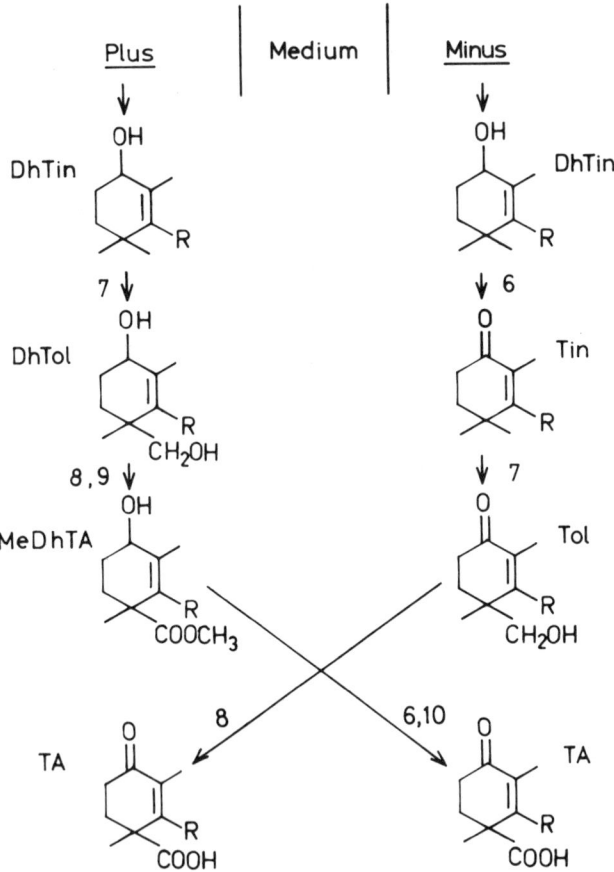

Figure 5 Suggested major biosynthetic steps of trisporic acid in *plus* and *minus* hyphae. *plus*: 4-dihydrotrisporin (*DhTin*), 4-dihydrotrisporol (*DHTol*), methyl-4-dihydrotrisporate (*MeDhTA*), and trisporic acid (*TA*). *minus*: 4-dihydrotrisporin (*DhTin*), trisporin (*Tin*), trisporol (*Tol*), and trisporic acid (*TA*). The numbers refer to enzyme activities in Table 2.

3. *Phycomyces blakesleeanus,* which produces very small amounts and is poorly sensitive to exogenous trisporic acid, but which has been the subject of most genetic study in the Mucorales (Sutter, 1975; Cerdá-Olmedo, 1975).

Trisporic acids are C_{18} terpenoids, the two major ones being B (ketone) and C (alcohol) (Fig. 3). Both compounds elicit a complete switch from asexual to sexual differentiation in (+) and (−) mycelia.

Table 2 Mating-Type Specificity and Enzymology of Trisporic Acid Biosynthesis[a]

Metabolic step	Product	Probable enzyme	Site
1. β-Carotene cleavage	Retinal	Carotene 14,15′-oxygenase	*plus, minus*
2. Retinal oxidation	Retinoic acid	Dehydrogenase	*plus, minus*
3. Retinoate cleavage	β-C_{18} Ketone	Lyase via coenzyme A derivative	*plus, minus*
4. Oxidation at C-4 to alcohol	4-Hydroxy-β-C_{18} ketone	Monooxygenase	*plus, minus*
5. Reduction at C-11,12	4-Dihydrotrisporin	?	*plus, minus*
6. Oxidation at C-4 to ketone	Trisporin, trisporol, trisporate	Dehydrogenase, nicotinamide adenine dinucleotide linked	(*plus*) *minus*
7. Oxidation of 1-methyl to alcohol	4-Dihydrotrisporol, trisporol	Monooxygenase, *cyt P450*?	*plus, minus*
8. Oxidation of 1-CH_2-OH to 1-COOH	4-Dihydrotrisporate, trisporate	Dehydrogenases, via aldehyde	*plus*
9. Methylation of 1-carboxylate	Methyl 4-dihydrotrisporate (methyl trisporate)	?	*plus*
10. Demethylation	Trisporate	Esterase	(*plus*) *minus*
11. 13-oxo/13-hydroxyl interversion	Trisporate C/B	Dehydrogenase	*plus, minus*

[a] References in text.

β-Carotene (C_{40}) and its oxidation product retinal (C_{20}) are biosynthetic precursors of trisporic acid (Austin et al., 1970). Retinal has been demonstrated as a minor metabolite in *P. blakesleeanus* (Meissner and Delbrück, 1968).

Unmated cultures of (+) and (−) strains produce negligible trisporic acid. The carbon skeleton of trisporic acid is contributed, in approximately equal amounts, by each mating type in the form of sex-specific metabolites, which are converted to trisporic acid by the opposite mating type (Figs. 4 and 5 and Table 2; reviewed by van den Ende, 1976; Bu'Lock et al., 1976; Sutter, 1977; Gooday et al., 1979; Jones et al., 1981). As a general summary, (−) strains oxidize at C-4 to form the ketone, and hydrolyze the methyl ester; (+) strains oxidize the pro-S-methyl group on C-1 through to the methyl ester of the carboxyl group. Sutter and Whitaker (1981a,b) reported finding small amounts of methyl trisporates in unmated (+) cultures of *B. trispora*, showing that the (+) mating type does have some ability to oxidize the alcohol at C-4. They further found a C_{15} degradation product, trisporone (Fig. 3), which is biologically inactive, accumulating in these cultures, and suggested that its formation from trisporol, the C_{18} precursor of trisporates, is a regulatory mechanism preventing trisporic acid formation in unmated (+) cells.

The sex-specific intermediates are produced in only small amounts in the unmated strains, but trisporic acid greatly stimulates their production, so that the entire metabolic sequence becomes fully operational only as (+) and (−) approach one another—termed "cascade expression" by Bu'Lock (1975).

As common with most pathways of secondary metabolism, little is known of the enzymology of trisporate biosynthesis of the enzyme reactions in Table 2: Only those steps for 6 and 10 have been identified in *M. mucedo* (Werkman, 1976); steps 1 and 2 are characterized for mammals (see the review by Jones et al., 1981).

The morphological response to trisporic acid is a complete cessation of asexual differentiation and an initiation of sexual differentiation. Our knowledge of the biochemical responses is incomplete. Cyclic adenosine 5′-monophosphate (cAMP) levels rise transiently to a maximum at 4 hr after treatment of mycelium of *M. mucedo* with trisporic acid (Bu'Lock et al., 1976). Zygophore surfaces have a distinctive chemistry, as revealed by their specific straining with fluorescently labeled antibodies and lectins (Jones and Gooday, 1977, 1978). The best-characterized biochemical response is the marked increase in terpenoid biosyntheses.

Addition of trisporic acid to unmated mycelium of *M. mucedo* and *B. trispora* leads to a very large increase in carotene content, an increase in sterol content, proportionally smaller but much greater in total quantity, and increases in ubiquinone and polyprenol contents (reviewed by Jones et al., 1981). Similar increases are also found in the mating region where two compatible colonies meet, in rare sexually heterozygous diploids of *Mucor hiemalis*, and in mating-type heterokaryons of *P. blakesleeanus* (Gooday, 1978; Gooday and Gauger, 1975; Gauger et al., 1980). The resultant accumulated carotene is a precursor for sporopollenin, a component of the zygospore wall (Gooday, 1981).

The possible mechanisms for regulation of the increased terpenoid biosynthesis are discussed by Jones et al. (1981). Investigation of two enzymes, mevalonate kinase and 3-hydroxy-3-methylglutaryl-coenzyme A reductase, has shown that in *M. mucedo* neither has the behavior to be expected from a regulatory enzyme for this biosynthetic pathway (I. P. Williamson, personal communication).

In order to explain the apparent anomalies in sexuality of homothallic and heterothallic Mucorales, Schipper and Stalpers (1980) suggested that these fungi also may have their mating-type regulatory gene as part of a transposable "cassette" system as described for *S. cerevisiae*. Homothallic and heterothallic stains would all have the genetic *potential* for "*plus*ness" and "*minus*ness," but this would normally be under strict control. This model would allow such phenomena as a mating-type switch induced by mutagenetic treatment (Nielsen, 1978; Bu'Lock and Hardy, 1979) and the "Laniger" (+) strain of *M. mucedo*, which produces zygophores (Gooday, 1973).

PEPTIDES

α and *a* Factors

Saccharomyces species are ascosporogenous yeasts. Haploid cells are of two mating types, α and *a*. These will grow vegetatively by budding until brought together, when they will fuse to form the diploid zygote. The zygote in turn will grow vegetatively by budding until induced to sporulate. Meiosis then occurs, with the formation of four haploid ascospores. The ascospores germinate to give haploid yeast cells to complete the life cycle. Levi (1956) showed that mating in *S. cerevisiae* is regulated by diffusible sex hormones. When α and *a* cells are placed next to one another on agar, they cease budding, swell, elongate toward each other, and fuse. The resultant swollen, pear-shaped cells are called "shmoos," after the amorphous character in the Li'l Abner cartoon. The hormones produced by α and *a* cells are called α factor and *a* factor, respectively.

α Factor has been characterized as a mixture of four peptides (Fig. 6; Stötzler and Duntze, 1976; Stötzler et al., 1976; Sakurai et al., 1976); α*1* and α*3* are tridecapeptides, and α*2* and α*4* are dodecapeptides, lacking the N-terminal tryptophan residue; α*3* and α*4* are oxidation products containing a methionine sulfoxide instead of methionine. α Factor has been synthesized by several groups and the synthetic peptides are fully active in bioassays (Manney et al., 1981).

a Factor has proved much more intractable, as it is very hydrophobic and shows a strong propensity to associate with very high molecular weight mannans

1. α Factor
   - α1: NH_2–Trp–His–Trp–Leu–Gln–Leu–Lys–Pro–Gly–Gln–Pro–Met–Tyr–COOH
   - α2: NH_2–His–Trp–Leu–Gln–Leu–Lys–Pro–Gly–Gln–Pro–Met–Tyr–COOH
   - α3: NH_2–Trp–His–Trp–Leu–Gln–Leu–Lys–Pro–Gly–Gln–Pro–Met(SO)–Tyr–COOH
   - α4: NH_2–His–Trp–Leu–Gln–Leu–Lys–Pro–Gly–Gln–Pro–Met(SO)–Tyr–COOH

2. a Factor
   - NH_2–Tyr(Asx, Gly, Ala, Val, Ile, Ile, Phe, Lys, Trp, Pro)–COOH

3. Tremerogen A–10
   - NH_2–Glu–His–Asp–Pro–Ser–Ala–Pro–Gly–Asn–Gly–Tyr–Cys(farnesyl)–$COOCH_3$

4. Tremerogen a–13
   - NH_2–Glu–Gly–Gly–Gly–Asn–Arg–Gly–Asp–Pro–Ser–Gly–Val–Cys(farnesyl)–COOH

5. Rhodotorucine A
   - NH_2–Tyr–Pro–Glu–Ile–Ser–Trp–Thr–Arg–Asn–Gly–Cys(farnesyl)–COOH

Figure 6 Structures of peptide sex hormones.

during purification. Nevertheless, Betz et al. (1977) and Betz and Duntze (1979) have characterized it as an undecapeptide (Fig. 6).

α Factor (and presumably also a factor) is synthesized via messenger RNA and ribosomes, and presumably specific proteolysis of the resultant precursor protein, as its synthesis is inhibited by cycloheximide and in conditional mutants deficient in protein or RNA synthesis (Scherer et al., 1974). The yield is about 30 mM in the culture filtrate (Duntze et al., 1973).

The morphological response to α and a factor, "shmooing," is the basis of the simplest bioassay, the determination of the dilution end point where cells cease to schmoo. Because α factor has been more readily available than a factor, most workers have preferentially studied the effects of α factor on a cells, but the action of a factor on α cells appears to be identical.

When a cells are treated with α factor, they swell so that the mean volume nearly doubles over 4 hr (Manney et al., 1981). The wall of the elongating part of the cell acquires a diffuse outer surface and is thinner than that of the vegetative cell. The nucleus often migrates toward the tip. The wall becomes much more susceptible to lysis by glucanases. The ratio of glucan to mannan rises from 1:1 to 3:2, and the mannan contains an increased proportion of shorter side chains (Lipke et al., 1976). The apical portions of the elongating cells show a higher affinity for fluorescent concanavalin A (MacKay, 1978). Chitin synthase activity increases, and the elongating wall has a higher chitin content (Schekman and Brawley, 1979). Sexual agglutinins appear on the cell surface (Betz et al., 1978; Yanagishima and Yoshida, 1981). The cell cycle is arrested at G_1 "start," so that the cells accumulate as unbudded haploid cells, with initiation of DNA synthesis inhibited (Hartwell, 1978). The membrane-bound adenylate cyclase is also inhibited (Liao and Thorner, 1980).

All of these activities can be seen as courtship responses preparing the cells for mating. They are transient effects; the cells resume vegetative growth after a few hours if mating has not occurred. Thus 3 hr after the addition of 3.38 units/ ml of α factor to a cells, the proportion of budded cells had dropped from 50 to 0.2%, but 2 hr later had risen rapidly to 70% (Manney et al., 1981). The response was concentration dependent, so that equivalent figures were a drop from 50 to 20% after 1 hr and a rise to 40% after a further 1 hr for 0.22 units/ml, and a drop from 50 to 10% after 3 hr and a rise to 46% after a further 2 hr for 0.93 units/ml. To a large extent, the recovery is probably the result of specific degradation of the peptide hormone (Chan, 1977; Ciejek and Thorner, 1979; Finkelstein and Strausberg, 1979).

α Factor is not species specific, as α factors from *S. cerevisiae* and *Saccharomyces kluyveri* are both active on a cells of both species (McCullough and Herskowitz, 1979).

Saccharomyces strains are very amenable to genetic analysis, and great progress has been made in understanding genetic regulation in this organism. By analysis of the regulation of mating type in a range of strains, especially of the mating-type switches in homothallic strains (Klar et al., 1981; Nasmyth et al., 1981; Sprague et al., 1981), DNA sequences and hence the amino acid sequences of the mating-type gene products of *S. cerevisiae* have been characterized. At this rate of achievement, our knowledge of regulatory mechanisms in this organism will soon far outstrip that of other systems.

Tremerogens

Tremella mesenterica is a heterobasidiomycetous jelly fungus capable of growing vegetatively as a budding yeast. Bandoni (1965) has demonstrated its hormonal control of sexual reproduction by showing that yeast cells of each mating type, A and a, produce diffusible factors that cause cells of opposite mating type to cease budding and form conjugation tubes. Reid (1974) reported partial purification of the hormones and suggested that they could be peptides. This has been confirmed by the characterization of two peptides, termed tremerogen *A-10* (from A-type cells) and tremerogen *a-13* (from a-type cells) (Fig. 6; Sakagami et al., 1978a,b, 1979, 1981a,b; Yoshida et al., 1981). Tremerogen *A-10* is a lipopeptide, with the terpenoid farnesol linked to the sulfhydryl group of the C terminal cysteine residue, and with the carboxyl group of the cysteine methylated. Tremerogen *a-13* is also a lipopeptide, but with the C terminus of S-trans, trans-farnesylcysteine, that is, with a free carboxyl group. Synthetic tremerogen is fully active in the bioassay (Kitada et al., 1981), and Fujino et al. (1980) have presented biological activities of a range of analogs.

The morphological response to the tremerogens is the formation of conjugation tubes. This filamentous growth continues as long as the tremerogens are added, but the cells revert to budding when they are removed (Tsuchiya and Fukui, 1978a; Flegel, 1981). The conjugation tube wall is similar in composition to the yeast wall, but contains less α-glucan and more β-glucan and chitin (Reid and Bartinicki-Garcia, 1976). Tremerogen *A-10* arrests DNA synthesis in *a* cells (E. Tsuchiya, quoted in Flegel, 1981).

Related yeasts, namely, other species of *Tremella* (Reid, 1974) and *Sirobasidium magnum* (Flegel, 1981), appear to have very similar hormonal systems, but without cross-reactivity. Unexpectedly the hormones and hormone activities of *Tremella* species are very closely analogous to those of *Saccharomyces* species, but again, there is no cross-reactivity. (Note: The tremerogens must not be confused with the "tremorgens," a class of mycotoxins inducing tremors in animals.)

Rhodotorucines

Rhodosporidium is also a genus of heterobasidiomycetous yeasts, of the Ustilaginales. Abe et al. (1975) have demonstrated that conjugation tube formation in *Rhodosporidium toruloides* is regulated by diffusible hormones. Unlike the situation in *Saccharomyces* species and *Tremella* species, where both mating types constitutively produce their sex-specific hormones, only A cells of *R. toruloides* constitutively produce their hormone, rhodotorucine *A*; *a* cells then produce their hormone rhodotorucine *a* in response. This is analogous to the antheridiol-oogoniol system of *Achlya* species, but rhodotorucine *A* has proved to be much easier to work with.

Rhodotorucine *A* is a lipopeptide (Fig. 6) remarkably similar to the tremerogens, with a farnesyl side chain (Kamiya et al., 1977, 1978a,b; Sakurai et al., 1978). Synthesized rhodotorucine *A* is fully active in the bioassay (Kitada et al., 1979).

The responses to rhodotorucine *A* by *a* cells of *R. toruloides* include: (1) inhibition of budding, (2) formation of conjugation tubes, (3) inhibition of DNA synthesis (G_1 arrest), and (4) secretion of rhodotorucine *a* (Tsuchiya and Fukui, 1978b; E. Tsuchiya, quoted by Flegel, 1981). As for *Saccharomyces* species and *Tremella* species responding to their hormones, these responses are transient, and cells will revert to vegetative budding unless the rhodotorucine *A* is supplied continuously or the cells mate with *A* cells.

UNCHARACTERIZED HORMONES

As stated in the Introduction, there must be many additional fungal sex hormones waiting to be discovered, yet only a few have been isolated and are close to being characterized.

Among the oomycetes, we can expect the elucidation of further systems analogous to that of *Achlya* species, presumably also with steroid hormones. Thus Sherwood (1966) described the induction of oogonial formation in female mycelia of *Dictyuchus monosporus* by culture filtrates from mated cultures. *Achlya* and *Dictyuchus* are genera of the Saprolegniales, the members of which have fucosterol, cholesterol, 24-methylenecholesterol, and desmosterol as their major sterols (McCorkindale et al., 1969). The important plant pathogens, *Phytophthora* and *Pythium,* members of the Peronosporales, synthesize no detectable sterols. They require exogenous sterols, however, for sexual differentiation to give oospore production (Elliott et al., 1964). An appealing idea is that the added sterols provide precursors of specific sex hormones. Thus when cholesterol or cholestanol were added to cultures of *Phytophthora cactorum,* the numbers of mature oospores formed were 313 and 6, respectively, whereas the numbers of aborting oogonia were 53 and 224 (Elliott and Sansome, 1977). These authors suggested that there is a specific delimiting hormone controlling meiosis for which cholesterol is an effective precursor but cholestanol is not. Perhaps because this putative sterol hormone will only be produced in the presence of a large excess of exogenous precursor, attempts at characterizing it have not succeeded to date. Ko (1978, 1980), growing species of *Phytophthora* on a complex medium, described induction of oospores in heterothallic strains by diffusible hormones from compatible heterothallic strains and from homothallic strains. He termed these α hormones, $α^1$ being secreted by A^1 isolates and inducing oospore formation in A^2 isolates, and $α^2$ being secreted by A^2 isolates and inducing oospore formation in A^1 isolates.

Among the ascomycetes, cell-free extracts from mating cultures of *Neurospora crassa* can stimulate sexual morphogenesis when added to cultures of that fungus, by stimulating either fertility in poorly fertile crosses or development and even ascospore formation in protoperithecia of unmated strains (Islam, 1981; Vigfusson and Cano, 1974). The factor described by Islam is lipid in nature, perhaps a hydrocarbon, while that of Vigfusson and Cano behaves like a protein. Likewise, the estrogenic mycotoxin zearalenone can enhance perithecium production in its producing strains of *Gibberella zea* (*Fusarium roseum*); Wolf and Mirocha (1973) suggested that this metabolite may be an endogenous regulator of sexual differentiation. In *Glomerella cingulata,* extracts from strongly mating cultures also will stimulate mating in other cultures (Driver and Wheeler, 1955). Hormone systems which are ripe for reinvestigation have been described for *Bombardia lunata* (Zickler, 1952) and *Ascobolus stercorarius* (Bistis, 1957).

The sexual development of the cellular slime molds results in the production of large dormant structures, the macrocysts. This process is controlled by hormones which have yet to be characterized (O'Day and Lewis, 1981). *Dictyostelium discoideum* strain *NC-4* produces a volatile sex hormone which induces macrocyst formation in strain *V-12* (O'Day and Lewis, 1975; Lewis and O'Day, 1977); *Dictyostelium purpureum* strains *Dp6* and *Dp7* produce a sex hormone that induces macrocyst formation in strain *Dp2* (Lewis and O'Day, 1976). In each of these species, the reciprocal activity (e.g., effect of *V-12* on *NC-4*) has not been observed, and the hormones are species specific. Of the four mating types of *Dictyostelium giganteum,* each secretes its own sex hormone to which each of the other three strains responde by forming macrocysts. The four can be placed in hierarchical order, the one with the strongest ability to induce, *WS589*, being also the one with the weakest ability to respond (Lewis and O'Day, 1979).

DISCUSSION

The major senses of the microbe are chemical. Fungi use chemicals so that they can communicate to coordinate their sexual reproduction. Sexual reproduction is important to them, as it allows meiotic recombination, but it requires a diversion from vegetative growth and asexual reproduction. Thus it must be an

Table 3 Properties of Fungal Sex Hormones

Hormone	Molecular structure	Probable precursor	Site and specificity of synthesis	Optimal yield (M)	Sensitivity of bioassay (M)	Response
Sirenin	Sesquiterpene, $C_{15}H_{24}O_2$	Farnesyl pyrophosphate	♀ Gametes, *Allomyces* species, constitutive	10^{-6}	10^{-10}	Chemotaxis of ♂ gametes
Antheridiol	Sterol $C_{29}H_{42}O_5$	Fucosterol	♀ Cells, *Achyla* species, constitutive	10^{-8}	10^{-11}	Antheridia and oogoniol production by ♂
Oogoniol	Sterol ester, $C_{33}H_{54}O_6$	Fucosterol	♂ Cells, *Achlya* species, induced	—	10^{-6}	Oogonia by ♀
Trisporic acid	Apocarotenoid, $C_{18}H_{26}O_4$	Retinal	(+)/(−) Cells, Mucorales, collaborative	10^{-6} (*M. mucedo*); 10^{-3} (*B. trispora*)	10^{-8}	Zygophores by (+) and (−)
α Factor	Dodeca- and tridecapeptides	Protein	α Cells, *S. cerevisiae*, constitutive	10^{-8}	10^{-8}	Shmoos by *a*
a Factor	Undecapeptide	Protein	*a* Cells, *S. cerevisiae*, constitutive	—	—	Shmoos by α
Tremerogen *A*	Farnesyl dodecapeptide	Protein, farnesyl pyrophosphate	*A* Cells, *T. mesenterica*, constitutive	10^{-7}	10^{-9}	Conjugation tubes by *a*
Tremerogen *a*	Farnesyl tridecapeptide	Protein, farnesyl pyrophosphate	*a* Cells, *T. mesenterica*, constitutive	10^{-8}	—	Conjugation tubes by *A*
Rhodotorucine *A*	Farnesyl undecapeptide	Protein, farnesyl pyrophosphate	*A* Cells, *R. toruloides*, constitutive	—	—	Conjugation tubes and rhodotorucine production by *a*

efficient and orderly process. The specificity of the hormones ensures that sexual differentiation only occurs when there is a good chance of successful mating. Most are species specific. An exception is trisporic acid, which appears to be a hormone for all of the Mucorales. Perhaps there is sufficient ecological specialization among these fungi to make an unconsummated pairing a rare event.

General properties of these hormones are summarized in Table 3. All are produced in tiny amounts (except trisporic acid by *B. trispora*), as befits molecules with such powerful activities. Some are produced constitutively; the synthesis of others such as oogoniol and rhodotorucine *a* requires induction by the complementary hormone. They tend to be unstable molecules and to be destroyed in some manner by responding cells, both phenomena presumably helping to accentuate concentration gradients and responsiveness. It has been argued that in the biosynthesis of trisporic acid in the Mucorales it is the mating-type-specific precursors that should be considered as the hormones (for example, *P. blakesleeanus* is more responsive to these than to trisporic acid) (Sutter, 1977). These precursors are converted to trisporic acid by recipient mycelium; trisporic acid is active without further metabolism.

As two mating-type cells approach each other, there is increased hormone synthesis, as illustrated by the trisporic acid system. This is a "chemical display" analogous to animal mating displays, for example, the visual displays of birds during courtship.

All of these hormones are active in very tiny amounts. All are terpenoids or hydrophobic peptides, but no information is available on receptors in responding cells. Seeing the progress that has been made with the identification of receptors for mammalian hormones, the interaction of these hormones with cell membranes merits investigation. The evolution of these hormone systems required the acquisition of the hormone and its receptor, presumably in that order. If the organism was already producing the hormone molecule or a related molecule as a minor secondary metabolite, it would already have, in the biosynthetic enzyme, a protein with specificity for this molecule. The receptor in the present-day system might then have arisen from such a protein. With the current rate of progress of knowledge with *S. cerevisiae*, it is with this system that we can expect such questions to be answered first.

The fungal sex hormones have clear structural affinities with hormones and biologically active molecules in other systems: sirenin with the "paper factors" juvabione and dehydrojuvabione with insect juvenile hormone activity, antheridiol and oogoniol with ecdysone and other animal sterol hormones, trisporic acid with abscisic acid in plants and retinal and its metabolites in mammals, and the hydrophobic peptides with mammalian hormones such as luteinizing hormone releasing factor (which has sequences in common with them) (Kitada et al., 1979). In this last example, it would seem that all of these peptides have structural features necessary for interaction with particular types of receptors, probably membrane bound, so that similarities are unlikely to indicate any phylogenetic relationship between peptides of *Saccharomyces, Tremella,* and *Rhodosporidium* species.

ACKNOWLEDGMENTS

I thank the many authors that have provided me with reprints and preprints of their work.

REFERENCES

Abe, K., Kusaka, I., and Fukai, S. (1975). Morphological change in the early stages of the mating process of *Rhodosporidium toruloides.* J. Bacteriol. *122*:710-718.

Arsenault, G. P., Biemann, K., Barksdale, A. W., and McMorris, T. C. (1968). The structure of antheridiol, a sex hormone in *Achlya bisexualis.* J. Am. Chem. Soc. *90*:5635-5636.

Austin, D. J., Bu'Lock, J. D., and Drake, D. (1970). The biosynthesis of trisporic acids from β-carotene via retinal and trisporol. Experientia *26*:348-349.

Bandoni, R. J. (1965). Secondary control of conjugation in *Tremella mesenterica.* Can. J. Bot. *43*:627-630.

Barksdale, A. W., and Lasure, L. L. (1974). Production of hormone B by *Achlya heterosexualis.* Appl. Microbiol. *28*:544-546.

Barksdale, A. W., McMorris, T. C., Seshadri, R., Aranachalam, T., Edwards, J. A., Sundeen, J., and Green, J. M. (1974). Response of *Achlya ambisexualis* E87 to the hormone antheridiol and certain other steroids. J. Gen. Microbiol. *82*:295-299.

Betz, R., and Duntze, W. (1979). Purification and partial characterisation of a-factor, a mating hormone produced by mating-type-*a* cells from *Saccharomyces cerevisiae.* Eur. J. Biochem. *95*:469-475.

Betz, R., MacKay, V. L., and Duntze, W. (1977). a-Factor from *Saccharomyces cerevisiae*: Partial characterization of a mating hormone produced by cells of mating type *a.* J. Bacteriol. *132*:462-472.

Betz, R., Duntze, W., and Manney, T. R. (1978). Mating-factor-mediated sexual agglutination in *Saccharomyces cerevisiae.* FEMS Microbiol. Lett. *4*:107-110.

Bistis, G. (1957). Sexuality in *Ascobolus stercorarius.* II. Preliminary experiments on various aspects of the sexual process. Am. J. Bot. *44*:436-443.

Blakeslee, A. F. (1904). Sexual reproduction in the Mucorineae. Proc. Am. Acad. Arts Sci. *40*:205-319.

Blakeslee, A. F., and Cartledge, J. L. (1927). Sexual dimorphism in Mucorales. II Interspecific reactions. Bot. Gaz. Chicago *84*:51-57.

Bu'Lock, J. D. (1975). Cascade expression of the mating-type-locus in Mucorales. In *Proceedings of the Second International Symposium on Genetics of Industrial Micro-organisms*. 1974, K. D. Macdonald (Ed.). Academic, London, pp. 497-509.

Bu'Lock, J. D., and Hardy, T. M. (1979). Mating raaction between sexually altered strains of *Mucor pusillus*. Exp. Mycol. *3*:194-196.

Bu'Lock, J. D., and Osagie, A. U. (1973). Prenols and ubiquinones in single strain and mated cultures of *Blakeslea trispora*. J. Gen. Microbiol. *76*: 77-83.

Bu'Lock, J. D., Jones, B. E., and Winskill, N. (1976). The apocarotenoid system of sex hormones and prohormones in Mucorales. Pure Appl. Chem. *47*: 191-202.

Caglioti, L., Cainelli, G., Camerino, B., Mondelli, R., Prieto, A., Quilico, A., Salvatori, T., and Selva, A. (1964). Sulla constituzione degli acidi trisporici. Chim. Ind. Milan *46*:961-966.

Caglioti, L., Cainelli, G., Camerino, B., Mondelli, R., Prieto, A., Quilico, A., Salvatori, T., and Selva, A. (1967). The structure of trisporic acid C. Tetrahedron Suppl. *7*:175-187.

Carlile, M. J., and Gooday, G. W. (1978). Cell fusion in myxomycetes and fungi. In *Membrane Fusion*, G. Poste and G. L. Nicholson (Eds.). Elsevier, New York, pp. 219-265.

Carlile, M. J., and Machlis, L. (1965). A comparative study of the chemotaxis of the motile phases of *Allomyces*. Am. J. Bot. *52*:484-486.

Cerdá-Olmedo, E. (1975). The genetics of *Phycomyces blakesleeanus*. Genet. Res. *25*:285-296.

Chan, R. K. (1977). Recovery of *Saccharomyces cerevisiae* mating-type *a* cells from G_1 arrest by α-factor. J. Bacteriol. *130*:766-774.

Ciejek, E., and Thorner, J. (1979). Recovery of *Saccharomyces cerevisiae a* cells from G_1 arrest by α-factor pheromone requires endopeptidase action. Cell *18*:623-635.

Driver, C. H., and Wheeler, H. E. (1955). A sexual hormone in *Glomerella*. Mycologia *47*:311-316.

Duntze, W., Stötzler, D., Bucking-Throm, E., and Kaltitzer, S. (1973). Purification and partial characterization of α-factor, a mating-type specific inhibitor of cell reproduction from *Saccharomyces cerevisiae*. Eur. J. Biochem. *35*: 357-365.

Elliott, C. G., and Sansome, E. (1977). The influence of sterols on meiosis in *Phytophthora cactorum*. J. Gen. Microbiol. *98*:141-145.

Elliott, C. G., Hendrie, M. R., Knights, B. A., and Parker, W. (1964). A steroid growth factor requirement in a fungus. Nature *230*:427-428.

Emerson, R., and Wilson, C. M. (1954). Interspecific hybrids and the cytogenetics and cytotaxonomy of *Euallomyces*. Mycologia *46*:393-434.

Ende, H. van den (1976). *Sexual Interactions in Plants. The Role of Specific Substances in Sexual Reproduction*. Academic, London.

Finkelstein, D., and Strausberg, S. (1979). Metabolism of α-factor by *a* mating type cells of *Saccharomyces cerevisiae*. J. Biol. Chem. *254*:796-803.

Flegel, T. W. (1981). The pheromonal control of mating in yeasts and its phylogenetic implication: A review. Can. J. Microbiol. *27*:373-389.

Fujino, M., Kitada, C., Sakagami, Y., Isogai, A., Tamura, S., and Suzuki, A. (1980). Biological activity of synthetic analogs of tremerogen A-10, a mating hormone of the heterobasidiomycetous yeast, *Tremella mesenterica*. Naturwissenschaften *67*:406-407.

Gauger, W. L., Pelaez, M. I., Alvarez, M. I., and Eslava, A. P. (1980). Mating type heterokaryons in *Phycomyces blakesleeanus*. Exp. Mycol. *4*:56-64.

Gooday, G. W. (1973). Differentiation in the Mucorales. Soc. Gen. Microbiol. Symp. *23*:269-293.

Gooday, G. W. (1974). Fungal sex hormones. Annu. Rev. Biochem. *43*:35-49.

Gooday, G. W. (1978). Functions of trisporic acid. Philos. Trans. R. Soc. London B. *284*:509-520.

Gooday, G. W. (1981). Biogenesis of sporopollenin in fungal spore walls. In *The Fungal Spore: Morphogenetic Controls*, G. Turian and H. Hoch (Eds.). Academic, London, pp. 487-505.

Gooday, G. W., and Gauger, W. L. (1975). Control of sexuality in Mucorales. In *Proceedings of the First Intersectional Congress of IAMS*, T. Hasegawa (Ed.). Science Council of Japan, Tokyo, pp. 273-282.

Gooday, G. W., Jones, B. E., and Leith, W. H. (1979). Trisporic acid and the control of sexual differentiation in the Mucorales. In *Regulation of Secondary Product and Plant Hormone Metabolism*, M. Luckner and K. Schreiber (Eds.). Pergamon, Oxford, pp. 221-229.

Gwynne, D. J., and Brandhorst, B. P. (1980). Antheridiol-induced differentiation in the absence of detectable synthesis of new proteins. Exp. Mycol. *4*:251-259.

Harold, F. M. (1982). Pumps and currents: A biological perspective. Curr. Top. Membr. Transp. *16*:485-516.

Hartwell, L. H. (1978). Cell division from a genetic perspective. J. Cell. Biol. *77*:627-637.

Hatch, W. R. (1938). Conjugation and zygote germination in *Allomyces arbuscula*. Ann. Bot. N.S. *2*:583-614.

Horgen, P. A. (1981). The role of the steroid sex pheromone antheridiol in controlling the development of male sex organs in the water mold, *Achlya*. In *Sexual Interactions in Eukaryotic Microbes*, D. H. O'Day and P. A. Horgen (Eds.). Academic, New York, pp. 155-178.

Horgen, P. A., and Ball, S. F. (1974). Nuclear protein acetylation during hormone-induced sexual differentiation in *Achlya ambisexualis*. Cytobios *10*:181-185.

Islam, M. S. (1981). Sex pheromones in *Neurospora crassa*. In *Sexual Interactions in Eukaryotic Microbes*, D. H. O'Day and P. A. Horgen (Eds.). Academic, New York, pp. 131-154.

Jones, B. E., and Gooday, G. W. (1977). Lectin binding to sexual cells in fungi. Biochem. Soc. Trans. *5*:719-721.

Jones, B. E., and Gooday, G. W. (1978). An immunofluorescent investigation of the zygophore surface of Mucorales. FEMS Microbiol. Lett. *4*:181-184.

Jones, B. E., Williamson, I. P., and Gooday, G. W. (1981). Sex pheromones in *Mucor*. In *Sexual Interactions in Eukaryotic Microbes*, D. H. O'Day and P. A. Horgen (Eds.). Academic, New York, pp. 179-198.

Kamiya, Y., Sakurai, A., Tamura, S., Abe, K., Tsuchiya, E., and Fukui, S. (1977). Isolation and chemical characterization of the peptidyl factor controlling mating tube formation in *Rhodosporidium toruloides*. Agric. Biol. Chem. *41*:1099-1100.

Kamiya, Y., Sakurai, A., Tamura, S., Takahashi, N., Abe, K., Tsuchiya, E., and Fukui, S. (1978a). Isolation of rhodotoruine A, a peptidyl factor inducing the mating tube formation in *Rhodosporidium toruloides*. Agric. Biol. Chem. *42*:1239-1243.

Kamiya, Y., Sakurai, A., Tamura, S., Takahashi, N., Abe, K., Tsuchiya, E., and Fukui, S. (1978b). Amino acid sequences of rhodotoruine A, a peptidyl factor controlling mating tube formation in *Rhodosporidium toruloides*. Agric. Biol. Chem. *42*:209-211.

Kitada, C., Fujino, M., Tsuchiya, E., Abe, K., Fukui, S., Kamiya, Y., Sakurai, A., Tamura, S., Takahashi, N., Sakagami, Y., Isogai, A., and Suzuki, A. (1979). Synthesis of sex hormones of heterobasidiomycetous yeast. In *Peptide Chemistry 1978* N. Izumiya (Ed.). Protein Research Foundation, Osaka, pp. 195-200.

Kitada, C., Fujino, M., Sakagami, Y., Yoshida, M., Isogai, A., and Suzuki, A. (1981). Synthesis and structure-activity relations of tremergoen a-13, a peptidal sex hormone of *Tremella mesenterica*. Agric. Biol. Chem. *45*: 1049-1051.

Klapper, B. F., and Klapper, M. H. (1977). A natural inhibitor of sexual attraction in the water mold *Allomyces*. Exp. Mycol. *1*:352-355.

Klar, A. J. S., Strathern, J. N., Broach, J. R., and Hicks, J. B. (1981). Regulation of transcription in expressed and unexpressed mating type cassettes of yeast. Nature *289*:239-244.

Ko, W. H. (1978). Heterothallic *Phytophthora*: Evidence for hormonal regulation of sexual reproduction. J. Gen. Microbiol. *107*:15-18.

Ko, W. H. (1980). Hormonal regulation of sexual reproduction in *Phytophthora*. J. Gen. Microbiol. *116*:459-463.

Levi, J. D. (1956). Mating reaction in yeast. Nature *177*:753-754.

Lewis, K. E., and O'Day, D. H. (1976). Sexual hormone in the cellular slime mold *Dictyostelium purpureum*. Can. J. Microbiol. *22*:1269-1273.

Lewis, K. E., and O'Day, D. H. (1977). Sex hormone of *Dictyostelium discoideum* is volatile. Nature *268*:730-731.

Lewis, K. E., and O'Day, D. H. (1979). Evidence for a hierarchical mating system operating via pheromones in *Dictyostelium giganteum*. J. Bacteriol. *138*:251-253.

Liao, H., and Thorner, J. (1980). Yeast mating pheromone alpha-factor inhibits adenylate cyclase. Proc. Nat. Acad. Sci. USA *77*:1898-1902.

Lipke, P. N., Taylor, A., and Ballou, C. E. (1976). Morphogenic effects of α-factor on *Saccharomyces cerevisiae*. J. Bacteriol. *127*:610-618.
McCorkindale, N. J., Hutchinson, S. A., Pursey, B. A., Scott, W. T., and Wheeler, R. (1969). A comparison of the types of sterol found in species of the Saprolegniales and Leptomitales with those found in some other Phycomycetes. Phytochemistry *8*:861-867.
McCullough, J., and Herskowitz, I. (1979). Mating pheromones of *Saccharomyces kluyveri* and *Saccharomyces cerevisiae*. J. Bacteriol. *138*:146-154.
Machlis, L. (1958). A study of sirenin, the chemotactic sexual hormone from the water mold *Allomyces*. Physiol. Plant. *11*:845-854.
Machlis, L. (1973a). Factors affecting the stability and accuracy of the bioassay for the sperm attractant sirenin. Plant Physiol. *52*:524-526.
Machlis, L. (1973b). The chemotactic activity of various sirenins and analogues and the uptake of sirenin by the sperm of *Allomyces*. Plant Physiol. *52*:527-530.
Machlis, L., Nutting, W. H., William, M. W., and Rapoport, H. (1966). Production, isolation, and characterization of sirenin. Biochemistry *5*:2147-2152.
Machlis, L., Nutting, W. H., and Rapoport, H. (1968). The structure of sirenin. J. Am. Chem. Soc. *90*:1674-1676.
MacKay, V. L. (1978). Mating-type specific pheromones as mediators of sexual conjugation in yeast. In *Molecular Control of Proliferation and Differentiation*, J. Papaconstantinou (Ed.). Academic, New York, pp. 243-259.
McMorris, T. C. (1978). Antheridiol and the oogoniols, steroid hormones which control sexual reproduction in *Achlya*. Philos. Trans. R. Soc. London B. *248*:459-470.
McMorris, T. C., and Barksdale, A. W. (1967). Isolation of a sex hormone from the water mould *Achlya bisexualis*. Nature *215*:320-321.
McMorris, T. C., and White, R. M. (1977). The biosynthesis of the oogoniols, steroidal sex hormones of *Achlya*: The role of fucosterol. Phytochemistry *16*:359-362.
McMorris, T. C., Seshadri, R., Weiche, G. R., Arsenault, G. P., and Barksdale, A. W. (1975). Structures of oogoniol -1, -2, and -3, steroidal sex hormones of the water mold, *Achlya*. J. Am. Chem. Soc. *97*:2544-2545.
Manney, T. R., Duntze, W., and Betz, R. (1981). The isolation characterization, and physiological effects of the *Saccharomyces cerevisiae* sex pheromones. In *Sexual Interactions in Eukaryotic Microbes*, D. H. O'Day and P. A. Horgen (Eds.). Academic, New York, pp. 21-51.
Meissner, G., and Delbrück, M. (1968). Carotenes and retinal in Phycomyces mutants. Plant Physiol. *43*:1279-1283.
Mesland, D. A. M., Huisman, J. G., and Ende, H. van den (1974). Volatile sex hormones in *Mucor mucedo*. J. Gen. Microbiol. *80*:111-117.
Mullins, J. T. (1973). Lateral branch formation and cellulase production in the water molds. Mycologia *65*:1007-1014.
Musgrave, A., and Nieuwenhuis, D. (1975). Metabolism of radioactive antheridiol by *Achlya* species. Arch. Microbiol. *105*:313-317.

Nasmyth, K. A., Tatchell, K., Hall, B. D., Astell, C., and Smith, M. (1981). A position effect in the control of transcription of yeast mating type loci. Nature 289:244-250.

Nielsen, R. I. (1978). Sexual mutants of a heterothallic *Mucor* species, *Mucor pusillus*. Exp. Mycol. 2:193-197.

Nutting, W. H., Rapoport, H., and Machlis, L. (1968). The structure of sirenin. J. Am. Chem. Soc. 90:6434-6438.

O'Day, D. H. (1981). Modes of cellular communication and sexual interactions in eukaryotic microbes. In *Sexual Interactions in Eukaryotic Microbes*, D. H. O'Day and P. A. Horgen (Eds.). Academic, London, pp. 3-17.

O'Day, D. H., and Lewis, K. E. (1975). Diffusible mating type factors induce macrocyst development in *Dictyostelium discoideum*. Nature 254:431-432.

O'Day, D. H., and Lewis, K. E. (1981). Pheromonal interactions during mating in *Dictyostelium*. In *Sexual Interactions in Eukaryotic Microbes*, D. H. O'Day and P. A. Horgen (Eds.). Academic, New York, pp. 199-221.

Plattner, J. J., and Rapoport, H. (1971). The synthesis of *d*- and *l*- sirenin and their absolute configurations. J. Am. Chem. Soc. 93:1758-1761.

Plattner, J. J., Bhalerao, U. T., and Rapoport, H. (1969). Synthesis of *dl*-sirenin. J. Am. Chem. Soc. 91:4933.

Pommerville, J. (1977). Chemotaxis of *Allomyces* gametes. Exp. Cell. Res. 109:43-51.

Pommerville, J. (1978). Analysis of gamete and zygote motility in *Allomyces*. Exp. Cell Res. 113:161-172.

Pommerville, J. (1981). The role of sexual pheromones in *Allomyces*. In *Sexual Interactions in Eukaryotic Microbes*, D. H. O'Day and P. A. Horgen (Eds.). Academic, New York, pp. 53-92.

Pommerville, J., and Fuller, M. S. (1976). The cytology of the gametes and fertilization of *Allomyces macrogynus*. Arch. Microbiol. 109:21-30.

Popplestone, C. R., and Unrau, A. M. (1973). Major sterols of *Achlya bisexualis*. Phytochemistry 12:1131-1133.

Popplestone, C. R., and Unrau, A. M. (1974). Studies on the biosynthesis of antheridiol. Can. J. Chem. 52:462-468.

Raper, J. R. (1951). Sexual hormones in *Achlya*. Am. Sci. 39:110-120.

Raper, J. R. (1973). Fungal sex hormones. p. 665-668. In *Biology Data Book*, Vol. 2, P. L. Altman and D. S. Dittmer (Eds.). FASEB, Bethesda, Md.

Raper, J. R., and Haagen-Smit, A. G. (1942). Sexual hormones in Achlya. IV. Properties of hormone A of *Achlya bisexualis*. J. Biol. Chem. 143:311-320.

Reid, I. D. (1974). Properties of conjugation hormones (erogens) from the Basidiomycete *Tremella mesenterica*. Can. J. Bot. 52:521-524.

Reid, I. D., and Bartnicki-Garcia, S. (1976). Composition and structure of cell walls of *Tremella mesenterica*: Yeast cells vs. conjugation tubes. J. Gen. Microbiol. 96:35-50.

Rozek, C. E., and Timberlake, W. E. (1980). Absence of evidence for changes in messenger RNA populations during steroid hormone-induced cell differentiation in *Achlya*. Exp. Mycol. 4:33-47.

Sakagami, Y., Isogai, A., Suzuki, A., Tamura, S., Tsuchiya, E., and Fukui, S. (1978a). Isolation of a novel sex hormone, tremerogen A-10, controlling conjugation tube formation in *Tremella mesenterica* Fries. Agric. Biol. Chem. *42*:1093-1094.

Sakagami, Y., Isogai, A., Suzuki, A., Tamura, S., Tsuchiya, E., and Fukui, S. (1978b). Amino acid sequence of tremerogen A-10, a peptidal hormone, inducing conjugation tube formation in *Tremella mesenterica* Fr. Agric. Biol. Chem. *42*:1301-1302.

Sakagami, Y., Isogai, A., Suzuki, A., Tamura, S., Kitada, C., and Fujino, M. (1979). Structure of tremerogen A-10, a peptidal hormone inducing conjugation tube formation in *Tremella mesenterica*. Agric. Biol. Chem. *43*: 2643-2645.

Sakagami, Y., Yoshida, M., Isogai, A., and Suzuki, A. (1981a). Structure of tremerogen a-13, a peptidal sex hormone of *Tremella mesenterica*. Agric. Biol. Chem. *45*:1045-1047.

Sakagami, Y., Yoshida, M., Isogai, A., and Suzuki, A. (1981b). Peptidal sex hormones inducing conjugation tube formation in compatible mating-type cells of *Tremella mesenterica*. Science *212*:1525-1527.

Sakurai, A., Tamura, S., Yanagishima, N., and Shimoda, C. (1976). Structure of the peptidyl factor inducing sexual agglutination in *Saccharomyces cerevisiae*. Agric. Biol. Chem. *40*:1057-1058.

Sakurai, A., Tamura, S., Takahashi, N., Abe, K., Tsuchiya, E., Fukui, S., Kitada, C., and Fujino, M. (1978). Structure of rhodotorucine A, a novel lipopeptide, inducing mating tube formation in *Rhodosporidium toruloides*. Biochem. Biophys. Res. Commun. *83*:1077-1083.

Schekman, R., and Brawley, V. (1979). Localized deposition of chitin on the yeast cell surface in response to mating pheromone. Proc. Nat. Acad. Sci. USA *76*:645-649.

Scherer, G., Haag, G., and Duntze, W. (1974). Mechanism of α factor biosynthesis in *Saccharomyces cerevisiae*. J. Bacteriol. *119*:386-393.

Schipper, M. A. A., and Stalpers, J. A. (1980). Various aspects of the mating system in Mucorales. Persoonia *11*:53-63.

Sherwood, W. A. (1966). Evidence for a sexual hormone in the water mold *Dictyuchus*. Mycologia *58*:215-220.

Silver, J. C., and Horgen, P. A. (1974). Hormonal regulation of presumptive mRNA in the fungus *Achlya ambisexualis*. Nature *249*:252-254.

Sprague, G. F., Rine, J., and Herskowitz, I. (1981). Homology and nonhomology at the yeast mating type locus. Nature *289*:250-252.

Stötzler, D., and Duntze, W. (1976). Isolation and characterization of four related peptides exhibiting α-factor activity from *Saccharomyces cerevisiae*. Eur. J. Biochem. *65*:257-262.

Stötzler, D., Kiltz, H., and Duntze, W. (1976). Primary structure of α-factor peptides from *Saccharomyces cerevisiae*. Eur. J. Biochem. *69*:397-400.

Sutter, R. P. (1975). Mutations affecting sexual development in *Phycomyces blakesleeanus*. Proc. Nat. Acad. Sci. USA *72*:127-130.

Sutter, R. P. (1977). Regulation of the first stage of sexual development in *Phycomyces blakesleeanus* and in other mucoraceous fungi. In *Eukaryotic Microbes as Model Development Systems*, D. H. O'Day and P. A. Horgen (Eds.). Marcel Dekker, New York, pp. 251-272.

Sutter, R. P., and Whitaker, J. P. (1981a). Sex pheromone metabolism in *Blakeslea trispora*. Naturwissenschaften *68*:147-148.

Sutter, R. P., and Whitaker, J. P. (1981b). Zygophore-stimulating precursors (pheromones) of trisporic acids active in (–)-*Phycomyces blakesleeanus*. J. Biol. Chem. *256*:2334-2341.

Timberlake, W. (1976). Alterations in RNA and protein synthesis associated with steroid-hormone induced sexual morphogenesis in the water mold *Achlya*. Dev. Biol. *51*:202-214.

Trewavas, A. (1981). How do plant growth substances work? Plant Cell Environ. *4*:203-228.

Tsuchiya, E., and Fukui, S. (1978a). Biological activities of sex hormone produced by *Tremella mesenterica*. Agric. Biol. Chem. *42*:1089-1091.

Tsuchiya, E., and Fukui, S. (1978b). Binding of rhodotorucine A, a lipopeptidyl mating hormone to *a* cells of *Rhodosporidium toruloides* for induction of sexual differentiation. Biochem. Biophys. Res. Commun. *85*:473-479.

Turian, G. (1969). *Differenciation Fongique*. Masson, Paris.

Turian, G., and Ojha, M. H. (1969). Indications of an interspecific transformation in *Allomyces*. Experientia *25*:79-81.

Vigfusson, N. V., and Cano, R. J. (1974). Artificial induction of the sexual cycle of *Neurospora crassa*. Nature *249*:383-385.

Werkman, T. A. (1976). Localization and partial characterization of a sex-specific enzyme in homothallic and heterothallic Mucorales. Arch. Microbiol. *109*:209-213.

Wolf, J. C., and Mirocha, C. J. (1973). Regulation of sexual reproduction in *Gibberella zeae* (*Fusarium roseum* "Graminearum") by F-2 (zearalenone). Can. J. Microbiol. *19*:725-734.

Yanagishima, N., and Yoshida, K. (1981). Sexual interactions in *Saccharomyces cerevisiae* with special reference to the regulation of sexual agglutinability. In *Sexual Interactions in Eukaryotic Microbes*, D. H. O'Day and P. A. Horgen (Eds.). Academic, New York, pp. 261-295.

Yoshida, M., Sakagami, Y., Isogai, A., and Suzuki, A. (1981). Isolation of tremerogen *a*-13, a peptidal sex hormone of *Tremella mesenterica*. Agric. Biol. Chem. *45*:1043-1044.

Zickler, H. (1952). Zur Entwicklungsgeschichte des Askomyceten *Bombardia lunata*. Arch. Protistenkd. *98*:1-71.

10
Yeast/Mold Morphogenesis in *Mucor* and *Candida albicans*

Jim E. Cutler and Kevin C. Hazen* / Montana State University, Bozeman, Montana

INTRODUCTION

The phenomenon of yeast (Y) ⇆ mold (M) dimorphism (morphogenesis) has been observed for many years, but research in this area has intensified over the past 10 years. There are at least two major reasons for the current surge of interest. First, the nuclear complement of genetic material in a haploid fungal nucleus, such as in *Saccharomyces cerevisiae,* may be only five times that of a typical prokaryote (Hartwell, 1974). The relatively small number of genes in these primitive eukaryotes, coupled with the fact that Y ⇄ M dimorphism can be controlled in the laboratory and is a reversible event, makes fungi one of the simplest morphogenetic models of eukaryotes. It is hoped that clues to regulation and control of morphogenesis in mammalian and plant systems will be uncovered through studies on fungi. Second, several fungi that are capable of causing disease in man undergo a morphogenetic change upon invasion of host tissues. Organisms such as *Ajellomyces capsulatus* (*Histoplasma capsulatum*), *Ajellomyces* (*Blastomyces*) *dermatitidis, Paracoccidioides brasiliensis,* and *Sporothrix schenckii* grow as fluffy molds at room temperature on usual media, but predominate as yeasts when in host tissues or when grown in an appropriate medium at 37°C. *Candida* species, and most notably *Candida albicans,* usually grow as yeasts on common laboratory media at 24 and 37°C, but grow as a mixture of yeasts and hyphae (hyphae and pseudohyphae) upon tissue invasion in a susceptible animal host. Because a Y ⇄ M transformation of these important disease-producing agents of man seems to be a criterion for infectivity, an understanding of dimorphic mechanisms may lead to improved methods of controlling diseases by these fungi.

Many investigators have approached Y ⇄ M dimorphism in a descriptive manner by examining the effects of physical, chemical, and nutritional factors

*Present affiliation: Washington University School of Medicine, St. Louis, Missouri

on the inducement or repression of morphogenesis. These kinds of studies are needed and have yielded interesting and reproducible results. It is hoped that through such observations a common thread may appear which will give a clue to the underlying mechanisms and pathways that control and regulate fungal morphogenesis. Careful evaluation of the literature reveals a morass of data replete with conflicting conclusions. Some of the confusion may be because different fungal species have different morphogenetic regulatory mechanisms. Much of the problem, however, stems from the number of ways in which investigators study the influence of various factors on Y $\rightleftarrows$ M dimorphism. Some investigators use complex media, while others use media which are chemically defined. Dimorphic changes are induced in some laboratories by temperature shifts, and in other laboratories by nutritional changes. Investigations may be done on only one or two isolates within a given species, even though strain-specific characteristics within a single species of dimorphic fungi have been described. Studies are not always standardized with regard to the growth phase under which cells are analyzed. Many specific discrepancies in studies on *C. albicans* morphogenesis, for example, have been cited elsewhere (Odds, 1979) and by us in this chapter.

It is not our intent to give an exhaustive review of Y $\rightleftarrows$ M dimorphism. Rather, we have chosen to focus on work that has been done on *Mucor* species and on *C. albicans* because (1) systematic studies, and the rationale for such studies, done on *Mucor* over the past several years may serve as a basis or model for approaching morphogenesis in other fungi, and (2) the current level of interest in *C. albicans* as an opportunitistic human pathogen, along with the amount of work done on the morphogenesis of this organism, merits its inclusion in this chapter. Some very fine work on Y $\rightleftarrows$ M dimorphism of other organisms such as *H. capsulatum* has been done. However, studies on *H. capsulatum* and other dimorphic fungi have been reviewed elsewhere and the interested reader is encouraged to consult such reviews (Boguslawski and Stetler, 1979; Maresca et al., 1980, 1981; Rippon, 1980; San-Blas et al., 1980; Cole and Nozawa, 1981).

DIMORPHISM OF *Mucor*

Overview

Mucor is most widely known as a filamentous fungus. Under usual laboratory and natural conditions members of this genus do not produce yeasts, and, in fact, a yeast form is usually not depicted in illustrations on the life cycle of *Mucor* species (e.g., Swatek, 1967). Under anaerobic conditions, certain species of this genus are especially capable of producing a yeast form (Y) which reproduces asexually by bud or blastoconidial formation (Fig. 1). When environmental conditions are appropriately shifted, Y will germinate to initiate a hyphal or

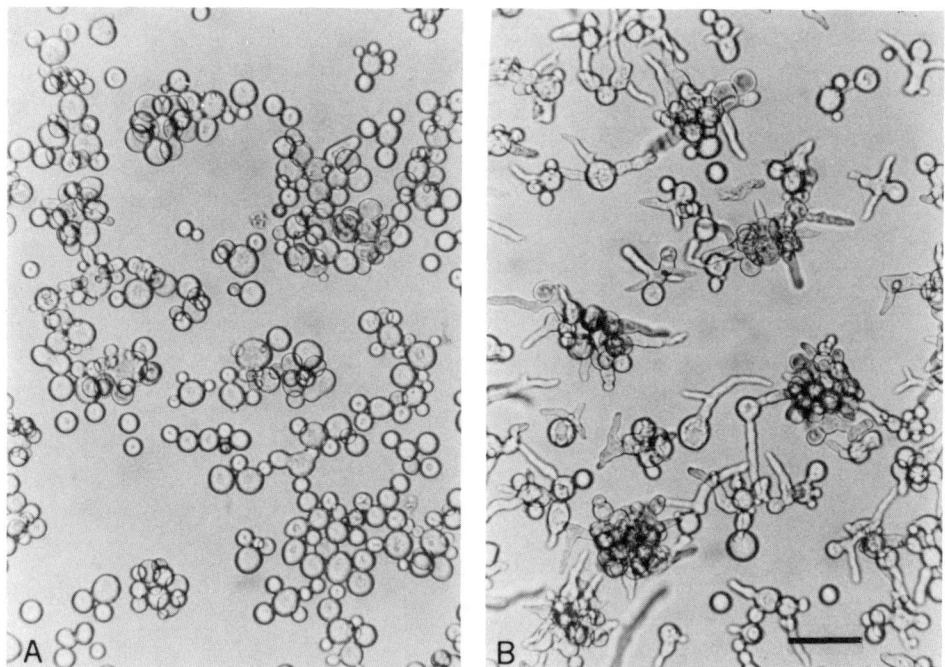

Figure 1 Yeast-mold morphogenesis in *Mucor racemosus*. The organism grows as a yeast form under CO_2 (A), but within 4 hr after a shift to air the yeast cells germinate to the mycelial form (B). The bar represents 20 μ m. (Photographs courtesy of J. Paznokas).

mycelial form (M) of the fungus. The fact that Y or M may be induced by controlling the environment makes *Mucor* a suitable model for learning about control and regulation of morphogenesis.

It has been known for over 140 years that *Mucor* may produce yeasts, but it was not until 1962 that generalities emerged from systematic studies on this phenomenon. Initially, investigations were designed to define which set of environmental situations would induce either Y → M or M → Y transformations. Induction of M → Y requires anaerobiosis and the presence of a hexose in the medium. A requirement for CO_2 in the anaerobic environment has been debated and is now considered not to be essential for M → Y. Differences in the cell walls of M and Y have been noted, and based upon these observations a mechanism of control of M → Y morphogenesis has been proposed.

Over the past 7 years studies have become geared to defining a physiological basis of M ⇄ Y morphogenesis. These studies have included examinations of enzymes which are important in the regulation of metabolic pathways, such as

Table 1 Effect of Various Factors on *Mucor* Y ⇄ M Morphogenesis

Factor(s) tested	*Mucor* species tested[a]	Direction of morphogenesis examined	Possible effect(s) of factor(s) on *Mucor* cells	References
Anaerobiosis + CO_2 + glucose	*M. indicus* (+) *M. subtilissimus* (+) *M. racemosus* (±) *M. mucedo* (+)	M → Y	Affects vesicle organization required for apical growth	Bartnicki-Garcia and Nickerson, 1962a, Farkas, 1979, Bartnicki-Garcia, 1968
Temperature	*M. indicus* (−)	M → Y	None	Bartnicki-Garcia and Nickerson, 1962a
CO_2	*M. indicus* (+)	M → Y	Incorporation of the carbon into cell-wall proteins	Bartnicki-Garcia and Nickerson, 1962d
Cysteine	*M. indicus* (−)	M → Y	None	Bartnicki-Garcia and Nickerson, 1962d
Cycloheximide, cyanide, acriflavin	*M. indicus* (−)	Y → M	Inhibits cytochrome oxidase	Haidle and Storck 1966a, b; Friedenthal et al., 1974

Phenethyl alcohol	*M. indicus* (−)	Y ↛ M	Inhibits oxidative phosphorylation	Terenzi and Storck, 1969
Cyclic guanosine 5′-monophosphate	*M. racemosus* (+)	Y ↛ M	None	Orlowski and Sypherd, 1977
Cyclic AMP	*M. racemosus* (+)	M ↛ Y	Inhibits β-glucosidase synthesis	Larsen and Sypherd, 1974
			Induces changes in chitin-chitosan synthesis	Domek and Borgia, 1981
Glucose	*M. racemosus* (+)	M ↛ Y	Affects activities of hexokinase, phosphofructokinase, or pyruvate kinase	Paznokas and Sypherd, 1977
Air	*M. racemosus* (+)	Y ↛ M	Increases ribosomal movement and protein synthesis	Orlowski and Sypherd, 1977
			Increases intracellular s-adenosylmethionine and and protein methylation	Garcia et al., 1980
			Increases activities of glutamate dehydrogenase and ornithine decarboxylase	Peters and Sypherd, 1979; Inderlied et al., 1980

[a] +, morphogenesis occurred; −, no morphological change; ±, some strains +, some −.

in glucose metabolism, oxidative phosphorylation, and posttranslational modifications of ribosomal proteins. Differences between M and Y were noted in some of the above studies, but they were found to follow, rather than precede, the onset of morphological evidence of morphogenesis. More recently it has been found that events which precede morphological evidence of Y $\rightleftarrows$ M include changes in the levels of cyclic adenosine 5'-monophosphate (cAMP), protein synthesis and protein methylation, rates of RNA synthesis, and activities of nicotinamide adenine dinucleotide (NAD) -dependent glutamate dehydrogenase and ornithine decarboxylase. Table 1 is a summary of some of the studies done on M $\rightleftarrows$ Y in *Mucor*. Albeit these studies do not mechanistically define dimorphism in *Mucor*, they are yielding results which may lead to an understanding of this process.

Factors Which Influence Morphogenesis

Growth Atmosphere Which Favors the M $\rightleftarrows$ Y Transition

It has been known for many years that an anaerobic environment promotes a M → Y transition (e.g., Bartnicki-Garcia and Nickerson, 1962a), but the necessary gas composition has been debated. Many experiments were done using 100% CO_2, and this gas was thought to be a requirement of M → Y morphogenesis. Although CO_2 is still used for Y induction, the need for this gas is diminished or absent when glucose is appropriately released in the medium (Bartnicki-Garcia, 1968). The amount of glucose required for this effect varies depending on the species and strain of *Mucor* examined. It does not seem that a high glucose concentration merely allows sufficient endogenous CO_2 production to override an exogenous need, because CO_2 is not produced at a lower rate by cells growing in 1% glucose than by cells in 10% glucose (Bartnicki-Garcia, 1968). Under conditions in which CO_2 favors a M → Y conversion, $^{14}CO_2$ becomes incorporated into aspartic acid, which may indicate an association with cell-wall protein (viz., mannan protein), since other yeasts have a high aspartic acid content in cell-wall proteins (Bartnicki-Garcia and Nickerson, 1962b). It was postulated that malate dehydrogenase may be important in the CO_2 uptake, but this idea has yet to be substantiated. None of these investigations, however, have explained the CO_2-hexose relationship in Y-form development. This interesting observation should be pursued.

Contrary to the above, CO_2 was reported as unnecessary for Y development, provided that an invert gas, N_2 or argon, flows through the incubation vessel at a relatively high rate (Mooney and Sypherd, 1976). It was interpreted that the free-flowing gas removes a volatile factor, termed a "morphogen," which is produced by *Mucor racemosus* and inhibits M → Y morphogenesis in this organism. Unfortunately, evidence for this factor is only indirect; it has not been isolated and it has not been confirmed by other investigators.

Nutritional and Other Physical Factors Which Affect Growth or a M → Y Transition

In general, the Y form is more fastidious in its growth requirements than the M form. Whereas an array of various mono- and disaccharides and other substances such as ethanol, glycerol, and tricarboxylic acid (TCA) intermediates may serve as the carbon source for growth of the M form of *Mucor indicus* (*Mucor rouxii*), the Y form requires a hexose (Bartnicki-Garcia and Nickerson, 1962c). The hexose requirement holds for all species of *Mucor* tested, and glucose seems to give the best results (Bartnicki-Garcia, 1968). Because of the hexose requirement for M $\rightleftarrows$ Y transition, studies have been done to determine how glucose metabolism relates to morphogenesis (Bartnicki-Garcia, 1968; Clark-Walker, 1972; Rogers et al., 1974; Borgia and Sypherd, 1977; Paznokas and Sypherd, 1977). However, it has not been established whether the M → Y transition mechanism requires glucose or if glucose is merely a nutritional requirement for yeast growth. In fact, some evidence suggests that Y-form cells may become impermeable to some potential C sources such as disaccharides (Borgia and Sypherd, 1977). Some investigators felt that the glucose requirement is not due to anaerobiosis, since *Mucor* will grow anaerobically in the M form if the glucose level in the medium is low (Bartnicki-Garcia, 1968; Friedenthal et al., 1974).

The type of nitrogen source represents the only other major nutritional factor which affects *Mucor* morphogenesis. Various organic and inorganic nitrogen compounds support M growth, but Y-form growth of *M. indicus* is favored by an inorganic nitrogen source (Bartnicki-Garcia and Nickerson, 1962c) and the Y form of *M. racemosus* requires organic nitrogen (Borgia and Sypherd, 1977). Factors such as cysteine and incubation temperature, which influence Y $\rightleftarrows$ M transitions in other genera of dimorphic fungi (Nickerson, 1954; Manning and Mitchell, 1980c; Maresca et al., 1981), have no effect on *M. indicus* (Bartnicki-Garcia and Nickerson, 1962c).

Physiological Basis of Morphogenesis

Cell Walls

M and Y cell walls of *M. indicus* have been isolated, hydrolyzed, and chemically analyzed to determine if profound differences exist between the two forms (Bartnicki-Garcia and Reyes, 1964; Dow and Rubery, 1977). In some studies cell walls from sporangiospores were determined to differ qualitatively in chemical constituents. A higher mannose concentration, presumably as mannan protein (Dow and Rubery, 1977), is associated with Y cell walls. The rate of chitin-chitosan synthesis, which is correlated with the apical growth of *M. indicus* hyphal cells (Bartnicki-Garcia, 1968), is three times greater in *M. racemosus* M cells than in Y cells (Domek and Borgia, 1981). Furthermore, electron microscopic

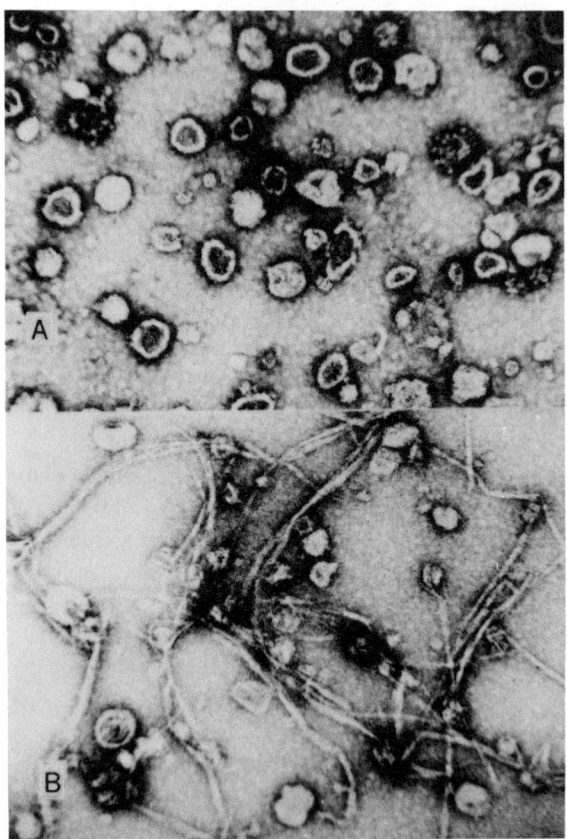

Figure 2 Production of chitin microfibrils by isolated vesicles (chitosomes) from *Mucor indicus*. (A) Chitosomes were isolated from yeast cells of *M. indicus* and photographed (×100,000). (B) Upon addition of appropriate activators and substrate for chitin synthesis, chitin microfibrils were produced in vitro (×80,000. Photographs courtesy of C. E. Bracker from Bartnicki-Garcia et al., 1978).

examinations of thin sections of whole cells revealed that Y cell walls are composed of two layers which together are about 10 times thicker than the single-layered M cell walls (Bartnicki-Garcia and Nickerson, 1962d).

Hyphal extension occurs by synthesis of new cell wall at the growing tip, and in *M. indicus* an "apical corpuscle" has been described which is believed to play a role in controlling apical growth (Bartnicki-Garcia et al., 1968). In some elegant studies, isolated vesicles ("chitosomes") from *M. indicus* were shown to contain chitin synthetase (Bracker et al., 1976; and Bartnicki-Garcia et al., 1978). In the presence of appropriate substrate and activators microfibrils develop within the chitosomes and the fibrils thicken and elongate with time (Fig. 2). Others have observed an association of cytoplasmic vesicles with the apical region of growing hyphal tips of several fungi, including the Zygomycetes (Grove and Bracker, 1970). If such vesicles in *Mucor* represent packaging of inactive chitin synthetase, then migration of these vesicles to the growing hyphal tip and protease activation of the synthetase (Ruiz-Herrera et al., 1977). would allow for site-specific cell-wall synthesis (McMurrough et al., 1971; Farkas, 1979). A breakdown in the organization of the vesicles to a specific area, perhaps due to a hexose metabolite (Bartnicki-Garcia, 1968), would result in random association of chitosomes with the cell membrane and hence yeast form (Farkas, 1979). Alternatively, vesicles (viz., chitosomes) may always be in random association with the cell membrane (Duran et al., 1979). Generalized activation of vesicle contents would result in yeast development, whereas selective activation would give rise to site-specific synthesis such as septum formation in *S. cerevisiae* (Duran et al., 1979) or, perhaps, hyphal extension as in *Mucor*. Though there may be localized activation of cell-wall synthesis for hyphal growth, protein synthesis is not restricted to apical regions (Orlowski and Sypherd, 1978a).

Respiration

The hexose-anaerobiosis requirement for induction of a M → Y transition in *Mucor* has led to the assumption that fermentative metabolism is required for Y-form development, and oxidative metabolism for the M form. This idea was supported by studies involving inhibitors of respiration. Yeasts taken from a CO_2 environment and placed in air germinated to the M form unless cycloheximide, acriflavin, or cyanide was added to the culture (Haidle and Storck, 1966a,b). Furthermore, KCN added to M-form cells induced a M → Y change even under aerobic conditions (Friedenthal et al., 1974).

It was also noted that a Y → M transition was correlated with the induction of cytochrome oxidase activity (Haidle and Storck, 1966b). The significance of cytochrome oxidase in morphogenesis has been questioned, however, because another inhibitor of respiration, phenethyl alcohol (PEA), prevents a Y → M conversion under aerobic conditions without affecting the oxidase activity (Terenzi and Storck, 1969). Although PEA does not prevent cellular uptake of O_2, it uncouples mitochondrial oxidative phosphorylation.

It is interesting that the PEA-induced suppression of Y → M is dependent on a high level of hexose in the medium, because this correlates with alcohol fermentation (Terenzi and Storck, 1969). Cells growing as Y in the absence of

O_2 or inhibitors also have high levels of activity of enzymes important in alcohol fermentation, such as pyruvate kinase and alcohol dehydrogenase (Zorzopulos et al., 1973). In conjunction with these observations, a respiratory deficient mutant of *Mucor bacilliformis* (which could not take up O_2, lacked cytochrome a-a_3 and cytochrome oxidase, and required hexose for growth) grew only in the Y form, even under aerobic conditions (Storck and Morrill, 1971).

The above studies suggest that morphogenesis is closely associated with mitochondrial activity, but this view has been challenged. Mitochondria have been found in anaerobically grown Y cells (Clark-Walker, 1972; Takeo, 1974) and a cyanide-insensitive respiratory pathway may be functioning in these cells (Clark-Walker, 1972). Also a M → Y morphogenesis under aerobic conditions will not occur upon addition of chloramphenicol, provided that the glucose level in the medium is kept low (Rogers et al., 1974). In some well-controlled studies it was reported that a Y → M transition would not occur under aerobic conditions if dibutyryl cAMP was added to the medium, yet these Y-form cells in air developed normal respiratory activity (Paznokas and Sypherd, 1975). In addition, M-form cells shifted to a N_2 anaerobic environment did not undergo M → Y morphogenesis if the flow rate of N_2 was kept low (Paznokas and Sypherd, 1975; Mooney and Sypherd, 1976), but these M-form cells growing anaerobically did not have measurable mitochondrial activity (Paznokas and Sypherd, 1975). Perhaps inhibition of Y → M transition by substances such as PEA is related more to shifts in levels of cAMP than to respiratory activity. Uncouplers of oxidative phosphorylation were found to increase levels of cAMP (Trevillyan and Pall, 1979), which is correlated with Y development (see below).

Cyclic AMP

Over the past several years interesting correlations have been noted in *Mucor* between levels of cAMP and morphogenesis. Addition of dibutyryl cAMP to M cells induced a M → Y transition under aerobic conditions, and this cyclic nucleotide prevented a Y → M transition when anaerobically grown Y cells were shifted to an aerobic environment (Larsen and Sypherd, 1974). Others obtained similar effects by addition of natural cAMP to the medium (Paveto et al., 1975). These findings agree well with endogenous intracellular levels of cAMP. Anaerobically grown Y-form cells have a greater cAMP level than M cells grown aerobically, and, more importantly, the level of cAMP in Y cells decreases prior to morphological evidence of a Y → M conversion (Larsen and Sypherd, 1974; Paznokas and Sypherd, 1975). On the other hand, cyclic guanosine 5′-monophosphate (GMP) does not influence *Mucor* morphogenesis (Orlowski and Sypherd, 1976).

Recently, studies were done to determine how cAMP interacts initially with the cell membrane of *Mucor* during morphogenesis. In work on *S. cerevisiae* four cell membrane proteins have been identified which bind cAMP (Jaynes et al., 1980). Studies similar to those on *S. cerevisiae* were done on *Mucor genevensis* and *M. racemosus* by Forte and Orlowski (1980) using 8-azido-[^{32}P]cAMP as a photoaffinity probe. Binding of cAMP to the cell membrane of either species decreased during a Y → M morphogenesis. These results support the experiments cited above which demonstrated decreasing intracellular cAMP levels during Y → M transition. Two proteins were isolated from the cell membrane of *M. genevensis* which accounted for all detectable binding of cAMP. Most cAMP binding to *M. racemosus* was also associated with two proteins which have molecular weights similar to those of the binding proteins from *M. genevensis*. The binding proteins of both species were detectable throughout the Y → M conversion and their binding affinities for cAMP remained constant. However, a qualitative reduction of binding proteins occurred during morphogenesis. These results imply that fewer cAMP-binding proteins are synthesized or are available for binding as Y becomes M.

In addition to having fewer cAMP-binding sites on M than on Y cells, the level of cAMP associated with each form of the fungus is controlled by other factors. Though the level of activity of adenylate cyclase is the same for Y and M cells, a four- to sixfold increase in cyclic adenosine 3′,5′-monophosphate phosphodiesterase activity was found in M cells (Paveto et al., 1975). These findings lend further support to the theory that the level of cAMP is important in the control of morphogenesis.

Changes in cAMP levels occur prior to the onset of morphological changes, suggesting a role for the cyclic nucleotide in regulation of the Y ⇌ M transition. However, dibutyryl cAMP is not sufficient for preventing a Y → M transition, because the effect was found to depend on the presence of hexose in the medium (Paznokas and Sypherd, 1975). The hexose dependency has not been explained.

Attempts have been made to understand how cAMP affects the ability of cells to utilize carbohydrates and synthesize cell-wall constituents. High levels of cAMP prevent β-glucosidase activity in *M. racemosus* (Borgia and Sypherd, 1977). These studies along with those cited above, which claim that Y cells have a high level of cAMP, allow a simplistic explanation as to why Y cells cannot grow in disaccharides. In the presence of high internal levels of cAMP β-glucosidase activity is repressed, thus preventing hydrolysis and utilization of disaccharides. The mechanism of decreased enzyme activity may be coarse rather than fine control, because cAMP apparently does not directly affect the enzyme in vitro (Borgia and Sypherd, 1977). Cyclic AMP may also affect the type of macromolecules synthesized for cell-wall formation. The rate of chitin-chitosan synthesis in Y growing anaerobically is one-third that of M cells growing aerobically. However, if dibutyryl cAMP is added to M cells growing in air, a reduction in the rate of chitin-chitosan synthesis occurs along with a M → Y transition (Domek and Borgia, 1981).

Hexose Metabolism

The importance of hexose in *Mucor* morphogenesis has been discussed earlier. The presence of glucose may override the need for CO_2 in the anaerobic environment; Y cells require a hexose carbon source; and, aerobic induction of a M → Y transition by cAMP is dependent on hexose. These observations have led investigators to look for enzyme differences associated with hexose metabolism between M- and Y-form cells. Although these studies have yielded important observations concerning glucose metabolism in *Mucor,* profound differences which can be linked to M → Y morphogenesis have not been found (Paznokas and Sypherd, 1977; Inderlied and Sypherd, 1978; Barrera and Corral, 1980; O' Connell and Paznokas, 1980).

Protein Synthesis

When yeasts are growing under CO_2 and then shifted to an aerobic environment, a burst in protein synthesis occurs prior to the emergence of germ tubes (Orlowski and Sypherd, 1977). The increased synthesis is not due to a shift to air, because it also occurs in a nitrogen atmosphere which favors a Y → M transition. Along with an increase in protein synthesis during a CO_2→ air shift, an increase in the synthesis of all classes of RNA occurs (Orlowski and Sypherd, 1978b). In some elaborate and interesting experiments Orlowski and Sypherd (1978b) found that the protein synthesis burst parallels an increase in the velocity of translation. These data, obtained by two different approaches, led to the conclusion that regulation of Y → M morphogenesis is correlated with or may even be dependent on an increase in ribosomal movement along mRNA.

In an attempt to understand regulation of the ribosomal movement, ribosomal proteins of Y and M were examined by two-dimensional gel electrophoresis (Larsen and Sypherd, 1979). Although it was thought that phosphorylation of a specific ribosomal protein (presumably protein S6 in the 40 S ribosomal subunit) greatly increases upon a Y → M morphogenesis, this apparently is not the case (Larsen and Sypherd, 1979, 1980). The phosphorylated S6 protein which is found in Y and M cells probably does not relate to a Y → M morphogenesis. However, protein modification by way of methylation seems to be closely associated with such a morphogenesis (Garcia et al., 1980). Intracellular pools of S-adenosylmethionine increase during a Y → M transition. This increase was not accompanied by increased spermidine levels (a polyamine dependent on S-adenosylmethionine for its production) but, rather, the increase was

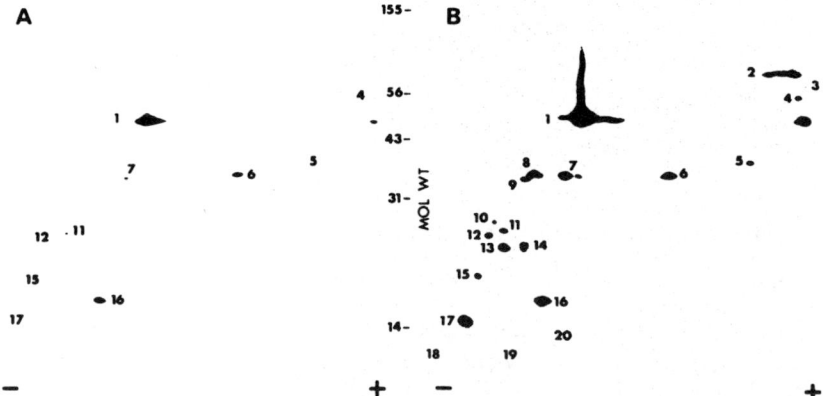

Figure 3 Comparison of methylated polypeptides by yeasts and mycelia of *Mucor racemosus*. Cells grown as yeasts or mycelia were labeled and proteins from each were analyzed by two-dimensional gel electrophoresis (Garcia et al., 1980). Note that there are fewer (A) yeast-methylated polypeptides than (B) polypeptides methylated by aerobic mycelial cells. Both gels received an equal load of total protein in the first dimension. (From Garcia et al., 1980).

correlated with profound changes in protein methylation occurring during the shift from CO_2 to air (Fig. 3). Protein methylation is important in other systems (e.g., Kondoh et al., 1979), but it is not known whether methylation is required for a Y → M transition in *Mucor*.

In other studies, attempts were made to compare total cytoplasmic proteins made in Y and M cells. Examination of over 400 cellular proteins by two-dimensional polyacrylamide gel electrophoresis showed apparent quantitative changes in some proteins between Y and M forms. In addition, 9 polypeptides appeared to be specific for Y cells and 10 were M-phase specific (Hiatt et al., 1980). By appropriate controls it was determined that these differences were associated with morphogenesis and were not due to a shift from CO_2 to air. Although such studies provide protein profiles for the purpose of comparing Y with M cells, they do not identify the proteins which differ and hence do not give clues about pathways involved in morphogenesis. Two enzymes have been found, however, which increase in activity during Y → M transition, and the increase occurs prior to the emergence of germ tubes. One of these enzymes, NAD-dependent glutamate dehydrogenase, increased about 10-fold during the transition (Peters and Sypherd, 1979). Furthermore, addition of dibutyryl cAMP, which prevents Y → M morphogenesis, also prevented the increase in dehydrogenase activity. The other enzyme, ornithine decarboxylase, increased

30- to 50-fold during morphogenesis (Inderlied et al., 1980). This enzyme, which is required for the synthesis of polyamines such as putrescine, is especially interesting, since its importance in cellular changes has also been implicated in other eukaryotic systems (Russell and Snyder, 1968; Russell et al., 1970; Hogan et al., 1974; Lembach, 1974).

DIMORPHISM OF CANDIDA ALBICANS

Background

The Y → M conversion by *C. albicans* cells has been the subject of an enormous number of investigations (for previous reviews see Fineman, 1921; Benham, 1931; Skinner, 1947; McClary, 1952; Johnson, 1954; Romano, 1966; Rippon, 1980). Most of the works prior to 1950 were concerned with the influence of the environment and nutritional factors on morphology. Physiological studies began in the late 1940s and, more recently, attempts have been made to understand the regulatory signals which dictate morphological change.

In spite of the number of studies done on *Candida* morphogenesis, the biochemical events important in the initiation of a Y → M transition have not been defined. In fact, evaluation of the literature in this area is a formidable task because of the conflicting results and conclusions reported by many different research groups. One major problem seems to be that not all strains of *C. albicans* behave in the same way under similar conditions. The solution to this may not be simply to distribute standardized strains to various workers in the field, because we have found that changes occur even within a single strain after repeated subculturing (our unpublished findings). Another problem is morphological interpretation. *Candida albicans* has the ability to exist in many morphological forms (Odds, 1979; Mackinnon, 1940; Hedden and Buck, 1980), such as budding yeasts, chains of yeasts, chains of elongated yeasts, germ tubes with a constricted base, germ tubes with a broad base, and, not infrequently, a yeast cell will develop two germ tubes of which one has a constricted base and the other a broad base (Fig. 4). To morphologically describe a Y → M conversion of least five terms have been used, including "hypha," "germ tube," "pseudohypha," "pseudo-germ tube," and "filament"—yet it is not always clear which morphology is actually observed by various investigators. In our studies we categorize the various forms into yeasts, germ tubes, and pseudohyphae. The latter category serves as a morphological receptacle, because cells are placed here which have not formed germ tubes (i.e., tubes with a length at least twice the diameter of the mother cell and with a broad base of attachment to the mother cell) or yeasts (i.e., ovoid cells with or without buds) (Fig. 4B-D). The various morphological forms make interpretation of experiments on Y → M conversion difficult. However, our observations and those of others (Odds, 1979; Cole and Nazawa, 1981)

Yeast/Mold Morphogenesis

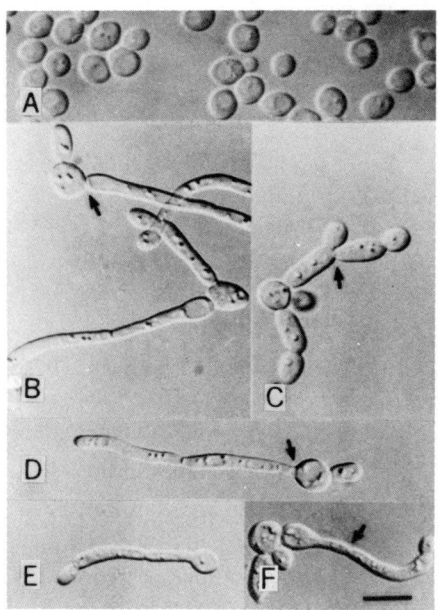

Figure 4 Some forms of *C. albicans* cells. (A) yeasts or Y form; (B,C) pseudohyphae (arrows denote a constricted base, which is a pseudohyphal characteristic); (D) hypha or germ tube (note the wide base of attachment, arrow); (E,F) characteristics of germ tubes and pseudohyphae (note the wide base of attachment, the buds at the ends of elongated cells, and constriction at the point of cell-cell juncture, arrow). The bar represents 6 μm.

have lead us to believe that a germ tube (as defined above) represents the end result of a Y → M transition and that interruption of pathways leading to the germ tube results in various types of pseudohyphae. Studies on yeast → pseudohyphal morphogenesis rather than yeast → germ tube development (which we term Y → M) may well be important in gaining an understanding of *Candida* Y → M morphogenesis.

Difficulty in interpreting and comparing various studies on *C. albicans* dimorphism also stems from a lack of uniformity of growth conditions for obtaining yeast cells (see Table 2). Various temperatures, germination media, and incubation times have been used. In spite of these problems, some consistent results have emerged among various laboratories.

Table 2 Factors Which Affect *Candida albicans* Y → M[a] Morphogenesis

Factor tested	Resultant morphology[b]	References
Carbohydrates		
Glucose	M	Land et al., 1975a, Shepherd and Sullivan, 1976
	No effect[c]	Evans et al., 1975
2-Deoxyglucose	Y	Shepherd et al., 1980b
Polysaccharides	M	Nickerson and Mankowski, 1953
Galactose	M	Shepherd and Sullivan, 1976; McClary, 1952
Fructose	M	Shepherd and Sullivan, 1976
Maltose	M	Shepherd and Sullivan, 1976
Sucrose	M	Shepherd and Sullivan, 1976
N-Acetyl-D-glucosamine	M	Simonetti et al., 1974; Chaffin and Sogin, 1976; Mattia and Cassone, 1979
	No effect	Wain et al., 1976a
Metals		
Co	M	Nickerson and Van Rij, 1949
	No effect	Shepherd et al., 1980b
Zn	Y	Yamaguchi, 1975
	No effect	Vaughn and Weinberg, 1978
Fe	M	Landau et al., 1964
	No effect	Vaughn and Weinberg, 1978
Cu	Y	Vaughn and Weinberg, 1978
Mn	M	Simonetti et al., 1974
Mg	No effect	Vaughn and Weinberg, 1978
Amino acids		
Cysteine	Y	Nickerson, 1948; Nickerson and Van Rij, 1949; Shepherd and Sullivan, 1976; Hazen and Cutler, 1979
	No effect	McClary, 1952; Wain et al., 1975; Nishioka and Silva-Hutner, 1974
Hydroxyproline	M	Johnson, 1954

Table 2 (Continued)

Factor tested	Resultant morphology[b]	References
Amino acids (contd)		
Proline	M	Dabrowa et al., 1976; Land et al., 1975a
Glutamate	Y	Mardon et al., 1969
	M	Nishioka and Silva-Hutner, 1974
Methionine	M	Mardon et al., 1969
Arginine	M	Land et al., 1975a
Alanine	M	Land et al., 1975a
Miscellaneous		
Phenylethyl alcohol	Y	Shepherd et al., 1980a
	No effect	Hazen and Cutler, 1979
L-α-Amino-n-butyric acid	M	Mardon et al., 1971
Acetate	M	Shepherd and Sullivan, 1976
Lactate	M	Shepherd and Sullivan, 1976
NH_4Cl	Y	Land et al., 1975a
Phosphate	Y	McClary, 1952; Nickerson and Mankowski, 1953; Land et al., 1975a; Dabrowa et al., 1976
	M	Chattaway et al., 1976
Polysorbate (Tween 80)	M	Schaar et al., 1974
	No effect	Shepherd et al., 1980a
Ethylenediamine tetraacetic acid	M	Nickerson, 1954
Cyclic AMP	M	Land et al., 1975a
Dibutyryl cAMP	M	Chattaway et al., 1981; Niimi et al., 1980
Oxine + cobalt	Y	Nickerson and Van Rij, 1949
Light	Y	Saltarelli and Coppola, 1979
Seminal fluid	M	Chattaway et al., 1980
Low biotin concentration	M	Yamaguchi, 1974
Specific antibody	Y	Grappel and Calderone, 1976
	M	Smith and Louria, 1972
Coculture with bacteria	Y	Auger and Joly, 1975, 1977; Purohit et al., 1977

Table 2 (Continued)

Factor tested	Resultant morphology[b]	Reference
Miscellaneous (contd)		
Penicillin	M	Nickerson and Mankowski, 1953
Aminopterin	M	Nickerson and Mankowski, 1953
5-Fluorocytosine	Y	Shepherd et al., 1980; Polak and Wain, 1979
Nystatin	Y	Shepherd et al., 1980a
2,4-Dinitrophenol	Y	Shepherd et al., 1980a
Iodoacetamide	Y	Shepherd et al., 1980a
Chloramphenicol	No effect	Shepherd et al., 1980a
Temperature	M	Lee et al., 1975; Vaughn and Weinberg, 1978; Manning and Mitchell, 1980a
NaCl	Y	Dabrowa et al., 1965

[a]M cells are defined as any elongated cellular appendage. Therefore pseudohyphae and hyphae are termed M cells in this table.
[b]In all germination assays, the inocula consisted of yeast cells; that is, investigators were considering a Y → M morphogenesis.
[c]No effect implies that the factor tested neither promoted nor repressed a Y → M morphogenesis.

Y and M Cell Walls and Cell Membranes

Chemical analysis of Y and M cell walls and membranes have shown quantitative but not qualitative differences between these morphological forms (Kessler and Nickerson, 1959; Chattaway et al., 1968; Phaff, 1971). These findings are like those obtained by others on the Y and M cells of *M. indicus* (Bartnicki-Garcia and Nickerson, 1962d) and *H. capsulatum* (Kobayashi and Guiliacci, 1967). *Candida* cell walls are composed chiefly of glucose as glucan, mannose as mannan (Ballou, 1976), protein, and a small amount of lipid (Kessler and Nickerson, 1959). Whereas Y and M cell walls have a similar content of glucose and mannose, M cell walls have one-third as much protein and three times as much glucosamine (presumably as chitin) as Y cell walls (Chattaway et al., 1968). The larger quantity of glucosamine in hyphal cell walls is correlated with the fact that the electron-transparent layer (layer 4 of the yeast cell wall in Fig. 5) which increases in width in the germ tube cell wall is composed of chitin (Cassone et al.,

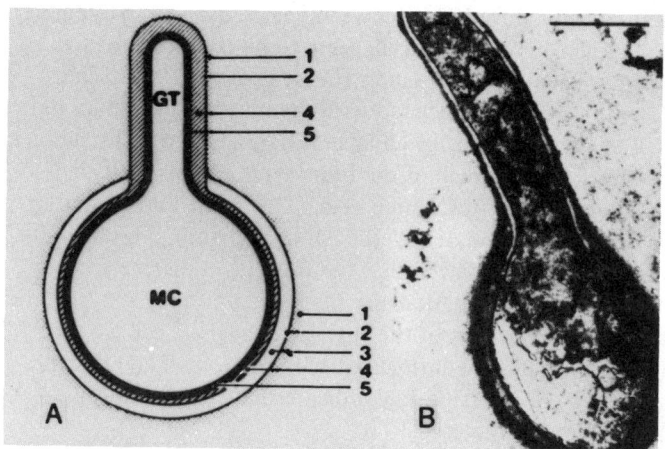

Figure 5 Schematic (A) of an electron micrograph (B) depicting layers of yeast and hyphal cell wall. Layer 4 is electron transparent and is continuous between the yeast and hyphal cell regions. The bar represents 1.5 μm. (From Scherwitz et al., 1978).

1973). Increased chitin synthesis by M cells of *C. albicans* was also supported by Braun and Calderone (1978), who found that hypha-forming cells take up more N-acetyl glucosamine than yeasts. In these studies, chitin deposition was found to occur at the site of budding for blastoconidium-forming cells, and at the apical region of hypha-forming cells.

Substantial differences in phospholipid and sterol content between cytoplasmic membranes of Y and M cells have been reported. Yeast cell membranes contain phosphotidyl serine, phosphotidyl inositol, sphingolipids, phosphotidyl ehtanolamine, and phosphotidyl choline. However, only phosphotidyl ethanolamine and phosphotidyl choline have been detected in membranes of exponential-phase M cells (Marriott, 1975). This is well correlated with other observations that the major lipids synthesized from [^{14}C]-acetate during Y → M conversion are phosphotidyl choline, phosphotidyl ethanolamine, and neutral lipids (Ballman and Chaffin, 1979). Overall lipid synthesis, however, is lower in M than in Y cells (Sundaram et al., 1981).

Only stationary-phase yeast cells are capable of germinating (Chaffin and Sogin, 1976; Soll and Bedell, 1978), yet stationary-phase yeast cell membranes more closely resemble membranes of exponential-phase mycelial cells than those of exponential-phase yeast cells (Kerridge, 1980). These data may imply that changes in the chemical makeup of the cell membrane mark the initiation of morphogenesis. Clearly, more work is needed to define this possible correlation.

At the ultrastructural level, differences between yeasts and young hyphae have been observed (see Fig. 5). Yeast cell walls seem to be composed of five layers, whereas germ tubes have only four layers (Cassone et al., 1973; Scherwitz et al., 1978). However, hyphae associated with the invasion of human tissues may be composed of six distinct cell-wall layers (Rajasingham and Cawson, 1980). There is disagreement as to which of the four layers of yeast cell wall disappear during hyphal development. Cassone et al. (1973) felt that dissolution of the second layer of the germinating yeast cell occurs during hyphal initiation, but Scherwitz et al. (1978) believed that the third layer disappears (see Fig. 5). The initiation of bud formation and that of germ tubes appear to be similar processes, in that the major part of the emerging daughter cell walls of both forms are composed of material continguous with the fourth layer of the mother cell (Scherwitz et al., 1978). This may indicate that the fourth cell-wall layer is composed of flexible material.

Fewer cell-wall layers are usually found in hyphal cells and a wall of this form of *C. albicans* is about half the thickness of a yeast cell wall (Cassone et al., 1973). Similar differences between yeast and mycelial cell walls have been observed in *H. capsulatum* (Domer et al., 1967) and *M. indicus* (Bartnicki-garcia and Nickerson, 1962d).

Exogenous Substances Which Affect Morphogenesis of *C. albicans*

Since the discovery that serum induces a Y → M conversion of *C. albicans* (Johnson, 1954) and that this reaction in serum is a distinguishing characteristic of this *Candida* species (Taschdjian et al., 1960) attempts have been made to define the simplest medium for induction of *C. albicans* morphogenesis. Although the medium of Lee et al. (1975) has received the most widespread acceptance, a modification of their recipe containing even fewer components has been reported to promote a *Candida* Y → M transition (Manning and Mitchell, 1980a). *Candida albicans* requires biotin and a carbon source, usually glucose, for growth and mycelial development. The effect and importance of various substances on morphogenesis, however, is a very controversial topic.

Table 2 lists some of the substances which prevent or promote M cell formation in various media. Unfortunately, most of these reports have been of a phenomenological nature, with little attempt having been made at defining possible mechanisms of inhibition or stimulation of morphogenesis and some of the reports are contradictory. For example, some workers obtained filamentous cells using high concentrations of glucose in an otherwise yeast-inducing medium (Land et al., 1975a), but others found that glucose does not enhance germination (Nickerson and Mankowski, 1953; Evans et al., 1975; Shepherd and Sullivan, 1976). Reports have also indicated that glucose is important in maintaining the yeast form because if glucose levels are decreased (Johnson, 1954) or if another carbon source such as galactose (McClary, 1952) or a less readily metabolizable carbohydrate is substituted for glucose (Fineman, 1921; Nickerson and Mankowski, 1953; Shepherd and Sullivan, 1976), germination is enhanced.

Similar confusion exists when considering the effects of many other substances on morphogenesis. Until more definitive studies utilizing standardized conditions are carried out, existing contradictions will not be resolved.

**Physiology and Biochemistry of *C. albicans*
Mycelial Formation**

Autoinhibitors of Morphogenesis

Some of the earlier physiological studies on the dimorphism of *C. albicans* were stimulated by the observation of Langeron and Guerra (1939), that cells located on apposed sides of two parallel streaks would not germinate when the distance between the streaks was sufficiently short. These workers believed that the effect might be due to production by *C. albicans* of a germination-inhibiting factor, but some felt that this so-called "Langeron effect" was merely due to nutritional depletion (Magni, 1948). Others, however, sought to determine the nature of the germination inhibitor and isolated two substances: phenylethyl alcohol and tryptophol (Lingappa et al., 1969). Because these substances supposedly inhibited growth of *C. albicans*, it is difficult to evaluate the effects on morphogenesis. Actually, others (Hazen and Cutler, 1979; Soll et al., 1981a) found that neither substance inhibits *Candida* Y → M conversion.

Some investigators have obtained evidence that *C. albicans* produces substances presumably other than tryptophol and phenethyl alcohol which affect morphogenesis (Saltarelli, 1973; Jillson and Nickerson, 1948; Hazen and Cutler, 1979; Pugh and Cawson, 1979). The identity of these substances and the manner in which they biochemically effect a Y → M conversion are unknown. The production of germination self-inhibitors by *C. albicans* yeast cells provides an explanation for the long-observed phenomenon. That is, germination will not occur during a germ tube test if the inoculum of the *C. albicans* yeast cells is too concentrated (Landau et al., 1965; Joshi et al., 1973; Shepherd and Sullivan, 1976; Hazen and Cutler, 1979).

Autoinhibitors of fungal germination have been reported before (Lingappa and Lingappa, 1966; Macko et al., 1970; Mooney and Sypherd, 1976), but an explanation for such production is not apparent. In the case of *C. albicans* it may be that while conditions are nutritionally optimal for cell growth, inhibition of germination keeps the fungus in that particular area. An environment which becomes unfavorable for cell growth may favor germination and thus allows the organism to extend into a new and possibly more hospitable environment.

Carbohydrate Utilization

Candida albicans contains most, and perhaps all, of the enzymes of the glycolytic pathway and the tricarboxylic acid (TCA) cycle (Rao et al., 1960, 1962). The dehydrogenase enzymes of the glycolytic pathway from yeast grown at 37°C differ from some organisms, however, in that they have a $NADP^+$, rather than a NAD^+, cofactor requirement (Rao et al., 1960). In studies using metabolic inhibitors, Hasilik (1973) reported that phosphofructokinase (PKF) and glucose-6-phosphate dehydrogenase (G6PD) are important for growth. Chattaway et al. (1973) studied some enzymes of glycolysis, of the oxidative pentose pathway, and the first enzyme of chitin synthesis, L-glutamine-D-fructose-6-phosphate aminotransferase. Phosphofructokinase activity was lowest when germination was at its peak, and cells which had germinated contained less PFK activity than yeast-form cells. The investigators felt that the low PFK activity would allow more fructose-6-phosphate to be available for chitin synthesis via L-glutamine-D-fructose-6-phosphate aminotransferase (which will be discussed in more detail later in this chapter). This idea is correlated with their earlier findings, that chitin is more abundant in mycelial form cells than in yeasts (Chattaway et al., 1968). Unfortunately, the mycelial extracts used for enzyme assays were obtained from cells grown at 40°C and the yeast extracts were from cells grown at 30°C. Therefore, it is impossible to know whether reduced PFK levels are required for Y → M morphogenesis or whether the reduction in activity is due to an increase in incubation temperature.

Cyclic AMP

The involvement of cyclic adenosine 3',5'-monophosphate (cAMP) in various metabolic events of prokaryotes is well established (Rickenberg, 1974) and studies have also been done in fungi (Pail, 1981). As was alluded to earlier, morphogenesis of *M. racemosus* is closely correlated with changes in cAMP levels. In *C. albicans,* such correlations are less clear. Exogenous addition of cAMP to *C. albicans* has been shown to decrease RNA and protein synthesis (Bhattacharya and Datta, 1977) and increase Y → M conversion (Land et al., 1975b). Although others found that cAMP would not promote M cell development (Niimi et al., 1980), the Y → M transition was enhanced by addition of the lipophilic dibutyryl derivative of cAMP (Niimi et al., 1980; Chattaway et al., 1981).

Attempts to correlate intracellular cAMP levels with germination revealed that the cAMP concentration increases prior to (Niimi et al., 1980) and during germination (Chattaway et al., 1981), but these changes are also correlated with a shift of the incubation temperature from 30 to 37°C. However, in accordance with correlating cAMP changes with morphogenesis, substances which either

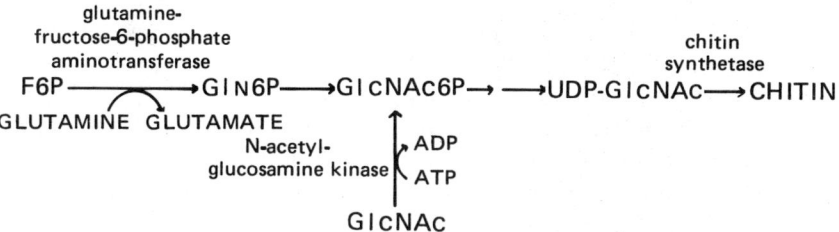

Figure 6 Chitin synthesis pathways.

increase or decrease intracellular cAMP levels increased or decreased germination, respectively (Chattaway et al., 1981).

Only one investigation has endeavored to integrate cAMP levels with metabolism. The addition of glucose to the *C. albicans* growth medium did not affect levels of cAMP or N-acetylglucosamine catabolic enzymes. Glucose, on the other hand, did repress the enzymes and cAMP levels in *S. cerevisiae*. These data suggest that *C. albicans* catabolic repression mechanisms differ from those of other fungi (Singh and Datta, 1978). Furthermore, the mechanisms by which increased glucose may prevent mycelial formation (see earlier discussion) may not involve cAMP.

Chitin Synthesis

Interest in chitin synthesis by *C. albicans* (Fig. 6) appears to have developed because of two observations: Mycelial cells contain three times more chitin than yeast cells (Chattaway et al., 1968), and N-acetyl-D-glucosamine (GlcNAc) in water induces germ tube formation (Simonetti et al., 1974). Although this latter observation has been disputed (Wain et al., 1976a), it is difficult to evaluate the controversy because of differences in experimental protocol. N-Acetyl-D-glucosamine has also been shown to induce germination more quickly (Chaffin and Sogin, 1976) and to induce Y → M conversion in cells which normally do not germinate (i.e., early and midlogarithmic-phase cells) (Mattia and Cassone, 1979).

Exogenous GlcNAc appears to induce synthesis of the chitin synthesis enzyme N-acetylglucosamine kinase by *C. albicans* (Bhattacharya et al., 1974). However, the induction kinetics are similar in both yeast and mycelial cells (Shepherd et al., 1980b), as are the specific activities. This suggests that N-acetyl glucosamine kinase is not important in Y → M conversion (Shepherd et al., 1980b), which is not surprising, because the organism can synthesize chitin from glucose (see Fig. 6).

The key enzyme in the production of chitin from glucose, L-glutamine-D-fructose-6-phosphate aminotransferase, has been demonstrated to have a higher specific activity in germinating cells. In addition, induction of the enzyme appears to be related to the appearance of germ tubes (Chiew et al., 1980).

Further evidence that chitin synthesis is involved in the germination of *C. albicans* cells is the observation that the last enzyme in the chitin synthesis pathway, chitin synthetase (synthase), has a higher specific activity in hyphal cells and hyphal cells incorporate 10 times more GlcNAc than yeasts (Braun and Calderone, 1978). Modeling their studies on those done on *S. cerevisiae* (Cabib and Keller, 1971; Duran et al., 1975; Keller and Cabib, 1971), Braun and Calderone (1978) suggested that the proteolytic activator of chitin synthetase binds more strongly to the enzyme of mycelial cells than to the enzyme of yeast cells. Although these observations strongly suggest a relationship between chitin synthesis and germ tube formation, it is unclear as to whether morphogenesis is initiated by increased chitin synthesis or whether increased chitin synthesis is the result of initiation of morphogenesis.

Respiration

Reduced oxygen tension or increased atmospheric CO_2 have been shown, in early studies, to favor Y → M conversion by *C. albicans* (Fineman, 1921; Skinner, 1947; Weld, 1952; McClary, 1952; Johnson, 1954; Mardon et al., 1969). Under anaerobic conditions (100% H_2 atmosphere), however, the percentage of germinating cells in serum was reduced (Davies and Denning, 1972). Although attempts have been made to correlate repressed respiratory activity (mitochondrial repression) with germ tube formation (Land et al., 1975a,b), the data obtained from using various respiratory inhibitors are not clear.

It is interesting to note that under conditions which are considered to enhance filamentation, for example, reduced glucose concentrations (see earlier discussion), the respiratory pathway that becomes dominant is resistant to cyanide. Under less limiting conditions, the major respiratory pathway is sensitive to cyanide (Kot et al., 1976). Clearly, more work is needed in this area to understand how O_2 tension relates to morphogenesis.

Cysteine

Extensive investigations into the role of cysteine in Y → M conversion by *C. albicans* were conducted by Nickerson because of several observations: First, *Trichophyton rubrum* was found to release a —SH containing compound which inhibited germination of *C. albicans* (Jillson and Nickerson, 1948; Nickerson and Jillson, 1948); second, 10^{-2} M cysteine inhibited germination of wild-type *C. albicans* and of a filamentous mutant of *C. albicans* (Nickerson and Van Rij, 1949; Nickerson, 1951); third, *C. albicans* cells accumulated cobalt when cobalt was present in the growth medium (Nickerson and Zerahn, 1949), and it was known that cysteine chelates cobaltous ions (Albert, 1952); and fourth, the presence of a readily utilizable carbon source such as glucose did not prevent Y → M morphogenesis (Nickerson and Mankowski, 1953). From these observations, Nickerson postulated that filamentation results when cell growth occurs without cell division (Nickerson, 1948; Nickerson, 1951) and that cell division occurs only in the presence of a sufficient number of —SH groups (provided, for instance, by cysteine) (Nickerson, 1951). In support of this concept, *C. albicans* was described as having reduced nicotinamide adenine dinucleotide (NADH)-dependent cystine reductase (Nickerson and Romano, 1952; Romano and Nickerson, 1954), of which the NADH is provided through glucose metabolism (Nickerson and Mankowski, 1953). Hence carbon sources such as polysaccharides would not yield enough NADH to power the cystine reductase and thus a Y → M conversion would result.

In related studies a filamentous mutant of *C. albicans*, designated as strain *806*, was induced to grow as a yeast when incubated in the presence of a substance released from wild-type cells (Nickerson and Chung, 1954). Furthermore, strain *806* and wild-type mycelial cells, but not yeasts, released protons into the medium surrounding the colony (Nickerson, 1954). Nickerson (1954) proposed that release of H^+ indicated that they were unavailable for reducing the appropriate substance(s) required for maintenance of yeast growth (Nickerson, 1953). Although the final proton acceptor molecule in yeasts was not identified, it was suggested that it may reside in the cell wall (Falcone and Nickerson, 1956; Nickerson and Falcone, 1956a,b). Furthermore, by accepting protons and subsequently forming more covalent bonds (possibly as —SH groups) the acceptor molecule in the wall would favor the prolate spheroid shape of the yeast form (Falcone and Nickerson, 1959). Unfortunately subsequent work to support or deny these ideas has not been done.

As in other investigations on *C. albicans* Y → M conversion, the observations of Nickerson and co-workers suffer because the studies were done on developed yeast and mycelial cells rather than on events which occur prior to morphological evidence of a transition. Therefore it is not possible to make definitive conclusions about the importance of cysteine in initiating *Candida* morphogenesis.

Other Amino Acids

The soluble pool of s-adenosylmethionine (SAM) has been reported to be greater in the M form of *C. albicans* than in the Y form (Balish and Svihla, 1966). High SAM levels are associated with the initiation of a Y → M morphogenesis in *M. racemosus* (Garcia et al., 1980). However, the SAM appears to increase in *C. albicans* after initiation of the Y → M conversion (Balish, 1973).

Several investigators have reported that proline promotes *Candida* morphogenesis (Johnson, 1954; Land et al., 1975a; Dabrowa et al., 1976). In attempts

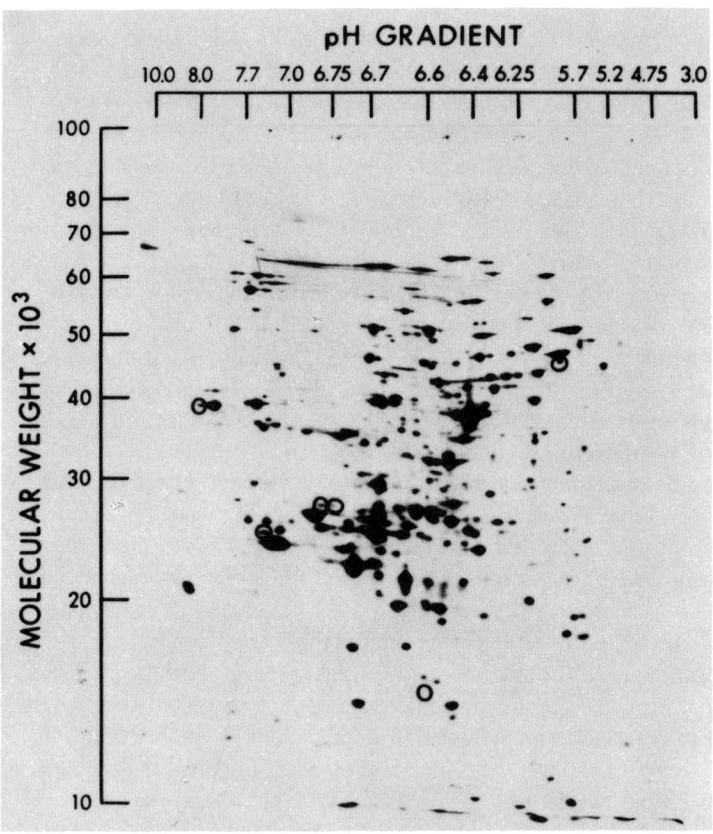

Figure 7 Two-diemsnional gel electrophoresis of cytoplasmic proteins of M-form cells of *C. albicans*. The cells were grown in a low-sulfate synthetic medium (Manning and Mitchell, 1980a) with $Na_2^{35}SO_4$ at $37°C$ for 3 hr before harvesting the cytoplasmic constituents. Under these conditions all of the cells germinated. Protein profiles of the M-form cytoplasmic proteins were compared with profiles obtained from cells grown as above, but at $24°C$ for 6 hr (resulting in pure Y-form cells). Although some proteins were found only in preparations from M cells (as indicated by the circles), evidence was obtained that these were modified yeast proteins and not proteins unique to the M form (Manning and Mitchell, 1980b,c).

to define a simple medium which supports the Y → M transition we also found proline an essential ingredient for maximal responses (our unpublished data). Although it has been suggested that proline enters into the TCA cycle via glutamate (Nishioka and Silva-Hutner, 1974; Land et al., 1975a), a mechanism for how this might influence morphogenesis has not been defined.

Nucleic Acid Synthesis

The relationship between nucleic acid synthesis and *Candida* morphogenesis is not clear. Recently evidence indicates that germination may be a sexual characteristic of *C. albicans* (Olaiya et al., 1980). However, this view is not universally accepted (Sarachek et al., 1981; Poulter et al., 1981).

During a Y → M conversion nuclear division occurs when the germ tube reaches a critical length (Bedell et al., 1980; Soll et al., 1978), whereas in budding cells nuclear division occurs when the bud attains a critical volume (Bedell et al., 1980; Soll et al., 1981b). Production of M-form cells was found to occur in the presence of an inhibitor of purine synthesis (Nickerson and Mankowski, 1953) and in adenine-deficient strains of *C. albicans* (Kwon-Chung and Hill, 1970). However, a Y → M transition was blocked by using other inhibitors of RNA synthesis (Shepherd et al., 1980a) and M-form cells were reported as having more RNA than the yeast form (Yamaguchi, 1975). Then again, others found DNA and RNA synthesis to be similar in Y and M forms (Dabrowa et al., 1970; Wain et al., 1976b). It may be more informative to evaluate RNA synthesis in terms of which species of RNA is synthesized during initiation of morphogenesis.

Protein Synthesis

Protein synthesis occurs at similar rates in Y and M forms of *C. albicans* (Dabrowa et al., 1970). Whereas cytoplasmic protein synthesis is required for morphogenesis, mitochondrial protein synthesis seems unnecessary (Shepherd et al., 1980a).

In an attempt to determine whether there are proteins unique to Y- or M-form cells, Dabrowa et al. (1970) used disk gel electrophoresis to analyze cell-free extracts of these forms. Two protein bands were reported to be unique to yeasts, and one band unique to mycelial cells. However, using two-dimensional gel electrophoresis mycelial cells were found to have fewer proteins than yeast cells and proteins unique to mycelial forms were not detected (Manning and Mitchell, 1980a,b,c; Brown and Chaffin, 1981) (Fig. 7). It was suggested that Y → M conversion occurs because of posttranslational protein modifications (Manning and Mitchell, 1980b,c) or differential gene expression (Brown and Chaffin, 1981). Evaluation of proteins synthesized within the first hour of morphogenesis induction is needed.

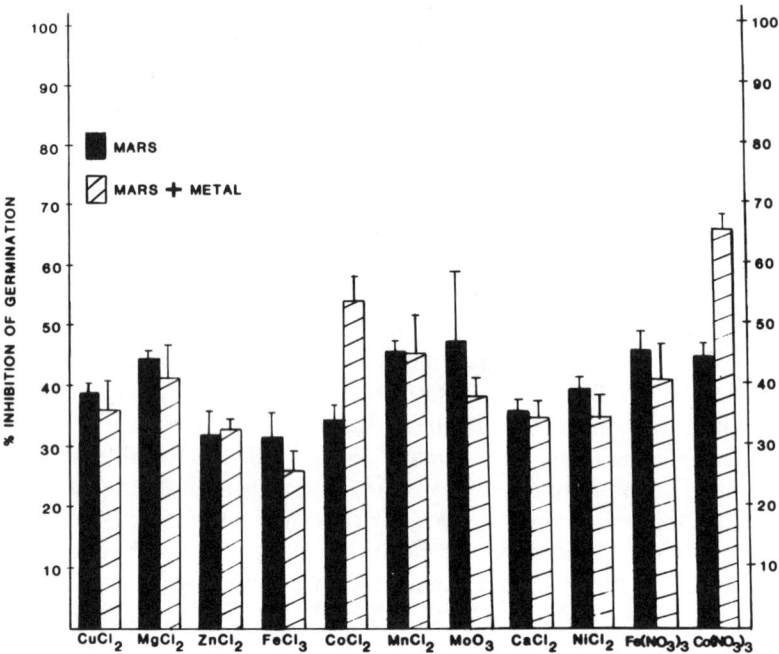

Figure 8 The effect of different metals on the enhancement of *C. albicans* morphogenetic autoregulatory substance (MARS). In all cases, the concentration of metal tested was 1 $\log_{10}$ less than the concentration determined to adversely affect germination. (Hazen and Cutler, 1983.)

Other Factors Which Influence Morphogenesis

Zinc may be involved in hyphal formation and yeast growth (Bedell and Soll, 1979; Soll et al., 1981a). Yeasts grew to a greater concentration in media supplemented with zinc, and zinc-grown Y cells initiated germination more quickly than Y-form cells grown in the absence of the metal.

In our laboratory *C. albicans* yeasts released a soluble factor, termed morphogenic autoregulatory substance (MARS), that prevents Y → M conversion (Hazen and Cutler, 1979). The inhibitory activity of MARS was enhanced when cobalt, but not zinc or other metals, was present in the medium (Fig. 8). The mechanism of action of MARS is unknown at this time, but we have found that (1) MARS-treated cells take up a greater quantity of cobalt than untreated cells, (2) MARS-treated cells take up fewer amino acids and uridine than untreated cells, and (3) MARS has characteristics which suggest it may be a proline derivative (our unpublished observations).

CONCLUSION

The ease with which Y → M or M → Y conversions can be induced or suppressed in *Mucor* sp. and *C. albicans* makes these organisms practical models for studies on eukaryotic morphogenesis. Indeed, through the carefully controlled systematic approaches currently used, a definition of morphogenesis at the molecular level in *M. racemosus* may be forthcoming. The multiplicity of factors which influence morphological transformation in *Mucor* and *Candida* suggests that a single Y ⇄ M pathway does not exist. Instead, the form a *Mucor* or *Candida* cell assumes may be the end result of several different pathways.

Although studies on *C. albicans* morphogenesis have not been as systematic as those of *Mucor*, it is interesting to compare certain morphogenetic features of both types of organisms. In some ways induction of a M → Y conversion in *M. racemosus* is similar to that of a Y → M transition in *C. albicans*. Anaerobiosis promotes the change in *M. racemosus* and low oxygen tension favors the transition in *C. albicans*. Increased levels of exogenous and endogenous cyclic AMP parallel the respective morphogenetic changes in each organism and neither fungus is affected by cyclic GMP. These observations indicate that changes in cell-wall structure which dictate form (i.e., Y or M form) are not directly dependent on decreased O_2 tension or increased cAMP levels because these parameters induce opposite forms in these two fungi.

In the case of *C. albicans* the Y → M transition is favored by poor growth conditions. Yeast cells, especially log-phase cells, tend to replicate as yeasts in an enriched medium at 25 and 37°C under high O_2 tension. In a nutritionally deprived medium, yeast-form stationary-phase cells readily undergo a shift to the M form at 37°C under low O_2 tension. Hence there may be some validity to the statements made many years ago by Weidman (1922), who wrote, "The rationale of filament production here is clear. Many yeasts develop mycelial forms only under adversity. Under optimum conditions they reproduce, reproduce, reproduce. . . . Under adversity they reach out by their mycelial means. . . to more removed, perchance more favorable positions." If one extrapolates these ideas to disease in man caused by *C. albicans*, the characteristic presence of yeasts and mycelial forms in tissue lesions may be rationalized on the basis of local nutritional conditions. Perhaps the yeast form may be favored during initial lesion development; but, as host defense cells arrive, an adverse environment for fungal growth evolves, thus promoting a Y → M morphogenesis and spread of the fungus to new tissue sites. If control of *Candida* morphogenesis in animal tissues becomes possible, the above hypothesis may be tested.

Investigations on *Mucor* and *Candida* morphogenesis should allow extrapolation to more developed eukaryotic systems and to disease states involving other dimorphic fungi. Future studies should focus on molecular events which occur during initiation, rather than completion, or morphogenesis.

ACKNOWLEDGMENTS

Some of the unpublished research cited in this chapter was supported by research grants to the senior author from the National Institutes of Health.

REFERENCES

Albert, A. (1952). Quantitative studies of the avidity of naturally occurring substances for trace metals. 2. Amino acids having three ionizing groups. Biochem. J. *50*:691-697.

Auger, P., and Joly, J. (1975). Etude de quelques aspects de la pathogénèse des infections à *Candida albicans*. Sabouraudia *13*:263-273.

Auger, P., and Joly, J. (1977). Factors influencing germ tube production in *Candida albicans*. Mycopathologia *61*:183-186.

Balish, E. (1973). Methionine biosynthesis and S-adenosylmethionine degradation during an induced morphogenesis of *Candida albicans*. Can. J. Microbiol. *19*:847-853.

Balish, E., and Svihla, G. (1966). Ultraviolet microscopy of *Candida albicans*. J. Bacteriol. *92*:1812-1820.

Ballman, G. E., and Chaffin, W. L. (1979). Lipid synthesis during reinitiation of growth from stationary phase cultures of *Candida albicans*. Mycopathologia *67*:39-43.

Ballou, C. (1976). Structure and biosynthesis of the mannan component of the yeast cell envelope. Adv. Microbiol. Physiol. *14*:93-158.

Barrera, C. R., and Corral, J. (1980). Effect of hexoses on the levels of pyruvate decarboxylase in *Mucor rouxii*. J. Bacteriol. *142*:1029-1031.

Bartnicki-Garcia, S. (1968). Control of dimorphism in *Mucor* by hexoses. Inhibition of hyphal morphogenesis. J. Bacteriol. *96*:1586-1594.

Bartnicki-Garcia, S., and Nickerson, W. J. (1962a). Induction of yeastlike development in *Mucor* by carbon dioxide. J. Bacteriol. *84*:829-840.

Bartnicki-Garcia, S., and Nickerson, W. J. (1962b). Assimilation of carbon dioxide and morphogenesis of *Mucor rouxii*. Biochem. Biophys. Acta *64*:548-551.

Bartnicki-Garcia, S., and Nickerson, W. J. (1962c). Nutrition, growth, and morphogenesis of *Mucor rouxii*. J. Bacteriol. *84*:841-858.

Bartnicki-Garcia, S., and Nickerson, W. J. (1962d). Isolation, composition, and structure of cell walls of filamentous and yeast-like forms of *Mucor rouxii*. Biochem. Biophys. Acta *58*:102-119.

Bartnicki-Garcia, S., and Reyes, E. (1964). Chemistry of spore wall differentiation in *Mucor rouxii*. Arch. Biochem. Biophys. *108*:125-133.

Bartnicki-Garcia, S., Nelson, N., and Cota-Robles, E. (1968). A novel apical corpuscle in hyphae of *Mucor rouxii*. J. Bacteriol. *95*:2399-2402.

Bartnicki-Garcia, S., Bracker, C. E., Reyes, E., and Ruiz-Herrera, J. (1978). Isolation of chitosomes from taxonomically diverse fungi and synthesis of chitin microfibrils *in vitro*. Exp. Mycol. *2*:173-192.

Bedell, G. W., and Soll, D. R. (1979). Effects of low concentrations of zinc on the growth and dimorphism of *Candida albicans*: Evidence for zinc-resistant and -sensitive pathways for mycelium formation. Infect. Immun. 26:348-354.

Bedell, G. W., Werth, A., and Soll, D. R. (1980). The regulation of nuclear migration and division during synchronous bud formation in released stationary phase cultures of the yeast *Candida albicans*. Exp. Cell Res. *127*:103-113.

Benham, R. W. (1931). Certain monilias parasitic to man. Their identification by morphology and by agglutination. J. Infect. Dis. *49*:183-215.

Bhattacharya, A., and Datta, A. (1977). Effect of cyclic AMP on RNA and protein synthesis in *Candida albicans*. Biochem. Biophys. Res. Commun. *77*: 1438-1444.

Bhattacharya, A., Banerjee, S., and Datta, A. (1974). Regulation of N-acetylglucosamine kinase synthesis in yeast. Biochim. Biophys. Acta *374*:384-391.

Boguslawski, G., and Stetler, D. A. (1979). Aspects of physiology of *Histoplasma capsulatum* (a review). Mycopathologia *67*:17-24.

Borgia, P., and Sypherd, P. S. (1977). Control of β-glucosidase synthesis in *Mucor racemosus*. J. Bacteriol. *130*:812-817.

Bracker, C. E., Ruiz-Herrera, J., and Bartnicki-Garcia, S. (1976). Structure and transformation of chitin synthetase particles (chitosomes) during microfibril synthesis *in vitro*. Proc. Nat. Acad. Sci. *73*:4570-4574.

Braun, P. C., and Calderone, R. A. (1978). Chitin synthesis in *Candida albicans*: Comparison of yeast and hyphal forms. J. Bacteriol. *135*:1472-1477.

Brown, L. A., and Chaffin, W. L. (1981). Differential expression of cytoplasmic proteins during yeast bud and germ tube formation in *Candida albicans*. Can. J. Microbiol. *27*:580-585.

Cabib, E., and Keller, F. A. (1971). Chitin and yeast budding. Allosteric inhibition of chitin synthetase by a heat-stable protein from yeast. J. Biol. Chem. *246*:167-173.

Cassone, A., Simonetti, N., and Strippoli, V. (1973). Ultrastructural changes in the wall during germ-tube formation from blastospores of *Candida albicans*. J. Gen. Microbiol. *77*:417-426.

Chaffin, W. L., and Sogin, S. J. (1976). Germ tube formation from zonal rotor fractions of *Candida albicans*. J. Bacteriol. *126*:771-776.

Chattaway, F. W., Holmes, M. R., and Barlow, A. J. E. (1968). Cell wall compsition of the mycelial and blastospore forms of *Candida albicans*. J. Gen. Microbiol. *51*:367-376.

Chattaway, F. W., Bishop, R., Holmes, M. R., Odds, F. C., and Barlow, A. J. E. (1973). Enzyme activities associated with carbohydrate synthesis and breakdown in the yeast and mycelial forms of *Candida albicans*. J. Gen. Microbiol. *75*:97-109.

Chattaway, F. W., O'Reilly, J., Barlow, A. J. E., and Aldersley, T. (1976). Induction of the mycelial form of *Candida albicans* by hydrolysates of peptides from seminal plasma. J. Gen. Microbiol. *96*:317-322.

Chattaway, F. W., Wheeler, P. R., and O'Reilly, J. (1980). Purification and properties of peptides which induce germination of blastospores of *Candida albicans*. J. Gen. Microbiol. *120*:431-437.

Chattaway, F. W., Wheeler, P. R., and O'Reilly. (1981). Involvement of adenosine 3':5'-cyclic monophosphate in the germination of blastospores of *Candida albicans*. J. Gen. Microbiol. *123*:233-240.

Chiew, Y. Y., Shepherd, M. G., and Sullivan, P. A. (1980). Regulation of chitin synthesis during germ-tube formation in *Candida albicans*. Arch. Microbiol. *125*:97-104.

Clark-Walker, G. D. (1972). Development of respiration and mitochondria in *Mucor genevensis* after anaerobic growth: Absence of glucose repression. J. Bacteriol. *109*:399-408.

Cole, G. T., and Nozawa, Y. (1981). Dimorphism. In *Biology of Conidial Fungi*, Vol. 1, G. T. Cole and B. Kendrick (Eds.). Academic Press, New York, pp. 97-133.

Dabrowa, N., Landau, J. W., and Newcomer, V. D. (1965). The antifungal activity of physiologic saline in serum. J. Invest. Dermatol. *45*:368-377.

Dabrowa, N., Howard, D. H., Landau, J. W., and Shechter, Y. (1970). Synthesis of nucleic acids and proteins in the dimorphic forms of *Candida albicans*. Sabouraudia *8*:163-169.

Dabrowa, N., Taxer, S. S. S., and Howard, D. H. (1976). Germination of *Candida albicans* induced by proline. Infect. Immun. *13*:830-835.

Davies, R. R., and Denning, T. J. (1972). Growth and form in *Candida albicans*. Sabouraudia *10*:180-188.

Domek, D. B., and Borgia, P. T. (1981). Changes in the rate of chitin-plus-chitosan synthesis accompany morphogenesis of *Mucor racemosus*. J. Bacteriol. *146*:945-951.

Domer, J. E., Hamilton, J. G., and Harkin, J. C. (1967). Comparative study of the cell walls of the yeast-like and mycelia phases of *Histoplasma capsulatum*. J. Bacteriol. *94*:466-474.

Dow, J. M., and Rubery, P. H. (1977). Chemical fractionation of the cell walls of mycelial and yeast-like forms of *Mucor rouxii*: A comparative study of the polysaccharide and glycoprotein components. J. Gen. Microbiol. *99*: 29-41.

Duran, A., Bowers, B., and Cabib, E. (1975). Chitin synthetase zymogen is attached to the yeast plasma membrane. Proc. Nat. Acad. Sci. *72*:3952-3955.

Duran, A., Cabib, E., and Bowers, B. (1979). Chitin synthetase distribution on the yeast plasma membrane. Science *203*:363-365.

Evans, E. G. V., Odds, F. C., Richardson, M. D., and Holland, K. T. (1975). Optimum conditions for initiation of filamentation in *Candida albicans*. Can. J. Microbiol. *21*:338-342.

Falcone, G., and Nickerson, W. J. (1956). Cell wall mannan-protein of baker's yeast. Science *124*:272-273.

Falcone, G., and Nickerson, W. J. (1959). Enzymatic reactions involved in cellular division of microorganisms. Proc. 4th Int. Congr. Biochem. Vienna 6: 65-70.

Farkas, V. (1979). Biosynthesis of cell walls of fungi. Microbiol. Rev. *43*:117-144.
Fineman, B. C. (1921). A study of the thrush parasite. J. Infect. Dis., *28*:185-200.
Forte, J. W., and Orlowski, M. (1980). Profile of cyclic adenosine 3',5'-monophosphate-binding proteins during the conversion of yeasts to hyphae in the fungus *Mucor*. Exp. Mycol. *4*:78-86.
Friedenthal, M., Epstein, A., and Passeron, S. (1974). Effect of potassium cyanide, glucose and anaerobiosis on morphogenesis of *Mucor rouxii*. J. Gen. Microbiol. *82*:15-24.
Garcia, J. R., Hiatt, W. R., Peters, J., and Sypherd, P. S. (1980). S-Adenosylmethionine levels and protein methylation during morphogenesis of *Mucor racemosus*. J. Bacteriol. *142*:196-201.
Grappel, S. F., and Calderone, R. A. (1976). Effect of antibodies on the respiration and morphology of *Candida albicans*. Sabouraudia *14*:51-60.
Grove, S. N., and Bracker, C. E. (1970). Protoplasmic organization of hyphal tips among fungi: Vesicles and Spitzenkörper. J. Bacteriol. *104*:989-1009.
Haidle, C. W., and Storck, R. (1966a). Inhibition by cycloheximide of protein and RNA synthesis in *Mucor rouxii*. Biochem. Biophys. Res. Commun. *22*:175-180.
Haidle, C. W., and Storck, R. (1966b). Control of dimorphism in *Mucor rouxii*. J. Bacteriol. *92*:1236-1244.
Hartwell, L. H. (1974). *Saccharomyces cerevisiae* cell cycle. Bacteriol Rev. *38*:164-198.
Hasilik, A. (1973). Perturbation of growth and metabolism on *Candida albicans* by 4-bromobenzyl isothiocyanate and iodoacetate. Z. Naturforsch. *28c*:21-31.
Hazen, K. C., and Cutler, J. E. (1979). Autoregulation of germ tube formation by *Candida albicans*. Infect. Immun. *24*:661-666.
Hazen, K. C., and Cutler, J. E. (1983). Effect of cobalt and morphogenic autoregulatory substance (MARS) on morphogenesis of *Candida albicans*. Exp. Mycology (in press).
Hedden, D. M., and Buck, J. D. (1980). A reemphasis-germ tubes diagnostic for *Candida albicans* have no constriction. Mycopathologia *70*:95-101.
Hiatt, W. R., Inderlied, C. B., and Sypherd, P. S. (1980). Differential synthesis of polypeptides during morphogenesis of *Mucor*. J. Bacteriol. *141*:1350-1359.
Hogan, G., Shields, R., and Curtis, D. (1974). Effect of cyclic nucleotides on the induction of ornithine decarboxylase in BHK cells by serum and insulin. Cell *2*:229-233.
Inderlied, C. B., and Sypherd, P. S. (1978). Glucose metabolism and dimorphism in *Mucor*. J. Bacteriol. *133*:1282-1286.
Inderlied, C. B., Cihlar, R. L., and Sypherd, P. S. (1980). Regulation of ornithine decarboxylase during morphogenesis of *Mucor racemosus*. J. Bacteriol. *141*:699-706.
Jaynes, P. K., McDonough, J. P., and Mahler, H. R. (1980). Identification of cAMP binding proteins associated with the plasma membrane of the yeast *Saccharomyces cerevisiae*. Biochem. Biophys. Res. Commun. *94*:16-22.

Jillson, O. F., and Nickerson, W. J. (1948). Mutual antagonism between pathogenic fungi. Inhibition of dimorphism in *Candida albicans*. Mycologia *40*: 369-385.

Johnson, S. A. M. (1954). *Candida (Monilia) albicans*. Arch. Dermatol. Syphilol. *70*:59-60.

Joshi, K. R., Bremner, D. A., Gavin, J. B., Herdson, P. B., and Parr, D. N. (1973). The formation of germ tubes by *Candida albicans* in sheep serum and trypticase soya broth. Am. J. Clin. Pathol. *60*:839-842.

Keller, F. A., and Cabib, E. (1971). Chitin and yeast budding. Properties of chitin synthetase from *Saccharomyces carlsbergensis*. J. Biol. Chem. *246*:160-166.

Kerridge, D. (1980). The plasma membrane of *Candida albicans* and its role in the action of anti-fungal drugs. In *The Eucaryotic Microbial Cell*, G. W. Gooday, D. Lloyd, and A. P. J. Trinci (Eds.). Cambridge University, Cambridge, pp. 103-127.

Kessler, G., and Nickerson, W. J. (1959). Glucomannan-protein complexes from cell walls of yeasts. J. Biol. Chem. *234*:2281-2285.

Kobayashi, G. S., and Guiliacci, P. L. (1967). Cell wall studies on *Histoplasma capsulatum*. Sabouraudia *5*:180-188.

Kondoh, H. C., Ball, C. B., and Adler, J. (1979). Identification of a methyl-accepting chemotaxis protein for the ribose and galactose chemoreceptors of *Escherichia coli*. Proc. Nat. Acad. Sci. *76*:260-264.

Kot, E. J., Olson, V. L., Rolewic, L. J., and McClary, D. O. (1976). An alternate respiratory pathway in *Candida albicans*. Antonie Van Leeuwen *42*:33-48.

Kwon-Chung, K. J., and Hill, W. B. (1970). Studies on the pink, adenine-deficient strains of *Candida albicans*. I. Cultural and morphological characteristics. Sabouraudia *8*:48-59.

Land, G. A., McDonald, W. C., Stjernholm, R. L., and Friedman, L. (1975a). Factors affecting filamentation in *Candida albicans*: Relationship of the uptake and distribution of proline to morphogenesis. Infect. Immun. *11*: 1014-1023.

Land, G. A., McDonald, W. C., Stjernholm, R. L., and Friedman, L. (1975b). Factors affecting filamentation in *Candida albicans*: Changes in respiratory activity in *Candida albicans* during filamentation. Infect. Immun. *12*:119-127.

Landau, J. W., Dabrowa, N., Newcomer, V. D., and Rowe, J. R. (1964). The relationship of serum transferrin and iron to the rapid formation of germ tubes by *Candida albicans*. J. Invest. Dermatol. *43*:473-482.

Landau, J. W., Dabrowa, N., and Newcomer, V. D. (1965). The rapid formation in serum of filaments by *Candida albicans*. J. Invest. Dermatol. *44*:171-179.

Langeron, M., and Guerra, P. (1939). Orientation de la filamentisation des champignons livuriformes cultivés sur lames gélosées. Ann. Parasitol. *17*: 580-589.

Larsen, A., and Sypherd, P. (1979). Robosomal proteins of the dimorphic fungus, *Mucor racemosus*. Mol. Gen. Genet. *175*:99-109.

Larsen, A., and Sypherd, P. S. (1980). Physiological control of phosphorylation of ribosomal protein S6 in *Mucor racemosus*. J. Bacteriol. *141*:20-25.

Larsen, A. D., and Sypherd, P. S. (1974). Cyclic adenosine 3′,5′-monophosphate and morphogenesis in *Mucor racemosus*. J. Bacteriol. *117*:432-438.

Lee, K. L., Buckley, H. R., and Campbell, C. C. (1975). An amino acid liquid synthetic medium for mycelial and yeast forms of *Candida albicans*. Sabouradia *13*:148-153.

Lembach, K. J. (1974). Regulation of growth *in vitro*. I. Control of ornithine decarboxylase levels in untransformed and transformed fibroblasts by serum. Biochim. Biophys. Acta *354*:88-100.

Lingappa, B. T., and Lingappa, Y. (1966). The nature of self-inhibition of germination of conidia of *Glomerella cingulata*. J. Gen. Microbiol. *43*:91-100.

Lingappa, B. T., Prasad, M., Lingappa, Y., Hunt, D. F., and Biemann, K. (1969). Phenylethyl alcohol and tryptophol: Autoantibiotics produced by the fungus *Candida albicans*. Science *163*:192-194.

McClary, D. O. (1952). Factors affecting the morphology of *Candida albicans*. Ann. Mo. Bott. Gard. *39*:137-165.

Mackinnon, J. E. (1940). Dissociation in *Candida albicans*. J. Infect. Dis. *66*: 59-77.

Macko, V., Staples, R. C., Gershon, H., and Renwick, J. A. A. (1970). Self-inhibitor of the bean rust uredospores: Methyl 3,4-dimethoxycinnamate. Science *170*:539-540.

McMurrough, I., Flores-Carreon, A., and Bartnicki-Garcia, S. (1971). Pathway of chitin synthesis and cellular localization of chitin synthetase in *Mucor rouxii*. J. Biol. Chem. *246*:3999-4007.

Magni, G. (1948). Biological significance of the pseudomycelium in asporogenous yeasts. I. Reciprocal inhibition of parallel colonies. Mycopathologia *4*:207-212.

Manning, M., and Mitchell, T. G. (1980a). Strain variation and morphogenesis of yeast- and mycelial-phase *Candida albicans* in low-sulfate, synthetic medium. J. Bacteriol. *142*:714-719.

Manning, M., and Mitchell, T. G. (1980b). Analysis of cytoplasmic antigens of the yeast and mycelial phases of *Candida albicans* by two-dimensional electrophoresis. Infect. Immun. *30*:484-495.

Manning, M., and Mitchell, T. G. (1980c). Morphogenesis of *Candida albicans* and cytoplasmic proteins associated with differences in morphology, strain, or temperature. J. Bacteriol. *144*:258-273.

Mardon, D., Balish, E., and Phillips, A. W. (1969). Control of dimorphism in a biochemical variant of *Candida albicans*. J. Bacteriol. *100*:701-707.

Mardon, D. N., Hurst, S. K., and Balish, E. (1971). Germ-tube production by *Candida albicans* in minimal liquid culture media. Can. J. Microbiol. *17*: 851-856.

Maresca, B., Kumar, B. V., Medoff, J., Medoff, G., and Kobayashi, G. S. (1980). Studies on dimorphism in *Histoplasma capsulatum*: Biochemical changes during the differentiation process. In *Medical Mycology*, H. J. Preusser (Ed.). Gustav Fischer Verlag, New York, pp. 17-22.

Maresca, B., Lambowitz, A. M., Kumar, V. B., Grant, G. A., Kobayashi, G. S., and Medoff, G. (1981). Role of cysteine in regulating morphogenesis and mitochondrial activity in the dimorphic fungus *Histoplasma capsulatum*. Proc. Nat. Acad. Sci. *78*:4596-4600.

Marriott, M. S. (1975). Isolation and chemical characterization of plasma membranes from the yeast and mycelial forms of *Candida albicans*. J. Gen. Microbiol. *86*:115-132.

Mattia, E., and Cassone, A. (1979). Inducibility of germ-tube formation in *Candida albicans* at different phases of yeast growth. J. Gen. Microbiol. *113*:439-442.

Mooney, D. T., and Sypherd, P. S. (1976). Volatile factor involved in the dimorphism of *Mucor racemosus*. J. Bacteriol. *126*:1266-1270.

Nickerson, W. J. (1948). Enzymatic control of cell division in microorganisms. Nature *162*:241-245.

Nickerson, W. J. (1951). Physiological basis of morphogenesis in animal disease fungi. Trans. N.Y. Acad. Sci. *13*:140-145.

Nickerson, W. J. (1953). Reduction of inorganic substances by yeasts. I. Extracellular reduction of sulfate by species of *Candida*. J. Infect. Dis. *93*:43-56.

Nickerson, W. J. (1954). Experimental control of morphogenesis in microorganisms. Ann. N.Y. Acad. Sci. *60*:50-57.

Nickerson, W. J., and Chung, C. W. (1954). Genetic block in the cellular division mechanism of a morphological mutant of a yeast. Am. J. Bot. *41*:114-120.

Nickerson, W. J., and Falcone, G. (1956a). Enzymatic reduction of disulfide bonds in cell wall protein of baker's yeast. Science *124*:318-319.

Nickerson, W. J., and Falcone, G. (1956b). Identification of protein disulfide reductase as a cellular division enzyme in yeasts. Science *124*:722-723.

Nickerson, W. J., and Jillson, O. F. (1948). Interaction between pathogenic fungi in culture. Considerations on the mechanism of cell division in the dimorphism of pathogenic fungi. Mycopathol. Mycol. Appl. *4*:279-283.

Nickerson, W. J., and Mankowski, Z. (1953). Role of nutrition in the maintenance of the yeast-shape in *Candida*. Am. J. Bot. *40*:584-592.

Nickerson, W. J., and Romano, A. H. (1952). Enzymatic reduction of cystine by coenzyme I (DPNH). Science *115*:676-678.

Nickerson, W. J., and Van Rij, N. J. W. (1949). The effect of sulhydryl compounds, penicillin and cobalt on the cell division mechanism of yeasts. Biochim. Biophys. Acta *3*:461-475.

Nickerson, W. J., and Zerahn, K. (1949). Accumulation of radioactive cobalt by dividing yeast cells. Biochim. Biophys. Acta *4*:476-483.

Niimi, M., Niimi, K., Tokunaga, J., and Nakayama, H. (1980). Changes in cyclic nucleotide levels and dimorphic transition in *Candida albicans*. J. Bacteriol. *142*:1010-1014.

Nishioka, Y., and Silva-Hutner, M. (1974). Dimorphism, sensitivity to nystatin and acriflavin uptake in a strain of *Candida albicans* grown with glutamate as sole nitrogen and carbon source. Sabouraudia *12*:295-301.

O'Connell, B. T., and Paznokas, J. L. (1980). Glyoxylate cycle in *Mucor racemosus*. J. Bacteriol. *143*:416-421.

Odds, F. C. (1979). *Candida and Candidosis*. University Park Press, Baltimore.
Olaiya, A. G., Steed, J. R., and Sogin, S. J. (1980). Deoxyribonucleic acid-deficient strains of *Candida albicans*. J. Bacteriol. *141*:1284-1290.
Orlowski, M., and Sypherd, P. S. (1976). Cyclic guanosine 3',5'-monophosphate in the dimorphic fungus *Mucor racemosus*. J. Bacteriol. *124*:1226-1228.
Orlowski, M., and Sypherd, P. S. (1977). Protein synthesis during morphogenesis of *Mucor racemosus*. J. Bacterioo. *132*:209-218.
Orlowski, M., and Sypherd, P. S. (1978a). Location of protein synthesis during morphogenesis of *Mucor racemosus*. J. Bacteriol. *133*:399-400.
Orlowski, M., and Sypherd, P. S. (1978b). RNA synthesis during morphogenesis of the fungus, *Mucor racemosus*. Arch. Microbiol. *119*:145-159.
Pall, M. L. (1981). Adenosine 3',5'-phosphate in fungi. Microbiol. Rev. *45*:462-480.
Paveto, C., Epstein, A., and Passeron, S. (1975). Studies on cyclic adenosine 3',-5'-monophosphate levels, adenylate cyclase and phosphodiesterase activities in the dimorphic fungus *Mucor rouxii*. Arch. Biochem. Biophys. *169*:449-457.
Paznokas, J. L., and Sypherd, P. S. (1975). Respiratory capacity, cyclic adenosine 3',5'-monophosphate, and morphogenesis of *Mucor racemosus*. J. Bacteriol. *124*:134-139.
Paznokas, J. L., and Sypherd, P. S. (1977). Pyruvate kinase isozymes of *Mucor racemosus*: Control of synthesis by glucose. J. Bacteriol. *130*:661-666.
Peters, J., and Sypherd, P. S. (1979). Morphology-associated expression of nicotinamide adenine denucleotide-dependent glutamate dehydrogenase in *Mucor racemosus*. J. Bacteriol. *137*:1134-1139.
Phaff, H. J. (1971). Structure and biosynthesis of the yeast cell envelope. In *The Yeasts*, Vol. 2, A. H. Rose and J. S. Harrison (Eds.). Academic, New York, pp. 135-210.
Polak, A., and Wain, W. H. (1979). The effect of 5-fluorocytosine on the blastospores and hyphae of *Candida albicans*. J. Med. Microbiol. *12*:83-97.
Poulter, R., Jeffery, K., Hubbard, M. J., Shepherd, M. G., and Sullivan, P. A. (1981). Parasexual genetic analysis of *Candida albicans* by spheroplast fusion. J. Bacteriol. *146*:833-840.
Pugh, D., and Cawson, R. A. (1979). The induction of germ tubes in *Candida albicans* by an intrinsic factor. Microbios. *24*:73-79.
Purohit, B. C., Joshi, K. R., Ramedo, I. N., and Bharadwaj, T. P. (1977). The formation of germ tubes by *Candida albicans* when grown with *Staphylococcus pyogene*, *Escherichia coli*, *Klebsiella pneumoniae*, *Lactobacilus acidophilus*, and *Proteus vulgaris*. Mycopathologia *62*:187-189.
Rajasingham, K. C., and Cawson, R. A. (1980). Cell wall and plasma membrane ultrastructure of the invasive hyphae of *Candida albicans*. Pathology *8*:430.
Rao, G. R., Ramakrishnan, T., and Sirsi, M. (1960). Enzymes in *Candida albicans*. I. Pathways of glucose dissimilation. J. Bacteriol. *80*:654-658.
Rao, G. R., Sirsi, M., and Ramakrishnan, T. (1962). Enzymes in *Candida albicans*. II. Tricarboxylic acid cycle and related enzymes. J. Bacteriol. *84*: 778-783.

Rickenberg, H. V. (1974). Cyclic AMP in prokaryotes. Annu. Rev. Microbiol. *28*:353-369.
Rippon, J. W. (1980). Dimorphism in pathogenic fungi. Crit. Rev. Microbiol. *8*:49-97.
Rogers, R. J., Clark-Walker, G. D., and Stewart, P. R. (1974). Effects of oxygen and glucose on energy metabolism and dimorphism of *Mucor genevensis* grown in continuous culture: Reversibility of yeast-mycelium conversion. J. Bacteriol. *119*:282-293.
Romano, A. (1966). Dimorphism. In *The Fungi*, Vol. 2, G. C. Ainsworth and A. S. Sussman (Eds.). Academic, New York, pp. 181-209.
Romano, A. H., and Nickerson, W. J. (1954). Cystine reductase of pea seeds and yeasts. J. Biol. Chem. *208*:409-416.
Ruiz-Herrera, J., Lopez-Romero, E., and Bartnicki-Garcia, S. (1977). Properties of chitin synthetase in isolated chitosomes from yeast cells of *Mucor rouxii*. J. Biol. Chem. *252*:3338-3343.
Russell, D. H., and Snyder, S. H. (1968). Amine synthesis in rapidly growing tissues: Ornithine decarboxylase activity in regenerating rat liver, chick embryos, and various tumors. Proc. Nat. Acad. Sci. *66*:1420-1427.
Russell, D. H., Synder, S. H., and Medine, V. J. (1970). Growth hormone induction of ornithine decarboxylase in rat liver. Endocrinology *86*:1414-1419.
Saltarelli, C. G. (1973). Growth stimulation and inhibition of *Candida albicans* by metabolic by-products. Mycopathol. Mycol. Appl. *51*:53-63.
Saltarelli, C. G., and Coppola, C. P. (1979). Effect of light growth and metabolite synthesis in *Candida albicans*. Mycologia *71*:773-785.
San-Blas, F., San-Blas, G., and Inlow, D. (1980). Dimorphism in *Paracoccidioides brasiliensis*. In *Medical Mycology*, H. J. Preusser (Ed.). Gustav Fischer Verlag, New York, pp. 23-28.
Sarachek, A., Rhoads, D. D., and Schwarzhoff, R. H. (1981). Hybridization of *Candida albicans* through fusion of protoplasts. Arch. Microbiol. *129*:1-8.
Schaar, G., Long, I., and Widra, A. (1974). A combination rapid and standard method for identification of *Candida albicans*. Mycopathol. Mycol. Appl. *52*:203-207.
Scherwitz, C., Martin, R., and Ueberberg, H. (1978). Ultrastructural investigations of the formation of *Candida albicans* germ tubes and septa. Sabouraudia *16*:115-124.
Shepherd, M. G., and Sullivan, P. A. (1976). The production and growth characteristics of yeast and mycelial forms of *Candida albicans* in continuous culture. J. Gen. Microbiol. *93*:361-370.
Shepherd, M. G., Chiew, Y. Y., Ram, S. P., and Sullivan, P. A. (1980a). Germ tube induction in *Candida albicans*. Can. J. Microbiol. *26*:21-26.
Shepherd, M. G., Ghazali, H. M., and Sullivan, P. A. (1980b). N-Acetyl-D-glucosamine kinase and germ-tube formation in *Candida albicans*. Exp. Mycol. *4*:147-159.
Simonetti, N., Strippoli, V., and Cassone, A. (1974). Yeast mycelial conversion induced by N-acetyl-D-glucosamine in *Candida albicans*. Nature *250*:344-346.

Singh, B. R., and Datta, A. (1978). Glucose repression of the inducble catabolic pathway for N-acetylglucosamine in yeast. Biochem. Biophys. Res. Commun. *84*:58-64.

Skinner, C. E. (1947). The yeast-like fungi: *Candida* and *Brettanomyces*. Bacteriol. Rev. *11*:227-274.

Smith, J. K., and Louria, D. B. (1972). Anti-*Candida* factors in serum and their inhibitors. II. Identification of a *Candida*-clumping factor and the influence of the immune response on the morphology of *Candida* and on anti-*Candida* activity of serum in rabbits. J. Infect. Dis. *125*:115-122.

Soll, D. R., and Bedell, G. W. (1978). Bud formation and the inducibility of pseudo-mycelium outgrowth during release from stationary phase in *Candida albicans*. J. Gen. Microbiol. *108*:173-180.

Soll, D. R., Stasi, M., and Bedell, G. (1978). The regulation of nuclear migration and division during pseudo-mycelium outgrowth in the dimorphic yeast *Candida albicans*. Exp. Cell. Res. *116*:207-215.

Soll, D. R., Bedell, G. W., and Brummel, M. (1981a). Zinc and the regulation of growth and phenotype in the infectious yeast *Candida albicans*. Infect. Immun. *32*:1139-1147.

Soll, D. R., Bedell, G., Thiel, J., and Brummel, M. (1981b). The dependency of nuclear division on volume in the dimorphic yeast *Candida albicans*. Exp. Cell. Res. *133*:55-62.

Storck, R., and Morrill, R. C. (1971). Respiratory-deficient, yeastlike mutant of *Mucor*. Biochem. Genet. *5*:140-147.

Sundaram, S., Sullivan, P. A., and Shepherd, M. G. (1981). Changes in lipid composition during starvation and germ-tube formation in *Candida albicans*. Exp. Mycol. *5*:140-147.

Swatek, F. E. (1967). *Textbook on Microbiology*. C. V. Mosby, St. Louis, Mo. p. 397.

Takeo, K. (1974). Ultrastructure of polymorphic *Mucor* as observed by means of freeze-etching. II. Vegetative yeast form grown under anaerobic conditions. Arch. Microbiol. *99*:91-98.

Taschdjian, C. L., Burchall, J. J., and Kozinn, P. J. (1960). Rapid identification of *Candida albicans* by filamentation on serum and serum substitute. J. Dis. Child. *99*:212-215.

Terenzi, H. F., and Storck, R. (1969). Stimulation of fermentation and yeast-like morphogenesis in *Mucor rouxii* by phenethyl alcohol. J. Bacteriol. *97*: 1248-1261.

Trevillyan, J. M., and Pall, M. L. (1979). Control of cyclic adenosine $3',5'$-monophosphate levels by depolarizing agents in fungi. J. Bacteriol. *138*: 397-403.

Vaughn, V. J., and Weinberg, E. D. (1978). *Candida albicans* dimorphism and virulence: Role of copper. Mycopathologia *64*:39-42.

Wain, W. H., Price, M. F., and Cawson, R. A. (1975). A re-evaluation of the effect of cysteine on *Candida albicans*. Sabouraudia *13*:74-82.

Wain, W. H., Brayton, A. R., and Cawson, R. A. (1976a). Variations in the response to N-acetyl-D-glucosamine by isolates of *Candida albicans*. Mycopathologia *58*:27-29.
Wain, W. H., Price, M. F., Brayton, A. R., and Cawson, R. A. (1976b). Macromolecular synthesis during the cell cycles of yeast and hyphal phases of *Candida albicans*. J. Gen. Microbiol. *97*:211-217.
Weidman, F. D. (1922). Resemblance of yeasts in cutaneous scrapings to hyphomycetes. Arch. Dermatol. Syphilol. *5*:325-328.
Weld, J. T. (1952). *Candida albicans*. Rapid identification in pure cultures with carbon dioxide on modified eosin-methylene blue medium. Arch. Dermatol. Syphilol. *66*:691-694.
Yamaguchi, H. (1974). Mycelial development and chemical alteration of *Candida albicans* from biotin insufficiency. Sabouraudia *12*:320-328.
Yamaguchi, H. (1975). Control of morphism in *Candida albicans* by zinc: Effect on cell morphology and composition. J. Gen. Appl. Microbiol. *86*:370-372.
Zorzopulos, J., Jobbagy, A. J., and Terenzi, H. F. (1973). Effects of ethylenediaminetetraacetate and chloramphenicol on mitochondrial activity and morphogenesis in *Mucor rouxii*. J. Bacteriol. *115*:1198-1204.

11
The Yeast Genome in Yeast Differentiation

Michael Breitenbach and Eva Lachkovics* / Institut für Allgemeine Biochemie and Ludwig-Boltzmann-Forschungsstelle für Biochemie, Vienna, Austria

INTRODUCTION

The life cycles of the unicellular eukaryote *Saccharomyces cerevisiae* are presented schematically in Fig. 1. They include several steps of physiological and morphological differentiation. There is a distinction between adaptation processes, which are in principle changes in the enzymatic composition of cells brought about by environmental stimuli (e.g., the transition from respiring to glucose-repressed cells or from vegetative to stationary cells) and cell differentiation processes which involve morphogenesis. Such morphogenetic steps are the following:

1. The cell cycle itself, which may occur in haploid or diploid cells [although according to some authors, "it is a matter of semantics whether the vegetative cell cycle is considered a type of differentiation" (Piggot, 1979)]
2. The mating process, which brings about sexual fusion of two competent haploid cells and the formation of a diploid zygote
3. Meiosis and sporulation of a competent diploid cell, resulting in the formation of an ascus containing four haploid ascospores
4. Germination of the spores and formation of stable haploid clones

The cell cycle will be reviewed briefly here, because the analysis of cell cycle mutants (mainly by Hartwell and his co-workers) has provided many insights into the other three differentiation processes.

Haploid as well as diploid cells proliferate by *budding* in a very similar way. However, the orientation of the growing bud relative to preexisting bud scars (or relative to the birth scar) is different in haploids and diploids. Haploids bud "equatorially" (i.e., the growing bud is near the older bud scars), and diploids

Present affiliation: IOCU Regional Office, Penang, Malaysia

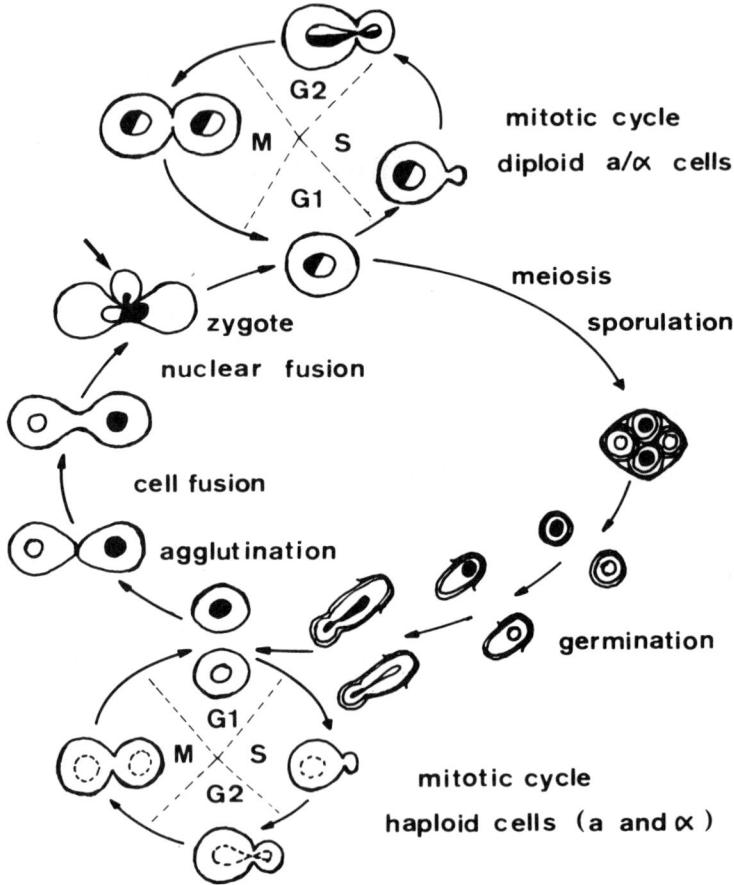

Figure 1 The life cycles of *S. cerevisiae* showing the haploid and diploid phases and also the differentiation processes occurring (i.e., mating, meiosis/sporulation, and germination).

do so in a "polar" way (i.e., the growing bud is as far away as possible from the older bud scars) (Freifelder, 1960).

The competence of a cell to undergo meiosis/sporulation or mating (which are mutually exclusive except for rare mutants) is governed by a complex cluster of regulatory genes, the mating-type locus (*MAT*). The two alleles of *MAT* defining the two mating types of yeast are called *a* and α. Cells containing only one type of *MAT* gene (haploids or homozygotes of higher ploidy) can mate, but not sporulate. Cells containing at least one *MATa* and one *MAT*α gene (i.e., heterozygotes) can sporulate, but are unable to mate. The mating-type locus seems to occupy a central role in the hierarchy of gene functions required for mating or meiosis/sporulation. The gene products of the *MAT* locus are synthesized constitutively. The two gene products of *MATα* are both needed for the regulation of mating and meiosis/sporulation. Presumably, one is an activator of α-specific genes, which are scattered throughout the genome. The other could be an inhibitor of *a*-specific genes. Only one gene product of *MATa* has been shown to be indispensable. It is not needed for the mating of haploid *a* cells, but possibly for the sporulation of *a/α* cells (Strathern et al., 1981; Klar et al., 1981; Nasmyth et al., 1981a,b).

Homothallic switching involves a replacement of *MATa* genetic information by *MATα* (or vice versa). In this way the haplophase of a homothallic strain is rendered unstable, because a haploid cell and its progeny (which carries the opposite mating type) will form a zygote ("diploidization"). The process has been considered as a model for differentiation by the transposition of genetic elements, which also occurs during lymphocyte differentiation and other steps of differentiation in higher animals (Herskowitz et al., 1980).

Germination is the transition from the dormant state of the spore to a haploid vegetative cell. It does not seem to be under the control of the mating-type locus. The morphogenetic pathway of germination resembles the corresponding pathway in *Bacillus subtilis*. Comparatively few research papers and no reviews have been published on the germination of yeast ascospores. Therefore we have attempted to give a comprehensive review of this topic.

For the other steps of differentiation mentioned above, recent reviews are available and will be cited at the appropriate points in this chapter. We have concentrated on the description of the methods and results of genetic experiments related to the cell cycle, mating, and sporulation.

Based on the many mutants specific for the differentiation processes of yeast, it seems to be clear that these processes are controlled by specific genes. However, they are also controlled by environmental stimuli. Nitrogen starvation in the presence of a nonfermentable carbon source (such as acetate) stimulates sporulation. The availability of growth medium containing a fermentable carbon source stimulates germination. Only if nutrients are available is a new cell cycle entered. Mating is initiated only in the presence of mating pheromones, mating partners, and nutrients. Therefore none of these processes is purely genetically or physiologically controlled. During the elucidation of the processes of differentiation the interrelationship of these two aspects will be shown.

Yeast has been studied as a model system for eukaryotic differentiation for several decades. Up to about 1970 a large number of research papers dealing with the physiological control of yeast differentiation (mostly sporulation) appeared in the literature (for a review, see Fowell, 1969). For example, the nutritional conditions allowing sporulation were studied extensively (Miller and Hoffmann-Ostenhof, 1964).

Around 1970, yeast genetics, which had been developed during the 1960s by Roman, Mortimer, Hawthorne, and others (Mortimer and Hawthorne, 1969), began to make their impact on the field of yeast differentiation. In the following decade this resulted in the isolation of differentiation-specific mutants, the discovery of "pathways" ordering the sequence of functions defined by the mutants, and the estimation of the number of genes involved. During this period it became clear that genetic methods which had been developed for the analysis of the morphogenesis of phages (Jarvik and Botstein, 1973) or bacteria (Piggot, 1979) are successfully applicable to eukaryotes.

The advantages of yeast compared to other eukaryotes, whose morphogenetic development is currently being studied, are enormous: The organism is easy to grow and the generation time is short (usually around 2 hr). The haplophase as well as the diplophase of heterothallic yeast is stable and sexual conjugation and meiosis can be induced and controlled. This enables the study of recessive genes, dominance relations, and complementation in an easy way. Meiotic recombination is frequent and all four products of a meiotic tetrad are viable and can be isolated by spore tetrad dissection. Therefore recombinational mapping of yeast genes (together with other mapping techniques) has led to a very detailed genetic map of *S. cerevisiae* (Mortimer and Schild, 1980). The construction of double mutants is straightforward. As yeast is heterotrophic and facultatively anaerobic, auxotrophic and many other kinds of mutants are easily gained, helping in the genetic analysis of differentiation. Having mentioned the great technical advantages of *S. cerevisiae* in the study of biological development, it has to be said that only the unicellular aspects of development can be studied in yeast. How much this can tell us about developmental processes in higher organisms will be discussed in the last section of this chapter.

GENETIC STRATEGIES

Determination of Pathways (Order-of-Function Maps)

Temperature-Sensitive Mutants

In almost all cases it is necessary to work with temperature-sensitive (*ts*) or other conditional mutants (less preferably with leaky mutants), because the block in differentiation conferred by the mutation precludes genetic analysis. The *terminal phenotype* of a mutant is the phenotype it obtains after prolonged time at the restrictive temperature (usually 36°C). By a series of temperature shift experiments the *execution point* of a mutation is determined; it is defined operationally as the latest point that allows completion of the process studied if it is traversed at the permissive temperature (usually 23°C). For example, if a cell cycle (*cdc*) mutant has passed the execution point at the permissive temperature and is then shifted to restrictive conditions, it will complete the ongoing cycle and enter a new one, stopping only at its terminal phenotype in the new cycle. If it is shifted to restrictive conditions before it has passed the execution point, it will not complete the ongoing cycle, but will stop at the terminal phenotype during the same cycle. The execution point of a mutant is allele specific, not gene specific (Pringle and Hartwell, 1981). This is one of the reasons why the determination of execution points alone is not sufficient to determine the order-of-function map. Several methods have to be combined to achieve this.

If pulses of varying length of permissive conditions are given to mutant cells, the exact time can be determined at which expression of the mutant gene is necessary during the course of development. It is advantageous to work with synchronized cells to improve the temporal resolution of this method (for details see Simchen, 1978). A comparison of the terminal phenotypes and execution points of a number of mutants gives a first impression of the order of functions defined by the mutations.

The Construction of Double Mutants

The terminal phenotype and execution point of a double mutant yields the order of dependence (*A-B* or *B-A*) of the two functions in question. If the two genes act in independent pathways, the phenotype of the double mutant may be different from either of the two single mutants.

The Method of Reciprocal Temperature Shifts

Originally introduced by Jarvik and Botstein (1973), the method of reciprocal temperature shifts requires both heat-sensitive and cold-sensitive mutants in the same developmental process. These are usually not available in the case of yeast differentiation. However, if an inhibitor is available that acts at a specific stage of differentiation, a similar experiment can be performed. An example is the inhibition of the elongation step of DNA synthesis during the cell cycle by hydroxyurea. Cell cycle mutants can be ordered with respect to this step by the following method: The synchronized mutant cells are first kept at restrictive temperature without inhibitor; then they are shifted to permissive temperature with inhibitor. In a second experiment, permissive temperature with inhibitor comes first, followed by restrictive temperature without inhibitor. The experiment that allows the completion of differentiation defines the order of the two functions investigated in a *dependent pathway.* If neither experiment allows completion of the process, the two functions are interdependent. If both experiments allow completion of the process, the two functions are independent of each other (for further discussion see Avers, 1980, p. 506). If the method is applicable, it yields unambiguous ordering of two events during differentiation.

These three methods usually give enough information to construct a developmental pathway. Very often branched pathways are found, as will be discussed below.

Determination of the Primary Defect of a Mutation

Many of the differentiation-defective mutants discussed in this chapter show a pleiotropic phenotype. The determination of the primary defect of such a mutation is an essential step in identifying the gene product of the gene observed:

1. *Partial suppression* (Piggot, 1979) by a second mutation often reduces the pleiotropy to a single phenotype, which is considered to be the primary one. The method has been applied in the field of *B. subtilis* sporulation, but should work with yeast as well.
2. Investigation of a *different type of differentiation* in the same mutant, for example, sporulation/meiosis in a cell cycle mutant (Simchen, 1978), often gives similar information. The defect which occurs in both conditions is probably closer to the primary one.
3. If a *ts* mutant is investigated at an *intermediary temperature,* the defect that occurs also at the intermediary temperature is probably the primary one.

Is the Mutation in a Structural Gene for a Protein?

The answer to this important question is a prerequisite for the identification of the gene product. The fact that a certain mutation is *ts* does not guarantee that the mutation is in a structural gene for a protein. Counterexamples may be given by *ts* petite mutations in the introns of the mitochondrial gene *COB* (R. J. Schweyen, personal communication). These mutations are located outside the protein-encoding genes of this region; it is concluded that they confer temperature sensitivity to nucleic acid conformation in the physiological temperature range.

    1. The introduction of *temperature-sensitive nonsense suppressors* enables mutations in structural genes to be identified (Hartwell, 1980). If a nonsense mutation is found, the time of translation of the gene can be determined by temperature shift experiments.

    2. In the very advanced stage of genetic analysis a *collinearity relationship* will be found between the position of nonsense mutations in the fine-structure genetic map of a gene and the mobility on SDS/polyacrylamide gels of the corresponding polypepeptides. This allows the unambiguous assignment of a polypeptide to a genetically defined locus. Interruption of the structural gene by introns disturbs this collinearity relationship (for an example see Haid et al., 1978). The method has not yet been applied to any developmental gene of yeast.

The Yeast Genome in Yeast Differentiation   313

3. If it is possible to "clone" the gene in question, the gene product can be identified by the methods to be described in the next section.

The Use of "Genetic Engineering" Techniques in the Analysis of Developmental Mutants

The finding that yeast can be transformed efficiently by DNA (Hinnen et al., 1978) started a new era and provided a host of methods in yeast genetics.

1. Transformation. The transformation of yeast spheroplasts is performed in the presence of polyethylene glycol and calcium, followed by regeneration of yeast cells and identification of clones carrying specific DNA sequences or being reconstituted with respect to specific functions. The complementation of a recessive mutation by the transforming DNA is the most usual assay for the isolation of a specific DNA clone. The method has been optimized by several laboratories, mostly improving on the "vector" (i.e., the circular DNA used for transformation, reviewed by Nasmyth, 1978). Nowadays sophisticated hybrid shuttle vectors (Petes, 1980) consist of DNA sequences from the following:

a. A bacterial plasmid (enabling multiplication in bacteria and containing one or more antibiotic resistance genes for selection of transformed bacterial clones)
b. Yeast DNA enabling stable multiplication in yeast (either 2-μm DNA, a a chromosomal "replicator," or a centromere sequence)
c. A yeast gene that can be selected because it complements an auxotrophy in the yeast acceptor strain

Hohn and Hinnen (1980) suggested the use of the *cos* sequence of lambda phage DNA as part of the vector for transformation experiments in yeast. By packaging into lambda phage heads, large pieces of DNA can be selected, which is essential for the analysis of eukaryotic genomes.

The vector contains a single site for a certain restriction enzyme, into which restriction fragments of the yeast genome can be inserted. Most often, a specific developmental mutant of yeast is transformed by a random collection of yeast DNA restriction fragments (inserted into the vector), and transformed clones (i.e., single colonies) are selected on plates. Among several thousand transformed clones usually some are found in which the original developmental defect of the host is healed. The vector is reisolated from these clones and pieces of DNA carrying the developmental gene are isolated in a highly purified form.

2. The isolated DNA can be used to characterize a gene product directly, either by injecting it into *Xenopus* oocytes (De Robertis and Gurdon, 1979) or by using a coupled transcription-translation system (Moorman et al., 1976). It would be highly desirable to use such a system from yeast. However, at present,

only a system from *Escherichia coli* is available which is not capable of expressing eukaryotic split genes correctly. These systems could also be used for the study of the maturation and processing steps of the primary transcripts and translation products. At present, the method of choice seems to be the isolation of the corresponding mRNA by hybridization with the DNA clone followed by translation in an in vitro translation system. The molecular weight and isoelectric point of the mature gene product can be determined in a two-dimensional electrophoretic system (O'Farrell, 1975).

3. The isolated DNA can be used as a probe to identify the corresponding mRNA from wild-type cells by DNA/RNA hybridization techniques (Thomas, 1980). In the same way, the time of transcription of the gene during differentiation can be determined. Eventually the question of transcriptional versus posttranscriptional control can be answered by the combined use of methods (2) and (3).

4. *Mutagenesis in vitro.* Single-base changes, deletions, and additions of varying lengths can be introduced into DNA by enzymatic methods before being inserted back into a mutant strain (a good example for this method is the analysis of the *MAT* locus by Nasmyth et al., 1981a). As the question of eukaryotic control regions (promotors) and the function of the large noncoding parts of eukaryotic DNA is still unclear, the method is used to identify regions essential for the regulation of gene expression.

5. *DNA sequencing* may lead to the identification of open reading frames and putative control regions. An example is provided by the sequence of the mating-type locus (Nasmyth et al., 1981a). Physical mapping, which is a prerequisite for sequencing, is often used together with other techniques for genetic mapping.

6. As a high copy number of the transforming DNA is attainable at will, *gene dosage effects* can easily be studied in this way.

Obviously the potential of the new "genetic engineering" techniques is great. A typical example for the present state of the art may be the isolation of several *CDC* and *SPO* genes in various laboratories (Nasmyth and Reed, 1980). These techniques have the great advantage of being widely applicable. However, they are not yet advanced enought to identify the function of a cloned gene in development. This has in some cases been achieved by classic genetic methods (e.g., for the *CDC21* gene; see the section on the cell cycle).

Determination of Specificity

A great number of biochemical parameters (metabolites, enzyme activities) have been measured during the differentiation processes in the wild type. In these cases two questions arise: Is the parameter (event) necessary, and is it specific for the differentiation process studied? The first of these questions can be

The Yeast Genome in Yeast Differentiation

answered if a mutant devoid of the metabolite (enzyme activity) in question is studied under differentiation conditions. The second question is best answered by measuring the parameter (event) under differentiation conditions in a cell that is genetically unable to differentiate (e.g., an α/α cell under sporulation conditions). In this way three classes of events can be distinguished. They are given below with examples:

1. "Not necessary and not specific." The event is only accidental to the differentiation process, for example, the peak of cellular respiration during yeast sporulation (Hartig and Breitenbach, 1980).
2. "Necessary but not specific." The event occurs also under conditions of vegetative growth. For example, gene functions necessary for DNA replication are needed during sporulation, but they are also needed during the vegetative cell cycle.
3. "Necessary and specific." An example is provided by the process of meiotic recombination. A block in this process always impedes the formation of normal ascospores. It does not occur in mitotic cells (mitotic recombination has a different mechanism and is 100 times less frequent).

THE CELL CYCLE AND ITS RELATION TO OTHER DIFFERENTIATION PATHWAYS

The order-of-function map of the yeast cell cycle depicted in Fig. 2 was established by several laboratories using a collection of *ts cdc* mutants and various combinations of the methods described earlier. A detailed list of all known *cdc* mutants containing information on terminal phenotypes, diagnostic landmarks, genetic mapping, and biochemical characterization has been published (Pringle and Hartwell, 1981). Among a few thousand *ts* lethal mutations, only those displaying first cycle arrest and a well-defined terminal phenotype at the restrictive temperature were selected. Complementation analysis of these mutants revealed 50 complementation groups representing the major part of all genes which can be found by this selection procedure. Included are the G_1-arrested mutants isolated by a specific selection procedure (Reed, 1980a,b). Suppressors of sterility (*ste*) mutations often confer a *cdc* phenotype (M. E. Katz, P. Kayne and S. I. Reed, personal communication), but a complete analysis of these mutations in relation to the previously known *cdc* mutations has not yet been performed. It is possible that many more *CDC* genes exist but cannot be found by the selection procedures used so far (discussed by Pringle and Hartwell, 1981).

In only three cases is the gene product of *cdc* mutants known. These are *CDC21* coding for thymidylate synthetase (Game, 1976), *CDC19-1* coding for pyruvate kinase (Kawasaki, 1979), and *CDC9-1* coding for DNA ligase (Johnston

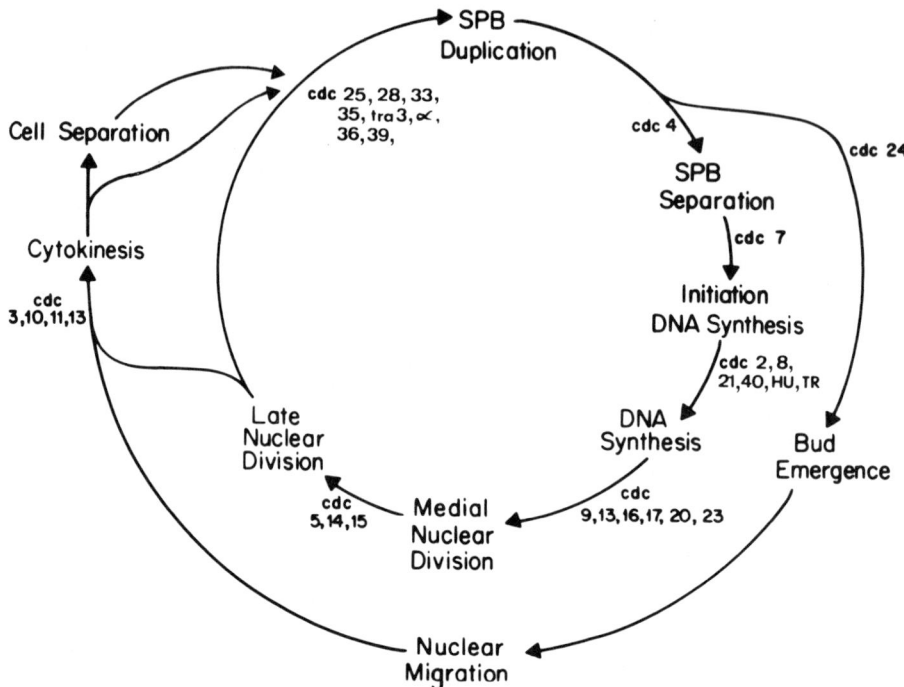

Figure 2 An outline of the function of genes during the cell cycle of yeast based on terminal phenotypes of *cdc* single and double mutants and shift experiments with α mating factor, hydroxyurea and trenimon (SPB, spindle pole body; α, HU, and TR, execution points of α mating factor, hydroxyurea, and trenimon, respectively; *tra3*, mutation causing derepression of enzymes for amino acid biosynthesis). (After Hartwell, 1978, with modifications.)

and Nasmyth, 1978). The phenotype of mutants *cdc21* and *cdc9-1*, which are both blocked in DNA synthesis and arrested in the "nuclear branch" of the map, is readily understood. Gene *CDC9* is allelic with *MMS8*, a gene needed for spontaneous mitotic recombination (Montelone et al., 1981). The relationship of pyruvate kinase to the early G_1 arrest of *cdc19-1* is less clear.

Three more *CDC* genes have been isolated on recombinant plasmids. The first-known start gene, *CDC28*, was isolated by Nasmyth and Reed (1980) and the molecular weight and isoelectric point of the corresponding protein were determined to be 27,000 and 6.4, respectively (S. I. Reed, J. C. Groppe and J. Ferguson, personal communication). The DNA corresponding to *CDC10* (Clarke and Carbon, 1980) and *CDC36* (T. A. Petersen and S. I. Reed, personal communication) has also been isolated. C. Kuo and J. L. Campbell (personal communication) have purified the *CDC8* protein, which is a component of the nuclear replication apparatus, using replication in permeabilized yeast cells as an assay of *CDC8* activity.

Byers and Sowder (1980) constructed hybrids containing a *cdc* mutant nucleus in wild-type cytoplasm. After a shift to the restrictive temperature, most of the hybrids were able to undergo several cell cycles, although the original mutants showed first-cycle arrest. The authors interpreted this to mean that in the wild type the gene products of most *CDC* genes are present in excess throughout the cell cycle. Therefore the order of functions in the cell cycle is not achieved by sequential synthesis of gene products, but probably by protein modification or availability of substrates. The only known exception is *CDC4*, probably coding for a component of the spindle pole body (SPB, the yeast centriole). The *CDC4* product seems not to be present in excess.

The branching of the cell cycle order-of-function map into two independent parallel pathways (the "nuclear pathway" and the "budding pathway") is documented by several experimental results. For example, in *cdc4*, which is blocked in the separation of the duplicated SPBs, multiple budding continues in the absence of DNA synthesis and nuclear division. In the mutant *cdc24* budding and chitin ring formation are impossible because the cell lacks the spatial organization of growth of the outer microtubuli determining the point of bud emergence in wild-type cells. The cells remain unbudded, but DNA synthesis and nuclear division go on, leading to inviable multinucleate cells. However, *cdc28*, which has to be executed before *cdc4* and *cdc24*, blocks both the nuclear and the budding pathway.

Therefore the branching off of the budding pathway occurs after *cdc28*, but before *cdc4*. The merging of the two pathways after nuclear division means that nuclear division is a prerequisite for cytokinesis. (No mutants budding off unnucleated cells have been found). After cell separation the cells have to return to the start event. Start (sometimes called G_0) is a period during G_1, after cytokinesis, but before SPB duplication. It is represented in Fig. 2 by genes *cdc28*, *cdc35*, and so on. The completion of the start event brings about *commitment* to a new cell cycle. Alternative routes of differentiation like mating or sporulation can only be entered by G_0 cells, that is, cells that have not yet completed start (see below).

The most up-to-date version of the order-of-function map (Pringle and Hartwell, 1981) is much more complicated than Fig. 2 (containing five parallel pathways) and is, in part, hypothetical. An important new feature is the independence of the development of the SPB from the initiation and progress of DNA synthesis. This is shown by the new mutant *cdc31*. The terminal phenotype of *cdc31* is a G_2 cell with a single bud, a doubled DNA content, and a short unipolar spindle with a single SPB twice as large as normal.

The Control of Start and of the G_1 Period

Nongrowing cells (e.g., stationary-phase cells) mostly remain in the G_1 period (before completion of start). These cells remain viable much longer than cells arrested at other points in the cell cycle. A number of mechanisms have been found which mediate the control of internal and external stimuli over G_1 cells.

During growth the *size* of the G_1 cell *determines* whether the start event can be completed. Mother cells as well as the smaller daughter cells have to grow in order to be able to enter a new cell cycle. The daughter cells stay in G_1 for a longer time than the mother cells and grow much more on a percentage basis (Hartwell and Unger, 1977). Apparently the variable length of G_1 is used for adjusting cells under various conditions to the size required for entering a new cell cycle (Johnston et al., 1977; Jagadish and Carter, 1977). Factors increasing the length of the G_1 period usually do so by changing the rate constant of a first-order rate process (a "probabilistic" process) leading to initiation of the cell cycle (Shilo et al., 1976, 1977; Samokhin et al., 1980). This means that both a deterministic (size-related) and a probabilistic element act in the control of start and of the G_1 period (Nurse, 1975, 1980).

The Influence of CDC Genes on Mating

Two questions are most interesting in this connection: (1) Which *CDC* genes are directly needed in mating and (2) at which point of the cell cycle can the mating pathway be entered? To answer these questions, several techniques have been devised. When unsynchronized *cdc* cells are shifted from permissive to restrictive temperature and at the same time are transferred to mating conditions, a certain fraction always mates. Only when the corresponding *CDC* gene product is essential for the mating process is this fraction zero. When the cells are synchronized by addition of mating hormone at the permissive temperature and then shifted to mating conditions at the restrictive temperature, the mating fraction reaches 100% of the wild-type control, but it is again zero when *CDC* genes are essential for mating.

The question of whether only one or more stages of the cell cycle allow the induction of mating is best answered by arresting *cdc* cells at the restrictive temperature and then transferring them to mating conditions, again at the restrictive temperature. Only *cdc* mutants arresting at those stages that allow mating directly will mate under these conditions (Reid and Hartwell, 1977). In those cases where a gene product has been shown to be essential for mating is it still possible to test if the stage where it arrests allows mating. The cells are arrested at the restrictive temperature and then shifted to mating conditions at the permissive temperature. If individual cells (as observed in the microscope) enter directly into mating, the stage does allow mating; if they first complete a ccell cycle, it does not (Simchen, 1978).

In this way it was found that *cdc28, cdc36, cdc37,* and *cdc39* allow mating directly. They all arrest at the same stage where α factor arrests the cells, namely, after SPB satellite formation but before SPB enlargement and duplication. They grow and show a "shmooing" phenotype at the restrictive temperature. (Therefore this step is regarded as separate in a dependent sequence within start; however, the ordering in this sequence is not unequivocal). It was shown by reciprocal shift experiments that the steps mediated by α factor and by *CDC 25, CDC28, CDC33,* and *CDC35* are independent.

The function of the *CDC* genes *28, 36, 37,* and *39* for the mating process can be further elucidated by testing their ability to suppress *ste* mutations. Hartwell (1980) isolated *ste* mutants belonging to 10 different complementation groups. Those in which mating is blocked because the cells do not arrest at start in the presence of α factor should be able to mate when combined in a double mutant with a *cdc* mutation enforcing the arrest at start. Double mutants with *cdc* mutant *28* or *37* were not able to mate in any case. Double mutants with *cdc36* or *cdc39* did suppress the *ste4* mutation in α cells and consequently were able to mate. There is no influence on three other phenotypically similar complementation groups of *ste* mutants (*ste7, ste11,* and *ste12*). It has been concluded that *STE4* in the wild type has the function of inactivating *CDC36* and *CDC39* in response to α factor. Genes *CDC28* and *CDC37* seem to act independently of (or prior to) the tested *STE* genes (J. R. Shuster, personal communication). For a detailed discussion of *ste* mutants, see the section on the mating process.

The genes *CDC1, CDC4, CDC5, CDC24, CDC33,* and *CDC34* are essential for mating. They are mostly involved in the budding pathway and in the development of the SPB. Ultrastructural investigations have shown that these functions are essential for mating in the wild type. The requirement for *CDC5*, which is involved in nuclear reorganization after nuclear division, is less well understood.

Several of the mutants mentioned (*cdc4, cdc28,* and *cdc37*) confer defects in *karyogamy* similar to the *kar1* mutation (Dutcher, 1980; Conde and Fink, 1976). In *cdc28* and *cdc37* only a fraction of the cells fail to undergo karyogamy. The *cdc4* defect is dominant in the mating reaction with wild-type cells (they fail to undergo karyogamy), but recessive in diploids of the configuration *cdc4/CDC4* (they carry out a normal cell cycle at the restrictive temperature). This is a strong argument for *cdc4* conferring a defect on a component of the SPB.

The influence of *cdc* genes on the induction of mating in *Schizosaccharomyces pombe* was studied by Nurse and Bissett (1981). Their methods and results are very similar to those described for *S. cerevisiae*. Only those *cdc* genes that arrest the cells at the start point allow direct entry into the mating pathway. One of these *CDC* genes has a second execution point in G_2, but cells arrested at this point are not capable of mating.

The Influence of *CDC* Genes on Sporulation

The influence of *CDC* genes on sporulation has been studied extensively by Simchen's group (Simchen, 1974, 1978; Hirschberg and Simchen, 1977; Shilo et al., 1978). The same questions as above (in relation to mating) have been posed and were answered by essentially the same methods. Only the genes *CDC3, CDC10, CDC11, CDC15,* and *CDC24* are not needed for meiosis/sporulation. They are involved in bud emergence (*CDC24*) or cytokinesis (*CDC3, CDC10,* and *CDC11*), functions which are obviously not needed during sporulation. Mutants *cdc25* and *cdc35* are special cases and will be discussed below. All other *CDC* genes are essential for meiosis/sporulation. This has led Simchen (1978) to the concept of meiosis being controlled by a modified program of mitosis. The steps after meiosis II (deposition of the spore wall structures around the haploid nuclei, maturation of the spores) are essentially unrelated to the mitotic cell cycle.

Three mutants, *cdc4, cdc25,* and *cdc35* allow sporulation directly. Mutants *cdc25* and *cdc35*, which are arrested before SPB satellite formation, allow sporulation at the restrictive temperature in vegetative growth medium containing acetate as a carbon source. It seems that these mutations give a signal to the cell which is normally only present under starvation conditions. The primary defect of these mutations could reside in the sensing mechanism for the presence of nutrients in the medium (Shilo et al., 1978).

Mutant *cdc4* poses a problem because it apparently contradicts the notion that only unbudded G_1 cells can enter sporulation. Arrested cells which are budded undergo sporulation without completing the ongoing cycle when shifted to sporulation medium and permissive temperature.

Similarly, not all *cdc* mutants which are arrested before SPB satellite formation allow the cells to enter meiosis/sporultation directly (*cdc19* and *cdc33* do not). We conclude that it is not only the arrest at a certain stage of the cell cycle that allows the induction of sporulation in *cdc* mutants, but also the nature of the specific gene product.

Some of the genes of the nuclear branch of the cell cycle have two execution points in meiosis, because two meiotic divisions have to occur (Simchen, 1978). An example was given by Schild and Byers (1980) who investigated mutants *cdc5* and *cdc14* in more detail under sporulation conditions. Diploids homozygous for *cdc5* and *cdc14*, which under vegetative growth conditions arrest at late nuclear division, were subjected to restrictive conditions in sporulation medium and to a series of shift experiments. DNA synthesis and commitment to recombination were normal, but the cells were arrested after SPB duplication and separation in meiosis I. The shift experiments showed that the gene products of *CDC5* and *CDC14* are needed in both meiosis I and meiosis II. If meiosis I is allowed but meiosis II is blocked by shifting the cells to restrictive temperature, a certain proportion of the cells develop asci containing two diploid spores. Genetic analysis shows that centromere-linked markers are mostly homozygous in these diploids, indicating that reduction division (meiosis I) has indeed occurred. The exact point of arrest in mitotic nuclear division, meiosis I, and meiosis II differs in these mutants. In mitosis they stop at the point of a very elongated spindle (*cdc14*) or even after nuclear division (*cdc5*); in meiosis I they arrest with separated SPBs, without forming a spindle; and in meiosis II they arrest at the stage of a very short spindle. Further analysis of this behavior should lead to identification of the primary defect of these mutants.

The Influence of *CDC* Genes on the Germination of Yeast Ascospores

No detailed investigation has yet been done. It seems to be clear that germinating spores, in order to form haploid colonies, eventually have to perform normal haploid cell cycles. Therefore all *CDC* genes are needed for germination. In addition to this, specific genes are expected to be responsible for the initial stages of germination. It would be interesting to know at what stages specific *cdc* genes interfere with germination.

The Influence of *CDC* Genes on Mitochondria and Extranuclear Genomes

Two questions have to be considered: (1) How do genes responsible for nuclear DNA synthesis influence the replication of extrachromosomal genomes such as the mitochondrial genome and the 2-μm plasmid, and (2) how do *CDC* genes influence the transfer of mitochondria and other cytoplasmic organelles during mating and their segregation during sporulation? The first of these questions has been studied in the wild type by Sena et al. (1975) and Cottrell and Lee (1981). They found that mitochondrial DNA synthesis is not restricted to a specific stage of the cell cycle. The investigation of mitochondrial DNA synthesis in *cdc* mutants (Newlon and Fangman, 1975) revealed that it is not influenced by two of the *cdc* genes blocking nuclear DNA synthesis (*cdc4* and *cdc7*). These genes are, however, needed for the replication of the 2-μm plasmid. Genes *CDC8* and *CDC21* are needed for mitochondrial, nuclear, and 2-μm DNA replication; *CDC21*, the structural gene for thymidylate synthetase, is obviously needed for the supply of precursors for every kind of DNA. Gene *CDC8* is a more complicated case. It is also essential for error-prone repair (Prakash et al., 1979). The mutant produced petites at a high frequency, as does *cdc21* (Newlon et al., 1979), and undergoes chromosome loss (Kawasaki, 1979). For the transfer of cytoplasmic organelles (most notably mitochondria) during mating *CDC* genes 5 and 27 are essential (Dutcher, 1980). Little is known about the segregation of mitochondria during sporulation (Brewer and Fangman, 1980). Apparently the influence of *CDC* genes on this process has not yet been tested.

THE MATING PROCESS

General Description

The process of mating is induced by the hormone-like oligopeptide *a* and α factors specifically secreted by the respective cells of *a* and α mating type (Thorner, 1980; Betz et al., 1981). Each mating type responds to the pheromone produced by the opposite mating type. Both sex factors act in a similar fashion (Betz et al., 1981). This response can be considered as a transient differentiation from a vegetative cell to a cell with the typical features of a gamete cell, although *S. cerevisiae* does not develop gametes in the usual sense of the word. The events taking place between the time of mixing of *a* and α cells and zygote formation have been termed "courtship" (Hartwell, 1973).

As a consequence of the pheromone action several distinct cell reactions can be observed. After binding to a hypothetical receptor sex factors seem to trigger several reactions in their target cells, which appear to be essentially independent of each other. These are synchronization of the cells by arresting them in the G_1 phase (Betz et al., 1981; Thorner, 1980) at a point termed start (Hartwell, 1974), induction of specific surface agglutinins (Yanagishima and Yoshida, 1981; Fehrenbacher et al., 1978; Shimoda et al., 1978) leading to sexual agglutination (Thorner, 1980), and localized lysis and new synthesis of the cell wall leading to cell fusion (Betz et al., 1981). It is feasible that a connection between these functions exists which is not yet known. Another possibility would be to assume the existence of three separate receptors for the pheromones. Genetic analysis might help clarify this problem (Betz et al., 1981).

The exact location of the execution point of α factor relative to cytological and biochemical cell cycle "landmarks" has been elucidated by Bücking-Throm et al. (1973). It is identical with the start point of the cell cycle (Fig. 2). Arrest at the "start" point is essential for mating. Only unbudded cells arrested at start are capable of mating efficiently (Bücking-Throm et al., 1973). Once a cell has started a new cell cycle, it will complete it and will be stopped only afterward by the respective pheromone (Thorner, 1980; Bücking-Throm et al., 1973). Protein, RNA, and polysaccharide synthesis (Throm and Duntze, 1970), as well as mitochondrial DNA synthesis (Cryer et al., 1973) continue after the arrest at start; only nuclear DNA replication is inhibited (Throm and Duntze, 1970). An increase in unbudded cells becomes visible in the mating mixture after one-fourth of the time needed for a mitotic cell cycle (Betz et al., 1981). The experiments of Udden and Finkelstein (1978) point to the fact that a single receptor site saturated in *a* cells by α factor might be sufficient to cause cell arrest.

Cell growth is not stopped by interruption of the cell division cycle. In fact, cells which do respond to sex factor but are somehow prevented from conjugation increase in size. They develop quite bizarre shapes, mostly elongated pearlike forms called "shmoos" (Thorner, 1980; Manney and Meade, 1977). "Shmoo" formation occurs at the same pole where the haploid cell would be expected to bud. After removal of α factor, cells develop new buds right at the growing tip of the shmoo (Schekman and Brawley, 1979). Shmoo formation provides a convenient assay for pheromone concentration and activity (Duntze et al., 1970; Betz et al., 1977).

Budding is delayed by sex factor beyond the normal lag time which occurs after a shift to new medium (Sena et al., 1973). The delay is not an "all or none" response but, rather, a dose-dependent lowering of the rate constant of cell cycle initiation (and therefore transition probability) for each separate cell (Samokhin et al., 1980). Mating efficiency is dependent on the age of the cells and the growth conditions. Mid-log-phase cultures mate at a frequency 10 times higher than late log- or early stationary-phase cultures (Lipke et al., 1976).

The sequence of steps in the mating process is not yet clear. The first reaction of a haploid cell to the pheromone might be cell cycle arrest, which is a prerequisite to all following steps (Betz et al., 1981, 1978; Campbell, 1973). Alternatively, the first response might be surface alteration and sexual agglutination after random collision in liquid media (Thorner, 1980; Fehrenbacher et al., 1978; Sakai and Yanagishima, 1972; Shimoda et al., 1978; Yanagishima and Inaba, 1981; Yanagishima, 1978). In this model cell cycle arrest would occur independently of, but generally later than, agglutination. In spite of this uncertainty, a sequence of events can be established. The time scale of these events as shown in Table 1 is remarkably reproducible in different strains and by different workers (Yanagishima et al., 1976; Betz et al., 1978; Fehrenbacher et al., 1978; Shimoda et al., 1978). If synchronized haploid cells are used, the steps of the mating process are shifted to somewhat earlier times as compared to unsynchronized cells (Sena et al., 1973).

Once conjugation pairs have been formed, autolytic activities are required to remove the separating cell walls (Shimoda and Yanagishima, 1972; Thorner, 1980), while cell-wall synthesis must occur at the same time in order to form a zygote. The SPB at this stage is single, but has an additional "half bridge" (satellite). Microtubules radiating from it on the outside of the nucleus seem to determine the site of cell fusion. The nuclei start migrating to the point of cell fusion even before fusion is complete. (In "shmoos" the nuclei migrate to the tip of the growing shmoo.) Once the cells have fused, the microtubules appear to pass through the aperture (Byers and Goetsch, 1973, 1975a) and eventually connect the two single SPBs of the two nuclei. Nuclear fusion (karyogamy) is now carried out near the junction at which the zygote will later normally start to bud

Table 1 The Timing of the Steps of the Mating Process

Time (min)	Steps of the mating process
0	a and α cells are mixed
20	Early aggregates that are easily separable
60	Strong and extensive sexual agglutination
60-140	Cell fusion
150	Cell fusion is complete, *zygotes*
170	First diploid buds
180	"Shmoos" in nonmated cells
170-350	First diploid cell cycle

Source: Data from Sena et al. (1973).

(Lipke et al., 1976). The two SPBs fuse laterally, forming a "half-cylinder" structure; then the SPB is duplicated and outer microtubules and the zygotic bud are formed (Byers and Goetsch, 1973, 1975a).

It is a striking feature of the conjugation of *S. cerevisiae* that multiple mating does not occur more frequently than 1 time in 10^4 (only triploids and tetraploids were scored, not diploid and haploid cytoductants; Rogers and Bussey, 1978). It is not yet known by what processes such events are prevented. One might think of some mechanism similar to the prevention of multiple fertilization (cortical reaction, fertilization membrane) in animals (Shapiro and Eddy, 1980). Yet the localized cell-wall alterations might provide a simpler answer to this problem. In the following section the steps of the mating process will be dealt with in more detail.

Cell-Wall Alterations

One of the responses of a *S. cerevisiae* cell to sex factor is the remarkably localized modification of the cell wall. At 1-2 hr after the beginning of treatment of *a* cells with α factor Lipke et al. (1976) determined an increasing glucanase sensitivity of the *a* cells subsequent to the morphological changes of the cell (shmoo formation). The extent of this response and of the elongation of the cells was dependent on the dose of α factor. It seems that the organization of glucan must have been altered to produce the higher glucanase sensitivity. In fact, a higher content of glucan and a lower content of mannan were found as compared to vegetative cells. The ratio of glucan to mannan is 1:1 in vegetative cells (Northcote and Horne, 1952), but 1.5:1 after sex factor treatment (Lipke et al., 1976).

The Yeast Genome in Yeast Differentiation 325

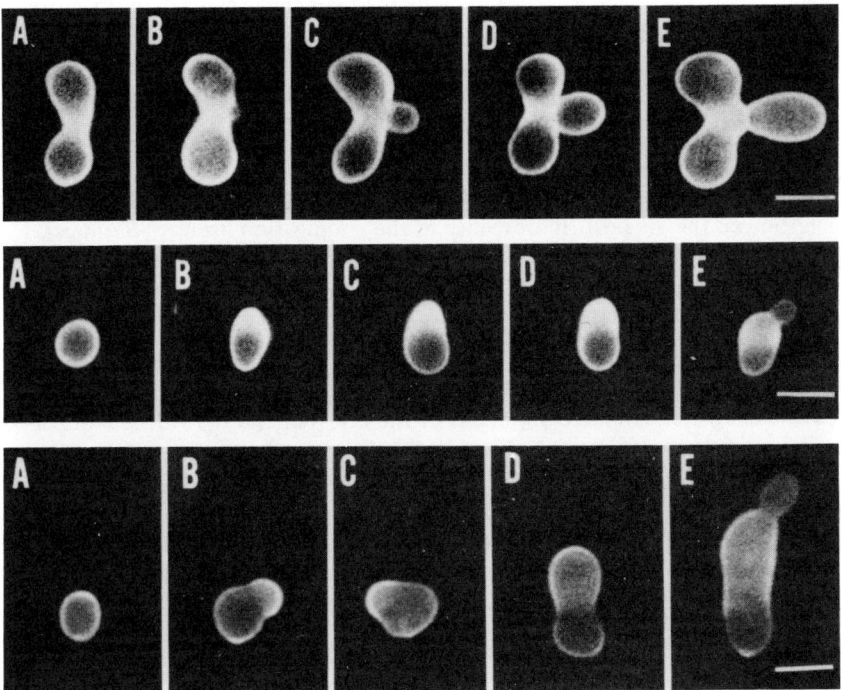

Figure 3 (Top.) Zygotes formed after conjugation between a and α cells of *S. cerevisiae* stained with FITC-ConA 4 hr after mixing of the cells. (A-E) Zygotes in progressive stages of development present in the mixture. Fluorescence was most intense in the wall of the conjugation bridge before bud initiation (A) as well as during all stages of bud development (B-E). The bar in this and subsequent figures represents 5 μ m. (Middle.) Fluorescence of a cells exposed to α factor and stained with FITC-ConA immediately after addition of α factor (A) or after incubation with α factor for 2 (B-D) or 4 hr (E). (Bottom.) Fluorescence of α cells exposed to a factor and stained with FITC-ConA immediately after addition of a factor (A) or after incubation with a factor for 2 (B-D) or 4 hr (E). (From Tkacz and MacKay, 1979).

Furthermore, the mannan of these a cells (shmoos) was found to contain a greater portion of short oligosaccharide side chains and unsubstituted backbone as compared to control cells. By electron microscopy it could be observed that the cell wall grew thinner and developed a more diffuse outer layer at the protuberant tip of the elongation, corresponding to observations of the conjugation bridge of zygotes (Osumi et al., 1974). The mannan coat obviously got thinner at this point. Cell-wall synthesis inhibitors (2-deoxy-D-glucose and other analogs of D-glucose) completely blocked morphogenesis at a comparatively low concentration where basic cellular metabolism was only partly inhibited.

A fluorescent derivative of concanavalin A (FITC-ConA) which specifically binds to mannan (Tkacz et al., 1971) was used to stain the cell wall of zygotes, shmoos, and budding cells. Asymmetrical staining was observed with zygotes which showed the greatest fluorescence at the conjugation bridge. "Shmoos" exhibited the most intense fluorescence at the growing tip, regardless of the mode of induction (sex factor alone or mixing with the opposite mating type). (Tkacz and MacKay, 1979; Fig. 3). Vegetative cells and even budding cells were stained uniformly after incubation with the appropriate pheromone. Taken together, these results suggest that mating involves specific cell surface alterations involving the mannan component of the cell wall and that the site of cell-wall alterations is identical with the site of cell fusion.

There seems to be an apparent contradiction between the decrease in total mannan (Lipke et al., 1976) upon exposure to sex factor and the increase in mannan-specific FITC-ConA fluorescence at the growing tip of the shmoo (Tkacz and MacKay, 1979). This can be explained by the assumption that mannan is more accessible at the growing tip of the shmoo than in the vegetative cell wall.

Another important and significant change in the cell-wall composition is the localized accumulation of chitin at the growing tip of the shmoo and at the site of cell fusion (Schekman and Brawley, 1979). Chitin synthetase activity in the plasma membrane increases 11-fold upon treatment of haploid cells with the appropriate pheromone, while the zymogen (precursor of chitin synthetase to be activated by proteolytic cleavage) in intracellular particles increases four-fold and chitin production increases threefold compared to resting cells (Schekman and Brawley, 1979). By staining pheromone-treated cells with primulin, which specifically binds to chitin (Cabib and Bowers, 1975), the sites of chitin accumulation became visible. The site of deposition is the tip of the growing shmoo. Acid phosphatase activity, a marker of cell-wall growth, is also detectable at this point (Schekman and Brawley, 1979).

For the sake of comparison, it may be mentioned here that chitin synthetase is also activated in the plasma membrane of budding cells, though in a very restricted area. Chitin is accumulated in the nascent division septum. After cell division the bud and birth scars left behind are composed mainly of chitin (Cabib, 1975).

Sloat and Pringle (1978) and Sloat et al. (1981) found an almost uniform chitin deposition on the surface of *cdc24* mutant cells which are mating deficient at the restrictive temperature. (Under these conditions the mutant is blocked in bud emergence and grows uniformly over the entire surface of the cell; Reid and Hartwell, 1977). This presents a parallel to the nonmaters staining uniformly with FITC-ConA after incubation with the appropriate sex factor (Tkacz and MacKay, 1979). Delocalization of mannan deposition was also observed in those mutants (Sloat et al., 1981). These cells fail to bud at the nonpermissive temperature, but instead grow over the whole surface. The *CDC24* gene may play a crucial role in determining asymmetric yeast growth at the different stages of yeast differentiation.

The observation in *Candida albicans* that the transition from budding to hyphal growth is accompanied by an increased rate of chitin synthesis and a shift in the location of chitin deposition from the division septum to the hyphal tip (Braun and Calderone, 1978) is another hint toward a specific role of asymmetric chitin synthesis in yeast differentiation.

An accumulation of small vesicles within the growing tips of shmoos was reported (Cortat et al., 1971; Byers and Goetsch, 1973) that was similar to the vesicle accumulation in budding cells at the budding site (Matile et al., 1969; Sentandreu and Northcote, 1969). These vesicles very likely take part in the various cell-wall alterations. The vesicles observed in budding cells carry cell-wall lytic enzymes required for local cell-wall lysis in the course of budding (Cortat et al., 1971). Exo-β-1,3-glucanase-containing vesicles arising from the endoplasmic reticulum could be isolated from budding cells; they also contained mannan and protein. It is assumed that the contents of these vesicles take part in cell-wall alterations after fusion with the cell membrane. The glucanase cleaves part of the glucan chains and thus renders the cell wall expandable. Concomitantly, the vesicles supply the growing bud with cell-wall components until the daughter cell is ready for separation (Cortat et al., 1971). The secretory vesicles could also contain chitin synthetase zymogen or chitin synthetase-activating protease (Schekman and Brawley, 1979). Independent evidence for the active participation of the vesicles in cell-wall synthesis has been gained by the examination of temperature-sensitive mutants which are deficient in secretion and consequently in cell-wall growth and exocytosis (Novick and Schekman, 1979). The vesicles observed in shmoos could play a role very similar to the ones discussed above during the formation of shmoos and the mating process.

Only speculation about the role of the cytoskeleton in mating is possible at present. Byers and Goetsch (1976) observed a specialized structure of the cytoskeleton in budding cells. It consists of filamentous rings, which are attached to the plasma membrane of the neck of the budding cell and which are presumably involved in the transport of vesicles to the bud and the formation of the primary septum. A similar structure would be expected in shmooing and mating cells, but this has not yet been detected.

Sexual Agglutination

Mating efficiency depends on random cell collisions (Thorner, 1980; Campbell, 1973). Up to 20-30 min after the mixing of "naive" (not preconditioned) cells, constitutive cell-wall components cause a relatively weak aggregation (Fehrenbacher et al., 1978). Clumps of many cells occur. This seems to be an effect that is separate from the inducible mating-specific adhesion described below. The agglutinins responsible for the weak constitutive agglutination are possibly different from those induced by the mating factors (Betz et al., 1978). Under nonnutrient conditions only the weak constitutive agglutination is observed (Campbell, 1973; Sena et al., 1973; Lipke et al., 1976). Approximately equal numbers of a and α cells are found in the clumps, regardless of the ratio of cell types in the original mating mixture (Kawanabe et al., 1979).

Strong sexual agglutination and the formation of conjugation pairs occur 60 min after mixing (Fehrenbacher et al., 1978; Shimoda et al., 1978). This is brought about by mating-type-specific a and α agglutinins becoming accessible on the cell surface of the haploid cells (Betz et al., 1978). These agglutinins were identified as two different glycoproteins (Betz et al., 1978; Sakai and Yanagishima, 1972; Yoshida et al., 1976). Both types of agglutinins could be isolated by snail gut enzymes (Shimoda et al., 1975) or by a special autoclaving method and subsequent purification (Yoshida et al., 1976). Claims that the agglutinins show univalent interaction (Yoshida et al., 1976; Yanagishima and Yoshida, 1981) could not be verified in the experiments of Fehrenbacher et al. (1978). These authors observed that pretreatment of cells with extracts of preconditioned cells of the opposite mating type completely blocked agglutination. However, when both a and α cells pretreated with extracts of preconditioned cells of the opposite mating types were mixed, some agglutination was still observed. It was found that preconditioning of only one cell type was sufficient to raise the agglutinability, though not to the maximum (Fehrenbacher et al., 1978; Betz et al., 1978). This means that the agglutinins react with ordinary constitutive cell-wall components also, but the reaciton is stronger with the agglutinin of the opposite mating type. The agglutinins are synthesized de novo in response to the pheromones in the so-called "inducible" strains (Betz et al., 1978; Sakai and Yanagishima, 1972; Yanagishima, 1978; Yanagishima and Yoshida, 1981; Yanagishima et al., 1976). (Most $MATa$ strains and some $MAT\alpha$ strains are inducible by mating factor for a surface agglutinin; Hartwell, 1980). The a agglutinin consists of at least two glycoprotein subunits (23,000 daltons); it contains 61% carbohydrate, its isoelectric point is 4.5, it has no hydrolase activity, and it does not destroy the complementary agglutinin of the opposite mating type. The α agglutinin has a molecular weight of 130,000, a carbohydrate content of 47%, and its isoelectric point is 4.3. Isolated a and α agglutinins are capable of forming complementary complexes at pH 5.5 in vitro which can be reseparated by exposure to pH 9.5 (Yoshida et al., 1976).

Constitutive (α^c, a^c) and inducible (α^i, a^i) strains with respect to sexual agglutination were found in both mating types (Sakai and Yanagishima, 1971). Inducible cells agglutinate only after a lag period and mixing with the opposite mating type, whereas constitutive cells show immediate strong agglutination (Sakurai et al., 1975). Obviously the cell surface of constitutive strains contains the agglutinin as a constitutive cell-wall component.

Temperature-sensitive conversion from constitutive (25°C) to inducible (36°C) agglutinability was shown to occur in most wild-type strains (Doi and Yoshimura, 1978). The synthesis of the agglutinins changes from being constitutive to inducible (as shown by shift experiments in the presence of inhibitors). After the shift to 36°C only the newly made daughter cells are inducible. The agglutinins seem to be fairly stable once they are incorporated in the cell wall (Betz et al., 1978; Doi and Yoshimura, 1978). This seems also to be the cause of the agglutinability of zygotes and their diploid daughter cells (Tohoyama et al., 1979).

A single recessive gene (*sag1*) regulating the conversion from constitutive to inducible sexual agglutination was found to be closely linked or identical with the mating-type locus (Doi and Yoshimura, 1978).

Yeasts Other Than *Saccharomyces*

Inducible agglutination has also been observed in *S. pombe* (Egel, 1971).

Sexual agglutination in *Hansenula wingei* is much stronger compared with that of *S. cerevisiae* (Crandall and Brock, 1968; Sakai and Yanagishima; 1971). The complementary agglutinins of the mating types *5* and *21* have been thoroughly studied (Crandall and Brock, 1968; Crandall et al., 1976; Yen and Ballou, 1974). These cell-wall factors determine the specificity of cellular recognition and constitute the first step in mating. Both factors could be isolated by snail gut enzyme or subtilisin digestion, purified (Taylor, 1964, 1965; Yen and Ballou, 1973, 1974; Crandall and Brock, 1968), and identified as mannoproteins (Taylor, 1965; Crandall and Brock, 1968). The *5* factor is a heterogeneous protein-mannan complex, apparently with six active site-binding fragments joined together by disulfide bonds (Taylor and Orton, 1971). The factor was well characterized as to composition (high content of serine and threonine) and structure (amino acids substituted by mannose and manno-oligosaccharides) (Yen and Ballou, 1973, 1974). It is inactivated and destroyed by mannanase or pronase treatment, which suggests that both components (protein and saccharide) are essential for the agglutinating function of the glycoprotein (Yen and Ballou, 1973, 1974). The *21* factor is also a mannoprotein, but a homogeneous one which is considered univalent (Crandall and Brock, 1968).

α-factor

H₂N-Trp-His-Trp-Leu-Gln-Leu-Lys-Pro-Gly-Gln-Pro-Met-Tyr-COOH

a-factor:

H₂N-Tyr-Ile-Ile-Lys-Gly-$\genfrac{}{}{0pt}{}{\text{Val}}{\text{Leu}}$- Phe-Trp-Ala-Asx-Pro-COOH

Figure 4 The structures of a and α factor according to Stötzler et al. (1976) and Betz et al. (1981). The structure of a factor is only preliminary; α factor usually consists of four components (Stötzler and Duntze, 1976), three of which are probably isolation artifacts leading to the loss of N-terminal tryptophan and/ or formation of a sulfoxide in methionine 12. The sequence given for a factor is that of the biologically most active component. In another component leucine is substituted for valine 6. In position 10 a modified form of asparagine with an unknown substituent on the γ-carboxylate group is found.

These recognition mechanisms involving cell surface glycoproteins rich in threonine and serine might represent structural analogies between yeast agglutination factors and substances responsible for animal cell adhesion (Watkins, 1972; Codington et al., 1972). Glycoproteins in general play an important role in cell recognition processes in animals, plants, and microorganisms.

Sex Factors and Their Reactions and Effects

The existence of diffusible mating-type factors (a and α) in *S. cerevisiae* was first reported by Levi (1956). The action of the sex factors is a prerequisite for all the subsequent mating-specific phenomena (Radin, 1976; Betz et al., 1981; Manney and Meade, 1977). The sex factors are small, diffusible, hormone-like polypeptides produced constitutively by the respective mating-type cells. The concentration of sex factor in early stationary culture media is 5-30 μg/liter. During the work on purification and elucidation of the structure of the sex factor, it was of crucial importance to develop quantitative assays for the factors. The bioassays used are based on serial dilutions and induction of shmooing in cells of opposite mating types (Duntze et al., 1973) or induction of agglutinability (Hartwell, 1980). A radioimmunoassay for α factor has been developed by Jones-Brown et al. (cited as a personal communication in Manney et al., 1981). The amino acid sequences of the sex factors have been determined (Stötzler et al., 1976; Betz et al., 1981) and are shown in Fig. 4. Various authors have confirmed the sequence data for α factor (Ciejek, 1980; Sakurai et al., 1976, 1977; Tanaka et al., 1977). Final proof for the structure of α factor as given in Fig. 4 has been provided by achieving complete synthesis (Ciejek et al., 1977; Khan et al., 1981; Samokhin et al., 1979; Masui et al., 1977). The product was identical with natural α factor in every respect, including biological activity. The dipeptide histidine-tryptophan (positions 2 and 3 of α factor) occurs only three times in all the known protein sequences. The other examples are luliberin (luteinizing hormone-releasing factor, also a hormone) and $E.\ coli$ β-galactosidase (Dayhoff et al., 1978). The first and the last amino acid of the α factor polypeptide seem to be dispensable for its biological activity. A large number of partial sequences of α factor have been tested by various authors and the residues of histidine 2, leucine 6, and lysine 7 have been found to be most significant for maintenance of the biological activity (for a review see Thorner, 1980). However, Samokhin et al. (1979) reported no effect on the biological activity of α factor after substitution of arginine for lysine 7.

α Substance I (isolated and sequenced by Yanagishima's group) is identical with the factor $α_2$ of Stötzler et al. (1976), which lacks the N-terminal tryptophan but is unmodified in methionine 12 (Sakurai et al., 1977).

a Factor was postulated by Hartwell (1973) and was shown to be present in the cell-free culture medium of a cells (Wilkinson and Pringle, 1974); it was shown to be a peptide (Yanagishima, 1978; Betz and Duntze, 1979). The identity of the two factors described by Yanagishima (a substance I) and Betz and Duntze (a factor) still has to be proven. A preliminary sequence has been published by Betz et al. (1981) (see Fig. 4). There are no obvious structural homologies between the two mating factors.

The appearance of α factor is inhibited by cycloheximide, which suggests that α factor is synthesized on ribosomes (Scherer et al., 1974). A larger polypeptide carrying six more C-terminal amino acids, which is a precursor of α factor, has been found by Tanaka and Kita (1977). Recently, J. Kurjan and I. Herskowitz (personal communication) isolated and sequenced a structural gene for α factor. They transformed a sterile, nonsecreting *matα2* mutant with a high copy number plasmid carrying yeast nuclear DNA and screened for transformants that had regained the ability to secrete α factor. The α factor sequence is repeated four times (separated by octapeptide spacers) in a large, uninterrupted, open reading frame. Specific proteolytic processing is thus necessary for the production of mature α factor. Protein sequencing will show whether the precursor isolated by Tanaka and Kita is indeed encoded by the DNA clone described above. The a factor sequence is not present in this region.

The secretion of the sex factors probably follows the pathway also used for the secretion of killer toxins and extracellular enzymes. This assumption is based on the fact that the *kex2* mutation (deficient in secreting killer toxin; Rogers et al., 1979) inhibits α factor secretion (Leibowitz and Wickner, 1976).

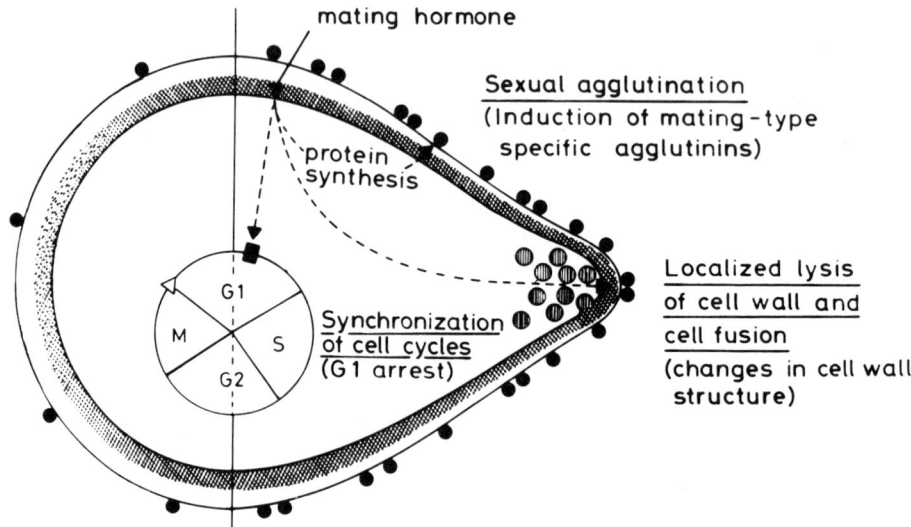

Figure 5 Schematic outline of events triggered by mating hormone in a haploid cell of *S. cerevisiae*. The mating hormone is assumed to interact with a surface receptor (not shown here). The pleiotropic response consists of the arrest of the cell cycle in G_1 phase, the induction of mating-type-specific surface agglutinins (full circles), and the accumulation of "glucanase vesicles" (crosshatched circles). (From Betz et al., 1981).

The effects of the two mating-type hormones on their respective target cells are nearly identical (Betz et al., 1981; Hartwell, 1973). Both mating types exhibit the same responses, including G_1 arrest, induction of specific surface agglutinins, localized cell-wall changes (including chitin synthetase activation; Schekman and Brawley, 1979) leading to shmoo formation, and readiness for conjugation. The response of α cells to *a* factor is less pronounced than the response of *a* cells to α factor (Manney and Woods, 1976).

Some evidence for the existence of a specific receptor for α factor on the surface of *a* cells has been provided by biochemical and genetic experiments. Thorner (1980) showed that after elimination of unspecific ionic and hydrophobic interactions, isotopically labeled α factor binds 10 times more efficiently to *a* cells than to α cells. In the absence of such specific precautions no difference in binding to *a* or α cells is found (Maness and Edelman, 1978). Experiments with FITC-labeled α factor added evidence to support the idea of specific recognition of α factor in *a* cells (Tanaka and Kita, 1978). A single gene (*STE2*) probably codes for α factor receptor; it is the only *a* specific gene in the collection of sterile mutants described by Hartwell (1980). It is defective for both of the inducible functions, cell cycle arrest and agglutination. *a* Cells are able to destroy α factor rapidly by limited and relatively specific proteolytic cleavage (Maness and Edelman, 1978; Ciejek and Thorner, 1979; Hicks and Herskowitz, 1976; Chan, 1977) during response to it. The transient effect of α factor on *a* cells (Throm and Duntze, 1970) is explained by α factor degradation in *a* cells. Protease inhibitors impede recovery from α factor effects (Thorner, 1980). α Cells do not degrade α factor under growing conditions; however, extracts from α cells (Finkelstein and Strausberg, 1979), α spheroplasts (Maness and Edelman, 1978), and even old stationary-phase α cells (Tanaka and Kita, 1977; Ciejek, 1980) do cleave the polypeptide, which suggests that the accessibility of the degrading system is responsible for the specificity of *a* and α cells with respect to α factor inactivation (Thorner, 1980).

A transient action was also observed in the case of *a* factor, which indicates that *a* factor is similarly degraded by α cells (Betz et al., 1977). The response to *a* factor is dose dependent (Lipke et al., 1976).

Liao and Thorner (1980) reported an inhibitory effect of α factor on membrane-bound adenylate cyclase activity, which would be an analogy ot the effects of peptide hormones. Presence of cAMP accelerates recovery from G_1 arrest. Moreover, a class of sterile, temperature-sensitive *a* cell mutants (*ste5*) unable to respond to α factor at the restrictive temperature possesses an adenylate cyclase activity insensitive to α factor inhibition at the restrictive temperature in vitro. These results seem to indicate that at least part of the response could be mediated by changes in the intracellular cAMP level (Liao and Thorner, 1980). This has led Finkelstein and McAlister (1981) to look for changes in protein phosphorylation as a result of the influence of α factor. They found a transient modification of a phosphoprotein in *a* cells treated with α factor. Although the protein is present in α and *a*/α cells also, the modification does not occur in these sexually unresponsive cells. The nature of the modification is unknown; it is not a change in protein phosphorylation. Diffusible polypeptides as intracellular signals triggering conjugation processes are quite a common feature among both eukaryotic and prokaryotic microorganisms (Thorner, 1980; Kamiya and Sakurai, 1981).

The optimal concentration of α factor to induce sexual agglutination is 1 ng/ml. The concentration necessary to inhibit DNA replication (as a consequence of G_1 arrest) is much higher (1 μg/ml), which suggests that sexual agglutination is the primary action, followed by inhibition of DNA synthesis (G_1 stop) (Shimoda et al., 1978). Agglutination is observed at a much lower hormone concentration than shmoo formation (Fehrenbacher et al., 1978).

Sterile Mutations

Sterile mutations have been used for genetic analysis of the mating process by a number of researchers. The mutants have to be either temperature sensitive, to allow genetic analysis at the permissive temperature, or leaky. In the latter case, diploids which occur at a low frequency have to be isolated by prototrophic selection and are used for tetrad analysis.

MacKay and Manney (1974a,b) isolated a large collection of nonmating a and α strains after mutagenesis with ultraviolet (UV) light. They mixed mutagenized haploids carrying a recessive antibiotic resistance marker with a large excess of haploids of the opposite mating type not carrying this marker. After 24 hr they plated the mixture on a selective agar medium that allowed growth only for the original haploid. In this way defects in most of the steps of the mating process could be found, with the exception of defects that occur only after nuclear fusion (i.e., mutants that form inviable zygotes). Sterile mutants which are arrested in the cell cycle and therefore do not respond to the opposite mating factor can hardly be detected (because they do not multiply during the 24-hr period of mixing with mating partners). Only one such mutant was found.

The mutants were tested for their ability to respond to their respective mating pheromones, to produce their own pheromone, to mate at a low frequency, and to produce living spores in the resulting diploids. They had lost one or more, and in some cases all, of the above-mentioned capabilities. The majority of the a and α mutants selected as nonmaters were simultaneously found to be unable to produce their mating pheromone and to respond to the opposite one.

The analysis of the nonmating mutants of either mating type led to the definition of 16 phenotypical classes of mutants representing at least 5 distinct genetic loci (*ste1, ste2, ste3, ste4,* and *ste5*, identified by recombination experiments). Of the 16 classes, 4 contain mutants closely linked to the mating-type loci. The α-linked classes 1, 2, and 10 define a hierarchy of functions associated with the α locus. Class 2 does not produce α factor, but retains the ability to respond to a factor (this class is represented by a single mutation, *ste1-5*). In addition, class 1 is unresponsive to a factor and class 10 shows the phenotype of preventing sporulation and at the same time allowing mating with α partners in diploids of the configuration a/α (class 10). This class has apparently lost the expression of all α specific information. Class 9 is analogous to class 10, but linked with the a locus. Analogs of classes 1 and 2 in the a locus have not been found, which points to a basic asymmetry between the a and α loci. (Compare also the discussion on the structure of the mating-type locus in the Introduction.) The remaining classes of mutants define functions unlinked to the mating-type locus, being either a specific, α specific, or nonspecific. A fourth group, which is a/α specific, might also lead to sterility, but will generally be associated with sporulation deficiency. For every mutant that is unlinked to *MAT* and α specific, an analogous *a* specific mutant could be found (and vice versa). The main phenotypes of the mutations unlinked to *MAT* were as follows:

1. Mating-type-specific mutations affecting the response to the opposite sex factor, but not the production of their own sex factor or the ability to sporulate in *a*/α diploids. Presumably the binding mechanism for the sex factor is affected. If so, these genes are regulated by their mating-type alleles.
2. Mutations that are not mating-type specific yet exhibit the same phenotype as the mutations mentioned above. Possibly there are genes involved which are responsible for aspects of the sex factor response common to both mating types.
3. All the other mutations unlinked to *MAT* exhibit neither the ability to produce their own sex factor nor the ability to respond to the opposite one. Apart from *MAT* there may be several genes conferring this phenotype in either mating type, which suggests that there are functional and regulatory interrelationships in addition to *MAT*.

Manney and Woods (1976) and Hartwell (1980) isolated *a* mutants resistant to the cell cycle-arresting activity of α factor on agar plates supplied with an excess of α factor. These were tested for mating with *MATa* and *MATα* testers on nonselective plates. All of them were found to be either nonconditional or temperature-sensitive nonmaters. Some of them were able to mate at low frequencies. Manney and Woods (1976) carried out random spore analyses with viable spores of the diploid progeny of low-frequency maters. They thereby proved that none of the 93 sterile *a* mutations examined were closely linked to mating-type locus. (All of them recombined with *MAT*.)

Hartwell (1980) defined eight complementation groups among temperature-sensitive *a* nonmaters isolated by the method described above. The complementation tests were carried out by mating the conditionally sterile *a* mutants to α strains, which were also conditionally sterile and which were obtained by recombination with the original *a* mutants at the permissive temperature (22°C). A recessive gene for cryptopleurine resistance was closely linked to *MATa*. Spontaneous gene conversion at the *MAT* locus produced *MATa/a* diploids which were selected on plates spread with cryptopleurine. They were presumed to be heterozygous for the two parental *ste* mutations. Being *MATa/a* diploids, they mated with α strains at the permissive temperature. If they did not mate at the restrictive temperature (34°C), the mutants were assigned to the same complementation group. A possible source of error would arise if the *ste* locus was linked to *MAT*. Yet none of the mutants studied by Hartwell (1980) were so linked. This result agrees with findings of Manney and Woods (1976) (see above).

The complementation groups thus obtained were termed *ste2, ste4, ste5, ste7, ste8, ste9, ste11,* and *ste12*. Mutants *ste2, ste4,* and *ste5* were allelic with the respective ones characterized by MacKay and Manney (1974b). Together with *ste1* and *ste3* (MacKay and Manney, 1974b) 10 sterility genes have thus been published so far.

Characterization of the α factor-resistant nonmaters isolated by Hartwell (1980) revealed that seven of the eight genes are not mating-type specific, whereas *ste2* was proven to be *a* specific, since it could not be transferred to an α strain by recombination. The *ste2* mutant is defective for both inducible functions, cell cycle arrest and agglutination. These properties are consistent with the possibility that *ste2* codes for the α factor receptor. In addition, this mutant shows a remarkable hyperproduction of *a* factor which cannot yet be explained.

Some nonconditional sterility mutants belonging to three of the eight complementation groups became temperature sensitive in a genetic background containing a temperature-sensitive amber nonsense suppressor (*SUP4-3*), which suggests that a nonsense mutation might be the cause of the nonmating phenotype. Evidently these three genes code for proteins. Presumably this is also the case for the rest of the described *STE* genes.

Experiments with double mutants containing *ste* genes and *cdc* genes conferring cell cycle arrest at the start point have already been discussed (see the section on the cell cycle). These experiments are very useful for the identification of *STE* genes which directly mediate cell cycle arrest by α factor.

None of Hartwell's mutants could be healed by the presence of wild-type cells of the same mating type providing extracellular components (mating hormones and agglutinins). The genetic defect in each of these mutants therefore appears to be in some cellular process intrinsic to the cell.

A newly isolated α specific *ste* mutant (*ste13-1*) apparently causes a defect in the processing of α factor precursor. This could be shown by transformation of the mutant cells with a high copy number plasmid carrying the structural gene for α factor. The transformants did not secrete mature α factor, but higher amounts of the misprocessed product. This product immunologically cross-reacts with α factor, but is 10 times less active (L. C. Blair, A. J. Brake, D. J. Julius, J. M. Lugovoy, and J. Thorner, personal communication).

Other Mutations Affecting the Process of Mating

Mutants which are supersensitive to G_1 arrest by α factor (*sst*) were analyzed by R. K. Chan and C. A. Otte (personal communication). Mutant *sst1* was found to be specifically deficient in α factor-degrading activity. It is allelic with *bar1* (a mutation which eliminates the ability of *a* cells to act as a barrier to the diffusion of α factor). Mutant *sst1* can be healed by the addition of wild-type *a* cells. It is *a* specific and is able to mate with *ste* mutations that are deficient in the production of active α factor, namely, *ste13* and *kex2*. Mutant *sst2* (unlinked to *sst1*) cannot be healed by the addition of wild-type cells. It grows slowly and the cells display a shmoo morphology. G_1 arrest by α factor is only very slowly reversible in *sst2*, even when excess α factor is removed by washing the cells. The defect of *sst2* is therefore intrinsic to the cell and is involved in the mechanism of α factor action at a step later than receptor binding or degradation.

The *kex2* gene, one of the two chromosomal genes (*kex1* and *kex2*) required for the secretion of killer toxin, which is coded for by a double-stranded RNA plasmid in "killer" yeast strains, is involved in the process of mating (Leibowitz and Wickner, 1976; Wickner and Leibowitz, 1976). Mutants *kex1* and *kex2* still carry the complete cytoplasmic killer genome (Wickner and Leibowitz, 1976). They were selected for failing to secrete killer toxin while still retaining resistance to the toxin (Wickner, 1974a). *kex2* mutants of α strains fail to secrete α factor, yet essentially retain the ability to respond to *a* factor, whereas *kex2* mutants of *a* strains show no defect in mating, pheromone secretion, and response to α factor.

Rogers et al. (1979) found that in *kex2* mutants the secretion of many extracellular proteins and glycoproteins is abnormal, probably at the level of protein processing, which suggests a common mode of secretion for all these products, including α factor. A radioimmunoassay also indicated that the α factor deficiency of α*kex2* mutants was due to lack of extracellular secretion and not to secretion of an inactive α factor-related product (Jones-Brown et al., 1980). The exact secretion mechanism of α factor remains to be elucidated, as well as the puzzling question of why *a* factor secretion is not affected at all by *kex2*.

In a series of papers by Schekman and co-workers (Novick et al., 1980; Esmon et al., 1981) the secretory pathway of yeast was characterized with the help of *ts* mutants falling into 23 complementation groups. These mutations have been termed *sec*. They are expected to be sterile, similar to the *kex* mutations.

Two further mutation sites affecting the mating reaction have been detected: *nul3*, which is defective in zygote formation (Mortimer and Hawthorne, 1973) and *tup1*, a gene controlling deoxythymidine 5′-monophosphate uptake. α Strains carrying the *tup1* mutation mated at a low efficiency with both *a* and α strains (Wickner, 1974b).

There are some mutations altering the structure of mannan (Ballou, 1976) and apparently affecting the intrinsic adhesiveness to cells of the opposite mating type (Radin, 1976). In spite of this, these mutants react to pheromone by showing enhancement of agglutinability (Thorner, 1980).

The *kar* mutations (Conde and Fink, 1976) leading to viable heterokaryons and haploid heteroplasmons prove that nuclear fusion is neither a necessary outcome of cell fusion nor an indispensable prerequisite for the viability of the zygotes and the daughter cells. Nuclear fusion is obviously a separate mating function. Mutants *cdc4*, *cdc28*, and *cdc37* also show the *kar* phenotype (Thorner, 1980; Pringle and Hartwell, 1981).

Temperature-sensitive conversion from constitutive to inducible sexual agglutinability in *a* cells was described by Doi and Yoshimura (1978). They found a recessive gene, *sag1*, closely linked to *MAT*, being responsible for the constitutivity of agglutination (see the discussion on sexual agglutination).

THE PROCESS OF SPORULATION

General Description

Most of the work in the field of yeast differentiation has been carried out on sporulation and the number of research papers appearing in the literature is enormous. We refer the reader to some of the reviews published during the last few years: Baker et al., (1976), Esposito and Esposito (1975, 1978), Haber et al., (1975), Haber and Halvorson (1975), Esposito and Klapholz (1981), and Tingle et al. (1973).

Sporulation is usually induced by shifting genetically competent yeast cells from early stationary phase on glucose-containing growth medium or from mid-log phase on acetate-containing growth medium (Roth and Halvorson, 1969) to 1% potassium acetate (sporulation medium). In some strains sporulation is greatly enhanced by the addition of small amounts of yeast extract and glucose to the 1% potassium acetate medium. Several methods are available for obtaining a high degree of *synchronous* sporulation which is advantageous for the biochemical analysis of the different steps in the sporulation process (Petersen et al., 1978; Hartig and Breitenbach, 1977; Watson and Berry, 1979).

The time course of sporulation differs widely in wild-type strains of different origins; however, in genetically defined laboratory strains, the completion of sporulation requires a minimum of 12 hr and a maximum of 20 hr (presporulation on acetate- or glucose-containing medium, respectively). In the latter case, premeiotic DNA synthesis starts at 6 hr after shift to sporulation medium and lasts for 2 hr (Croes, 1966; Roth and Lusnak, 1970; Piñon et al., 1974). Histones are synthesized only at the beginning of premeiotic DNA synthesis (Marian and Wintersberger, 1980). Synaptonemal complexes (synaptons) can be observed during the pachytene part of meiotic prophase I (Engels and Croes, 1968; Moens and Rapport, 1971a; Zickler and Olson, 1975; Byers and Goetsch, 1975b; Olson and Zimmerman, 1978; Horesh et al., 1979). The meiotic karyotype determined at this stage (Byers and Goetsch, 1975b) is in reasonable agreement with genetic investigations of the number of chromosomes of yeast (Mortimer and Schild, 1980). Experiments with synchronized cells showed that meiotic recombination occurs after DNA synthesis (G. Simchen, personal communication). The ensuing two meiotic divisions are similar to those observed in higher organisms, with the exception that the nuclei do not divide before spore maturation (they are formed by a process called "nuclear budding") (Moens and Rapport, 1971b; Moens, 1971). Figure 6 shows nine stages of meiosis/sporulation. Figure 7 shows

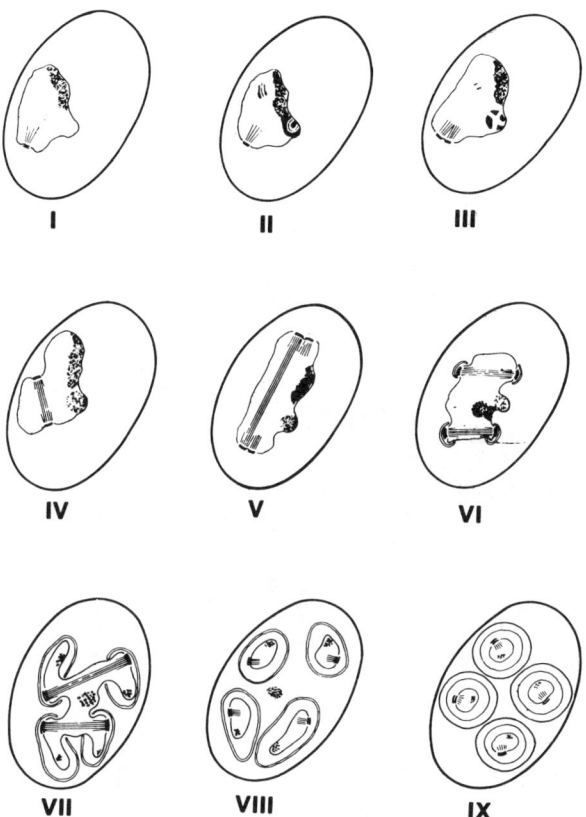

Figure 6 Nine stages of meiosis/sporulation. Stage I: During premeiotic DNA synthesis the SPB is single. The dotted area demarcates the nucleolus. Stage II: At the end of premeiotic DNA synthesis the "polycomplex body" (containing multiple synaptonemal complex-like elements) appears. Syanptons occur at the chromosomes. Stage III: The SPB is duplicated. Stage IV: The spindle of meiosis I is complete. The polycomplex body begins to degenerate. Stage V: The SPBs duplicate again. Stage VI: The two spindles of meiosis are complete. The SPBs form "outer plaques" which induce the formation of prospore walls. Stage VII: The nucleus forms four lobes each of which is surrounded by prospore walls. Stage VIII: The prospore walls are closed. The nucleolus seems to be excluded from the prospore. Each of the prospore nuclei contains a single SPB. Stage IX: The spores form mature spore walls. (From Esposito and Esposito, 1975).

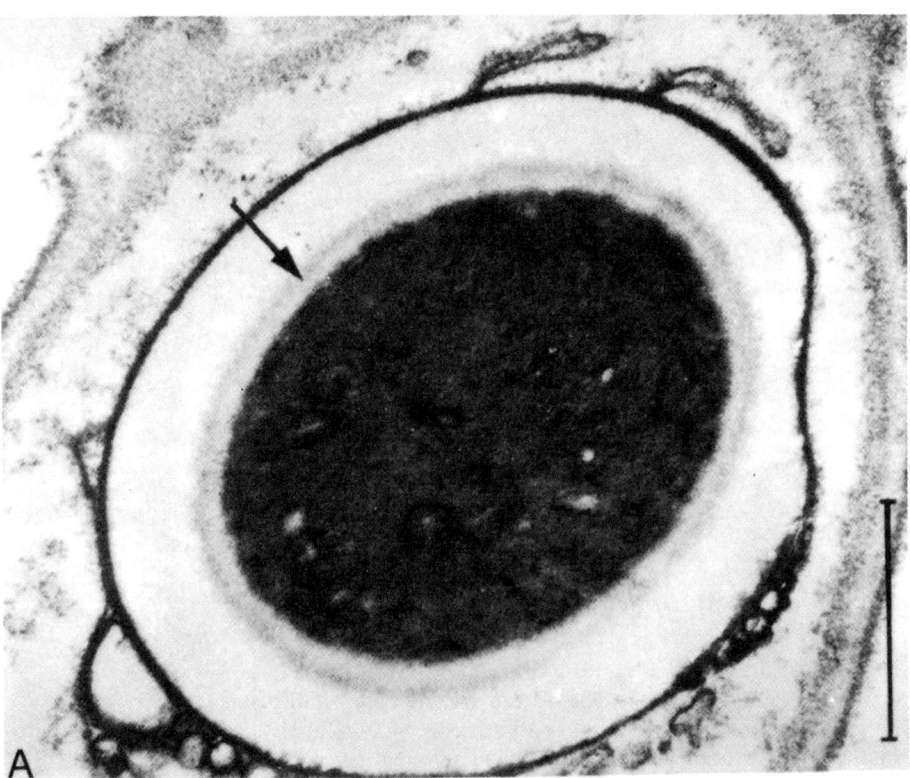

Figure 7 (A) A mature spore within an ascus. Three layers of the spore wall can be discerned: an outer osmiophilic dark layer, a wide light layer, and an inner greyish layer surrounding the protoplast. The length of the bar is 0.5 μm. (B) The spore wall at higher magnification. The outer osmiophilic part of the spore wall now seems to consist of two layers (GL: greyish inner layer of the spore wall). (From Kreger-Van Rij, 1978).

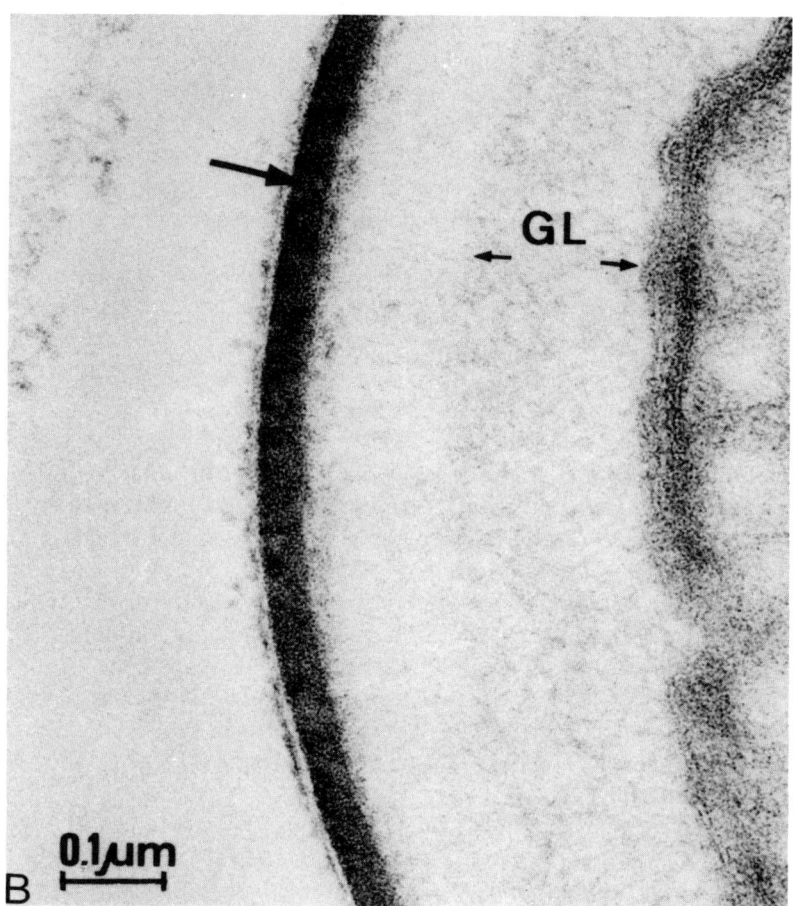

Figure 7 (Continued).

the ultrastructure of the mature spore wall. The outer osmiophilic layer seems to be derived from the prospore wall. The wide light layer seems to be formed during spore maturation (Kreger-Van Rij, 1978). The vegetative cell wall of the sporulating cell becomes the ascus wall after spore maturation; it is, however, not necessary for sporulation, since protoplasts can be induced to sporulate (Kopecka, 1974; Tsuboi, 1981).

In order to sporulate, the cells have to adapt (either during presporulation or after the shift to sporulation medium) to metabolic pathways, for example, gluconeogenesis, that are needed for the utilization of acetate. Some other nonfermentable substrates, such as dihydroxyacetone or pyruvate (Miller and Hoffmann-Ostenhof, 1964), also support sporulation and require gluconeogenesis. Freese et al. (1981) have investigated this problem in more detail using media on which yeast cells, after reaching stationary phase, sporulate efficiently without shift to sporulation medium. They found that under these conditions the exhaustion of either the carbon, nitrogen, or phosphate source may induce sporulation. By this technique cells can be induced to sporulate in the presence of excess nitrogen, indicating that sporulation is not necessarily affected by nitrogen repression. Nitrogen repression of sporulation was also investigated by genetic methods (Vezinhet et al., 1979; Croes et al., 1978). The influence of nitrogen repression on intracellular catabolic enzymes during sporulation was studied by Opheim (1979). Conversely, an excess of glucose always prevents sporulation.

A few other aspects of the *metabolism of sporulating cells* will be discussed.

1. Tracer experiments have shown that acetate is incorporated into lipids, macromolecules, and carbon dioxide during sporulation (Illingworth et al., 1973; Henry and Halvorson, 1973; Esposito et al., 1969). The inositol phospholipids synthesized during sporulation were studied with the help of inositol auxotrophic mutants and one sporulation-specific lipid was found (Schroeder and Breitenbach, 1981a,b).

2. During sporulation the pH of the medium rises from 6 to about 9; unknown basic end products of metabolism are excreted. If the sporulation medium is buffered at pH 6, sporulation is usually prevented. McCusker and Haber (1977), however, found that certain buffer substances allow sporulation at pH 6. Under these conditions the cells remain capable of taking up low molecular weight substances.

3. Carbohydrate metabolism during sporulation was studied by Kane and Roth (1974). Glycogen is synthesized during the early stages of sporulation, but degraded later on. The degradation of glycogen and the induction of the corresponding enzymes do not occur in α/α cells under sporulation conditions. This is one of the few known instances of a sporulation-specific enzyme (Colonna and Magee, 1978; Fonzi et al., 1979; Del Rey et al., 1979, 1980).

4. Surprisingly, the comparison of two-dimensional protein maps of sporulating cells and α/α cells under sporulating conditions revealed almost no new proteins (Wright and Dawes, 1979; Petersen et al., 1979; Trew et al., 1979; Kraig and Haber, 1980). This might result from label being primarily taken up by cells not entering meiosis/sporulation. Alternatively, it might be argued that specific changes occur only in less abundant proteins which are not seen on the map or in the classes of proteins that are not extracted or separated by the standard procedure of O'Farrell (e.g., basic or hydrophobic proteins). When the proteins associated with folded chromosomes were investigated (Piñon, 1979), sporulation-specific changes could indeed be seen (Petersen, 1981; Petersen and Piñon, 1979).

5. Degradation and reassembly of the macromolecular components of sporulating cells (RNAs, proteins, ribosomes) were studied by a number of workers (Wejksnora and Haber, 1974; Mills, 1974; Hopper et al., 1974; Magee and Hopper, 1974; Frank and Mills, 1978; Pearson and Haber, 1980; Emanuel and Magee, 1981). Spores contain a specific set of polyadenylated and capped mRNAs (Harper et al., 1980).

6. As yeast does not depend on exogenous nitrogen sources for sporulation, the necessary synthesis of new proteins depends on the degradation of pre-existing cellular protein. Around 50% of the cellular protein (and 70% of the RNA) is catabolized in sporulating cells (Chen and Miller, 1968; Zubenko and Jones, 1981). The process is only partially specific for sporulation (Hopper et al., 1974). The proteinases A and B increase several fold during sporulation (Klar and Halvorson, 1975; Chen and Miller, 1968), but no sporulation specific proteinases have been found. The influence of sporulation of several mutations conferring deficiencies in proteinases have been investigated (Wolf and Ehmann, 1978; Betz, 1979; Zubenko et al., 1979; Zubenko and Jones, 1981). It was found that defects in proteinases A, S, and Y do not influence sporulation. A deficiency in proteinase B leads to structurally abnormal but viable spores. Mutation *pep4-3* (conferring deficiencies in proteinases A, B, and Y, ribonuclease, and alkaline phosphatase) completely blocks sporulation (Zubenko and Jones, 1981).

7. The resistance of yeast spores to adverse conditions is less pronounced than that of bacterial spores. Yeast spores are cryptobiotic and resistant to starvation and dryness; however, they show very little heat resistance (Ho and Miller, 1978). The substances responsible for spore resistance are unknown, but the high content of free proline and the low internal pH of the spores has been correlated with resistance (Ho and Miller, 1978; Barton et al., 1980). Yeast spores are resistant to acid (Ho and Miller, 1978; Seigel and Miller, 1971) and to diethylether (Dawes and Hardie, 1974).

Commitment to sporulation occurs after DNA synthesis, but before SPB separation. This means that cells which are shifted to vegetative growth medium after commitment will nevertheless complete sporulation; but the exact

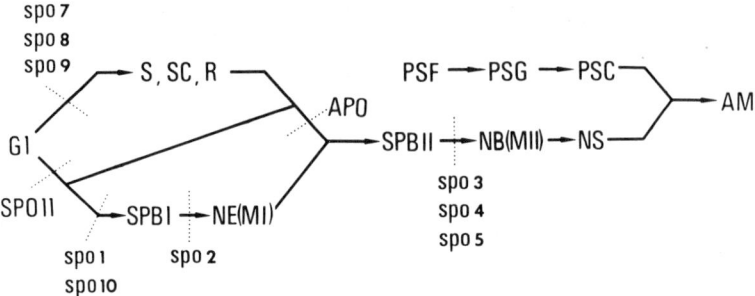

Figure 8 Working model for the order-of-function map of meiosis/sporulation. The dotted lines indicate the stage of sporulation at which cells of a given genotype are blocked or first exhibit abnormal development (APO, gene for apomixis; S, DNA synthesis; SC synaptonemal complex formation; R, recombination; SPB, spindle pole body; NE, nuclear elongation; NB, nuclear budding; NS, nuclear separation; PSF, prospore wall formation; PSG, prospore wall growth; PSC, prospore wall closure; AM, ascospore maturation). The model is discussed in the text. (From Esposito and Esposito, 1978).

time of commitment depends on the composition of the medium to which the cells are shifted. The same is true for bacterial sporulation (Cooney et al., 1977). The age of a cell determines its sporulation capability. Mother cells sporulate much more efficiently than daughter cells (Yanagita et al., 1970).

Mutants Specific for Meiosis/Sporulation

These mutants define a subset of all the genes needed for the process of meiosis/sporulation. Many other genes are also necessary, for instance, *CDC* genes and the *RAD* genes discussed below. Therefore a complete order-of-function map for meiosis/sporulation will only be available when combinations of all the classes of mutants involved have been tested for their sporulation properties. Three groups of mutants will be discussed here: the *spo* mutants (Esposito and Esposito, 1975, 1978; Esposito et al., 1974; Esposito and Klapholz, 1981), the *mei* and *con* mutants (Roth and Fogel, 1971; Fogel and Roth, 1974; Roth, 1973), and the *apo* mutants (Grewal and Miller, 1972; Klapholz and Esposito, 1980a,b). The current map, based on the *spo* and *apo* mutants, is presented in Fig. 8.

The *spo* mutants were isolated by mutagenizing the spores of a homothallic strain and plating them on vegetative growth medium. The spores form diploid clones on this medium (see also the Introduction). As these diploids are homozygous for all genes except *MAT*, they express both recessive and dominant mutations. They were screened for mutants that are *ts* during meiosis/sporulation but normal during the vegetative cell cycle and which retain the ability to utilize acetate during growth. The genetic analysis of the mutants was possible because at the permissive temperature they form normal spores which can be mated to heterothallic mating partners. A total of 3 dominant mutations and 11 complementation groups (genetic loci) were found. Statistical analysis showed that a maximum of 50 sporulation-specific genes can be expected. Recently, a more advanced and more complicated mutant selection procedure has been developed (N. Marmiroli, personal communication) which avoids the possible shortcomings of mutagenizing spores. In this procedure heterothallic haploid cells are mutagenized and then mated to an excess of homothallic haploids. The resulting diploids are induced to sporulate, clones derived from homothallic spores are selected, and their sporulation properties tested.

By means of another mutant selection system (Roth and Fogel, 1971; Fogel and Roth, 1974; Roth, 1973, 1976) mutants specifically deficient in premeiotic DNA synthesis (*mei* mutants) or in meiotic recombination and /or gene conversion (*con* mutants) were isolated. In this method, a strain disomic for chromosome III (but otherwise haploid), heterozygous at *MAT*, and heteroallelic at the *leu1* locus is mutagenized and exposed to sporulation conditions. Such strains sporulate, but the spores are aneuploid and inviable. Intragenic recombinants (or convertants) can be detected in a colony by replica plating on leucine-free selective medium. Growth on the leucine-free medium indicates a normal amount of meoitic recombination; no growth indicates a mutant. The method is practical because the cells, when shifted to vegetative growth conditions, return to mitotic cycles even after commitment to meiotic recombination (Esposito and Esposito, 1974; Olson and Zimmerman, 1978). Mutants influencing recombination will generally be expressed because a large part of the genome is haploid. Recessive mutants on chromosome III cannot be detected by this method. Normal haploids could be derived from the disomic mutants by mitotic segregation and used for genetic analysis. Results obtained with this method can be summarized as follows:

1. One class of mutants was defective in premeiotic (but not mitotic) DNA synthesis. All of these mutants were also defective in meiotic recombination. Three complementation groups (termed *mei1*, *mei2*, and *mei3*) were found (Roth, 1973).
2. Another class of mutants synthesized DNA under sporulation conditions, but was defective in meiotic recombination/conversion. Mitotic recombination (both spontaneous and UV induced) was normal. Again, three complementation groups (termed *con1*, *con2*, and *con3*) were found (Fogel and Roth, 1974).
3. A third class of mutants was deficient in both meiotic and UV-induced mitotic recombination (Roth and Fogel, 1971).

All of the mutants discussed above are recessive. The authors confirmed the results obtained with the disomic strains by constructing and testing diploids homozygous for these mutations. The formation of spores was blocked in most of the mutants obtained in classes (1), (2), and (3). However, the mutants *con2* and *con3* could proceed to the formation of inviable spores. In all the cases found so far, a block in meiotic recombination prevents the formation of viable haploid spores. Apparently meiotic recombination functions are needed for reductional division (Kassir and Simchen, 1978). In the presence of a mutation suppressing meiosis I (*spo12* and *spo13*, to be discussed below), mutants specifically defective in meiotic recombination but still forming viable spores can be selected (M. S. Esposito, C. V. Bruschi, D. T. Maleas, K. A. Bjornstad, and J. E. Golin, personal communication).

Genes leading to *apoximis* (absence of mixing of genomes, asexual sporulation) were studied in natural isolates of yeast by Miller's group (Grewal and Miller, 1972; Ashraf and Miller, 1978; Bilinski and Miller, 1980) and by Moens (1974). Later these mutants were analyzed genetically (Klapholz and Esposito, 1980a,b). On sporulation media they produced two-spored asci containing diploid spores. The diploid strains derived from the spores are themselves capable of sporulation. Four-spored asci can be induced in the strains by zinc chloride or by amitrole (Ashraf and Miller, 1978; Bilinski and Miller, 1980), but they were not analyzed genetically. Ultrastructural investigations (Moens, 1974) show that synaptons are formed, but consequently meiosis I is skipped and a single equational division (similar to meiosis II) ensues. This was corroborated by genetic analysis (Klapholz and Esposito, 1980b) investigating the segregation of centromere-linked markers in the dyads (two-spored products). No increased frequency of homozygosity for centromere-linked markers was found. The strains contained two independent genes (*spo12-1* and *spo13-1*), each of which confers the ability to form two-spored asci containing diploid spores (Klapholz and Esposito, 1980a).

Other conditions under which two-spored asci are formed include the following:

1. The presence of a mutant *cdc5* or *cdc14* gene at a semipermissive temperature (see the section on the cell cycle). In this case, however, meiosis II is skipped (Schild and Byers, 1980).
2. Interrupted sporulation in the wild type (Davidow et al., 1980; Srivastava et al., 1981) leads to two-spored asci containing haploid as well as aneuploid spores owing to disturbed meiosis II.
3. The presence of the mutant gene *spo3* at a semipermissive temperature leads to two-spored asci arising from random inclusion of normal haploid nuclei in prospore walls (Esposito et al., 1974).

The order-of-function map represented in Fig. 8 (Esposito and Esposito, 1978) is based on techniques discussed in the section on genetic strategies. As the temperature-sensitive defects of the *spo* mutants are generally not reversible, the techniques involving temperature shift could not be fully applied to the analysis of the mutants. Most of the mutants confer abnormal meiotic behavior rather than the blocking of meiosis/sporulation at a certain stage. The map was constructed with the help of double mutants. The positions of mutants in the map are those where they first induce abnormal behavior. The pathway of prospore wall formation is independent of DNA synthesis and the completion of meiosis, but dependent on the presence of SPBs, as is documented by the behavior of *spo2, spo3*, and other mutants. Anucleated prospores but no mature spore walls are formed in *spo2* cells (Moens et al., 1974). The formation of the mature spore wall seems to depend on the genome of the maturing spore.

The two pathways represented by *SPO11* and *APO* are blocked by the wild-type gene products, but opened up in the mutants (*APO* consists of the two genes *SPO12* and *SPO13*, as has been discussed above). Meiosis/sporulation in the wild type proceeds via DNA synthesis, recombination, meiosis I, and meiosis II to ascospore maturation. In the *spo11* mutant meiosis I is entered without completion of DNA synthesis, leading to aneuploid inviable products. In *apo* mutants the 4N nucleus directly enters an equational division.

An interesting pleiotropic phenotype is displayed by *spo7*, which is an *antimutator* mutant (Esposito et al., 1975) pointing to function in meiotic recombination and DNA repair.

The Influence of Genes Conferring Radiation Sensitivity on Meiosis/Sporulation

Among a great number of radiation-sensitive mutants of yeast defining essentially three pathways of DNA repair, several mutants have been found to confer specific defects during meiosis/sporulation. Mutants in the first of these pathways are sensitive to x-rays but not to UV light. Under sporulation conditions they synthesize DNA but do not undergo commitment to recombination and form only inviable spores. Examples are given by *rad50, rad52,* and *rad57* (Game et al., 1980; Resnick et al., 1981). Analysis of the DNA formed under sporulation conditions revealed that single-strand breaks occur in the parental DNA. The fragments, which are normal recombination intermediates (Jacobson et al., 1975) cannot be reannealed and accumulate in these strains, rendering the spores inviable (Resnick et al., 1981). Genes *RAD50* and *RAD57* act only in meiotic recombination, but *RAD52* is possibly involved in both meiotic and mitotic recombination.

Some of the mutants in the second pathway ("error-prone" repair, sensitive to both x-rays and UV light) double their DNA under sporulation conditions, but do not form spores. Examples are provided by *rad6* and *rad18* (Game et al., 1980). It is very probable that the gene *RAD6* codes for two different functions in DNA repair represented by two domains of a polyprotein (Tuite and Cox, 1981). Nonsense mutations in *RAD6A* have a polar effect on the *RAD6B* function. The gene *RAD6A*, which is involved in "error-prone" repair, seems to be necessary for sporulation. The gene *RAD6B* seems to code for a function in "error-free" repair and is not needed for sporulation, as has been documented by the analysis of different allelic mutants in *rad6* (Game et al., 1980).

The third repair pathway ("error-free" repair), represented by mutants sensitive only to UV light (*rad1, rad2, rad3, rad4, rad10,* and *rad16*) is involved in the excision of UV-induced pyrimidine dimers in DNA. The loss of these functions does not influence the process of sporulation. Radiation-sensitive mutations conferring a defect in meiotic recombination have also been found in higher eukaryotes (Baker et al., 1976). The situation in yeast is complicated by the fact that a very large number of mutations have been isolated, but is has not been possible so far to establish the exact relationships between all the functions involved (UV sensitivity, x-ray sensitivity, spontaneous mutability, meiotic and mitotic DNA synthesis and recombination, gene conversion, formation of spores, etc.). For instance, the ordering of *SPO* and *RAD* functions in a common order-of-function map has not been undertaken.

Other Mutations Affecting the Sporulation Process

The role of *CDC* and *RAD* genes in sporulation has already been discussed.

What is of interest here is the influence of mutations conferring auxotrophies or membrane transport deficiencies on sporulation. Only a few examples are given. The locus *TUP1* (enabling the cell to utilize exogenous deoxythymidine monophosphate) has already been discussed, because it influences α specific mating functions (Wickner, 1974b). Lemontt et al. (1980) showed that this locus maps near *MAT* on chromosome III and also influences UV-induced mutagenesis. Homozygotes for *tup1* do not sporulate.

Mutant *cho1*, which is blocked in the methylation of lipids and requires exogenous choline and ethanolamine, is blocked in sporulation on acetate medium even in the presence of choline and ethanolamine (Atkinson et al., 1980). D-Glucosamine auxotrophs are defective in ascospore wall formation (Whelan and Ballou, 1975).

THE PROCESS OF SPORE GERMINATION

General Description and Definitions

Rousseau et al. (1972) and Rousseau and Halvorson (1973a-c) investigated the transformation of dormant yeast spores into haploid vegetative cells. Three

The Yeast Genome in Yeast Differentiation

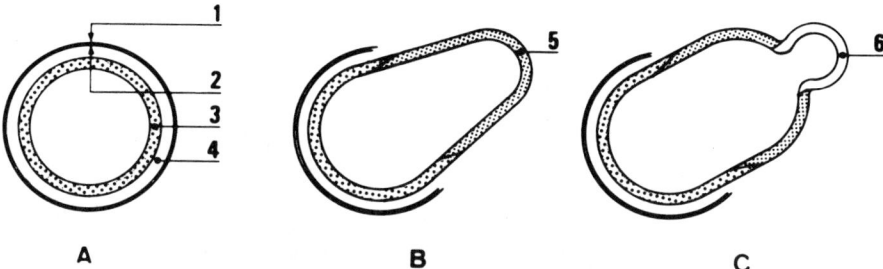

Figure 9 Schematic outline of spore germination. (A) Mature ascospore. In the wall the following layers may be distinguished: (1) a thin, dark outer layer which is the original outer membrane of the prospore wall, (2) a broader, dark layer under the outer layer, (3) the greyish part of the inner layer, and (4) the light part of the inner layer. (B) Outgrowing ascospore. The osmiophilis outer parts of the spore wall are ruptured. The greyish parts of the spore wall form the wall of the outgrowing spore (5). (C) Development of the first bud. The wall of the bud (6) arises from under the wall of the outgrowing spore (from Kreger-Van Rij, 1978).

sequential processes can be discriminated on a morphological and physiological basis; they are depicted schematically in Fig. 9.

1. "Germination proper." During the first 60-90 min after suspending the spores in vegetative growth medium the suspension decreases gradually in optical density (OD) by 10-30% of the initial value. At the same time a small loss of dry weight (approximately 2%) is observed. Synthesis of new biological macromolecules could not be detected (Rousseau and Halvorson, 1973b). The process depends on energy, since it cannot be initiated by water alone; however, it is not blocked by inhibitors of glycolysis (Rousseau and Halvorson, 1973a).

2. "Outgrowth" (Figs 9B and 10). During this phase (60-80 min after suspending the spores in vegetative growth medium) the outer layer of the spore wall breaks and a "germ tube" is formed. The OD increases again. Protein synthesis starts before RNA synthesis (Rousseau and Halvorson, 1973b; Steele and Miller, 1977). (Resting spores contain stable mRNAs; Harper et al., 1980.) The newly growing wall of the outgrowing spore is discussed in the legend in Fig. 9. Scanning electron microscopy of the outgrowing ascospores of an apomictic strain (Steele and Miller, 1974) and of a normal heterothallic strain (Rousseau et al., 1972) leads to conclusions similar to those of Kreger-Van Rij (1978). The newly formed wall is an extension of the innermost layer of the spore wall. The outgrowing spore become gradually sensitive to glusulase (Savarese, 1975), as are vegetative cells.

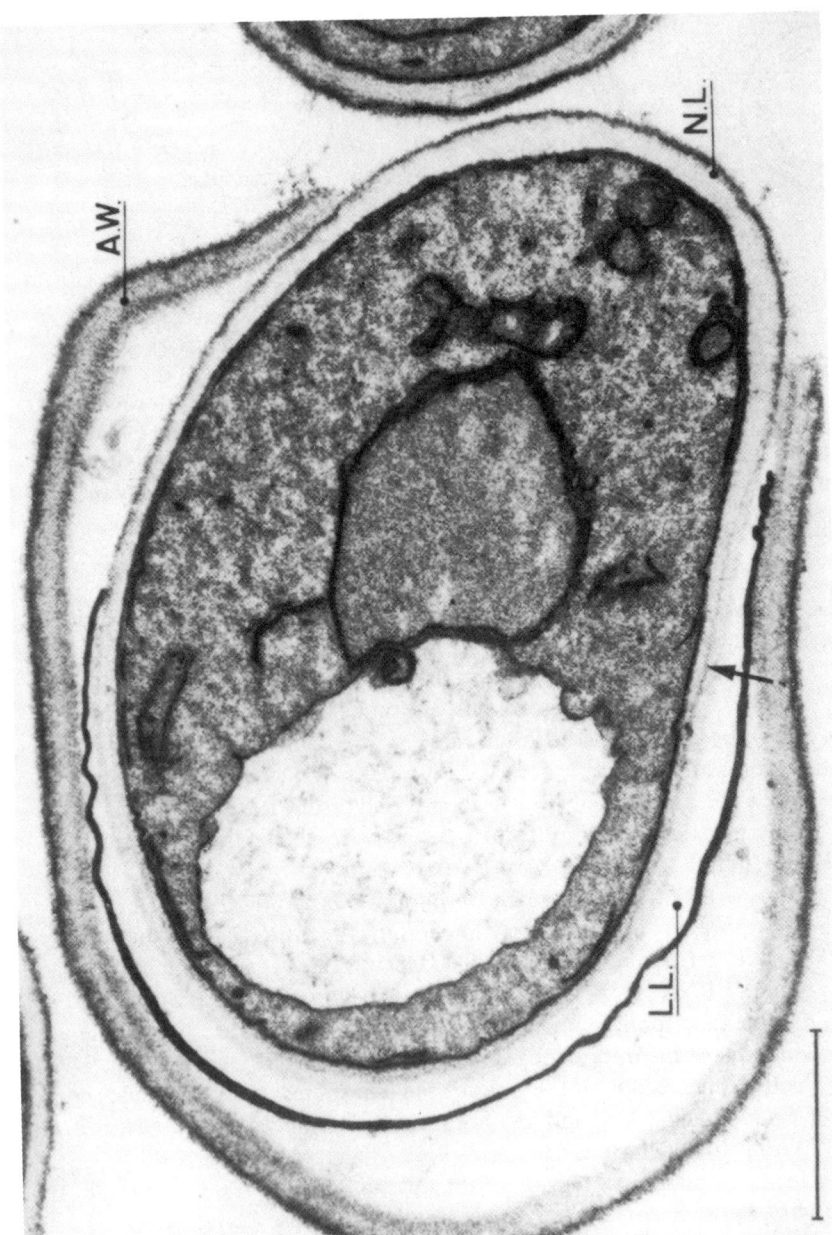

Figure 10 Outgrowing ascospore partly within the ascus. The ascus wall (AW) and the osmiophilic layers of the spore wall are ruptured. The new wall (NL) is formed by an extension of the innermost layer of the spore wall (compare Fig. 9B). The light part of the original spore wall (LL) does not extend to the outgrowing part of the spore. (From Kreger-Van Rij, 1978).

3. "Budding." The outgrowth phase is followed by the first vegetative cell cycle. DNA synthesis and the events observed in the nucleus are similar to the events during the normal vegetative cell cycle. The newly formed cell wall, however, is different from a normal vegetative cell wall for up to three cell cycles (Kreger-Van Rij, 1978; Hartig et al., 1981).

For biochemical studies of ascospore germination it is essential to work with pure yeast ascospores. Several methods have been published which allow the isolation of 90-95% pure spores (Emeis and Gutz, 1958; Rousseau and Halvorson, 1969; Savarese, 1974). The use of Percoll density gradients for the separation of ascospores from the small vegetative cells remaining after glusulase treatment improves the yield to 99.9% pure ascospores (Hartig et al., 1981). A similar method has been employed for the preparation of pure spores from *Dictyostelium discoideum* (Killick, 1981).

The use of vegetative growth medium (other than specialized germination media; see Rousseau and Halvorson, 1973b) is preferred for biochemical studies of germination. Effects which are due to adaptation to succinate (a compound included in specialized germination media) can be avoided in this way and germinating spores can be directly compared with the isogenic haploid strains growing on vegetative growth medium. For instance, large differences in the two-dimensional peptide maps were observed between vegetative cells and spores labeled during the outgrowth phase (M. Breitenbach, unpublished results). Similar results were obtained with outgrowing spores of *D. discoideum* (Dowbenko and Ennis, 1980).

Over a period of 6 hr the outgrowing spores lose resistance to dryness. The appearance of mitochondrial cytochromes requires up to 10 hr in germination medium. In the cytochrome spectra of outgrowing yeast spores, cytochrome c is observed first, followed by cytochromes b and a/a_3 (Hartig et al., 1981). The mitochondria of spores are in a nonrespiring undifferentiated state (Brewer and Fangman, 1980). During outgrowth, normal mitochondria begin to reappear (Hashimoto et al., 1958). It is not surprising, therefore, that fermentable carbon sources (preferably glucose) are needed for the induction of germination (Palleroni, 1961; Banerjee, 1971). The spores of most strains do not germinate on acetate-containing vegetative growth media.

A specialized role has been ascribed to S-adenosylmethionine (SAM) in the metabolism of germinating yeast ascospores (Choih et al., 1977). S-Adenosylmethionine may be essential for the initiation of the first cell cycle in these experiments (Brawley and Ferro, 1980).

Anaerobic germination and outgrowth of yeast ascospores was studied by Sando et al. (1980).

The sensitivity of outgrowing yeast ascospores to chemical mutagens was tested by Redshaw (1975). The action spectrum of ICR-170 (an alkylating acridine derivative and a potent mutagen for yeast cells) was very unusual. The induction of mitochondrial petite mutations was maximal 5 hr after suspension in germination medium. The nuclear genome was much less sensitive to mutagenesis than that of vegetative cells.

Genetic Studies of Germination

If one takes into account the complicated morphological and physiological changes occurring during the transition of yeast spores to vegetative cells, the existence of germination specific genes could be expected. However, no detailed genetic studies of germination have been published so far. A germination-deficient mitochondrial mutant (Hartig et al., 1981) will be discussed in the next section. Many of the mutants already discussed produce abnormal or inviable spores (e.g., some of the *spo* mutants), but those genes are normally needed during sporulation and not during germination. M. S. Esposito (personal communication) suggested a method by which truly germination-specific *ts* mutants can be distinguished from sporulation-specific *ts* mutants which produce dead spores. If the spores are *ts* during germination, irrespective of the temperature during sporulation, they carry a germination-specific mutant gene.

Germination Studies in Other Yeasts

The preparation of pure ascospores of *S. pombe* and their behavior during germination and outgrowth were studied by several groups (Johnke and Padilla, 1979; Nishi et al., 1978; Padilla et al., 1974, 1975; Shimoda, 1980). The essential features of these processes were very similar to the corresponding ones in *S. cerevisiae*.

THE ROLE OF MITOCHONDRIA IN YEAST DIFFERENTIATION

Petite mutants of yeast which have lost large parts or all of the mitochondrial genome (rho^- and rho^0 mutants, respectively) retain the ability to conjugate, but are unable to sporulate (Ephrussi et al., 1949). Therefore a specific role of the mitochondrial genome can be ruled out for conjugation, but it is left open for sporulation and germination.

Physiological Studies

The mitochondria undergo a remarkable differentiation process during sporulation and spore germination. Mitochondrial functions have to be activated at the onset of sporulation. At this time a large peak of respiration (which is, however, neither necessary nor specific for sporulation) and a large amount of cytochromes is observed in the wild type. Oxygen is required for sporulation (Hansen, 1902; Hartig and Breitenbach, 1980), but the exact amount of oxygen needed is not known. In the later stages of spore formation, however, 50% of the mitochondrial mass is excluded from the spores and the mature spores contain a single, large mitochondrion (Diala and Wilkie, 1977; Brewer and Fangman, 1980) which is devoid of cristae and looks "fermentation-like" (Hashimoto et al., 1958; Sando et al., 1980; Brewer and Fangman, 1980). It is completely inactive with respect to respiration and does not contain cytochromes. Thus it seems likely that these mitochondrial components (and possibly others) have to be "degraded" or excluded in some way during sporulation so that the mitochondria reach the extremely repressed state typical of mature yeast ascospores. Virtually nothing is known about this process. During germination and outgrowth the repressed mitochondria resume respiration and slowly and sequentially synthesize cytochromes c, b, and a/a_3 (see also the section on germination; Hartig et al., 1981). The sequential synthesis of cytochromes during spore germination is virtually identical in *S. cerevisiae* and *Botryodiplodia theobromae* (Brambl, 1975, 1977, 1980; Brambl and Handschin, 1976; Brambl and Josephson, 1977; Josephson and Brambl, 1980). It requires de novo protein synthesis, for which stable mRNA contained in the spores is probably used (Brambl and Van Etten, 1970). Possibly repression of respiration during sporulation and its reinitiation during spore germination is a common feature of fungal spores.

Studies with inhibitors of mitochondrial replication (Newlon and Hall, 1978) and mitochondrial protein synthesis (Puglisi and Zennaro, 1971; Marmiroli et al., 1980, 1981) point to the fact that the mitochondrial *genome* is needed during sporulation. Inhibition of mitochondrial protein synthesis partially inhibits the outgrowth of yeast spores (Rousseau and Halvorson, 1973a).

Genetic Studies

For a review of the mitochondrial genome of yeast, see Borst and Grivell (1978). The sporulation of respiration-deficient, mitochondrially inherited point mutants (*mit* mutants) has been studied by two groups (Hartig and Breitenbach, 1980; Pratje et al., 1979). It was found that sporulation of *mit*⁻ mutants is not caused by leakiness or reversion of the mutants. In the case of *oxi3 mit*⁻ mutants, it could be shown that defects in specific parts of the *oxi3* mosaic gene block sporulation, while others do not. This was interpreted as an indication of the requirement of specific mitochondrial genes (e.g., genes represented by the open reading frames in the introns of *oxi3*) for sporulation, although mitochondrial respiration itself was not required (Hartig and Breitenbach, 1980). Sporulation- and germination-specific products of mitochondrial transcription and translation could be shown to exist in a subsequent investigation (Schroeder and

Breitenbach, 1981a). To further corroborate these findings, a mutant selection system was devised (Hartig et al., 1981) and $MnCl_2$ mutagenesis (Baranowska et al., 1977; Putrament et al., 1973) applied to a diploid a/α strain. The method yielded predominantly mutants that were respiration-competent and germination deficient and which showed a pleiotropic phenotype lacking carbon catabolite repression of cytochromes during vegetative growth. No sporulation-deficient respiration-competent mutants have been found so far. The cytoplasmic mode of inheritance of the germination-deficient mutants was shown by deriving haploids and crossing them with the *kar1* mutant (Conde and Fink, 1976; Nagley and Linnane, 1978). The mitochondrial genome of one of the mutants was mapped by petite deletion mapping and by physical mapping (Hartig et al., 1981). Its location is between the loci *cob* and *oli2* in the segment of the map also containing the replication origins *ori2* and *ori7* (De Zamaroczy et al., 1981). The mutant spores are blocked during the formation of the first vegetative bud and do not form viable haploids on germination medium. They synthesize cytochrome c under these conditions, but not the cytochromes b and a/a_3, which are in part encoded in the mitochondrial genome. We interpret these findings as evidence for a regulatory gene that is needed for the reactivation of the mitochondrial genetic system during spore germination.

Küenzi et al. (1974) and Tingle et al. (1974) put forward the idea that respiration, but not the mitochondrial genome, is necessary for sporulation and spore germination. In their experiments fully respiratorily adapted yeast cells were treated with ethidium bromide for 2 hr before the start of the sporulation experiment. Consequently the cells were washed and resuspended in sporulation medium. At this time mitochondrial DNA could not be detected on cesium chloride gradients. After almost normal sporulation, tetrads were analyzed. All of the haploid progeny were petite, but many of them contained mitochondrial DNA. It can be argued that the presence and function of mitochondrial DNA during sporulation and germination has not been rigorously excluded in these experiments (Newlon and Hall, 1978; Hartig and Breitenbach, 1980).

The combined genetic and biochemical evidence available so far indicates that mitochondrial genes are indeed necessary for the processes of sporulation and germination.

PERSPECTIVES

The general aim of developmental biology is to understand the genetic programs of development (Grant, 1978). At present, this aim can best be followed by the identification of the genes and/or gene products involved in differentiation-specific processes. Only a few of these gene products have been identified so far, but the recently developed genetic techniques will bring about the identification of many more differentiation-specific gene products in the near future. Yeast offers the opportunity of combining classical genetics with molecular genetics, ultrastructural research, and biochemical analysis.

Some of the differentiation processes in yeast closely resemble analogous processes in higher organisms. Four of these aspects will be discussed here: (1) Meiosis is an outstanding example. A detailed molecular analysis of meiosis in yeast is to be expected earlier than in any other eukaryote. (2) Possibly molecular analysis of the yeast mating process will contribute to the understanding of fertilization processes. (3) The differentiation of mitochondria, which was discussed in the preceding section, is also an essential feature of animal cell differentiation (Pollak and Sutton, 1980; Vallejo et al., 1979; Boell and Greenfield, 1975; Neuberth et al., 1975). (4) The role of the spindle pole body (centriole) in yeast differentiation is being studied intensively with genetic and biochemical methods (Pringle and Hartwell, 1981; Byers, 1981). The functions of centrioles in animal cells seem to be similar to those in yeast.

Other aspects of yeast differentiation have little similarity with differentiation in higher organisms, but closely resemble bacterial differentiation processes. Many of the features of yeast germination discussed above can be found in bacterial germination (Keynan, 1974). Bacterial sporulation (Szulmajster, 1979) is transcriptionally regulated through promoter selection by differently modified RNA polymerases (Losick, 1981). Transcriptional regulation also occurs in eukaryotes, but the mechanism is only poorly understood. It can be expected that the analysis of the gene products of the yeast *MAT* locus will bring us nearer to an understanding of transcriptional regulation in eukaryotes, because the products of *MAT* seem to be transcriptional regulators for *a*-, α-, and *a*/α-specific genes.

The analysis of yeast differentiation has already shown that some genes necessary for differentiation are also used for a variety of other vital functions (e.g., the *RAD* genes). In this way, the picture of a complicated network of interrelated functions is slowly emerging.

ACKNOWLEDGMENTS

We want to express our gratitude to the colleagues who have made available to us manuscripts and useful information prior to publication, especially to W. Duntze and to J. R. Pringle and L. H. Hartwell; to H. Ruis, J. Hirsch, and K. Hartmuth for critically reading the manuscript; and to K. Hartmuth for typing most of the manuscript.

REFERENCES

Ashraf, M., and Miller, J. J. (1978). Sporulation in single spore isolates from amitrole induced multispored asci of *Saccharomyces cerevisiae*. Can. J. Microbiol. *24*:1614-1615.

Atkinson, K. D., Jensen, B., Kolat, A. I., Storm, E. M., Henry, S. A., and Fogel, S. (1980). Yeast mutants auxotrophic for choline or ethanolamine. J. Bacteriol. *141*:558-564.

Avers, C. J. (1980). *Genetics*. Van Nostrand, New York.

Baker, B. S., Carpenter, A. T. C., Esposito, M. S., Esposito, R. E., and Sandler, L. (1976). The genetic control of meiosis. Annu. Rev. Genet. *10*:53-134.

Ballou, C. E. (1976). Structure and biosynthesis of the mannan component of the yeast cell envelope. Adv. Microbiol. Physiol. *14*:93-158.

Banerjee, M. (1971). The role of carbohydrate in the germination of yeast ascospores. Ph.D. thesis, McMaster University, Hamilton, Ontario.

Baranowska, H., Ejchart, A., and Putrament, A. (1977). Manganese mutagenesis in yeast. V. On mutation and conversion induction in nuclear DNA. Mutat. Res. *42*:343-348.

Barton, J. K., Den Hollander, J. A., Lee, T. M., MacLaughlin, A., and Shulman, R. G. (1980). Measurement of internal pH of yeast spores by $31P$ nuclear magnetic resonance. Proc. Nat. Acad. Sci. USA *77*:2470-2473.

Betz, H. (1979). Loss of sporulation ability in a yeast mutant with low proteinase A levels. FEBS Lett. *100*:171-174.

Betz, R., and Duntze, W. (1979). Purification and partial characterization of a-factor from *Saccharomyces cerevisiae* Eur. J. Biochem. *95*:469-475.

Betz, R., MacKay, V. L., and Duntze, W. (1977). a-Factor from *Saccharomyces cerevisiae*. Partial characterization of a mating hormone produced by cells of mating type a. J. Bacteriol. *132*:462-472.

Betz, R., Duntze, W., and Manney, T. R. (1978). Mating-factor-mediated sexual agglutination in *Saccharomyces cerevisiae*. FEMS Lett. *4*:107-110.

Betz, R., Manney, T. R., and Duntze, W. (1981). Hormonal control of gametogenesis in the yeast *Saccharomyces cerevisiae*. Gamete Res. *4*:571-584.

Bilinski, C. A., and Miller, J. J. (1980). Induction of normal ascosporogenesis in two-spored *Saccharomyces cerevisiae* by glucose, acetate, and zinc. J. Bacteriol. *143*:343-348.

Boell, E. J., and Greenfield, P. C. (1975). Mitochondrial differentiation during animal development. In *Animal Cell Differentiation*, Vol. 3, R. Weber (Ed.). Plenum, New York, pp. 337-385.

Borst, P., and Grivell, L. A. (1978). The mitochondrial genome of yeast. Cell *15*:705-723.

Brambl, R. (1975). Characteristics of developing mitochondrial genetic and respiratory functions in germinating fungal spores. Biochim. Biophys. Acta *396*:175-186.

Brambl, R. (1977). Mitochondrial biogenesis during fungal spore germination. Development of cytochrome c oxidase activity. Arch. Biochem. Biophys. *182*:273-281.

Brambl, R. (1980). Mitochondrial biogenesis during fungal spore germination. J. Biol. Chem. 255:7673-7680.

Brambl, R., and Handschin, B. (1976). Mitochondrial biogenesis during fungal spore germination: Products of mitochondrial protein synthesis in vivo. Arch. Biochem. Biophys. 175:606-617.

Brambl, R., and Josephson, M. (1977). Mitochondrial biogenesis during fungal spore germination: Respiratory cytochromes of dormant and germinating spores of Botryodiplodia. J. Bacteriol. 129:291-297.

Brambl, R. M., and Van Etten, J. L. (1970). Protein synthesis during fungal spore germination V. Evidence that the ungerminated conidiospores of Botryodiplodia theobromae contain messenger ribonucleic acid. Arch. Biochem. Biophys. 137:442-452.

Braun, P. C., and Calderone, R. A. (1978). Chitin synthesis in Candida albicans: Comparison of yeast and hyphal forms. J. Bacteriol. 133:1472-1477.

Brawley, J. V., and Ferro, A. J. (1980). Stimulation of yeast ascospore germination and outgrowth by S-adenosyl methionine. J. Bacteriol. 142:608-614.

Brewer, B. J., and Fangman, W. L. (1980). Preferential inclusion of extrachromosomal genetic elements in yeast meiotic spores. Proc. Nat. Acad. Sci. USA 77:5380-5384.

Bücking-Throm, E., Duntze, W., Hartwell, L. H., and Manney, T. R. (1973). Reversible arrest of haploid yeast cells at the initiation of DNA synthesis by a diffusible sex factor. Exp. Cell Res. 76:99-110.

Byers, B. (1981). Multiple roles of the spindle pole bodies in the life cycle of Saccharomyces cerevisiae. In Molecular Genetics in Yeast. Alfred Benzon Symposium 16. D. von Wettstein, J. Friis, M. Kiellandt-Brandt, and A. Stenderup (Eds.). Munksgard, Copenhagen, pp. 119-133.

Byers, B., and Goetsch, L. (1973). Duplication of spindle plaques and integration of the yeast cell cycles. Cold Spring Harbor Symp. Quant. Biol. 28:123-131.

Byers, B., and Goetsch, L. (1975a). The behavior of spindles and spindle plaques in the cell cycle and conjugation of Saccharomyces cerevisiae. J. Bacteriol. 124:511-523.

Byers, B., and Goetsch, L. (1975b). Electron microscopic observations on the meiotic karyotype of diploid and tetraploid Saccharomyces cerevisiae. Proc. Nat. Acad. Sci. USA 72:5056-5060.

Byers, B., and Goetsch, L. (1976). A highly ordered ring of membrane-associated filaments in budding yeast. J. Cell Biol. 69:717-721.

Byers, B., and Sowder, L. (1980). Gene expression in the yeast cell cycle. J. Cell Biol. 87:6a.

Cabib, E. (1975). Molecular aspects of yeast morphogenesis. Annu. Rev. Microbiol. 29:191-214.

Cabib, E., and Bowers, B. (1975). Timing and functions of chitin synthesis in yeast. J. Bacteriol. 124:1586-1593.

Campbell, D. A. (1973). Kinetics of the mating-specific aggregation in Saccharomyces cerevisiae. J. Bacteriol. 116:323-330.

Chan, R. K. (1977). Recovery of *Saccharomyces* mating type *a*-cells from G1 arrest by α-factor. J. Bacteriol. *130*:766-774.

Chen, A. W., and Miller, J. J. (1968). Proteolytic activity of intact yeast cells. Can. J. Microbiol. *14*:957-964.

Choih, S. J., Ferro, A. J., and Shapiro, S. K. (1977). Function of S-adenosylmethionine in germinating yeast ascospores. J. Bacteriol. *131*:63-68.

Ciejek, E. (1980). Ph.D. thesis, University of California, Berkeley.

Ciejek, E., and Thorner, J. (1979). Recovery of *Saccharomyces a*-cells from G1 arrest by α-factor pheromone requires endopeptidase action. Cell *18*: 623-635.

Ciejek, E., Thorner, J., and Geier, M. (1977). Solid phase peptide synthesis of α-factor, a yeast mating pheromone. Biochem. Biophys. Res. Commun. *78*:952-961.

Clarke, L., and Carbon, J. (1980). Isolation of a yeast centromere and construction of small circular chromosomes. Nature *287*:504-509.

Codington, J. F., Sanford, B. H., and Jeanloz, R. W. (1972). Glycoprotein coat of the TA3 cell. Isolation and partial characterization of a sialic acid containing fraction. Biochemistry *11*:2559-2564.

Colonna, W. J., and Magee, P. T. (1978). Glycogenolytic enzymes in sporulating yeast. J. Bacteriol. *134*:844-853.

Conde, J., and Fink, G. R. (1976). A mutant of *Saccharomyces cerevisiae* defective for nuclear fusion. Proc. Nat. Acad. Sci. USA *73*:3651-3655.

Cooney, P. H., Whiteman, P. F., and Freese, E. (1977). Media dependence of commitment in *Bacillus subtilis.* J. Bacteriol. *129*:901-907.

Cortat, M., Matile, P., and Wiemken, A. (1971). Isolation of glucanase-containing vesicles from budding yeast. Arch. Mikrobiol. *82*:189-205.

Cottrell, S. F., and Lee, L. H. (1981). Evidence for the synchronous replication of mitochondrial DNA during the yeast cell cycle. Exp. Cell Res. *101*:1350-1356.

Crandall, M. A., and Brock, T. D. (1968). Molecular basis of mating in the yeast *Hansenula wingei.* Bacteriol. Rev. *32*:139-163.

Crandall, M., Lawrence, L. M., and Saunders, R. M. (1976). Molecular complementarity of yeast glycoprotein mating factors. Proc.Nat. Acad. Sci. USA *71*:26-29.

Croes, A. F. (1966). Duplication of DNA during meiosis in baker's yeast. Exp. Cell Res. *41*:452-454.

Croes, A. F., Steijns, J. M. J. M., DeVries, G. J. M. L., van der Putte, T. M. J. A. (1978). Inhibition of meiosis in *Saccharomyces cerevisiae* by ammonium ions: Interference of ammonia with protein metabolism. Planta *141*:205-209.

Cryer, D. R., Goldthwaite, C. D., Zinker, S., Lam, K. B., Storm, E., Hirschberg, R., Blamire, J., Finkelstein, D. B., and Marmur, J. (1973). Studies on nuclear and mitochondrial DNA of *Saccharomyces cerevisiae.* Cold Spring Harbor Symp. Quant. Biol. *38*:17-29.

Davidow, L., Goetsch, L., and Byers, B. (1980). Preferential occurrence of nonsister spores in two-spored asci of *Saccharomyces cerevisiae*. Genetics *94*: 581-595.

Dawes, I. W., and Hardie, I. D. (1974). Selective killing of vegetative cells in sporulated yeast cultures by exposure to diethyl ether. Mol. Gen. Genet. *131*:281-290.

Dayhoff, M. O., Hunt, L. T., Barker, W. C., Schwartz, R. M., and Orcutt, B. C. (1978). *Protein Sequence Dictionary 78*, National Biomedical Research Foundation, Silver Springs, Md.

Del Rey, F., Santos, T., Garcia-Acha, I., and Nombela, C. (1979). Synthesis of 1,3-β-glucanase in *Saccharomyces cerevisiae* during the mitotic cycle, mating and sporulation. J. Bacteriol. *139*:924-931.

Del Rey, F., Santos, T., Garcia-Acha, I., and Nombela, C. (1980). Synthesis of β-glucanases during sporulation in *Saccharomyces cerevisiae*: Formation of a new, sporulation-specific 1,3-β-glucanase. J. Bacteriol. *143*:621-627.

De Robertis, E. M., and Gurdon, J. B. (1979). Gene transplantation and the analysis of development. Sci. Am. *241*(6):60-68.

De Zamaroczy, M., Marotta, R., Faugeron-Fonty, G., Goursot, R., Mangin, M., Baldacci, G., and Bernardi, G. (1981). The origins of replication of the yeast mitochondrial genome and the phenomenon of suppressivity. Nature *292*: 75-78.

Diala, E. S., and Wilkie, E. (1977). Mitochondria-cell relationships in *Saccharomyces cerevisiae*: Aspects of sporulation. In *Mitochondria 1977*. W. Bandlow, R. J. Schweyen, K. Wolf, and F. Kaudewitz (Eds.). W. de Gruyter, Berlin, pp. 563-570.

Doi, S., and Yoshimura, M. (1978). Temperature-dependent conversion of sexual agglutinability in *Saccharomyces cerevisiae*. Mol. Gen. Genet. *162*:251-257.

Dowbenko, D. J., and Ennis, H. D. (1980). Regulation of protein synthesis during spore germination in *Dictyostelium discoideum*. Proc. Nat. Acad. Sci. USA *77*:1791-1795.

Duntze, W., MacKay, V. L., and Manney, T. R. (1970). *Saccharomyces cerevisiae*: A diffusible sex factor. Science *168*:1472-1473.

Duntze, W., Stötzler, D., Bücking-Throm, E., and Kolbitzer, S. (1973). Purification and partial characterization of α-factor, a mating-type specific inhibitor of cell reproduction from *Saccharomyces cerevisiae*. Eur. J. Biochem. *35*: 357-365.

Dutcher, S. K. (1980). Genetic control of karyogamy in *Saccharomyces cerevisiae*. Ph.D. thesis, University of Washington, Seattle.

Egel, R. (1971). Physiological aspects of conjugation in fission yeast. Planta *98*: 89-96.

Emanuel, J. R., and Magee, P. T. (1981). Timing of ribosome synthesis during ascosporogenesis of yeast cells: Evidence for early function of haploid daughter genomes. J. Bacteriol. *145*:1342-1350.

Emeis, C. C., and Gutz, H. (1958). Eine einfache Technik zur Massenisolation von Hefesporen. Z. Naturforsch. *13b*:647-650.

Engels, F. M., and Croes, A. F. (1968). The synaptonemal comples in yeast. Chromosoma *25*:104-106.

Ephrussi, B., Hottinguer, H., and Tavlitzki, J. (1949). Action de l'acriflavine sur les levures. II. Etude génètique du mutant "petite colonie." Ann. Inst. Pasteur *76*:419-450.

Esmon, B., Novick, P., and Schekman, R. (1981). Compartmentalized assembly of oligosaccharides on exported glycoproteins in yeast. Cell *25*:451-460.

Esposito, M. S., and Esposito, R. E. (1974). Genetic recombination and commitment to meiosis in *Saccharomyces*. Proc. Nat. Acad. Sci. USA *71*:3172-3176.

Esposito, M. S., and Esposito, R. E. (1975). Mutants of meiosis and ascospore formation. In *Methods in Cell Biology XI. Yeast Cells,* D. M. Prescott (Ed.). Academic, New York, pp. 303-326.

Esposito, M. S., and Esposito, R. E. (1978). Aspects of the genetic control of meiosis and ascospore development inferred from the study of *spo* (sporulation deficient) mutants of *Saccharomyces cerevisiae*. Biol. Cell. Paris *33*: 93-102.

Esposito, R. E., and Klapholz, S. (1981). Meiosis and spore formation in *Saccharomyces,* J. N. Strathern, E. W. Jones, and J. R. Broach (Eds.). Cold Spring Harbor Laboratory, Cold Spring Harbor, New York, pp. 211-287.

Esposito, M. S., Esposito, R. E., Arnaud, M., and Halvorson, H. O. (1969). Acetate utilization and macromolecular synthesis during sporulation of yeast. J. Bacteriol. *100*:180-186.

Esposito, M. S., Esposito, R. E., and Moens, P. B. (1974). Genetic analysis of two-spored asci produced by the *spo3* mutant of *Saccharomyces.* Mol. Gen. Genet. *135*:91-95.

Esposito, M. S., Bolotin-Fukuhara, M., and Esposito, R. E. (1975). Antimutator activity during mitosis by a meiotic mutant of yeast. Mol. Gen. Genet. *139*: 9-18.

Fehrenbacher, G., Perry, K., and Thorner, J. (1978). Cell-cell recognition in *Saccharomyces cerevisiae*: Regulation of mating-specific adhesion. J. Bacteriol. *134*:893-901.

Finkelstein, D. B., and McAlister, L. (1981). α-Factor-mediated modification of a ^{32}P-labeled protein by *MATa* cells of *Saccharomyces cerevisiae*. J. Biol. Chem. *256*:2561-2566.

Finkelstein, D. B., and Strausberg, S. (1979). Metabolism of α-*factor by a* mating type cells of *Saccharomyces cerevisiae.* J. Biol. Chem. *254*:796-803.

Fogel, S., and Roth, R. (1974). Mutations affecting meiotic gene conversion in yeast. Mol. Gen. Genet. *130*:189-201.

Fonzi, W. A., Shanley, M., and Opheim, D. J. (1979). Relationship of glycolytic intermediates, glycolytic enzymes and ammonia to glycogen metabolism during sporulation in the yeast *Saccharomyces cerevisiae*. J. Bacteriol. *137*: 285-294.

Fowell, R. R. (1969). Sporulation and hybridization of yeasts. In *The Yeasts*, Vol. 1, A. H. Rose and J. S. Harrison (Eds.). Academic, London, pp. 303-376.
Frank, K. R., and Mills, D. (1978). Ribosome activity and degradation in meiotic cells of *Saccharomyces cerevisiae*. Mol. Gen. Genet. *160*:59-65.
Freese, E. B., Chu, M. I., and Freese, E. (1981). Initiation of yeast sporulation by partial carbon nitrogen or phosphate deprivation. J. Bacteriol. *149*: 840-851.
Freifelder, D. (1960). Bud position in *Saccharomyces cerevisiae*. J. Bacteriol. *80*: 567-568.
Game, J. (1976). Yeast cell-cycle mutant *cdc* 21 is a temperature-sensitive thymidylate auxotroph. Mol. Gen. Genet. *146*:313-315.
Game, J. C., Zamb, T. J., Braun, R. J., Resnick, M., and Roth, R. M. (1980). The role of radiation (*rad*) genes in meiotic recombination in yeast. Genetics *94*: 51-68.
Grant, P. (1978). *Biology of Developing Systems*, Holt, Rinehart, and Winston, New York.
Grewal, N. S., and Miller, J. J. (1972). Formation of asci with two diploid spores by diploid cells of *Saccharomyces*. Can. J. Microbiol. *18*:1897-1905.
Haber, J. E., and Halvorson, H. O. (1975). Methods in sporulation and germination of yeasts. In *Methods in Cell Biology*, D. M. Prescott (Ed.). Academic, New York, pp. 45-69.
Haber, J. E., Esposito, M. S., Magee, P. T., and Esposito, R. E. (1975). Current trends in genetic and biochemical study of yeast sporulation. In *Spores VI*, P. Gerhardt, R. N. Costilow, and H. L. Sadoff (Eds.). American Society for Microbiology, Washington, D.C., pp. 132-137.
Haid, A., Schweyen, R. J., Bechmann, H., Kaudewitz, F., Solioz, M., and Schatz, G. (1978). The mitochondrial *COB* region in yeast codes for apocytochrome *b* and is mosaic. Eur. J. Biochem. *94*:451-464.
Hansen, E. C. (1902). Compt. Rend. Trav. Lab. Carlsberg *5*:68-107. Citation after: Fowell, R. R., Sporulation and hybridization of yeasts. In *The Yeasts*, Vol. 1, A. H. Rose and J. S. Harrison (Eds.). Academic, London, pp. 303-376.
Harper, J. F., Clancy, M. J., and Magee, P. T. (1980). Properties of polyadenylate-associated ribonucleic acid from *Saccharomyces cerevisiae* ascospores. J. Bacteriol. *143*:958-965.
Hartig, A., and Breitenbach, M. (1977). Intracellular cAMP in synchronous sporulation of *Saccharomyces cerevisiae*. FEMS Lett. *1*:79-82.
Hartig, A., and Breitenbach, M. (1980). Sporulation of mitochondrial *oxi*3 mutants of *Saccharomyces cerevisiae*. A correlation with the genetic map. Curr. Genet. *1*:97-102.
Hartig, A., Schroeder, R., Mucke, E., and Breitenbach, M. (1981). Isolation and characterization of yeast mitochondrial mutants defective in spore germination. Curr. Genet. *4*:29-36.
Hartwell, L. H. (1973). Synchronization of haploid yeast cell cycles, a prelude to conjugation. Exp. Cell Res. *76*:111-117.

Hartwell, L. H. (1974). *Saccharomyces cerevisiae* cell cycle. Bacteriol. Rev. *38*: 164-168.

Hartwell, L. H. (1978). Cell division from a genetic perspective. J. Cell Biol. *77*: 631.

Hartwell, L. H. (1980). Mutants of *Saccharomyces cerevisiae* unresponsive to cell division control by polypeptide mating hormone. J. Cell Biol. *85*:811-822.

Hartwell, L. H., and Unger, M. W. (1977). Unequal division in *Saccharomyces cerevisiae* and its implication for the control of cell division. J. Cell Biol. *75*:422-435.

Hashimoto, T., Conti, S. F., and Naylor, H. B. (1958). Fine structure of microorganisms III. Electron microscopy of resting and germinating ascospores of *Saccharomyces cerevisiae*. J. Bacteriol. *76*:406-416.

Henry, S. A., and Halvorson, H. O. (1973). Lipid synthesis during sporulation of *Saccharomyces cerevisiae*. J. Bacteriol. *114*:1158-1163.

Herskowitz, I., Blair, L., Forbes, D., Hicks, J., Kassir, Y., Kushner, P., Rine, J., Sprague, G., Jr., and Strathern, J. (1980). Control of cell type in the yeast *Saccharomyces cerevisiae* and a hypothesis for development in higher eukaryotes. In *The Molecular Genetics of Development*, T. Leighton (Ed.). Academic, New York, pp. 79-117.

Hicks, J. B., and Herskowitz, I. (1976). Evidence for a new diffusible element of mating pheromones in yeast. Nature *260*:246-248.

Hinnen, A., Hicks, J. B., and Fink, G. R. (1978). Transformation of yeast. Proc. Nat. Acad. Sci. USA *75*:1929-1933.

Hirschberg, J., and Simchen, G. (1977). Commitment to the mitotic cell cycle in yeast in relation to meiosis. Exp. Cell Res. *105*:245-252.

Ho, K. H., and Miller, J. J. (1978). Free proline content and sensitivity to desiccation and heat during yeast sporulation and spore germination. Can. J. Microbiol. *24*:312-320.

Hohn, B., and Hinnen, A. (1980). Cloning with cosmids in *E. coli* and yeast. In *Genetic Engineering II*. J. K. Setlow and A. Hollaender (Eds.). Plenum, New York, pp. 169-183.

Hopper, A. K., Magee, P. T., Welch, S. K., Friedman, M., and Hall, B. D. (1974). Macromolecule synthesis and breakdown in relation to sporulation and meiosis in yeast. J. Bacteriol. *119*:619-628.

Horesh, O., Simchen, G., and Friedmann, A. (1979). Morphogenesis of the synapton during yeast meiosis. Chromosoma *75*:101-115.

Illingworth, R. F., Rose, A. H., and Beck, A. (1973). Changes in the lipid composition and fine structure of *Saccharomyces cerevisiae* during ascus formation. J. Bacteriol. *113*:373-386.

Jacobson, G., Piñon, R., Esposito, R. E., and Esposito, M. S. (1975). Single strand scissions of chromosomal DNA during commitment to recombination at meiosis. Proc. Nat. Acad. Sci. USA *72*:1887-1891.

Jagadish, M. N., and Carter, B. L. A. (1977). Genetic control of cell division in yeast cultured at different growth rates. Nature *269*:145-147.

Jarvik, J., and Botstein, D. (1973). A genetic method for determining the order of events in a biological pathway. Proc. Nat. Acad. Sci. USA 70:2046-2050.
Johnke, R., and Padilla, G. M. (1979). Germination and outgrowth of *Schizosaccharomyces pombe* spores isolated by a simple batch centrifugation technique. J. Gen. Microbiol. 115:255-258.
Johnston, L. H., and Nasmyth, K. A. (1978). *Saccharomyces cerevisiae* cell cycle mutant *cdc* 9 is defective in DNA ligase. Nature 274:891-893.
Johnston, G. S., Pringle, J. R., and Hartwell, L. H. (1977). Coordination of growth with cell division in the yeast *Saccharomyces cerevisiae*. Exp. Cell Res. 105:79-98.
Jones-Brown, Y. R., Thorner, J., and Blair, L. C. (1980). submitted to J. Biol. Chem., citation after Thorner, J. (1980). Intercellular Interactions of the yeast *Saccharomyes cerevisiae*. In *The Molecular Genetics of Development*. T. Leighton (Ed.). Academic, New York, pp. 119-178.
Josephson, M., and Brambl, R. (1980). Mitochondrial biogenesis during fungal spore germination. Purification, properties and biosynthesis of cytochrome c oxidase from *Botryodiplodia theobromae*. Biochim. Biophys. Acta 606: 125-137.
Kamiya, Y., and Sakurai, A. (1981). Mating pheromones of heterobasidiomycetous yeasts. Naturwissenschaften 68:128-133.
Kane, S., and Roth, R. (1974). Carbohydrate metabolism during ascospore development in yeast. J. Bacteriol. 118:8-14.
Kassir, Y., and Simchen, G. (1978). Meiotic recombination and DNA synthesis in a new cell cycle mutant of *Saccharomyces cerevisiae*. Genetics 90:41-68.
Kawanabe, Y., Yoshida, K., and Yanagishima, N. (1979). Sexual cell agglutination in relation to the formation of zygotes in *Saccharomyces cerevisiae*. Plant Cell Physiol. 20:423-433.
Kawasaki, G. (1979). Karyotypic instability and carbon source effects in cell cycle mutants of *Saccharomyces cerevisiae*. Ph.D. thesis, University of Washington, Seattle.
Keynan, A. (1974). The transformation of bacterial endospores into vegetative cells. In *Microbial Differentiation*, J. M. Ashworth and J. E. Smith (Eds.). Cambridge University, London, pp. 85-123.
Khan, S. A., Merkel, G. J., Becker, J. M., and Naider, F. (1981). Synthesis of the dodecapeptide-α-factor of *Saccharomyces cerevisiae*. Int. J. Pept. Protein Res. 17:219-230.
Killick, K. A. (1981). Purification of *Dictyostelium discoideum* spores by centrifugation in Percoll density gradients with retention of morphological and biochemical integrity. Anal. Biochem. 114:46-52.
Klapholz, S., and Esposito, R. E. (1980a). Isolation of *spo* 12-1 and *spo* 13-1 from a natural variant of yeast that undergoes a single meiotic division. Genetics 96:567-588.
Klapholz, S., and Esposito, R. E. (1980b). Recombination and chromosome segregation during the single division meiosis in *spo* 12-1 and *spo* 13-1 diploids. Genetics 96:589-611.

Klar, A. J. S., and Halvorson, H. O. (1975). Proteinase activities of *Saccharomyces cerevisiae* during sporulation. J. Bacteriol. *124*:863-869.

Klar, A. J. S., Strathern, J. N., Broach, J. R., and Hicks, J. B. (1981). Regulation of transcription in expressed and unexpressed mating type cassettes of yeast. Nature *289*:239-244.

Kopecka, M. (1974). Sporulation of protoplasts of the yeast *Saccharomyces cerevisiae*. J. Gen. Microbiol. *83*:171-178.

Kraig, E., and Haber, J. (1980). Messenger ribonucleic acid and protein metabolism during sporulation of *Saccharomyces cerevisiae*. J. Bacteriol. *144*: 1098-1112.

Kreger-Van Rij, N. J. W. (1978). Electron microscopy of germinating ascospores of *Saccharomyces cerevisiae*. Arch. Microbiol. *117*:73-77.

Küenzi, M. T., Tingle, M. A., and Halvorson, H. O. (1974). Sporulation of *Saccharomyces cerevisiae* in the absence of a functional mitochondrial genome. J. Bacteriol. *117*:80-88.

Leibowitz, M. J., and Wickner, R. B. (1976). A chromosomal gene required for killer plasmid expression, mating and spore maturation in *Saccharomyces cerevisiae*. Proc. Nat. Acad. Sci. USA *73*:2061-2065.

Lemontt, J. F., Fugit, D. R., and MacKay, V. L. (1980). Pleiotropic mutations at the *TUP1* locus that affect the expression of mating-type-dependent functions in *Saccharomyces cerevisiae*. Genetics *94*:899-920.

Levi, J. D. (1956). Mating reaction in yeast. Nature *117*:753-754.

Liao, H., and Thorner, J. (1980). Yeast mating pheromone α-factor inhibits adenylate cyclase; Proc. Nat. Acad. Sci. USA *77*:1898-1902.

Lipke, P. N., Taylor, A., and Ballou, C. E. (1976). Morphogenic effects of α-factor on *Saccharomyces cerevisiae a* cells. J. Bacteriol. *127*:610-618.

Losick, R. (1981). Sigma factors, stage 0 genes, and sporulation. In *Sporulation and Germination, Proceedings of the Eighth International Spores Conferences*, H. S. Levinson, A. L. Sonenshein, and D. J. Tipper (Eds.). ASM Publications, Washington, D.C., pp. 48-56.

McCusker, J. H., and Haber, J. E. (1977). Efficient sporulation of yeast in media buffered near pH 6. J. Bacteriol. *132*:180-185.

MacKay, V., and Manney, T. R. (1974a). Mutations affecting sexual conjugation and related processes in *Saccharomyces cerevisiae* I. Isolation and phenotypic characterization of nonmating mutants. Genetics *76*:255-271.

MacKay, V., and Manney, T. R. (1974b). Mutations affecting sexual conjugation and related processes in *Saccharomyces cerevisiae* II. Genetic analysis of nonmating mutants. Genetics *76*:273-288.

Magee, P. T., and Hopper, A. K. (1974). Protein synthesis in relation to sporulation and meiosis in yeast. J. Bacteriol. *119*:952-960.

Maness, P. F., and Edelman, G. M. (1978). Inactivation and chemical alteration of mating factor α by cells and spheroplasts of yeast. Proc. Nat. Acad. Sci. USA *75*:1304-1308.

Manney, T. R., and Meade, J. (1977). Cell-cell interaction during mating in *Saccharomyces cerevisiae*. In *Microbial Interactions, Receptors and Recognition* J. L. Reissig (Ed.). Chapman and Hall, London, pp. 283-321.

Manney, T. R., and Woods, V. (1976). Mutants of *Saccharomyces cerevisiae* resistant to the α mating factor. Genetics *82*:639-644.
Manney, T. R., Duntze, W., and Betz, R. (1981). The isolation, characterization and physiological effects of the *Saccharomyces cerevisiae* sex pheromones. In *Sexual Interactions in Eukaryotic Microbes*, D. H. O'Day and P. A. Horgen (Eds.). Academic, New York, pp. 21-51.
Marian, B., and Wintersberger, U. (1980). Histone synthesis during sporulation of yeast. FEBS Lett. *117*:63-67.
Marmiroli, N., Ferri, M., Puglisi, P. P., and Bruschi, C. (1980). Effects of erythromycin on sporulation associated processes in *Saccharomyces cerevisiae*— Premeiotic DNA synthesis, genetic recombination, nuclei division, respiratory adaptation readiness and spores development. Atti. Assoc. Genet. Ital. *26*:206-209.
Marmiroli, N., Tassi, F., Bianchi, L., Algeri, A. A., Puglisi, P. P., and Esposito, M. S. (1981). Erythromycin and cycloheximide sensitivities of protein and RNA synthesis in sporulating cells of *Saccharomyces cerevisiae*: Environmentally induced modifications controlled by chromosomal and mitochondrial genes. Curr. Genet. *4*:51-62.
Masui, Y., Chino, N., Sakakibara, S., Tanaka, T., Murakami, T., and Kita, H. (1977). Synthesis of the mating factor of *Saccharomyces cerevisiae* and its truncated peptides: The structure-activity relationship. Biochem. Biophys. Res. Commun. *78*:534-538.
Matile, P., Moor, H., and Robinow, C. F. (1969). Yeast cytology. In *The Yeasts* A. Rose and J. S. Harrison (Eds.). Academic, New York, pp. 219-297.
Miller, J. J., and Hoffmann-Ostenhof, O. (1964). Spore formation and germination in *Saccharomyces*. Z. Allg. Mikrobiol. *4*:273-294.
Mills, D. (1974). Isolation of polyribosomes from yeast during sporulation and vegetative growth. Appl. Microbiol. *27*:944-948.
Moens, P. B. (1971). Fine structure of ascospore development in the yeast *Saccharomyces cerevisiae*. Can. J. Microbiol. *17*:507-510.
Moens, P. B. (1974). Modification of sporulation in yeast strains with two-spored asci. J. Cell Sci. *16*:519-527.
Moens, P. B., and Rapport, E. (1971a). Synaptic structures in the nuclei of sporulating yeast, *Saccharomyces cerevisiae* (Hansen). J. Cell Sci. *9*:665-667.
Moens, P. B., and Rapport, E. (1971b). Spindles, spindle plaques and meiosis in the yeast *Saccharomyces cerevisiae* (Hansen). J. Cell Biol. *50*:344-361.
Moens, P. B., Esposito, R. E., and Esposito, M. S. (1974). Aberrant nuclear behavior at meiosis and anucleate spore formation by sporulation-deficient (spo) mutants of *Saccharomyces cerevisiae*. Exp. Cell Res. *83*:166-174.
Montelone, B. A., Prakash, S., and Prakash, L. (1981). Spontaneous mitotic recombination in *mms* 8-1, an allele of the *CDC* 9 gene of *Saccharomyces cerevisiae*. J. Bacteriol. *147*:517-525.
Moorman, A. F. M., Lamie, F., and Grivell, L. A. (1976). A coupled transcription-translation system derived from *Escherichia coli*: The use of immobilized deoxyribonuclease to eliminate endogenous DNA. FEBS Lett. *71*:67-72.

Mortimer, R. K., and Hawthorne, D. C. (1969). Yeast genetics. In *The Yeasts*, Vol. 1, A. M. Rose and J. S. Harrison (Eds.). Academic, London, pp. 385-460.

Mortimer, R. K., and Hawthorne, D. C. (1973). Genetic mapping in *Saccharomyces*. Genetics *74*:33-54.

Mortimer, R. K., and Schild, D. (1980). Genetic map of *Saccharomyces cerevisiae*. Microbiol. Rev. *44*:519-571.

Nagley, P., and Linnane, A. W. (1978). Expression of mitochondrial DNA in *Saccharomyces cerevisiae*: The construction of sets of isonuclear haploid strains containing different specified mitochondrial genomes. Biochem. Biophys. Res. Commun. *85*:585-592.

Nasmyth, K. (1978). Eukaryotic gene cloning and expression in yeast. Nature *274*:741-743.

Nasmyth, K. A., and Reed, S. I. (1980). Isolation of genes by complementation in yeast: Molecular cloning of a cell-cycle gene. Proc. Nat. Acad. Sci. USA *77*:2119-2123.

Nasmyth, K. A., Tatchell, K., Hall, B. D., Astell, C., and Smith, M. (1981a). Physical analysis of mating-type loci in *Saccharomyces cerevisiae*. Cold Spring Harbor Symp. Quant. Biol. *45*:961-981.

Nasmyth, K. A., Tatchell, K., Hall, B. D., Astell, C., and Smith, M. (1981b). A position effect in the control of transcription at yeast mating type loci. Nature *289*:244-250.

Neuberth, D., Gregg, C. T., Bass, R., and Merker, H. J. (1975). Occurrence and possible functions of mitochondrial DNA in animal development. In *Animal Cell Differentiation*, Vol. 3, R. Weber (Ed.). Plenum, New York, pp. 387-464.

Newlon, C. S., and Fangman, W. L. (1975). Mitochondrial DBA synthesis in cell cycle mutants of *Saccharomyces cerevisiae*. Cell *5*:423-428.

Newlon, C. S., Ludescher, R. D., and Walter, S. K. (1979). Production of petites by cell cycle mutants of *Saccharomyces cerevisiae* defective in DNA synthesis. Mol. Gen. Genet. *169*:189-194.

Newlon, M. C., and Hall, B. D. (1978). Inhibition of yeast sporulation by ethidium bromide. Mol. Gen. Genet. *165*:113-114.

Nishi, K., Shimoda, C., and Hayashibe, M. (1978). Germination and outgrowth of *Schizosaccharomyces pombe* ascospores isolated by urografin density gradient centrifugation. Can. J. Microbiol. *24*:893-897.

Northcote, D. H., and Horne, R. W. (1952). Chemical composition and structure of the yeast cell wall. Biochem. J. *51*:232-236.

Novick, P., and Schekman, R. (1979). Secretion and cell-surface growth are blocked in a temperature-sensitive mutant of *Saccharomyces cerevisiae*. Proc. Nat. Acad. Sci. USA *76*:1858-1862.

Novick, P., Field, C., and Schekman, R. (1980). Identification of 23 complementation groups required for posttranslational events in the yeast secretory pathway. Cell *21*:205-216.

Nurse, P. (1975). Genetic control of cell size at cell fission in yeast. Nature *256*:547-551.

Nurse, P. (1980). Cell cycle control—Both deterministic and probabilistic? Nature *286*:9-10.

Nurse, P., and Bissett, Y. (1981). Gene required in G1 for commitment to cell cycle and in G2 for control of mitosis in fission yeast. Nature 292:558-560.
O'Farrell, P. H. (1975). High resolution two-dimensional electrophoresis of protein. J. Biol. Chem. 250:4007-4021.
Olson, L. W., and Zimmerman, F. K. (1978). Meiotic recombination and synaptonemal complexes in Saccharomyces cerevisiae. Mol. Gen. Genet. 166: 151-159.
Opheim, D. J. (1979). Effect of ammonium ions on activity of hydrolytic enzymes during sporulation of yeast. J. Bacteriol. 138:1022-1025.
Osumi, M., Shimoda, C., and Yanagishima, N. (1974). Mating reaction in Saccharomyces cerevisiae V. Changes in the fine structure during mating reaction. Arch. Microbiol. 97:27-38.
Padilla, G. M., Creanor, J., and Frazer, R. S. S. (1974). Early events in the germination of Schizosaccharomyces pombe ascospores. In Cell Cycle Controls, G. M. Padilla, I. L. Cameron, and A. M. Zimmermann (Eds.). Academic, New York, pp. 167-180.
Padilla, G. M., Carter, B. L. A., and Mitchison, J. M. (1975). Germination of Schizosaccharomyces pombe spores separated by zonal centrifugaion. Exp. Cell Res. 93:325-330.
Palleroni, N. J. (1961). The nutritional requirements for the germination of yeast spores. Phyton 16:117-128.
Pearson, N. J., and Haber, J. E. (1980). Changes in regulation of ribosomal protein synthesis during vegetative growth and sporulation of Saccharomyces cerevisiae. J. Bacteriol. 143:1411-1419.
Petersen, J. G. L. (1981). SDS polyacrylamide gel electrophoresis of chromatin proteins from yeast during vegetative growth and sporulation. Carlsberg Res. Commun. 46:107-119.
Petersen, J. G. L., and Piñon, R. (1979). In vivo labeling of proteins associated with folded chromosomes of yeast. Carlsberg Res. Commun. 44:395-402.
Petersen, J. G. L., Olson, L. W., and Zickler, D. (1978). Synchronous sporulation of Saccharomyces cerevisiae at high cell concentrations. Carlsberg Res. Commun. 43:241-253.
Petersen, J. G. L., Kielland-Brandt, M. C., and Nilsson-Tillgren, T. (1979). Protein patterns of yeast during sporulation. Carlsberg Res. Commun. 44:149-162.
Petes, T. D. (1980). Molecular genetics of yeast. Annu. Rev. Biochem. 49: 845-876.
Piggot, P. J. (1979). Genetic strategies for studying bacterial differentiation. Biol. Rev. 54:347-367.
Piñon, R. (1979). Folded chromosomes in meiotic yeast. J. Mol. Biol. 129: 433-447.
Piñon, R., Salts, Y., and Simchen, G. (1974). Nuclear and mitochondrial DNA synthesis during yeast sporulation. Exp. Cell Res. 83:231-238.
Pollak, J. K., and Sutton, R. (1980). The differentiation of animal mitochondria during development. Trends Biochem. Sci. 5:23-27.
Prakash, L., Hinkle, O., and Prakash, S. (1979). Decreased UV mutatgenesis in

cdc 8, a DNA replication mutant of *Saccharomyces cerevisiae*. Mol. Gen. Genet. *172*:249.

Pratje, E., Schulz, R., Schnierer, S., and Michaelis, G. (1979). Sporulation of mitochondrial respiratory deficient *mit-* mutants of *Saccharomyces cerevisiae*. Mol. Gen. Genet. *176*:411-415.

Pringle, J. R., and Hartwell, L. H. (1981). The *Saccharomyces cerevisiae* cell cycle. In *The Molecular Biology of the Yeast Saccharomyces*, J. N. Strathern, E. W. Jones, and J. R. Broach (Ed.). Cold Spring Harbor Laboratory, Cold Spring Harbor, New York, pp. 97-142.

Puglisi, P. P., and Zennaro, E. (1971). Erythromycin inhbition of sporulation in *Saccharomyces cerevisiae*. Experientia *27*:963-964.

Putrament, A., Baranowska, H., and Prazmo, W. (1973). Induction by manganese of mitochondrial antibiotic resistance mutations in yeast. Mol. Gen. Genet. *126*:357-366.

Radin, D. N. (1976). Ph.D. thesis, University of California, Berkeley.

Redshaw, P. A. (1975). Induction of petite mutations during germination and outgrowth of *Saccharomyces cerevisiae* ascospores. J. Bacteriol. *124*:1411-1416.

Reed, S. I. (1980a). The selection of *Saccharomyces cerevisiae* mutants defective in the start event of cell division. Genetics *95*:561-577.

Reed, S. I. (1980b). The selection of amber mutations in genes required for completion of start, the controlling event of the cell division cycle in *Saccharomyces cerevisiae*. Genetics *95*:579-588.

Reid, B. J., and Hartwell, L. H. (1977). Regulation of mating in the cell cycles of *Saccharomyces cerevisiae*. J. Cell Biol. *75*:355-366.

Resnick, M. A., Kasimos, J. N., Game, J. C., Braun, R. J., and Roth, R. M. (1981). Changes in DNA during meiosis in a repair-deficient mutant (*rad* 52) of yeast. Science *212*:543-545.

Rogers, D., and Bussey, H. (1978). Fidelity of conjugation in *Saccharomyces cerevisiae*. Mol. Gen. Genet. *162*:173-182.

Rogers, D. T., Saville, D., and Bussey, H. (1979). *Saccharomyces cerevisiae* killer expression mutant *kex* 2 has altered secretory proteins and glycoproteins. Biochem. Biophys. Commun. *90*:187-193.

Roth, R. (1973). Chromosome replication during meiosis: Identification of gene functions required for premeiotic DNA synthesis. Proc. Nat. Acad. Sci. USA *70*:3087-3091.

Roth, R. (1976). Temperature-sensitive yeast mutants defective in meiotic recombination and replication. Genetics *83*:675-686.

Roth, R., and Fogel, S. (1971). A system selective for yeast mutants deficient in meiotic recombination. Mol. Gen. Genet. *112*:259-305.

Roth, R., and Halvorson, H. O. (1969). Sporulation of yeast harvested during logarithmic growth. J. Bacteriol. *98*:831-832.

Roth, R., Lusnak, K. (1970). DNA synthesis during yeast sporulation: Genetic control of an early development event. Science *168*:493-494.

Rousseau, P., and Halvorson, H. O. (1969). Preparation and storage of single spores of *Saccharomyces cerevisiae*. J. Bacteriol. *100*:1426-1427.

Rousseau, P., and Halvorson, H. O. (1973a). Effect of metabolic inhibitors on germination and outgrowth of *Saccharomyces cerevisiae* ascopores. Can. J. Microbiol. *19*:1311-1318.

Rousseau, P., and Halvorson, H. O. (1973b). Macromolecular synthesis during the germination of *Saccharomyces cerevisiae* spores. J. Bacteriol. *113*: 1289-1295.

Rousseau, P., and Halvorson, H. O. (1973c). Physiological changes following the breaking of dormancy of *Saccharomyces cerevisiae* ascospores. Can. J. Microbiol. *19*:547-555.

Rousseau, P., Halvorson, H. O., Bulla, L. A., and Julian, G. S. (1972). Germination and outgrowth of single spores of *Saccharomyces cerevisiae* viewed by scanning electron and phase contrast microscopy. J. Bacteriol. *109*: 1232-1238.

Sakai, K., and Yanagishima, N. (1971). Mating reaction in *Saccharomyces cerevisiae*. I. Cell agglutination related to mating. Arch. Mikrobiol. *75*: 260-265.

Sakai, K., and Yanagishima, N. (1972). Mating reaction in *Saccharomyces cerevisiae* II. Hormonal regulation of agglutinability of *a* type cells. Arch. Mikrobiol. *84*:191-198.

Sakurai, A., Tamura, S., Yanagishima, N., and Shimoda, C. (1975). Isolation of a peptidyl factor controlling sexual agglutination in *Saccharomyces cerevisiae*. Proc. Jpn. Acad. *51*:291-294.

Sakurai, A., Tamura, S., Yanagishima, N., and Shimoda, C. (1976). Structure of the peptidyl factor inducing sexual agglutination in *Saccharomyces cerevisiae*. Agric. Biol. Chem. *40*:1057-1058.

Sakurai, A., Tamura, S., Yanagishima, N., and Shimoda, C. (1977). Structure of a peptidyl factor, α-substance I_A, inducing sexual agglutinability in *Saccharomyces cerevisiae*. Agric. Biol. Chem. *41*:395-398.

Samokhin, G. P., Lizlova, L. V., Bespalova, J. D., Titov, M. I., and Smirnov, V. N. (1979). Substitution of lys^7 by arg does not affect biological activity by α-factor, a yeast mating pheromone. FEMS Lett. *5*:435-438.

Samokhin, G. P., Lizlova, L. V., Bespalova, J. D., Titov, M. I., and Smirnov, V. N. (1980). The effect of α-factor on the rate of cell cycle initiation in *Saccharomyces cerevisiae*. Exp. Cell Res. *131*:267-275.

Sando, N., Oguchi, T., Nagano, M., and Osumi, M. (1980). Morphological changes in ascospores of *Saccharomyces cerevisiae* during aerobic and anaerobic germination. J. Gen. Appl. Microbiol. *26*:403-412.

Savarese, J. J. (1974). Germination studies on pure yeast ascospores. Can. J. Microbiol. *20*:1517-1522.

Savarese, J. J. (1975). Evidence for change in the ascospore wall during yeast germination. J. Gen. Appl. Microbiol. *21*:123-126.

Schekman, R., and Brawley, V. (1979). Localized deposition of chitin on the yeast cell surface in response to mating pheromone. Proc. Nat. Acad. Sci. USA *76*:645-649.

Scherer, G., Haag, G., and Duntze, W. (1974). Mechanism of α-factor biosynthesis in *Saccharomyces cerevisiae*. J. Bacteriol. *119*:386-393.

Schild, D., and Byers, B. (1980). Diploid spore formation and other meiotic effects of two cell-division-cycle mutants of *Saccharomyces cerevisiae*. Genetics *96*:859-876.

Schroeder, R., and Breitenbach, M. (1981a). The role of the mitochondrial genome in yeast differentiation. In *Sporulation and Germination, Proceedings of the Eighth International Spores Conference*, ASM Publications, Washington, D.C., pp. 305-308.

Schroeder, R., and Breitenbach, M. (1981b). The metabolism of *myo*-inositol during sporulation of *myo*-inositol requiring yeast. J. Bacteriol. *146*:775-783.

Seigel, J. L., and Miller, J. J. (1971). Observations on acid-fastness and respiration of germinating yeast ascospores. Can. J. Microbiol. *17*:837-845.

Sena, E. P., Radin, D. N., and Fogel, S. (1973). Synchronous mating in yeast. Proc. Nat. Acad. Sci. USA *70*:1371-1377.

Sena, E. P., Welch, J. W., Halvorson, H. O., and Fogel, S. (1975). Nuclear and mitochondrial deoxyribonucleic acid replication during mitosis in *Saccharomyces cerevisiae*. J. Bacteriol. *123*:497-504.

Sentandreu, R., and Northcote, D. H. (1969). The formation of buds in yeast. J. Gen. Microbiol. *55*:393-398.

Shapiro, B. M., and Eddy, E. M. (1980). When sperm meets egg: Biochemical mechanisms of gamete interaction. Int. Rev. Cytol. *66*:257-302.

Shilo, B., Shilo, V., and Simchen, G. (1976). Cell cycle initiation in yeast follows first order kinetics. Nature *264*:767-770.

Shilo, B., Shilo, V., and Simchen, G. (1977). Transition probability and cell cycle initiation in yeast. Nature *267*:648-649.

Shilo, V., Simchen, G., and Shilo, B. (1978). Initiation of meiosis in cell-cycle initiation mutants of *Saccharomyces cerevisiae*. Exp. Cell Res. *112*:241-248.

Shimoda, C. (1980). Differential effect of glucose and fructose on spore germination in the fission yeast, *Schizosaccharomyces pombe*. Can. J. Microbiol. *26*:741-745.

Shimoda, C., and Yanagishima, N. (1972). Mating reaction in *Saccharomyces cerevisiae* III. Changes in autolytic activity. Arch. Microbiol. *85*:310-318.

Shimoda, C., Kitano, S., and Yanagishima, N. (1975). Mating reaction in *Saccharomyces cerevisiae* VII. Effects of proteolytic enzymes on sexual agglutination ability and isolation of crude sex-specific substances responsible for sexual cell agglutination. A. V. Leeuwenhoek J. Microbiol. Serol. *41*, 513-519.

Shimoda, C., Yanagishima, N., Sakurai, A., and Tamura, S. (1978). Induction of sexual agglutinability of a mating-type cells as the primary action of the peptidyl sex factor from α mating-type cells in *Saccharomyces cerevisiae*. Plant Cell Physiol. *19*:513-517.

Simchen, G. (1974). Are mitotic functions required in meiosis? Genetics *76*:745-753.

Simchen, G. (1978). Cell cycle mutants. Annu. Rev. Genet. *12*:161-191.

Simchen, G., Piñon, R., and Salts, Y. (1972). Sporulation in *Saccharomyces cerevisiae*: Premeiotic DNA synthesis, readiness, and commitment. Exp. Cell Res. *75*:207-218.

Sloat, B. F., and Pringle, J. R. (1978). A mutant of yeast defective in cellular morphogenesis. Science *200*:1171-1173.
Sloat, B. F., Adams, A., and Pringle, J. R. (1981). Roles of the *cdc* 24 gene product in cellular morphogenesis during the *Saccharomyces cerevisiae* cell cycle. J. Cell Biol. *89*:395-405.
Srivastava, P. K., Harashima, S., and Oshima, Y. (1981). Formation of 2-spored asci by interrupted sporulation in *Saccharomyces cerevisiae*. J. Gen. Micorbiol. *123*:39-48.
Steele, S. D., and Miller, J. J. (1974). Ultrastructural changes in germinating spores of *Saccharomyces cerevisiae*. Can. J. Microbiol. *20*:923-933.
Steele, S. D., and Miller, J. J. (1977). Amino acid uptake and protein synthesis in germinating spores of *Saccharomyces cerevisiae*. Can. J. Microbiol. *23*: 407-412.
Stötzler, D., and Duntze, W. (1976). Isolation and characterization of four related peptides exhibiting α factor activity from *Saccharomyces cerevisiae*. Eur. J. Biochem. *65*:257-262.
Stötzler, D., Kiltz, H., and Duntze, W. (1976). Primary structure of α-factor peptides from *Saccharomyces cerevisiae*. Eur. J. Biochem. *69*:397-400.
Strathern, J., Hicks, J., and Herskowitz, I. (1981). Control of cell type in yeast by mating type locus. The α1-α2 hypothesis. J. Mol. Biol. *147*:357-373.
Szulmajster, J. (1979). Is sporulation a simple model for studying differentiation? Trends Biochem. Sci. *4*:18-22.
Tanaka, T., and Kita, H. (1977). Degradation of mating factor by *a*-mating type cells of *Saccharomyces cerevisiae*. J. Biochem. Tokyo *82*:1689-1693.
Tanaka, T., and Kita, H. (1978). Site of action of mating factor in *a*-mating type cell of *Saccharomyces cerevisiae*. Biochem. Biophys. Res. Commun. *83*:1319-1324.
Tanaka, T., Kita, H., Murakami, T., and Narita, K. (1977). Purification and amino acid sequence of mating factor from *Saccharmoyces cerevisiae*. J. Biochem. Tokyo *82*:1681-1687.
Taylor, N. W. (1964). Specific, soluble factor involved in sexual agglutination of the yeast *Hansenula wingei*. J. Bacteriol. *87*:863-866.
Taylor, N. W. (1965). Purification of sexual agglutination factor from the yeast *Hansenula wingei* by chromatography and gradient sedimentation. Arch. Biochem. Biophys. *111*:181-186.
Taylor, N. W., and Orton, W. L. (1971). Cooperation among the active binding sites in the sex-specific agglutinin from the yeast *Hansenula wingei*. Biochemistry *10*:2043-2049.
Thomas, P. S. (1980). Hybridization of denatured RNA and small DNA fragments transferred to nitrocellulose. Proc. Nat. Acad. Sci. USA *77*:5201-5205.
Thorner, J. (1980). Intercellular interactions of the yeast *Saccharmoyces cerevisiae*. In *The Molecular Genetics of Development* T. Leighton (Ed.). Academic, New York, pp. 119-178.
Throm, E., and Duntze, W. (1970). Mating-type-dependent inhibition of dexoyribonucleic acid synthesis in *Saccharomyces cerevisiae*. J. Bacteriol. *104*: 1388-1390.

Tingle, M., Klar, A. J. S., Henry, S. A., and Halvorson, H. O. (1973). Ascospore formation in yeast. In *Microbial Differentiation*, J. M. Ashworth and J. E. Smith (Eds.). Cambridge University, London, pp. 209-243.

Tingle, M. A., Küenzi, M. T., and Halvorson, H. O. (1974). Germination of yeast spores lacking mitochondrial deoxyribonucleic acid J. Bacteriol. *117*:89-93.

Tkacz, J. S., and MacKay, V. L. (1979). Sexual conjugation in yeast. Cell surface changes in response to the action of mating hormones. J. Cell Biol. *80*: 326-333.

Tkacz, J. S., Cybulska, E. B., and Lampen, J. O. (1971). Specific staining of wall mannan in yeast cells with fluorescein-conjugated concanavalan A. J. Bacteriol. *105*:1-5.

Tohoyama, H., Hagiya, M., Yoshida, K., and Yanagishima, N. (1979). Regulation of the production of the agglutination substance responsible for sexual agglutination in *Saccharomyces cerevisiae*. Changes associated with conjugation and temperature shift. Mol. Gen. Genet. *174*:269-280.

Trew, B. J., Friesen, J. D., and Moens, P. B. (1979). Two-dimensional protein pattern during growth and sporulation in *Saccharomyces cerevisiae*. J. Bacteriol. *138*:60-69.

Tsuboi, M. (1981). Sporulation of products of protoplast fusion without regeneration in *Saccharomyces cerevisiae*. Mol. Gen. Genet. *182*:1-7.

Tuite, M. F., and Cox, B. S. (1981). $RAD\ 6^+$ Gene of *Saccharomyces cerevisiae* codes for two mutationally separable deoxyribonucleic acid, repair functions. Mol. Cell Biol. *1*:153-157.

Udden, M. M., and Finkelstein, D. B. (1978). Reaction order of *Saccharomyces cerevisiae* α-factor-mediated cell cycle arrest and mating inhibition. J. Bacteriol. *133*:1501-1507.

Vallejo, C. G., Günther Sillero, M. A., and Marco, R. (1979). Mitochondrial maturation of the process. Cell. Mol. Biol. *25*:113-124.

Vezinhet, F., Kinnaird, J. H., and Dawes, I. W. (1979). The physiology of mutants derepressed for sporulation in *Saccharomyces cerevisiae*. J. Gen Microbiol. *115*:391-402.

Watkins, W. M. (1972). Blood-group specific substances. In *Glycoproteins, Their Composition, Structure and Function*, A. Gottschalk (Ed.). American Elsevier, New York, pp. 830-891.

Watson, D. C., and Berry, D. R. (1979). Use of an exchange filtration technique to obtain synchronous sporulation in an extended batch germination. Biotechnol. Bioeng. *21*:213-220.

Wejksnora, P. J., and Haber, J. E. (1974). Methionine dependent synthesis of ribosomal ribonucleic acid during sporulation and vegetative growth in *Saccharomyces cerevisiae*. J. Bacteriol. *120*:1344-1355.

Whelan, W. L., and Ballou, C. E. (1975). Sporulation in D-glucosamine auxotrophs of *Saccharomyces cerevisiae*: Meiosis with defective ascospore wall formation. J. Bacteriol. *124*:1545-1557.

Wickner, R. B. (1974a). Chromosomal and nonchromosomal mutations affecting the "killer character" of *Saccharomyces cerevisiae*. Genetics 76:423-432.

Wickner, R. B. (1974b). Mutants of *Saccharomyces cerevisiae* that incorporate deoxythymidine-5'-monophosphate into deoxyribonucleic acid in vivo. J. Bacteriol. *117*:252-260.
Wickner, R. B., and Leibowitz, M. J. (1976). Two chromosomal genes required for killing expression in killer strains of *Saccharomyces cerevisiae*. Genetics *82*:429-442.
Wilkinson, L. E., and Pringle, J. R. (1974). Transient G1 arrest of *Saccharomyces cerevisiae* cells of mating type α by a factor produced by cells of mating type *a*. Exp. Cell Res. *89*:175-187.
Wolf, D. H., and Ehmann, C. (1978). Carboxypeptidase S from yeast: Regulation of its activity during vegetative growth and differentiation. FEBS Lett. *91*:59-62.
Wright, J. F., and Dawes, I. W. (1979). Sporulation specific protein changes in yeast. FEBS Lett. *104*:183-186.
Yanagishima, N. (1978). Sexual cell agglutination in *Saccharomyces cerevisiae*. Sexual cell recognition and its regulation. Bot. Mag. Tokyo Spec. Issue *1*: 61-81.
Yanagishima, N., and Inaba, R. (1981). Induction of sexual agglutinability by the absorbed α peptidyl factor in *a* mating type cells of *Saccharomyces cerevisiae*. Plant Cell Physiol. *22*:317-321.
Yanagishima, N., and Yoshida, K. (1981). Sexual interactions in *Saccharomyces cerevisiae* with special reference to the regulation of sexual agglutinability. In *Sexual Interactions in Eukaryotic Microbes* D. H. O'Day and P. A. Horgen (Eds.). Academic, New York, pp. 261-295.
Yanagishima, N., Yoshida, K., Hamada, K., Hagiya, M., Kawanabe, Y., Sakurai, A., and Tamura, S. (1976). Regulation of sexual agglutinability in *Saccharomyces cerevisiae* of *a* and α types by sex-specific factors produced by their respective opposite mating types. Plant Cell Physiol. *17*:439-450.
Yanagita, T., Yagisawa, M., Oishi, S., Sando, N., and Suto, T. (1970). Sporogenic activities of mother and daughter cells in *Saccharomyces cerevisiae*. J. Gen. Appl. Microbiol. *16*:347-350.
Yen, P. H., and Ballou, C. E. (1973). Composition of a specific intercellular agglutination factor. J. Biol. Chem. *248*:8316-8318.
Yen, P. H., and Ballou, C. E. (1974). Partial characterization of the sexual agglutination factor from *Hansenula wingei* Y-2340 type 5 cells. Biochemistry *13*:2428-2437.
Yoshida, K., Hagiya, M., and Yanagishima, N. (1976). Isolation and purification of the sexual agglutination substance of mating type *a* cells in *Saccharomyces cerevisiae*. Biochem. Biophys. Res. Commun. *71*:1085-1094.
Zickler, D., and Olson, L. W. (1975). The synaptonemal complex and the spindle plaque during meiosis in yeast. Chromosoma *50*:1-23.
Zubenko, G. S., and Jones, E. W. (1981). Protein degradation, meiosis and sporulation in proteinase-deficient mutants of *Saccharomyces cerevisiae*. Genetics *97*:45-64.
Zubenko, G. S., Mitchell, A. P., and Jones, E. W. (1979). Septum formation, cell division, and sporulation in mutants of yeast deficient in proteinase B. Proc. Nat. Acad. Sci. USA *76*:2395-2399.

BIBLIOGRAPHY

The manuscript was completed in November 1981. Following is a selected list of important contributions to this field published between that time and March 1983.

Barton, J. K., Den Hollander, J. A., Hopfield, J. J., and Shulman, R. G. (1982). ^{13}C Nuclear magnetic resonance study of trehalose mobilization in yeast spores. J. Bacteriol. *151*:177-185.

Botstein, D., and Maurer, R. (1982). Genetic approaches to the analysis of microbial development. Annu. Rev. Genet. *16*:61-83.

Byers, B., and Goetsch, L. (1982). Reversible pachytene arrest of *Saccharomyces cerevisiae* at elevated temperature. Mol. Gen. Genet. *187*:47-53.

Chan, R. K., and Otte, C. A. (1982). Isolation and genetic analysis of *Saccharomyces cerevisiae* mutants supersensitive to G1 arrest by a-factor and α-factor pheromone. Mol. Cell. Biol. *2*:11-29.

Goetsch, L., and Byers, B. (1982). Meiotic cytology of *Saccharomyces cerevisiae* in protoplast lysates. Mol. Gen. Genet. *187*:54-60.

Kundu, S. C., and Moens, P. B. (1982). The ultrastructural meiotic phenotype of the radiation sensitive mutant *rad6-1* in yeast. Chromosoma *87*:125-132.

Kurjan, J., and Herskowitz, I. (1982). Structure of a yeast pheromone gene (*MFα*): a putative α-factor precursor contains four tandem copies of mature α-factor. Cell *30*:933-943.

Loumaye, E., Thorner, J., and Catt, K. J. (1982). Yeast mating pheromone activates mammalian gonadotrophs: evolutionary conservation of a reproductive hormone? Science *218*:1323-1325.

Matsumoto, K., Uno, I., Ishikawa, T. (1983). Initiation of meiosis in yeast mutants defective in adenylate cyclase and cyclic AMP-dependent protein kinase. Cell *32*:417-423.

Nasmyth, K. A. (1982). Molecular genetics of yeast mating type. Annu. Rev. Genet. *16*:439-500.

Shuster, J. R. (1982). Mating defective *ste* mutations are suppressed by cell division cycle start mutations in *Saccharomyces cerevisiae*. Mol. Cell. Biol. *2*:1052-1063.

Simchen, G., Kassir, Y., Horesh-Cabilly, O., and Friedmann, A. (1981). Elevated recombination and pairing structures during meiotic arrest in yeast of the nuclear division mutant *cdc5*. Mol. Gen. Genet. *184*:46-51.

Sprague, G. F., Jensen, R., and Herskowitz, I. (1983). Control of yeast cell type by the mating type locus: positive regulation of the α-specific *STE3* gene by the *MATα1* product. Cell *32*:409-415.

Uchida, A., Takano, A., and Suda, K. (1982). Distribution of ultraviolet light irradiated mitochondrial genomes during meiosis in yeast. Curr. Genet. *6*:99-103.

Wagstaff, J. E., Klapholz, S., and Esposito, R. E. (1982). Meiosis in haploid yeast. Proc. Nat. Acad. Sci. USA *79*:2986-2990.

Ecological Considerations

12
Phytoalexins

H. Grisebach / Biological Institute II, University of Freiburg, Freiburg, Federal Republic of Germany

INTRODUCTION

The story of phytoalexins began with the classical investigations of K. O. Müller and others (Müller and Börger, 1940; Müller, 1956) at the Biologische Reichsanstalt für Land- und Forstwissenschaft in Berlin on the resistance of potatoes (*Solanum tuberosum*) against *Phytophthora infestans*, the causative agent of potato rot ("late blight"). The essential facts uncovered by Müller's investigations were the following: (1) Inhibition of development of the fungus on tubers of resistant varieties is due to the presence of an active principle which is formed or activated only by the host-parasite interaction. (2) The plant defends itself against the parasite by sacrificing the tissue layers infected by the fungus. During this reaction, which was termed "defense necrosis," a defense substance is formed which was called "phytoalexin" from *phyton* (Greek for "plant") and *alexin* (Greek for "a warding-off compound"). (3) The phytoalexin is a chemical substance with fungitoxic properties. (4) The defense reaction is confined to the tissue colonized by the fungus and its immediate neighborhood. (5) The basic response that occurs in resistant hosts is similar to that occurring in susceptible hosts. The basis of differentiation between resistant and susceptible hosts is the speed of formation of the phytoalexin. (6) The sensitivity of the host cell that determines the speed of the host reaction is specific and genotypically determined.

It speaks for the excellence of Müller's work that after about 25 years and several hundred publications on phytoalexins these conclusions are still valid. Müller did not know the chemical nature of the potato phytoalexin(s). It was not until 1960 that a pure compound with fungitoxic properties was isolated by the drop-diffusate technique (see p. 381) from seed pods of peas (*Pisum sativum*) which were infected with spores of *Monilinia fructicola* (Cruikshank and Perrin, 1960). The compound was named pisatin. In 1962 Perrin and Bottomley (1962) showed its chemical structure to be 3,6a-dihydroxy-8,9-

methylenedioxypterocarpan (Table 1). Today the structures of about 100 phytoalexins are known and the number is still increasing.

Phytoalexins can thus be defined as substances with antimicrobial properties which are produced by the host after infection. They are contrasted to preformed internal chemical compounds (Schlösser, 1980) which are already present in sufficient concentration in noninfected tissue to prevent infections.

The role of phytoalexins in the resistance mechanism of plants is still a matter of controversy. The main question is whether phytoalexins contribute to the resistance of plants against primary infections or whether they are only effective against secondary infections (Király, 1980). The latter view is held mainly by van der Planck (1975).

The field of phytoalexins has been studied intensively in the past years and a number of recent reviews are available (Grisebach and Ebel, 1978; Gross, 1977; Kuć, 1976; Van Etten and Pueppke, 1976; Kuć et al., 1976; Deverall, 1976). This chapter is not intended to be a comprehensive review. After an overview of some important aspects of phytoalexins research, I will present in more detail a few case studies on the role of phytoalexins in plant-parasite infections, from which the reader should get an impression of the present state of research in this field.

STRUCTURE AND DISTRIBUTION OF PHYTOALEXINS

Phytoalexins are not a uniform class of substances, but belong to various groups of natural products. Phytoalexins include isoflavonoids, sesquiterpenes (Stoessl et al., 1976), diterpenes, furanoterpenoids, polyacetylenes, dihydrophenanthrenes, stilbenes, and miscellaneous other compounds. The structures of some representative examples of phytoalexins are shown in Table 1. The only criterion for classification of these substances as phytoalexins is the definition given above, namely, that they accumulate in high concentrations only after infection and that they inhibit the growth of certain microorganisms. Most plants produce several, often structurally related phytoalexins. A comprehensive survey of the nature and distribution of postinfectional defense substances in plants does not exist, but there is a clear relationship between the chemical nature of the phytoalexins and the plant families (Ingham and Harborne, 1976). The legumes generally produce isoflavonoids; the Solanaceae, diterpenes; the Compositae, acetylenes; and the Orchidaceae, dihydrophenanthrenes. Known exceptions are the formation of the furanoacetylene wyeronic acid (Table 1) in broad beans (*Vicia faba*) and the simultaneous accumulation of rishitin (Table 1) and polyacetylenes in tomato (Solanaceae) (de Wit and Kodde, 1981a). The study of phytoalexin formation can thus be used as a new dynamic approach to the chemotaxonomy of higher plants.

Phytoalexins

Table 1 Structure and Occurrence of Representative Phytoalexins

Structure	Trivial nane	Source
Isoflavonoids		
	Sativan	Clover
	Kievitone	Bush bean
Pterocarpanoids		
	Pisatin	Pea
	Phaseollin	Bush bean
Sesquiterpenoid		
	Rishitin	Potato, tomato

Table 1 (Continued)

Structure	Trivial name	Source
Diterpenes		
	Momilactone A	Rice
	Momilactone B	Rice
Polyacetylene		
$C_2H_5-CH\overset{c}{=}CH-C\equiv C-\underset{\underset{O}{\|}}{C}-\underset{O}{\text{furan}}-CH\overset{t}{=}CH-CO_2H$		
	Wyerone acid	Broad bean

INDUCTION OF PHYTOALEXIN FORMATION

General Observations

Plants are usually resistant to a large number of potential pathogens. Plant-parasite interactions are called specific if the parasite causes disease in certain plants but not in others. Graduation in specificity exists: There are parasites attacking many plant species, only one species, or only a few cultivars of a given species. It is the latter kind of specificity, namely, cultivar-specific resistance, which is most useful for investigating phytoalexin induction.

Cultivar-specific resistance is observed with many parasites. The gene-for-gene concept introduced by Flor (1942, 1971) states that for each gene conditioning avirulence in the parasite there is a corresponding gene conditioning resistance in the host plant. The gene-for-gene concept implies that specificity is associated with the incompatible (resistance) interaction. However, as will be discussed later (p. 419), specificity could at least in some cases be associated with the compatible interaction.

Defense necrosis (hypersensitivity), callose deposition, lignification, and phytoalexin accumulation are responses of the plant which are associated with the resistance reaction. Phytoalexins, however, as Müller already found, are also formed in compatible (susceptible) interactions, and it is only the degree and time course of phytoalexin accumulation which is different in compatible and incompatible interactions.

The accumulation of phytoalexins is induced not only by living microorganisms, but also by cell-free culture filtrates and compounds released from the cell walls of microbial cells (elicitors). Furthermore, different types of treatment which damage or poison plant tissue can stimulate phytoalexin formation.

Induction by Infection

Tissues from all parts of plants can produce phytoalexins in response to infection, but the age of the tissue can have a pronounced influence on the level of response (Paxton and Chamberlain, 1969; Lazarovits et al., 1981). Phytoalexin formation is induced by infection with fungi, bacteria, viruses, and nematodes. In the case of fungi a piece of mycelium can be introduced into a slash wound, for instance, in the hypocotyl. A more physiological mode of infection is the use of fungal spore suspensions which can be placed as a droplet on hypocotyls (Ward et al., 1979) or leaves (Ingham, 1978). Following inoculation part of the phytoalexin diffuses into the infection droplet and can be extracted from the combined drops with an organic solvent like ethyl acetate. Since only a fraction of phytoalexins are obtained by this method, quantitative evaluation requires additional extraction of a piece of tissue around the infection site.

Induction by Biotic Elicitors

Compounds of microbial origin which stimulate phytoalexin synthesis in plants have been named elicitors (Keen et al., 1971). The term "biotic elicitor" for such compounds and "abiotic elicitors" for compounds like mercuric chloride are also used (Yoshikawa, 1978). Elicitors which have been studied in more detail are listed in Table 2. Depending on the source and purity of the elicitor preparation, its capacity to induce phytoalexin accumulation can be equal or even higher than that of the organisms from which it is isolated. For example, an elicitor from *Cladosporium fulvum* (see Table 2) induced 10-50 times as much rishitin in tomato fruit tissue as a live conidial suspension of this fungus

Table 2 Source and Chemical Nature of Purified Elicitors

Source	Chemical nature	Phytoalexin assay system	Leading references
Phytophthora megasperma f. sp. *glycinea* Culture filtrate and cell wall	Branched β-glucan with predominantly 3- and 3,6-linked glucosyl residues	Glyceollins in *Glycine max*	Albersheim and Valent (1978)
Colletotrichum lindemuthianum Culture filtrate and cell wall	Glucan with predominantly 3- and 4-linked glucosyl residues	Phaseollin in *Phaseolus vulgaris*	Anderson-Prouty and Albersheim (1975) Anderson (1978)
Phytophthora infestans	Glucan	Rishitin in *Solanum*	Lisker and Kuć (1977)
Phytophthora cinnamoni Cell wall and mycelia			
P. infestans Mycelia	Eicosapentanoic acid, arachidonic acid	Rishitin, lubimin in *Solanum tuberosum*	Bostock et al. (1981)
Saccharomyces cerevisiae Yeast extract	Glucan	Glyceollin in *G. max* Rishitin in *S. tuberosum*	Albersheim and Valent (1978), Hahn and Albersheim (1978)
Rhizopus stolonifer	Polygalacturonase	Casbene synthetase in *Ricinus communis*	Stekoll and West (1978), Lee and West (1981a,b)
Cladosporium fulvum Culture filtrate, mycelia cell walls	Glycoprotein, peptidogalactoglucomannan	Rishitin in *Solanum lycopersicum*	De Wit and Kodde (1981b)
Fusarium solani f. sp. *phaseoli* Cell walls	Chitosan (polymer of β-1,4-linked glucosamine)	Pisatin in *Pisum sativum*	Hadwiger and Beckmann (1980)

containing 5×10^6 conidia per milliliter (de Wit and Roseboom, 1980; see also p. 409).

Most investigators have reported that elicitors are neither race- nor cultivar-specific with respect to phytoalexin accumulation. This means that phytoalexin accumulation is also induced in nonhost plants and that elicitor preparations from an incompatible race have the same activity as elicitors from a compatible race. An exception are the reports of Keen et al. on race-specific elicitors from *Phytophthora meagasperma* f. sp. *glycinea* (Keen, 1975) and *Pseudomonas glycinea* (Bruegger and Keen, 1979). Total solubilization of cell envelopes of *P. glycinea*, the causal agent of bacterial blight of soybeans, was achieved by sodium dodecylsulfate treatment. Extraction of lipids from these solubilized fractions yielded water-soluble preparations that contained both protein and carbohydrate. It was claimed that these preparations had the same elicitor specificity as different races of the live bacteria in the elicitation of glyceollin accumulation in soybean cultivars Harosoy and Acme. These results must be interpreted with caution because of the impurity of the preparations, the small differences found between preparations from differential bacterial races, and the analytical method used for glyceollin determination (which is subject to large errors. A later, more extended study on the glycoproteins of *P. megasperma* f. sp. *glycinea* also did not show conclusively that glycoproteins could act as race-specific elicitors (Keen and Legrand, 1980).

Wade and Albersheim (1979) have isolated extracellular glycoprotein preparations (ECGPs) from incompatible and compatible races of *P. megasperma* f. sp. *glycinea*. Partially purified ECGPs from the incompatible races, but not those from the compatible races, could protect soybean seedlings from attack by compatible races. However, the ECGPs themselves were very poor elicitors.

Further work with highly purified compounds is necessary to solve the problem of the "specificity factors." Also, no homogeneous elicitor is yet available to study the question of the interaction site of the elicitor with the host plant.

A new aspect of elicitor research was initiated by the work of Hargreaves and Bailey (1978) on phytoalexin production by hypocotyls of *Phaseolus vulgaris*. These authors found that phaseollin (Table 1) and phaseollidin accumulation occurred in hypocotyl sections adjacent to parts of the same hypocotyl which were kept at -20°C for 10 min and then incubated at 25°C. Accumulation also occurred when dead and live bean tissues were incubated together or when living tissue was treated with an aqueous extract of hypocotyl tissue. Phytoalexins were produced by live cells and seemed to be accumulated in dead cells. On the basis of these results, it was proposed that during the hypersensitivity reaction a preformed plant metabolite is released which stimulates phytoalexin synthesis. Such an *endogenous elicitor** could, for example, be released from the plant cell wall (see below). A study on phytoalexin accumulation in chloroform (Bailey and Berthier, 1981) or triton-treated cotyledons of *P. vulagris* (Hargreaves, 1981) also supports the suggestion that accumulation of phytoalexins can be a direct consequence of the damage of cells. Release of endogenous plant elicitors could also explain phytoalexin formation in bean tissues infected with viruses (Bailey and Ingham, 1971) or bacteria or exposed to ultraviolet radiation or salts of heavy metals (see p. 386).

An endogenous elicitor from soybean hypocotyls which elictis phytoalexin accumulation in soybean tissue appears to be a fragment of the pectic polysaccharide (Hahn et al., 1981). The partially purified elicitor preparation contained 96% galacturonyl, 2.4% rhamnosyl, and 1.7% xylosyl residues. Preparations with good elicitor activity were also obtained by partial acid hydrolysis from cell walls of tobacco, sycamore, and wheat.

If fragments of the plant cell wall function as endogenous elicitors, the question arises as to how these elicitors are released from the wall. One possibility is suggested by the recent discoveries that enzymes degrading pectic polysaccharide are phytoalexin elicitors. The elicitor from *Rhizopus stolonifer* (Table 2) has been identified as an endopolygalacturonase (Lee and West 1981a,b). The polygalacturonase elicitor was purified to apparent homogeneity by a series of gel filtration and ion-exchange chromatography steps, and it was shown that polygalacturonase activity and elicitor activity for casbene synthase in castor bean (*Ricins communis* L.) seedlings copurify in all steps. Examination of the types of products formed by the action of the enzyme suggests that it is an endohydrolase. The purified polygalacturonase elicitor is a glycoprotein with approximately 20% carbohydrate composed of 92% mannose and 8% glucosamine.

The soft rot-causing bacterium *Erwinia carotovora* produces a polygalacturonic acid-lyase when grown on pectin as the sole carbon source (G. Lyon and P. Albersheim, unpublished). This enzyme can function as an elicitor of phytoalexin accumulation in soybean tissue.

An explanation for this finding would be that the pectin-degrading enzymes secreted by these pathogens release the endogenous elicitor from the plant cell wall. Alternatively, a plant polysaccharide degradative enzyme could be activated by the pathogens and then cause solubilization of the endogenous elicitor from the walls surrounding these cells. These reactions are outlined in Scheme 1.

---

*Hahn (1981) has pointed out that the term "constitutive elicitor" used by Hargreaves et al. implies that the elicitor is present and active in the plant at all times. Since the elicitor is very probably not present in healthy untreated tissue, the term "endogenous elicitor" was proposed and is used in this article.

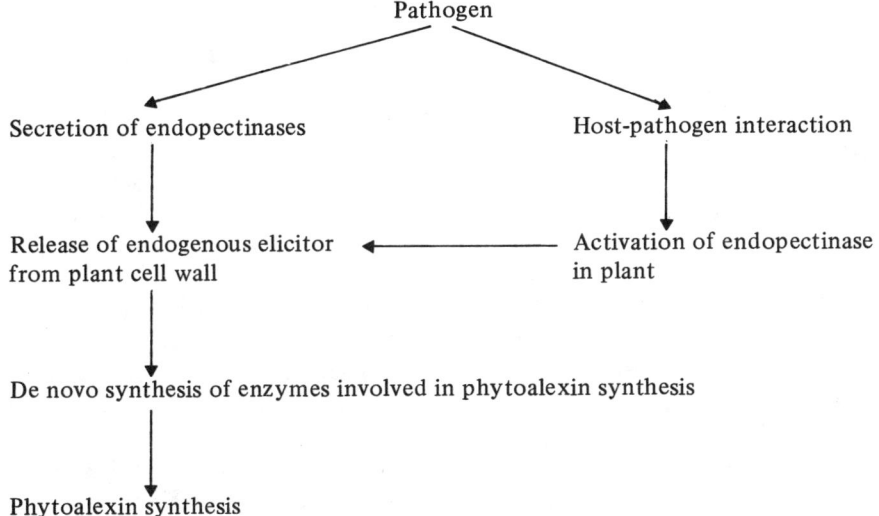

Scheme 1 Hypothetical reaction sequences for the induction of phytoalexin synthesis via an endogenous elicitor

Recently Yoshikawa et al. (1981) reported on the release of a soluble elictior of glyceollin accumulation from insoluble mycelial walls of *P. megasperma* f. sp. *glycinea* after incubations of the walls with soybean cotyledon tissue for only 2 min. Cell-free extracts from soybean cotyledons or hypocotyls also released soluble elicitors from the fungal cell walls. This observation is in agreement with results of a detailed study on the properties of β-glucosylases from soybean cell walls which were reported to hydrolyze the glucan elicitor from *P. megasperma* f. sp. *glycinea* (Cline and Albersheim, 1981a,b). While Albersheim et al., assumed that a possible function of these enzymes could be their ability to reduce the size of glucan elicitors and thereby expedite passage of elicitors through the cell wall, Yoshikawa et al. postulated a direct release of the fungal elicitor by the plant enzymes. These suggestions are still highly speculative.

Induction of phytoalexin formation by the plant hormone ethylene was also reported (Chalutz and Stahmann, 1969). Infection of bean leaves with *Uromyces phaseoli* leads to a cultivar-specific ethylene production which can also be induced by elicitor preparations from this fungus (Paradies et al., 1979). However, in further investigations it was shown that ethylene formation in soybean cotyledons can be influenced independently of phytoalexin production (Paradies and Elstner, 1980). Ethylene therefore seems to be an indicator of elicitor recognition, but it does not function as a signal between the initial interaction and phytoalexin accumulation.

Induction by Abiotic Elicitors and Stress Conditions

A whole range of chemicals like heavy-metal salts, polyamines, ribonuclease, antibiotics, and metabolic inhibitors can induce phytoalexin synthesis. Furthermore, stress conditions like cold injury (see p. 383) and ultraviolet (UV) light can trigger phytoalexin accumulation. Responses to wounds are less clear and can probably in some cases be attributed to contamination by bacteria. A special situation does exist in cut storage tissue like sweet potato root tissue. Uritani and others have compared the effects of wounds and infection on such tissues in detail (Uritani and Ôba, 1978).

Yoshikawa (1978) claimed that the primary mechanisms of action of biotic and abiotic elicitors are distinctly different. According to his results, biotic elicitors stimulate synthesis of phytoalexins but have no effect on degradation, whereas abiotic elicitors have only a slight effect on synthetic activity but strongly inhibit degradation. Since the steady-state concentration of a given phytoalexin is determined by its rates of synthesis and degradation, the final effect would be the accumulation of phytoalexin with either type of elicitor. The results of Yoshikawa are in contradiction to those from our experiments with soybean, in which we found that both the glucan elicitor from *P. megasperma* f. sp. *glycinea* and mercuric chloride stimulate mainly the rate of glyceollin synthesis (Moesta and Grisebach, 1980, 1981a). This work will be described in more detail on p. 406. Our findings are in agreement with the assumption that mercuric chloride causes the release in the plant of endogenous phytoalexin elicitors (Hargreaves and Bailey, 1978; Hargreaves, 1979).

BIOSYNTHESIS AND METABOLISM OF PHYTOALEXINS

For studies on the molecular mechanism of phytoalexin induction it is necessary to know the biosynthetic pathways to phytoalexins and the enzymes involved in them. Most studies which have been carried out on the biosynthesis of phytoalexins were in vivo investigations with labeled precursors. Knowledge about enzymes participating in phytoalexin biosynthesis is still meager.

Isoflavonoids

Biosynthesis

The majority of the phytoalexins belong to the group of isoflavonoids which includes the pterocarpanoids. Comprehensive reviews on the biosynthesis of flavonoids are available (Hahlbrock and Grisebach, 1975; Wong, 1976) and only a short outline is presented here.

Cinnamoyl-coenzyme A (cinnamoyl-CoA) thioesters and malonyl-CoA are precursors for all flavonoids. 4-Coumaroyl-CoA is formed from L-phenylalanine

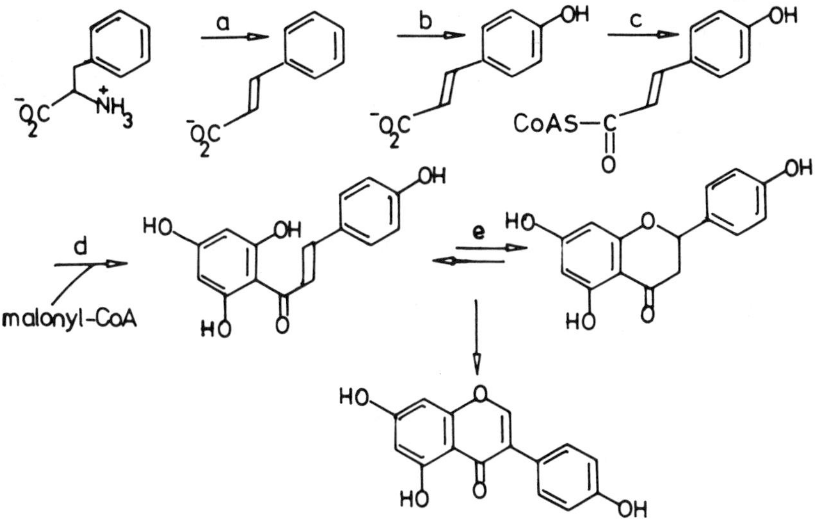

Figure 1 Biosynthetic pathway to isoflavones. Enzymes: (a) phenylalanine ammonia-lyase, (b) cinnamate-4-hydroxylase, (c) 4-courmarate:CoA ligase, (d) chalcone synthase, and (e) chalcone isomerase.

by the three reactions shown in Fig. 1, which are catalyzed by the enzymes phenylalanine ammonia-lyase, cinnamate 4-hydroxylase, and 4-courmarate: CoA ligase. Malonyl-CoA is supplied by the acetyl-CoA carboxylase reaction (Egin-Bühler et al., 1980). 4,2',4',6'-Tetrahydroxychalcone (naringenin chalcone) is the central intermediate for all flavonoids with a metadioxygen pattern in ring A, and 4,4',6'-trihydroxychalcone is the intermediate for all flavonoids with only a 7-oxygen substituent in this ring (flavonoid numbering). The formation of naringenin chalcone from one molecule of 4-coumaroyl-CoA and three molecules of malonyl-CoA is catalyzed by the enzyme chalcone synthase (Heller and Hahlbrock, 1980). The corresponding synthase for the formation of trihydroxychalcone is unknown. Isoflavonoids arise by a 1,2-aryl shift from the chalcone or the isomeric flavanone. The latter compound is formed by action of the enzyme chalcone isomerase.

Details concerning the biosynthesis of the pterocarpan phytoalexins in red clover (*Trifolium pratense*) are known. After fungal infection (Higgins and Smith, 1972) or in the presence of heavy-metal ions (e.g., Cu^{2+}) synthesis of medicarpin [Dewick, 1975; Fig. 2 (6)] and maackiain is induced in the seedlings. Incorporation experiments with ^{14}C-labeled precursors showed that 7-hydroxyl-

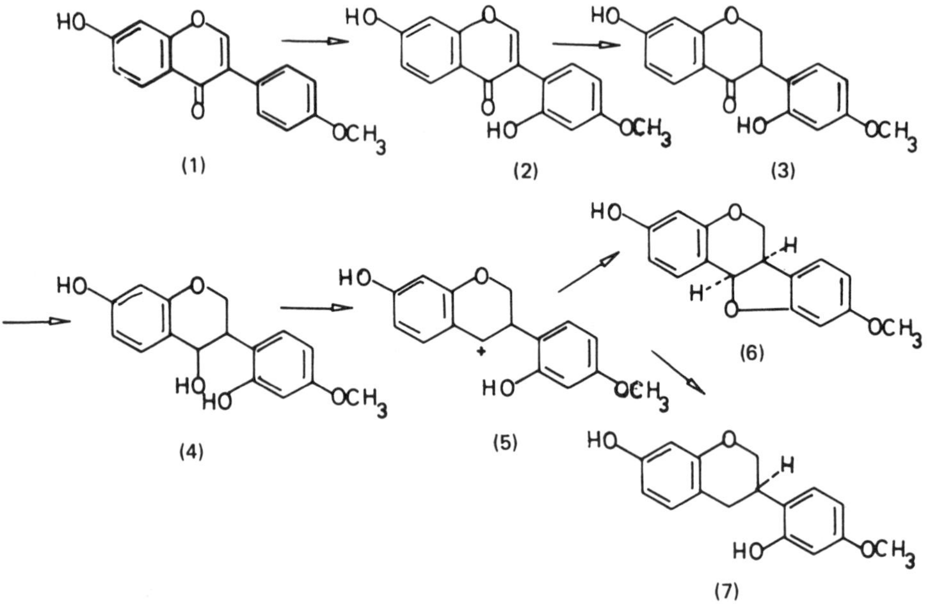

Figure 2 Biosynthetic pathway for medicarpin (6).

4′-methoxyisoflavone (formononetin) (1), the isoflavone (2), and the isoflavanone (3) were very good precursors for (6) but not for vestitol (7). Compounds (2) and (3) showed higher ^{14}C incorporation than compound (1) (Dewick, 1977). On the basis of these findings, the biosynthetic pathway for medicarpin (6) shown in Fig. 2 has been postulated. According to this, formononetin (1) is first hydroxylated to (2). The 2′-hydroxylation of isoflavones has been demonstrated in investigations on the biosynthesis of coumestrol in *Phaseolus aureus* (Dewick et al., 1970). The isoflavone could then be reduced to the isoflavanone (3), which would, on further reduction, give the isoflavanol (4). Loss of the hydroxyl group could lead to the carbenium ion (5), which would be converted into medicarpin (6) by cyclization and release of a proton. Vestitol (7) could be produced by addition of a hydride ion to the carbenium ion. The existence of the intermediate carbenium ion is strongly supported by the interconversion of (6) and (7) in lucerne (*Medicago sativa*) (Dewick and Martin, 1976). Further support for the above reaction sequence is the isolation of formononetin, 2′-hydroxyformononetin (2), and vestitone (7,2′-dihydroxy-4′-methoxyisoflavanone) (3) from clover leaves inoculated with a spore suspension of *M. fructicola* (Woodward, 1981).

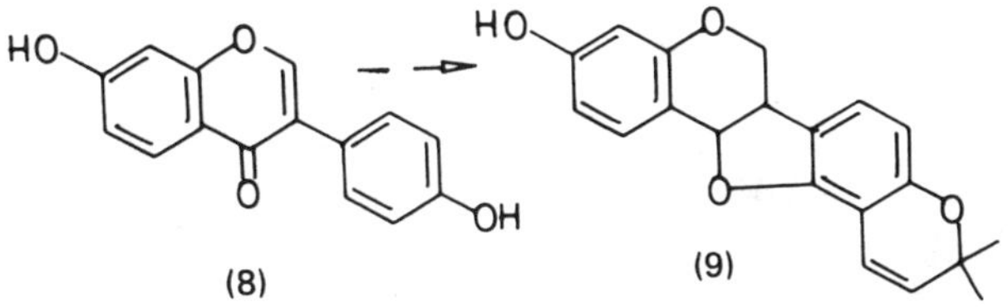

Figure 3 Conversion of daidzein (8) to phaseollin (9).

The incorporation of [^{3}H] daidzein (8) into phaseollin (9) which takes place in bean pods (Hess et al., 1971) is in agreement with the biosynthetic pathway for medicarpin (Fig. 3).

In accordance with the phytoalexin theory, it could be shown in experiments with beans that the precursors of phaseollin originate only from the host plant and not from the parasite. These experiments were performed with [^{14}C] glucose-labeled spores of *Sclerotinia fructicola* and with [^{14}C] glucose-labeled host tissue (van den Ende and Müller, 1964).

The relationship between phytoalexin induction and the increase in the activity of enzymes which are involved in their synthesis will be discussed later.

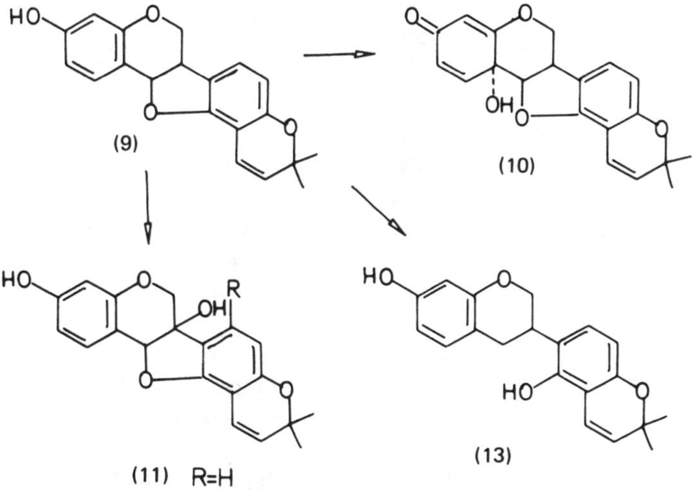

(11) R=H
(12) R=OH

Figure 4 Metabolites of phaseollin (9) by phytopathogenic fungi.

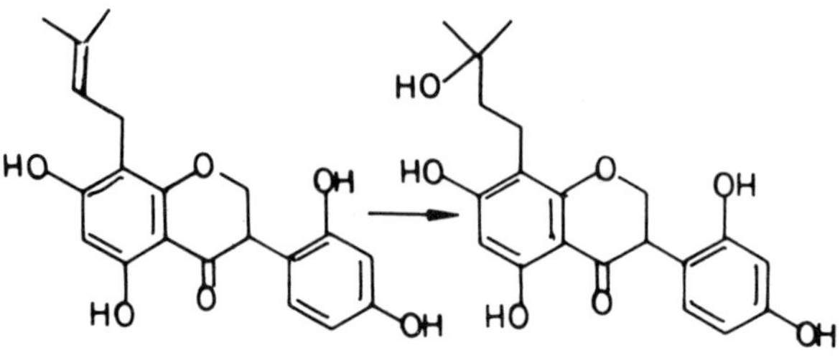

Figure 5 Conversion of kievitone (14) to the hydrated derivative (15).

Metabolism

Phytoalexins can be modified and probably also catabolized by fungi. These reactions may be a defense reaction of the fungi against the phytoalexins, since some of these metabolites have a lower fungitoxicity than the parent phytoalexins. Phaseollin (9) is oxidized to the more weakly fungitoxic 1a-hydroxyphaseollon (10) by the fungus *Fusarium solani,* which is pathogenic for beans (van den Heuvel et al., 1974; Van Etten and Smith, 1975) (Fig. 4). Hydroxylation of (9) to the products (11) and (12) (Burden et al., 1974) and the reductive opening of the benzyl phenyl ether bond to give phaseollin isoflavane (13) (Higgins et al., 1974) are other reactions which are carried out by phytopathogenic fungi (for other literature concerning these reaction see Gross, 1977; Van Etten and Pueppke, 1976; Fuchs et al., 1980). The modified products shown in Fig. 4 can probably be catabolized by the fungi to simple substances; the details of these processes, however, still have to be elucidated.

More metabolic products of the pterocarpanes are known (Gross, 1977; Van Etten and Pueppke, 1976; Ingham, 1976; Bell, 1981). It has recently been demonstrated that kievitone (14) is converted into product (15) by *F. solani* in a reaction which formally corresponds to the addition of water (Fig. 5) (Kuhn et al., 1977).

A similar reaction was found for detoxification of phaseollidin (Smith et al., 1980). A partially purified enzyme was obtained from *F. solani* f. sp. *phaseoli,* which catalyzes the hydration of kievitone and phaseollidin (Kuhn and Smith, 1979). This hydratase is apparently an extracellular enzyme.

The O-demethylation of pisatin (Table 1) has been observed with strains of *Ascochyta stemphylium* and *Fusarium* (Van Etten and Pueppke, 1976; Barz, 1978; Fuchs et al., 1980). A comparison of the fungitoxic properties of pisatin with its demethylation product shown in Table 3 demonstrate that the latter is a detoxification product. Cleavage of the O-methyl ether group is frequently the first reaction in a degradative pathway (Barz, 1978; Lappe and Barz, 1978).

Table 3 Inhibition of Fungal Growth by Pisatin and its Demethylation Product

Fungus	Percentage of inhibition[a]	
	Pisatin	3-O-Demethylpisatin
Fusarium solani sp. *phaseoli*	67	6
Helminthosporium turcicum	75	35
Neurospora crassa	78	28
Penicillium expansum	36	6
Rhizopus stolonifer	74	6
Stemphylium botyrosus	66	3

[a]Determined by inhibition of radial mycelial growth on solid medium containing 3×10^{-4} M of the respective compound.
Source: Van Etten and Pueppke (1976).

The metabolism of phytoalexins is not limited to the phytopathogenic fungi. *Septoria nodorum* and other fungi which are not pathogenic toward bean plants can convert phaseollin (9) into the much less toxic cis- and trans-12,13-dihydrodihydroxyphaseollin (16) (Fig. 6) (Bailey et al., 1977). Accordingly, the distinction between pathogenic and nonpathogenic fungi, in these cases at least, is not connected with their ability to catabolize phytoalexins.

Because of this metabolic capacity of fungi, the question has been raised as to whether all compounds isolated from infected plant material are of plant origin (Barz, 1978). An experimental approach to this question is a comparison with the compounds produced by induction with UV light or elicitors.

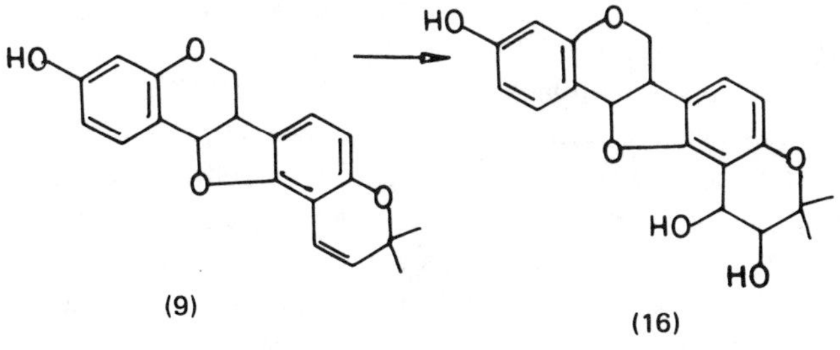

Figure 6 Conversion of phaseollin (9) to cis- or trans-12,14-dihydrodihydroxyphaseollin (16).

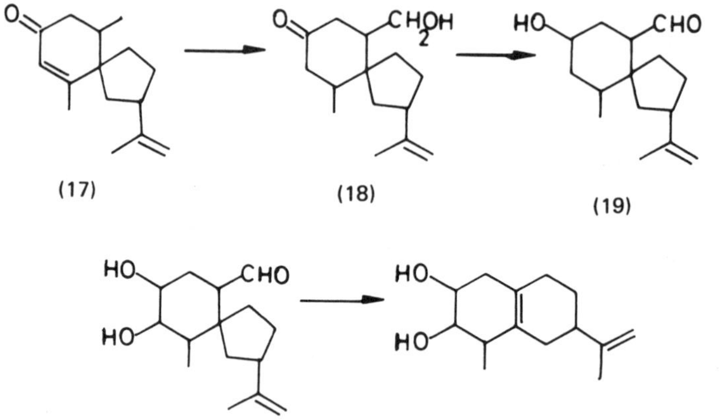

Figure 7 Postulated biosynthesis of rishitin (21) from the spirovetivadienone (17). Lubimin (19) and 4-hydroxylubimin (20) are possible intermediates. On conversion of (20) to rishitin, C-15(—CHO) is lost.

Sesquiterpenes

In this section only the experiments on biosynthesis of rishitin and furanoterpenoids will be discussed. Some other examples can be found in the review of Grisebach and Ebel (1978).

Several reports on the biosynthesis of rishitin in potatoes have appeared (Gross, 1977; Kuć et al., 1976; Grisebach and Ebel, 1978). The biosynthetic pathway shown in Fig. 7 was postulated on the basis of these results. The vetispira compounds isolubimim, lubimin (19), and 4-hydroxylubimin (20) have been proposed as intermediates and are also found to occur as phytoalexins in infected potatoes.

On comparison of the incorporation of ^{14}C-labeled acetate and mevalonate into rishitin incorporation up to 5.9-fold higher was found in resistant potato varieties as compared to susceptible ones (Shih and Kuć, 1973).

The changes in the activities of enzymes which are involved in the biosynthesis of isopentenyl pyrophosphate were investigated after infection of the root tissue of sweet potato (*Ipomoea batatas*) by spores of the fungus *Ceratocystis fimbriata*. Figure 8 shows the increase in activity of 3-hydroxy-3-methylglutaryl-coenzyme-A reductase (HMG-CoA reductase) after infection. This increase is followed by an increase in furanoterpenoids such as ipomeamarone (Suzuki et al., 1975). In root tissue which is damaged but not infected, however, neither

Phytoalexins

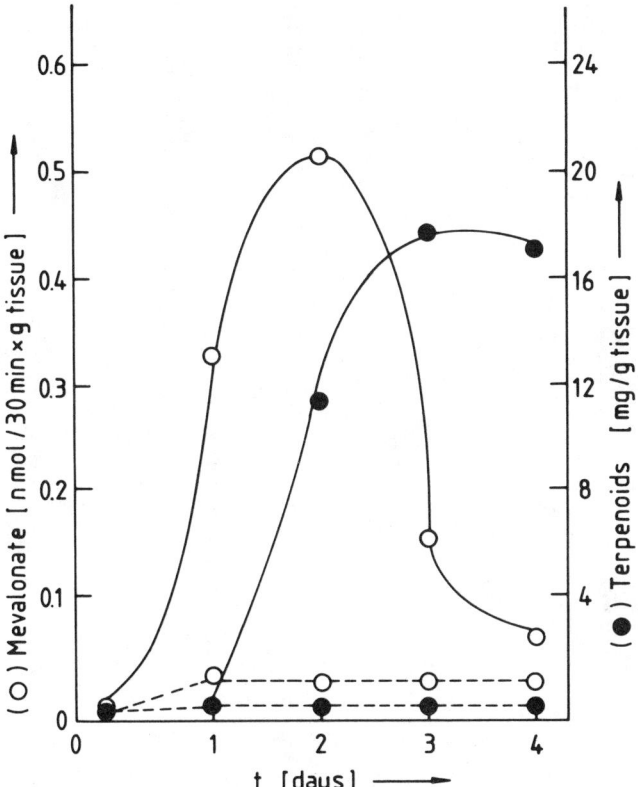

Figure 8 Changes in the activity of HMG-CoA reductase (o ——— o) and terpene content (• ——— •) after infection of the root tissue of sweet potato by spores of *Ceratocystis fimbriata*. The dashed curves are controls with wounded noninfected tissue.

the enzyme activity nor the furanoterpenoid content increases. These results are thus similar to those during induction of glyceollin formation in the soybean (see p. 405).

It is not known whether HMG-CoA reductase is the rate-determining enzyme in furanoterpenoid synthesis as is the case in the biosynthesis of cholesterol in rat liver. Increases in the activity of other enzymes of the pathway from mevalonate to isopentenylpyrophosphate (e.g., pyrophosphomevalonate decarboxylase) have been observed after infection of sweet potato (Oba et al., 1976).

Table 4 Antifungal activities of stress metabolites from *Vitis* spp. in various assay systems[a]

	TLC plate bioassay[b]	Spore germination assay[c]			*Plasmopara viticola* zoospores release/motility[d]	Mycelial growth on agar[e]			Germ tube growth assay[f]	
	C.c.	C.c.	B.c.	P.o.		B.c.	P.o.	P.s.	B.c.	P.o.
Pterostilbene	0.5	9	18	9	4.5/2.3	27.5	3	6.3	12.5	<6
Resveratrol	>60	>200	>200	>200	>200/>200	100	63	—	158 (331)	—
ε-Viniferin	2.5	37	100	—	19/12.5	100	54.4	19.3	79.4 (126)	—
α-Viniferin	1	47	49	28	35/11	—	—	—	72.4 (155)	—

[a] C.c. = *Cladosporium cucumerinum*; B.c. = *Botrytis cinerea*; P.o. = *Piricularia oryzae*; P.s. = *Pellicularia sasakii*
[b] Minimum amount (μg) detectable.
[c] Concentrations (μg/ml) causing 50% inhibition of spore germination.
[d] Concentrations (μg/ml) causing 50% inhibition of the release of zoospores from sporangia of *Plasmopara viticola* or of the motility of the zoospores after their release.
[e] Concentration (μg/ml) causing 50% inhibition of radial growth on agar media.
[f] Concentration (μg/ml) causing 50% inhibition of hyphal growth of sporelings (ED$_{90}$ values in parentheses).
Source: Langcake (1981).

BIOLOGICAL ACTION SPECTRUM OF PHYTOALEXINS

Bioassays

The common bioassays make use of the fungitoxic action of phytoalexins on spore germination, germ tube growth, radial mycelial growth on a solid nutrient medium (Pierre and Bateman, 1967), or the growth of fungal mycelium in liquid culture.

In the frequently used thin-layer chromatography (TLC) fungitoxicity assay the thin-layer plate is sprayed with a conidial suspension of *C. fulvum* or *Cladosporium cucumerinum* and the inhibition zones are observed after 2-3 days of incubation (Klarman and Stanford, 1968; Keen et al., 1971).

Germination as well as germ tube and mycelial growth can be observed on microscopic slides which contain conidia of the test organisms in a suitable medium and a concentration series of the respective phytoalexin (de Wit and Kodde, 1981a).

Since fungitoxic properties measured in vitro can vary depending on the assay conditions (Bailey, 1974), it is difficult to draw conclusions from these results that are applicable to the in vivo activity of phytoalexins.

It is advisable to use several different assay methods in parallel, as is shown in Table 4 for the case of phytoalexins from *Vitis* spp., in which five different assays and four different test fungi were used (Langcake, 1981).

Furthermore, the establishment of a dose-response curve for inhibitory activity increases the significance of a given assay. Such dose-response curves for the inhibitory activity of momilactones A and B toward germ tube growth of *Piricularia oryzae* sporelings (Cartwright et al., 1980) are shown in Fig. 9.

Antifungal Activity

Phytoalexins inhibit a large number of pathogenic and nonpathogenic fungi in the above-mentioned assays in concentrations of $10^{-5} - 5 \times 10^{-4}$ mol/liter (references see Grisebach and Ebel, 1978).

Antibacterial Activity

The antibacterial properties of phytoalexins have not been as widely investigated as their antifungal activities. Wyman and Van Etten (1978) assayed the effect of six isoflavonoids on bacteria in the semisolid medium bioassay. Phaseollin, pisatin, phaseollinisoflavan, and kievitone possessed some antibacterial properties, whereas coumestrol and formononetin lacked significant activity. No clear correlation between pathogenicity in leguminous species and tolerance of the bacteria to the isoflavonoids was evident. Although some isolates were very sensitive to several of the phytoalexins produced by the plant species with which they

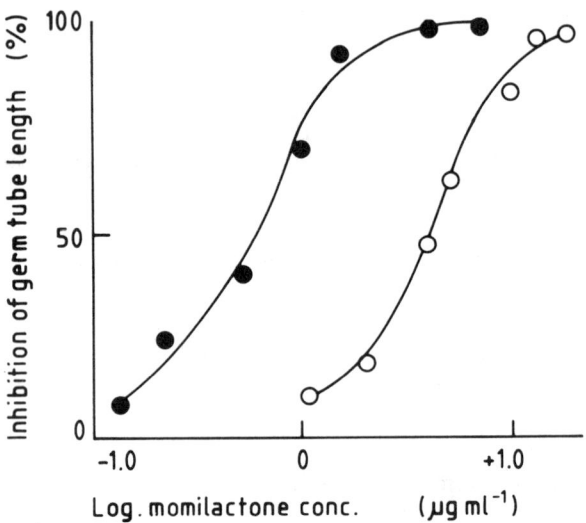

Figure 9 Dose-response curves for the inhibitory activity of momilactone A (o) and B (●) toward the germ tube growth of *P. oryzae* sporelings.

are normally associated, it is not known whether they are also inhibited by these compounds in vivo.

A selective toxicity of isoflavonoid phytoalexins for gram-positive bacteria was reported (Gnanamanickam and Smith, 1980). A total of 10-50 μg of kievitone or phaseollin inhibited the growth of all seven gram-positive but none of the eight gram-negative bacteria tested in a paper disk bioassay. Among the bacteria inhibited were *Corynebacterium fascians*, *Bacillus subtilis*, *Micrococcus luteus*, and *Streptomyces griseus*. In further studies (Gnanamanickam and Mansfield, 1981) it was found that wyerone, rishitin, and some hydroxyflavanes also have a selective toxicity for gram-positive bacteria.

Phytotoxicity

That phytoalexins have a toxic effect on plant cells has been shown in several studies. Cell suspension cultures of *P. vulgaris* and tobacco are sensitive to phaseollin (Glazener and Van Etten, 1978; Skipp et al., 1977). Evidence was also presented that exogenously added phaseollin was metabolized by the plant cells.

Studies on the mechanism of the phytotoxicity indicated that the first effects of phaseollin on beetroot protoplasts and cultured bean cells are associated with the tonoplast (Hargreaves, 1980). Disruption of the tonoplast membrane could lead to the release of harmful compounds of hydrolytic enzymes into the cytoplasm.

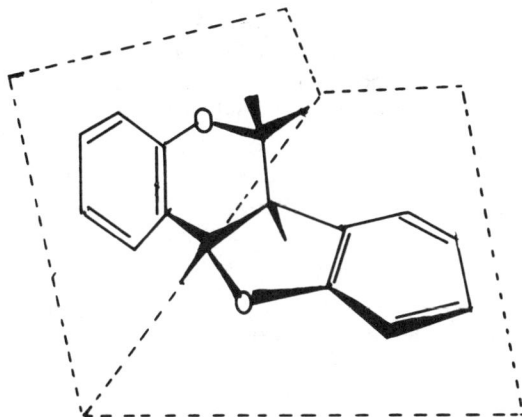

Figure 10 Conformation of 6aR, 11aR pterocarpans. The cis configuration of the 6a-11a junction has been established. If a 6a-hydroxyl group is present like in glyceollin, the sequence rule is reversed and the structure is designated as 6aS,11aS.

According to these results, necrosis of plant cells, which is always associated with the defense reaction, could occur as a consequence of the accumulation of phytoalexins.

Mode of Action of Phytoalexins

The exact mechanism by which phytoalexins inhibit the growth of microorganisms or kill them is unknown.

An electron microscopic study of the effect of phaseolin on zoospores of the soil fungus *Aphanomyces euteiches* showed that 10^{-4} mol/liter phaseollin caused rapid and marked degenerative changes in encysted and nonencysted zoospores (H. C. Hoch, H. D. van Etten, and P. S. Matthews, unpublished data). An effect on membranes has also been found in other investigations (Van Etten and Pueppke, 1976; Smith, 1976).

Since the concentration of glyceollin needed to inhibit mitochondrial respiration in soybean mitochondria was found to be 10-100 times smaller than the concentrations of isoflavonoid phytoalexins required to affect whole cells and inhibit fungal growth, it was suggested that inhibition of mitochondrial respiration could be an important aspect in the mode of action of phytoalexins against plant pathogens (Kaplan et al., 1980). This aspect needs further investigation.

Structure-activity relationship of pterocarpans have been investigated. The conformation of the pterocarpan molecule is one in which the two aromatic rings are almost perpendicular to each other (Perrin and Cruickshank, 1969) (Fig. 10). Mainly on the basis of the observation that 6a,11a-dehydropterocarpans (6a,11a-dehydropisatin and coumestans), which have an almost planar structure, lack antifungal activity, it was proposed that antifungal activity of the pterocarpans may be associated with two factors: (1) the aromatic rings A and B do not lie in the same plane and (2) small oxygen-containing substituents are present at the periphery of the molecule (Perrin and Cruickshank, 1969). However, in a later study it was found that two 6a,11a-dehydropterocarpans which were more water soluble than dehydropisatin exhibited antifungal activity (Van Etten, 1976). The hypothesis that the three-dimensional shape is a critical factor for antifungal activity of pterocarpans is therefore no longer tenable.

INVESTIGATIONS OF PHYTOALEXINS IN SELECTED PLANT-PARASITE INTERACTIONS

To investigate the importance of any phytoalexin in disease resistance it is necessary to relate its toxicity in vitro and the levels, sites, and time at which it accumulates in infected tissue and to relate the concentration of phytoalexin produced to the time at which fungal growth is inhibited in vivo. This combined approach has been followed only in a few cases. Besides investigating the role of phytoalexins in resistance, studies on the mechanism of phytoalexin accumulation caused by infection or by the formation of elicitors are directed toward answering the question of how the plant-parasite interaction ultimately leads to such an accumulation.

Soybean-*Phytophthora megasperma* f. sp. *glycinea*

General Observations

The soybean [*Glycine max* (L.) *Merrill*], a food and feed crop of great importance, can be parasitized by over 50 different fungi. A very destructive disease in susceptible soybean cultivars is *Phytophthora* rot caused by *P. megasperma* f. sp. *glycinea* (Kuan and Erwin, 1980) formerly *P. megasperma* var. *sojae*). *Phytophthora megasperma* exists in different races which are distinguishable only by their host-parasite interactions. Resistance of soybean cultivars to races 1 and 2 is believed to be controlled by a single dominant gene pair designated RPs. Additional loci could be involved in resistance to other races.* A number

---

Compendium of Soybean Diseases, American Phytopathological Society, St. Paul, Minn., 1975.

Table 5 Responses of Various Soybean Cultivars Toward *Phythophthora megasperma* f. sp. *glycinea*

Soybean	Races of *P. megasperma* f. sp. *glycinea*[a]						
	1	2	3	5	6	12	16
Harosoy	S	S	S				
Harosoy 63	R	R	S				
Wayne	S	S	S				
Sanga	R	S	R				
Amsoy 71	R	R	S				
Altona	R	R	R	S	S		
Williams	S	S	S			R	R
Williams 79	R	R	R		R		
Union	R	R	S				

[a]R = resistant, S = susceptible

of soybean cultivars and their reactions to different races of *P. megasperma* are listed in Table 5. A typical virulence assay (Wade and Albersheim, 1979) is shown in Fig. 11.

Soybean Phytoalexins

Glyceollin I [Fig. 12, (23)] (Burden and Bailey, 1975), erroneously described as 6-hydroxyphaseollin in earlier work (Sims et al., 1972), is the major phytoalexin which accumulates in inoculated soybean hypocotyls, cotyledons, or leaves, as well as in cell suspensions cultures of this plant (Ebel et al., 1976).

Analysis of the soybean phytoalexins by high-performance liquid chromatography (HPLC) led, in addition, to the isolation of glyceollins II-IV (24-26) (Lyne et al., 1976; Lyne and Mulheirn, 1978), 3,6a,9-trihydroxypterocarpan (22) (Lyne and Mulheirn, 1978; Weinstein et al., 1981), and 2-dimethylallyl (27) and 4-dimethylallyl-3,6a,9-trihydroxypterocarpan (28) (Zähringer et al., 1981; Ingham et al., 1981). The two dimethylallyl isomers are biosynthetic intermediates in the formation of glyceollins (see p. 408). It is interesting to note that trihydroxypterocarpan (THP) showed no inhibition (up to 10 μg) in the TLC fungitoxicity assay with *C. cucumerinum*. In contrast, 2-dimethylallyl-THP produced a clear inhibition zone which, however, was smaller than the inhibition zone obtained with a mixture of glyceollin isomers. Since THP has weak antifungal properties in other assays (Ingham, 1980) and also bacteriostatic properties (Weinstein et al., 1981), it can be classified as a phytoalexin.

Figure 11 Virulence assay of soybean cultivar Harosoy 63 with *P. megasperma*. Five-day-old seedlings are placed on steel needles and a small piece of mycelium is inserted into the slit wound. Shown are the results after 24 hr of incubation: left inoculation with race 3 (compatible); right innoculation with race 1 (incompatible).

In addition to these seven phytoalexins a number of isoflavones and pterocarpans occur in inoculated soybean plants. Daidzein accumulates in high concentrations in hypocotyls (Moesta and Grisebach, 1981b). It is therefore evident that quantitative analysis of the soybean phytoalexins poses a difficult problem. Whereas it is sufficient to determine the sum of the glyceollin isomers, one must differentiate between these strong fungitoxic compounds and the only rather weakly fungitoxic THP, which can accumulate to high concentrations in induced cotyledons (Moesta and Grisebach, 1980). Furthermore, in the TLC-UV method usually employed for separation of compounds (Yoshikawa et al., 1978) daidzein overlaps with glyceollins and this can lead to an overestimation of glyceollins.

Induction of Phytoalexin Synthesis with
P. megasperma f. sp. *glycinea*

Since race specificity is lost in elicitor preparations from *P. megasperma*, inoculation with spores or mycelium is necessary if the specificity of the host-parasite

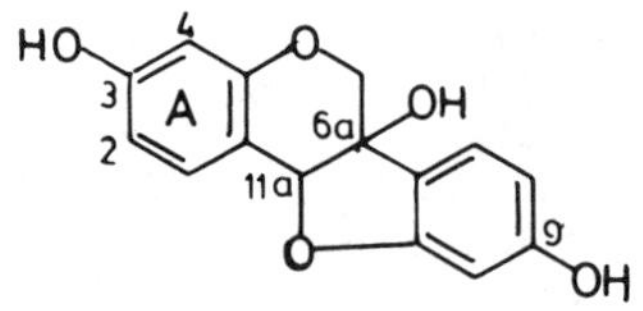

(22)

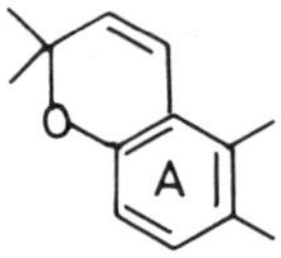

(23)

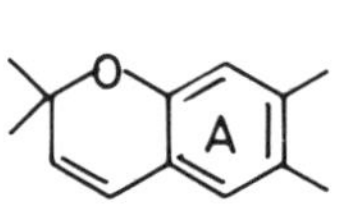

(24)

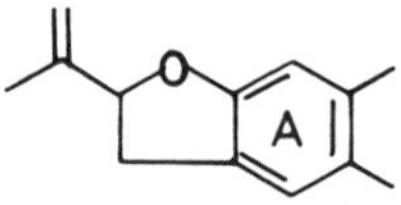

(25)

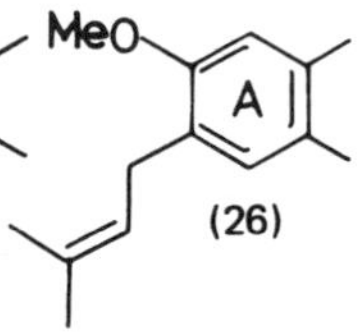

(26)

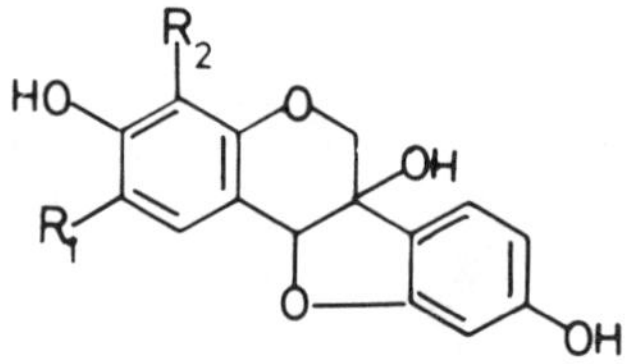

(27) R_1=dimethylallyl, R_2=H
(28) R_1=H, R_2=dimethylallyl

Figure 12 Structures of glyceollin isomers I (23), II (24), III (25), and IV (26), the dimethylallyltrihydroxypterocarpan isomers (27) and (28), and 3,6a,9-trihydroxypterocarpan (22).

interaction is to be explored. According to investigations with *P. megasperma* (race was not specified) and hypocotyls of the cultivars Harosoy (compatible) and Harosoy 63 (incompatible), differences in these host-parasite interactions occur between 4 and 8 hr (Frank and Paxton, 1970).

Early differences in phytoalexin accumulation would therefore be important. These conclusions were corroborated by studies of Yoshikawa et al. (1978) with the same host-parasite system (race 1 of *P. megasperma* was used) in which mycelial growth was correlated with glyceollin accumulation. With the above mentioned reservation concerning the analytical method employed (TLC with about 40% recovery), the results showed a high accumulation of glyceollins above the ED_{90} value for *P. megasperma* (approximately 200 μg/ml) at 8 hr after inoculation in tissue layers of Harosoy 63 hypocotyls which

Table 6 Glyceollin Concentrations in Tissue Sections of Soybean Hypocotyls of Harosoy (H) and Harosoy 63 (H 63) at Various Times After Inoculation with Race 1 of *P. megasperma* f. sp. *glycinea*

Time after inoculation (hr)	Host	Glyceollins (μg/g fresh weight) in tissue layer (0.25 mm thick)[a]				
		1	2	3	4	5
8	H 63	364**	75*	< 10	< 10	0
	H	10**	18*	0	0	0
9	H 63	589**	106**	24	0	0
	H	25**	46**	43	17	0
10	H 63	752**	261**	73	20	0
	H	20**	18**	19*	25	0
24	H 63	3889**	1820**	515	117	71
	H	557**	420**	273**	272**	155**

[a]Tissue layers with two asterisks were extensively colonized by the fungus; layers with one asterisk were only slightly colonized by the margins of advancing hyphae.
Source: Yoshikawa et al. (1978).

contained advancing hyphae. In contrast, such a high and localized accumulation of glyceollins was not detected in inoculated hypocotyls of the susceptible cultivar Harosoy (Table 6). The data indicate that phytoalexin concentrations surrounding the margins of the growing hyphae determine development of the pathogen. Delayed phytoalexin accumulation seems not to be effective in inhibiting fungal growth, since high levels (>ED$_{90}$) of glyceollin also accumulated in the susceptible hypocotyls at 24 hr after infection.

While the results of these investigations would suggest a very important role of glyceollins in the defense reaction of the soybean, similar studies in which a virulent and an avirulent race of *P. megasperma* were used in combination with Harosoy 63 did not give such an optimistic picture. We used races 1 and 3 of *P. megasperma* f. sp. *glycinea* in combination with Harosoy 63 to determine glyceollin accumulation as well as the rates of synthesis and metabolism of glyceollins in the incompatible and compatible interactions (Moesta and Grisebach, 1981b). Five-day-old whole seedlings were inoculated by inserting a piece of mycelium into a wound 1 cm below the cotyledonary node. Glyceollins were extracted around the infection site and analyzed by HPLC. For determination of the rates of synthesis and degradation, pulse and pulse-chase experiments with $^{14}CO_2$ were carried out.

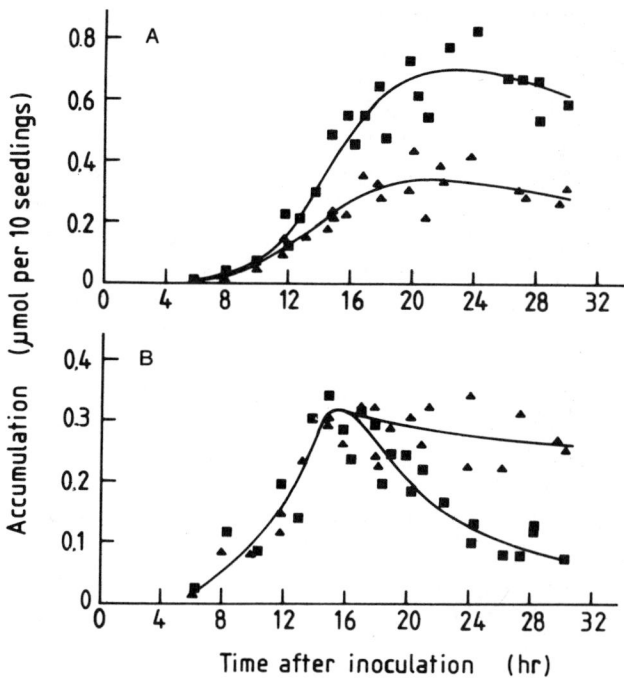

Figure 13 Time course of accumulation of glyceollin (A) and daidzein (B) in hypocotyls of Harosoy 63 after infection with race 1 (■) or race 3 (▲) of *P. megasperma* f. sp. *glycinea*.

The accumulation of glyceollin isomers and daidzein, a potential biosynthetic precursor of these compounds (see p. 408), is shown in Fig. 13. The incompatible combination leads to a higher accumulation of glyceollin, but a significant difference from the compatible reaction is not apparent until about 14 hr after inoculation. This is in agreement with the work of Ayers et al. (1976), in which it was found that glyceollin accumulation in Harosoy 63 hypocotyls is the same with races 1 and 3 up to 15 hr after infection. 3,6a,9-Trihydroxypterocarpan accumulates in cotyledons, but not in hypocotyls. We have consistently found only very low amounts (0.004-0.03 μmol/g fresh weight) of glyceollins in hypocotyls that were only wounded.

In agreement with the accumulation curves (Fig. 13) were the rates of synthesis of glyceollin, which were similar with both fungal races up to about 12 hr after inoculation. With race 1 the synthetic rate then reached a higher value and showed a subsequent slower decline than that with race 3.

From pulse-chase experiments the apparent half-life of glyceollin metabolism with race 1 was found to be 28 ± 7 hr, while for infection with race 3 no metabolism was observed.

Our data are not in agreement with the assumption, by Yoshikawa et al. (1979), that the differences in levels of glyceollin in Harosoy 63 hypocotyls upon infection with compatible and incompatible races of *P. megasperma* are determined by differential glyceollin-degrading activity of the hypocotyls. Our data instead show that the higher accumulation of glyceollin in the incompatible interaction is due to a longer persistence of synthetic activity. Since metabolism of glyceollin is slow, levels of glyceollin are determined mainly, if not exclusively, by its rate of synthesis.

We did not find more than about twice the amount of glyceollin in the incompatible interaction as in the compatible one. A more differentiated accumulation of glyceollin might be produced by infection with zoospores of *P. megasperma* instead of mycelium (Ward et al., 1979, 1980; Lazarovits et al., 1981). In the virulence assay (Wade and Albersheim, 1979) Harosoy 63 seedlings in the compatible interaction start to topple over about 14 hr after infection. The question can therefore be raised as to whether the difference in glyceollin accumulation between the compatible and incompatible interactions observed at about this time is responsible for the resistance of Harosoy 63 against race 1. To answer this question an analysis of glyceollin concentration at the cellular level is desirable.

It had been claimed that rates of synthesis of soybean phytoalexins are not regulated by "activation" of initial enzymes in flavonoid biosynthesis (Partridge and Keen, 1977). Infection of hypocotyls from the soybean cultivar Amsoy 71, which has the same response toward infection with *P. megasperma* races 1 and 3 as Harosoy 63, was used to assay enzyme levels in the compatible and incompatible interactions (Börner and Grisebach, 1982). The change in specific activity of phenylalanine ammonia-lyase (PAL) (compare Fig. 1) after infection of the hypocotyls in intact seedlings is shown in Fig. 14. Contrary to the results of Partridge and Keen (1977) wounded controls had very low levels of PAL, which did not change up to 21 hr after wounding. Infection with mycelium, however, induced a strong increase in PAL activity which was not different for the two races up to about 12 hr after infection. After this time PAL activity in the compatible interaction started to decline, whereas in the incompatible interaction it still increased up to about 16 hr after infection and then declined more slowly than in the compatible interaction. It should be noted that the difference in PAL levels occurred at the same time at which a difference in glyceollin accumulation was observable (see Fig. 13).

Chalcone synthase (see Fig. 1) also showed a strong induction effect after infection, but no significant difference between the two *P. megasperma* races was found. Again, wounding did not cause an increase in enzyme activity.

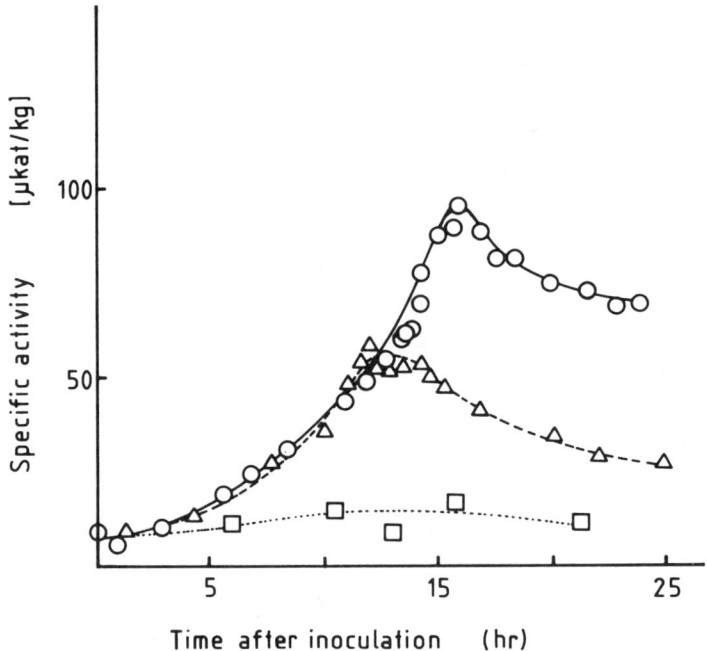

Figure 14 Changes in the specific activity of phenylalanine ammonia-lyase in soybean hypocotyls (Amsoy 71) after infection with race 1 (○) or race 3 (△) of *P. megasperma* f. sp. *glycinea*. Wounded controls (□).

In addition to the two enzymes which belong to the flavonoid biosynthetic pathway, the enzymes glucose 6-phosphate dehydrogenase and glutamate dehydrogenase were assayed. Again both enzymes showed a strong increase of activity after infection, but not after wounding. Infection therefore not only induces the phytoalexin pathway but also has a pronounced effect on other enzyme levels. The infected tissue obviously undergoes a dramatic change in metabolism which is very different from a wounding response.

Experiments in which the incorporation of L-^{35}S methionine into the immunoprecipitable PAL was determined have shown that the rise in PAL activity is due to a de novo synthesis of the enzyme (Börner and Grisebach, 1982).

Induction by Elicitors

Induction of glyceollin synthesis with an elicitor preparation from *P. megasperma* was studied in soybean cotyledons (Zähringer et al., 1978). As described in the case of infection with mycelium, large increases and subsequent decreases

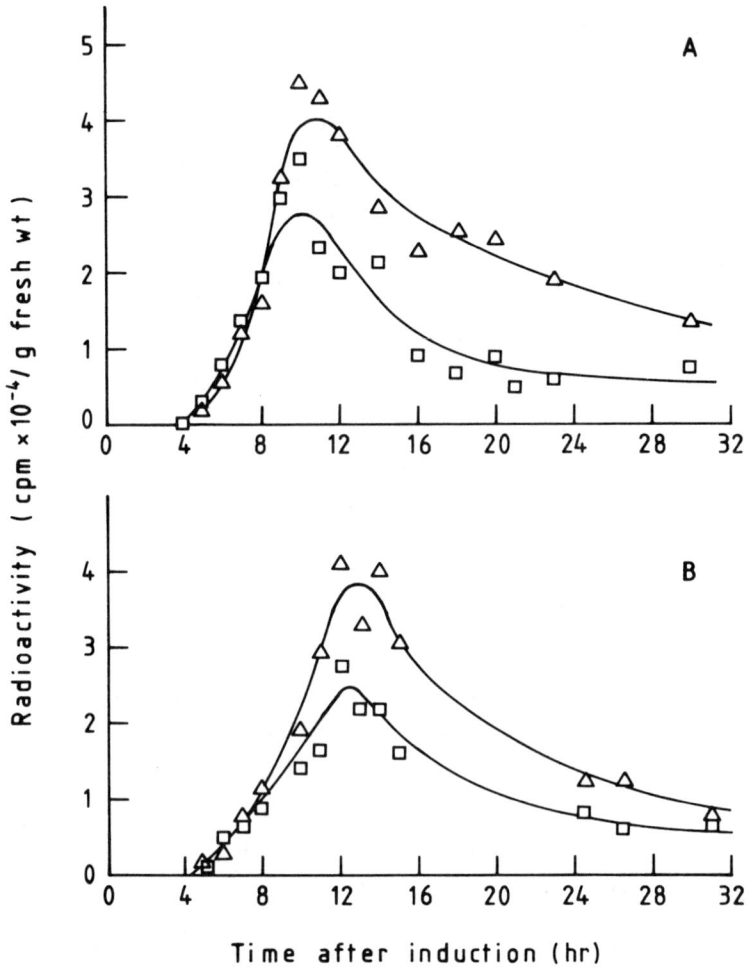

Figure 15 (A) ^{14}C Incorporation from $^{14}CO_2$ into glyceollin (□) and trihydroxypterocarpan (△) in Harosoy 63 cotyledons at various times after induction with glucan elicitor and (B) with mercuric chloride. A single 1-hr pulse of $^{14}CO_2$ was given before sampling.

in the activities of phenylalanine ammonia-lyase and chalcone synthase were found after elicitor treatment. Wounded controls showed no change in enzyme activities.

A detailed study on the mechanism of phytoalexin accumulation in Harosoy 63 cotyledons induced by either glucan elicitor or mercuric chloride was carried

Table 7 Apparent Half-Lives of Metabolism for Soybean Phytoalexins After Induction with Glucan or $HgCl_2$

Elicitor	Glyceollin	Trihydroxypterocarpan
Glucan	112 ± 63 hr	39 ± 8 hr
$HgCl_2$	91 ± 36 hr[a]	14 ± 2 hr
Total radioactivity in cotyledons after treatment with glucan elicitor	72 ± 11 hr	

[a] A second value from an independent experiment was 89 ± 53 hr.

out (Moesta and Grisebach, 1980, 1981a). As in the infection experiments described before, rates of synthesis and metabolism were determined by pulse or pulse-chase experiments with $^{14}CO_2$. The ^{14}C incorporation from $^{14}CO_2$ into glyceollin and trihydroxypterocarpan (see Fig. 12) at various times after induction with glucan elicitor or mercuric chloride is shown in Fig. 15. Very similar incorporation kinetics were found with the two types of elicitors. Apparent half-lives for glyceollin and trihydroxypterocarpan obtained from pulse-chase experiments are listed in Table 7. The half-life of glyceollin is not significantly different with the two elicitors, whereas trihydroxypterocarpan seems to turn over more rapidly after induction with $HgCl_2$.

These results are not in agreement with Yoshikawa's (1978) conclusion mentioned on p. 386, that the action of biotic elicitors is different from that of abiotic elicitors. Rather, they show that with both types of elicitors levels of glyceollins in soybean cotyledons are mainly, if not exclusively, controlled by their rates of synthesis. Since trihydroxypterocarpan is a precursor for glyceollins (Zähringer et al., 1979), the apparently shorter half-life of this compound with $HgCl_2$ induction is not relevant for the above argument.

The very similar effects of biotic and abiotic elicitors would be compatible with the hypothesis mentioned on p. 386, that both types of elicitors act by effecting the release of endogenous elicitors.

The glucan elicitor from *P. megasperma* also stimulates the accumulation of glyceollin and the activity of PAL in suspension-cultured soybean cells (Ebel et al., 1976).

With respect to elicitor action, it is important that the glucan elicitor caused large and rapid increases of PAL and 4-coumarate:CoA ligase in suspension-cultured parsley cells (a nonhost of *P. megasperma* f. sp. *glycinea*) (Hahlbrock et al., 1981). However, in contrast to the response of soybean cell cultures to treatment with elicitor and to the effect of irradiation of the parsley cells, treatment of the parsley cells with the elicitor did not induce the enzymes of the flavonoid glycoside pathway, as was demonstrated for acetyl-CoA carboxylase and chalcone synthase.

Biosynthesis of Glyceollins

The biosynthesis of pterocarpans has already been discussed on p. 387. According to the scheme shown in Fig. 2, daidzein (7,4'-dihydroxyisoflavone) would be a precursor for glyceollin, and in fact this isoflavone accumulates after infection of soybean hypocotyls. 3,6a,9-Trihydroxypterocarpan [Fig. 12,(22)], which is the presumed immediate precursor of the glyceollin isomers, was found to be a substrate for a dimethylallyltransferase. This enzyme is present only in elicitor-treated soybean cotyledons (Zähringer et al., 1979) or soybean cell cultures (Zähringer et al., 1981). Cotyledons which were only wounded did not display transferase activity. The enzyme preparation catalyses the transfer of the dimethylallyl residue from dimethylallyl pyrophosphate to the 2 and 4 positions, respectively, of trihydroxypterocarpan, leading to the two isomeric dimethylallyltrihydroxypterocarpans, which were also found to occur in low concentrations in elicitor-treated cotyledons. Probably two different transferases are present in this preparation. Enzyme activity is located in the 40,000g particulate fraction and depends on Mn^{2+} ions.

The enzymatic reactions leading from the two dimethylallyl isomers to the glyceollin isomers are still unknown.

Tomato-*Cladosporium fulvum*

General Observations

Cladosporium fulvum Cooke, the causal pathogen of leaf mold of the tomato (*Lycopersicum esculentum* Mill.), exists in many physiological races. In the cultivated tomato several genes for resistance are present which condition different reactions toward the pathogen. Tomato cultivars carrying the gene *Cf2* are susceptible to races 1, 2, and 3 and resistant to race 4 of *C. fulvum*. Plants with the *Cf4* gene are resistant to races 1, 2, and 3 and susceptible to race 4. Light and scanning electron microscopic studies of infection of tomato plants by virulent and avirulent races of *C. fulvum* were carried out be de Wit (1977). No differences were observed in growth between compatible and incompatible combinations during germination, the subsequent formation of runner hyphae, and stomatal penetration. In incompatible combinations fungal growth was arrested 1-2 days after penetration and confined to stomata and surrounding cells. The host cells in contact with the fungus deposited extensive amounts of callose.

Phytoalexins of Tomato

The antifungal sesquiterpene rishitin (see Table 1) accumulates in tomato fruit tissue after inoculation with *C. fulvum* or *C. cucumerinum* (de Wit and Flach,

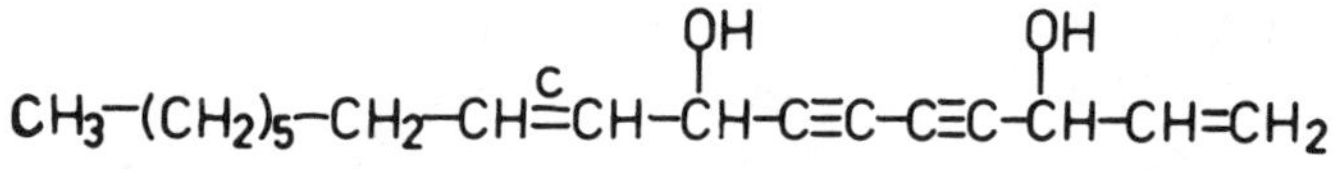

$$CH_3-(CH_2)_5-CH_2-CH\overset{c}{=}CH-\underset{OH}{CH}-C\equiv C-C\equiv C-\underset{OH}{CH}-CH=CH_2$$
(29)

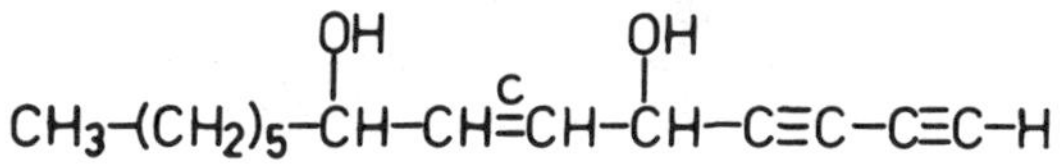

$$CH_3-(CH_2)_5-\underset{OH}{CH}-CH\overset{c}{=}CH-\underset{OH}{CH}-C\equiv C-C\equiv C-H$$
(30)

Figure 16 Structure of polyacetylenes from the tomato.

1979). Additional antifungal compounds which were all absent in healthy tissue were identified as new polyacetylenic phytoalexins (de Wit and Kodde, 1981a). Tomato fruits and leaves accumulate cis-heptadeca-1,9-diene-4,6-diyne-3,8-diol [Fig. 16, (29)] (falcarindiol) and probably the corresponding 3-ol (falcarinol); in addition, cis-tetradeca-6-ene-1,3-dyne-5,8-diol (30) was found in the fruits.

Falcarindiol has ED_{50} values of 24 and 12 µg/ml for conidial germination and germ tube elongation, respectively, of *C. fulvum* and *C. cucumerinum*. Mycelial growth of both fungi was inhibited completely at about 6 µg/ml. In tomato leaves phytoalexins accumulated earlier and to a higher level in incompatible than in compatible interactions (de Wit and Flach, 1979). The level to which falcarindiol accumulated nearly reached the ED_{50} value for hyphal and mycelial growth in vitro.

Contrary to results on phytoalexin elicitation in a number of other plants, mercuric chloride in concentrations up to 10 mM did not cause phytoalexin accumulation in leaves or fruits.

Elicitor from *Cladosporium fulvum*

High molecular weight elicitor preparations from culture filtrates, mycelial extracts, and cell walls of *C. fulvum* are very potent inducers of rishitin accumulation in tomato fruit tissue. They induced 10-50 times as much rishitin as a live conidial suspension of *C. fulvum* (de Wit and Flach, 1979).

Elicitors isolated from two races of *C. fulvum* were not host-specific and could also elicit the accumulation of pisatin and glyceollin in the pea and soybean, respectively (de Wit and Roseboom, 1980).

Since pronase and proteinase K destroyed most of the rishitin-inducing activity of a purified elicitor, a glycoprotein structure was assumed for the elicitor. This assumption was confirmed in further studies (de Wit and Kodde, 1981b). A glycoprotein produced at the end of the growth cycle of *C. fulvum* is probably derived from the peptido galactoglucomannan present on the surface of the mycelial wall. The carbohydrate moiety of this elicitor-active glycoprotein has a mannose:galactose ratio of 1.21:1 and contains only traces of glucose. The carbohydrate is O-glycosidically linked to serine and threonine. A positive correlation between the mannose and galactose content and its rishitin and necrosis-inducing activity was found.

However, this peptido galactomannan appeared to be neither race nor cultivar specific with respect to the accumulation of rishitin and polyacetylenes in tomato fruits and leaves.

Vine-*Botrytis cinerea/Plasmopara viticola*

General Observations

The grapevine (*Vitis vinifera* L.) pathogens *Botrytis cinerea* and *Plasmopara viticola* are the causal agents of grey mold and downy mildew, respectively.

In response to infection *V. vinifera* and *Vitis riparia* accumulate resveratrol (3,4',5-trihydroxystilbene) [Fig. 17, (31)], α-viniferin (33), ϵ-viniferin (34), and pterostilbene (32). The latter three compounds are fungitoxic (Langcake and Pryce, 1977; Langcake et al., 1979).

Relationship Between Resistance of Vine and Accumulation of Phytoalexins

A careful study of the relationship between the resistance of *V. vinifera* and *V. riparia* to *B. cinerea* and *P. viticola* has been carried out by Langcake (1981). Since cultivars of grapevine with resistance to the downy mildew disease are not readily available, the closely related species *V. riparia*, which is naturally resistant to *P. viticola*, was used.

The accumulation of phytoalexins surrounding the lesion area on leaves of *V. vinifera* infected by *B. cinerea* is shown in Fig. 18. α-Viniferin is the predominant compound after 2 days of infection. An important result was the inverse linear relationship between log α-viniferin concentration and susceptibility. The latter was determined by measurement of the amount of fungal glucosamine per lesion which is directly related to growth of *B. cinerea* at the infection site. From the regression line a concentration of α-viniferin in the necrotic zone when susceptibility approached zero was found to be 569 µg/g fresh weight. The observed maximum value was 606 µg/g fresh weight in a lesion of very low susceptibility and a minimum value of 8.4 µg/g fresh weight in the most susceptible lesions. Qualitatively, the response of *V. riparia* was similar to that of *V. vinifera*, but the concentration of phytoalexins in *V. riparia* was up to 10 times greater than in *V. vinifera*.

These results strongly support the assumption that α-viniferin plays an important role in resistance to *B. cinerea*. Langcake pointed out that while the

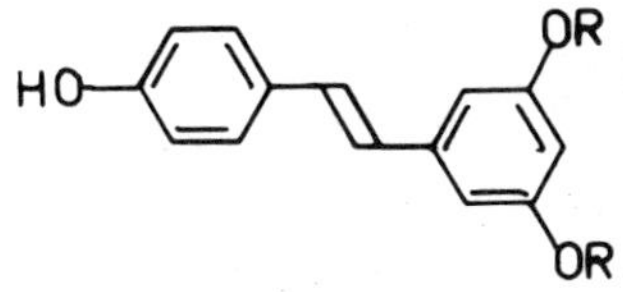

Figure 17 Structure of compounds which accumulate in *Vitis* spp. upon infection: (31) Resveratrol, (32) pterostilbene, (33) α-viniferin, and (34) ε-viniferin.

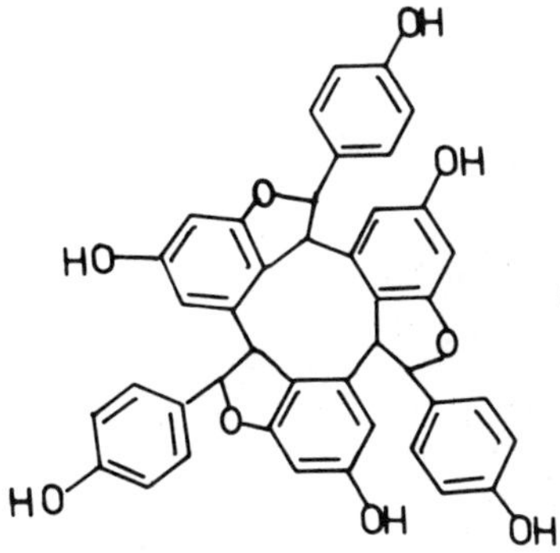

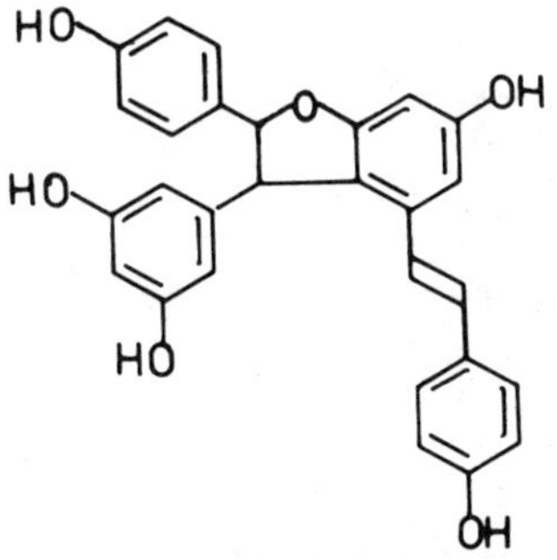

concentration of the phytoalexins may be sufficient to inhibit the growth of *B. cinerea* directly, these compounds may also inhibit the activity of fungal pectic enzymes and other hydrolases which are important for the pathogenicity of *B. cinera* (Verhoeff, 1972).

After infection of *V. riparia* by *P. viticola* resveratrol was the predominant compound during the first 24 hr. ε-Viniferins and α-viniferins then started to accumulate. In the susceptible *V. vinifera*, on the other hand, the concentrations of phytoalexins remained low.

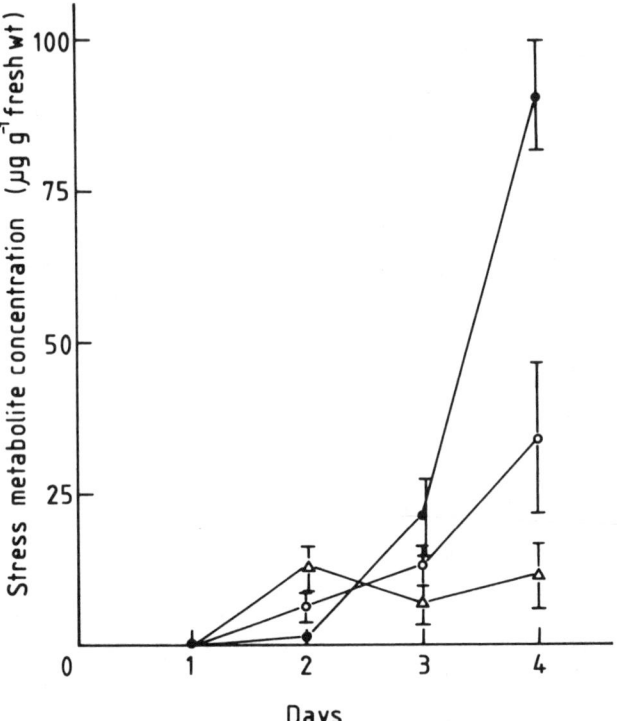

Figure 18 Accumulation of resveratrol (△), ε-viniferin (○), and α-viniferin (●) in tissue surrounding the lesion area on leaves of *V. vinifera* infected by *B. cinerea*.

It was estimated that the local concentrations of ε- and α-viniferins in *V. riparia* reach 1000-2000 µg/g fresh weight in the vicinity of the infection sites. It is therefore likely that their concentrations are sufficient to inhibit growth of the pathogen within the tissue.

Broad Bean-*Botrytis* Species

General Observations

Botrytis fabae, the causal agent of chocolate spot disease in broad bean (*Vicia faba* L.), colonizes the leaves and pod endocarp tissue of *V. faba*. These tissues are, however, resistant to *B. cinerea* Pers. and restrict the growth of invading hyphae to limited lesions. By contrast, cotyledons of *V. faba* are resistant to both fungi.

Me.CH₂CH=CH.C≡C.CO-[furan]-CH=CH.COOR

(35) R=Me
(36) R=H

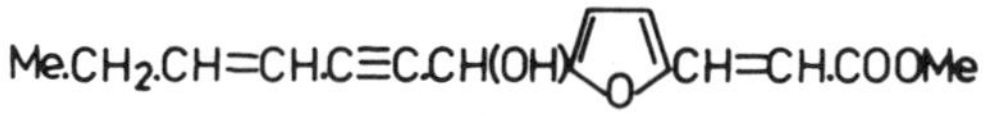

Me.CH₂.CH=CH.C≡C.CH(OH)-[furan]-CH=CH.COOMe

(37)

Me.CH₂CH₂CH₂.C≡C.CO-[furan]-CH=CH.COOR

(38) R=Me
(39) R=H

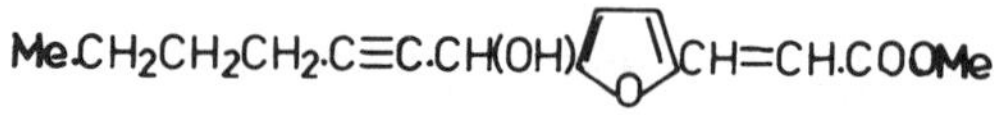

Me.CH₂CH₂CH₂.C≡C.CH(OH)-[furan]-CH=CH.COOMe

(40)

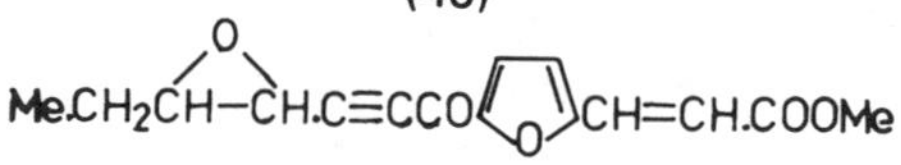

Me.CH₂CH–CH.C≡C.CO-[furan]-CH=CH.COOMe (epoxide on CH–CH)

(41)

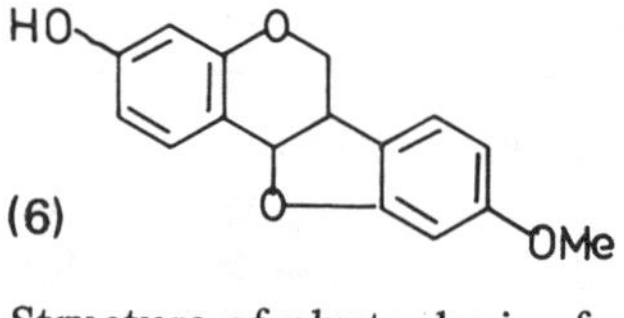

(6) [medicarpin structure: HO-...-OMe pterocarpan]

Figure 19 Structure of phytoalexins from *V. faba*: (35) wyerone, (36) wyerone acid, (37) wyerol, (38) dihydrowyerone, (39) dihydrowyerone acid, (40) dihydrowyerol, (41) wyerone epoxide, and (6) medicarpin.

Changes in Phytoalexin Concentrations After Inoculation

The broad bean shows a typical multicomponent phytoalexin response to fungal infection. The seven furanoacetylenic phytoalexins and the pterocarpan derivative medicarpin shown in Fig. 19 were isolated from bean tissue undergoing resistance reactions against *B. cinerea* or *B. fabae* (Hargreaves et al., 1977; Mansfield et al., 1980). The antifungal activity of some of the phytoalexins from *V. faba* against germ tubes of *B. cinerea* and *B. fabae* is shown in Table 8. Dihydro

Table 8 Antifungal Activity of Phytoalexins from *Vicia faba* Against Germ Tubes of *B. cinerea* and *B. fabae*

Phytoalexins	B. cinerea ED_{50}[a]	B. fabae ED_{50}
Wyerone	10.1	25.0
Wyerone acid[b]	3.5	6.8
Wyerone epoxide	2.7	5.6
Wyerol	85.0	140
Medicarpin	14.0	17.5

[a]Concentration of phytoalexin (μg/ml).
[b]Assay at pH 4.0 which reduced germ tube growth by 50%.
Source: Hargreaves et al. (1977).

derivatives possessed antifungal activity similar to that of their unsaturated analogues.

In cotyledons the pattern of phytoalexin accumulation was qualitatively similar in limited lesions caused by *B. cinerea* and *B. fabae*. The only compound which accumulated in large amounts in cotyledons was wyerone (35) (Hargreaves et al., 1977).

In leaf and pod endocarp tissue inoculated with *B. cinerea* wyerone acid (36) was the dominant phytoalexin. It accumulated rapidly and reached maximum concentration within 3 days after inoculation. In pod and leaf tissue infected with *B. fabae*, on the other hand, an initial small increase in concentration of wyerone acid was followed by a decrease as the susceptible tissues became blackened and invaded by the pathogen.

The time course of medicarpin (6) accumulation and its very low concentration during the resistance reaction suggests that compared to the wyerone derivatives this compound is not of much significance in the defense reaction.

Microscopic observations showed that during the development of limited lesions the events responsible for the initial restrictions of invading hyphae must be essentially complete before the third day of inoculation. According to its time course of accumulation, wyerone acid therefore seems to be the most important inhibitor produced by leaf and pod tissues. A total of two days after inoculation wyerone acid reached a concentration that was several times greater than that needed for the complete inhibition of germ tube growth of *B. cinerea*.

It can therefore be concluded that the rapid accumulation of wyerone or wyerone acid alone accounts for the restriction of fungal growth in the cotyledons or leaves and pods of *V. faba*, respectively.

ENHANCEMENT OF PHYTOALEXIN PRODUCTION BY FUNGICIDES

Dichlorocyclopropanes

A number of dichlorocyclopropanes provide a highly specific systemic protection against the rice blast disease caused by *Piricularia oryzae* Cav. The most active of these compounds found so far is 2,2-dichloro-3,3-dimethylcyclopropane carboxylic acid (WL 28325). On the basis of studies on the mode of action of WL-induced resistance to blast, it seemed unlikely that the fungicide has a direct action on the parasite or that fungitoxic metabolites of WL are produced within the plant (Langcake and Wickins, 1975). It was also discovered that rice (*Oryza sativa*), in response to infection by the blast fungus, produces the diterpene momilactones A and B (Cartwright et al., 1977, 1981) (Table 1), which are the first phytoalexins from a member of the *Gramineae*.

In further studies (Cartwright et al., 1980) on the action of WL 28325 wounded leaves of rice plants treated with fungicide and untreated controls were inoculated with a spore suspension of *P. oryzae* (2×10^6 spores/ml). After various incubation times a quantitative analysis of the momilactones in and around the inoculated area was made. Concomitantly the growth of the fungus was evaluated by histochemical techniques. Whereas germination, germ tube growth, and appresorium formation were similar on leaves of WL-treated and untreated plants, a necrotic reaction developed within 3 hr only in the WL-treated leaves. This reaction appeared to coincide with an almost immediate arrest of fungal development. As can be seen from Fig. 20, the accumulation of momilactones was far greater and more rapid in infected and WL-treated leaves than in infected but untreated leaves. It is also remarkable that the two phytoalexins not only accumulated at the infection site but were also present at some distance (at least 2 mm) from the infection site. In contrast, momilactones A and B were absent at all times in WL-treated but uninfected leaves. It is also noteworthy that phytoalexin accumulation could be induced by UV irradiation but that in this case WL-treatment had no influence on the level of momilactones (Cartwright et al., 1977). Antifungal activity of the momilactones was assayed by their effect on the germination of *P. oryzae* spores. For the momilactones A and B, ED_{50} values averaged 4.8 and 0.9 µg/ml, respectively. Momilactone B is therefore a very active antifungal compound.

It has been reported that certain fungicides are themselves capable of eliciting phytoalexin production (Reilly and Klarman, 1972). The mode of action of WL 28325, however, is clearly different, in that the compound itself does not stimulate momilactone production but, rather, increases the capacity of the rice plant to synthesize phytoalexins in response to infection. The biochemical mechanism for the enhancement of momilactone production is unknown. It has been

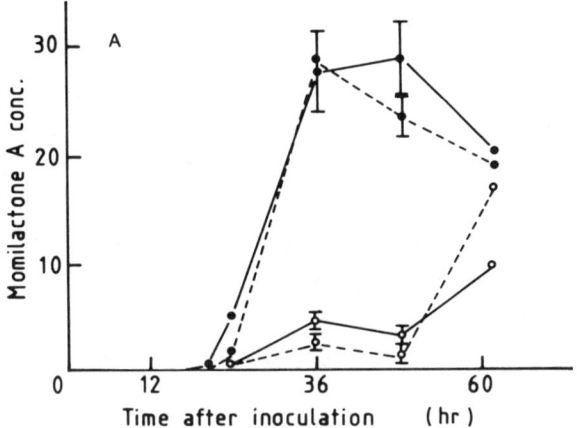

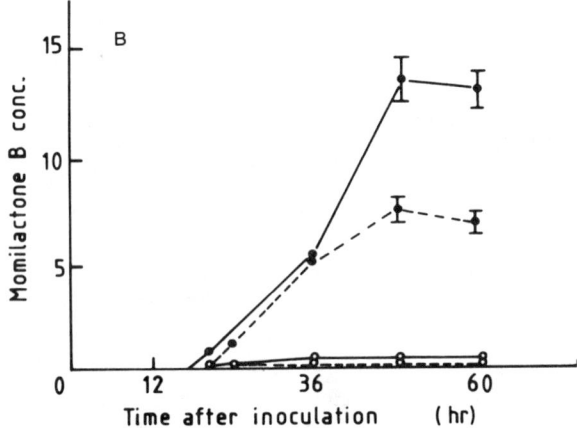

Figure 20 Changes in the concentration of (A) momilactone A and (B) momilactone B in the wounds (—) and wound surroundings (---) of WL-treated (●) and untreated (○) rice leaves following inoculation with *P. oryzae*. Momilactone concentration expressed as µg/g fresh wt. tissue.

suggested that although WL 28325 is not fungitoxic in vitro at the concentrations applied, it could damage the fungus in vivo and effect, for example, the release of fungal elicitors.

Since the early cessation of postpenetrative growth of *P. oryzae* in WL-treated leaves coincides with the accumulation of highly toxic concentrations of momilactones, these phytoalexins can be regarded as a major factor in the suppression of parasite development in WL-treated tissue.

Activation of the natural resistance mechanism could be an important method for plant disease control.

Metalaxyl

Reilly and Klarman (1972) reported that several fungicides in nonfungitoxic concentrations stimulated glyceollin production in hypocotyls of the soybean. The most effective compound was the dithiocarbamate maneb. This effect differed from the action of the chlorocyclopropanes, because the fungicides themselves stimulated phytoalexin accumulation.

An effect similar to that of WL 28325 was, however, found in later studies by Ward et al. (1980) on the control of *Phytophthora* rot of soybeans by the systemic fungicide metalaxyl (N-[2,6-dimethylphenyl]-N-[methoxyacetyl] alanine methyl ester). In this investigation hypocotyls of 5-day-old etiolated soybean seedlings (cultivar Altona) were inoculated with a suspension of zoospores (1×10^5 spores/ml) from races 4 (incompatible) and 6 (compatible) of *P. megasperma* f. sp. *glycinae*. The glyceollin (sum of isomers) concentration was then determined in metalaxyl-treated and untreated seedlings.

The response of hypocotyls inoculated with the incompatible race consisted of a typical hypersensitive reaction with limited lesions and was not influenced by metalaxyl treatment. In contrast, metalaxyl had a pronounced effect in the compatible combination. With increasing metalaxyl concentration the size of the lesions decreased and the glyceollin concentration increased from about 2 μg/ml to about 60% of the glyceollin concentration in the incompatible interaction. Unfortunately, the authors did not determine the increase between 0 and 12 hr after inoculation, which is a decisive period in the resistance response. It is important to note that metalaxyl treatment itself did not stimulate glyceollin accumulation in noninfected plants. The similarity to the effect of dichlorocyclopropanes of rice is therefore evident.

The results suggest that during a compatible plant-fungus interaction metalaxyl duplicates a typical incompatible interaction. As one of the possible mechanisms it has been proposed that the fungus, killed by the fungicide, could release cell-wall elicitors which then induce glyceollin accumulation. Experimental evidence for this hypothesis is not available.

DISCUSSION AND OUTLOOK

In assessing the role of phytoalexins in cultivar-specific resistance of plants against primary infections, one must keep in mind the current view of plant pathologists, that resistance is usually the result of multiple factors. "The key to disease resistance in plants is the functioning of multiple mechanisms for resistance and the key concept in understanding their interactions is one of coordinated defence" (Kuć and Caruso, 1977).

For example, resistance of tomato fruits to fungi can be associated with the saponine tomatine, the phytoalexin rishitin, and inhibitors of fungal hydrolases (Schlösser, 1980). Defense against the spread of wilt fungi in vascular tissue was reported to involve entrapment of conidia at the end walls of vessels followed by the formation, in sequence, of gels, tyloses, phytoalexins, tannins, and new xylem vessels (Bell, 1980).

According to Uritani et al. (1976), resistance of sweet potato root to nonpathogenic races of *C. fimbriata* is based on at least two mechanisms. In the initial stage of infection, agglutinating factors (Kojima and Uritani, 1978) in the host cells could agglutinate spores and possibly also advancing hyphae, thereby localizing the fungus in the host. In the second stage the accumulation of phytoalexins (furanoterpenoids) takes place. It could be that the speed of development of the pathogenic strain exceeds the rate of accumulation of fungitoxic levels of furanoterpenoids owing to a lack of agglutination, whereas spread of the nonpathogenic strain is limited by agglutination, which allows the phytoalexins to accumulate to fungitoxic levels at the infection site. Agglutinins have also been associated with the systemically induced resistance of the cucumber to *C. cucumerinum* and to *Cladosporium lagenarium* (Kuć and Caruso, 1977; compare Slusarenko and Wood, 1981).

The present state of research on phytoalexins taken together with the above views allows the conclusion that phytoalexins contribute as one of several factors to host defense in incompatible interactions.

In several cases it has been shown that phytoalexins accumulate at the infection site in concentrations high enough to restrict fungal growth. It would be desirable to obtain more precise information at the cellular level on the initial phase of phytoalexin accumulation around advancing hyphae and about the concentration gradient around the infection site. Suitable methods to obtain such information must be devised.

Mutants of plants which have a block in the biosynthesis of a phytoalexin could give clues as to the importance of phytoalexins in defense. At present no such mutants are available. The chain of events which follow the host-parasite interaction and finally lead to phytoalexin accumulation is not understood in any detail, but an overall picture is now emerging. The host-parasite interaction triggers by an unknown mechanism the de novo synthesis of enzymes which are involved in the biosynthesis of phytoalexins. This has, for example, been shown for phenylalanine ammonia-lyase in elicitor-treated *P. vulgaris* cell suspension cultures (Lawton et al., 1980), as well as in soybean hypocotyls inoculated with *P. megasperma* f. sp. *glycinea* (Börner and Grisebach, 1982) and in soybean cell cultures treated with elicitor from *P. megasperma* (Hille et al., 1982).

The elicitor concept is somewhat confusing at the moment. Such diverse substances as glucans, chitosan, glycoproteins, fatty acids, and hydrolytic enzymes have been reported to function as elicitors. Do receptors for these elicitor molecules exist in the plant, for example, at the plasma membrane? Are all signals mediated by endogenous elicitors? What are the specificity factors? What is the signal that initiates de novo protein synthesis? These are some of the open questions.

Many investigators in the phytoalexin field hold the view that "the most logical mechanism for conferral of specificity in gene-for-gene host-parasite systems would be for resistance genes to code for constitutive elements recognizing certain specific molecules of avirulent parasite races" (Keen and Legrand, 1980). This assumption is based on the gene-for-gene concept (Flor, 1942, 1971), which associates specificity with the incompatible interaction. It is possible, however, that at least in some host-parasite interactions the disease reaction is controlled by an active mechanism for susceptibility in which race-cultivar specificity may be due to a specific suppression of the hypersensitive response of host cells by fungal components. Doke et al. (1980) reported that a glucan isolated from the germination fluid of compatible races of *P. infestans* suppressed hypersensitive cell death and the accumulation of rishitin in potato tuber tissue infected with an incompatible race of *P. infestans*. In the same line is a working hypothesis expressed by E. Ziegler which states that glyceollin accumulation in soybean is initiated unspecifically by elicitors released from the walls of *P. megasperma* f. sp. *glycinea*. The quantitative differences in phytoalexin accumulation between the incompatible and compatible interaction (see Fig. 13) are caused by mannans (from cell walls or from extracellular glycoprotein of *P. megasperma*), which race-specifically inhibit the effect of the elicitors. This means that in the case of the compatible interaction inhibition of the elicitor effect occurs, leading to a lower phytoalexin accumulation. In support of this hypothesis, it was found that the effect of the glucan elicitor on glyceollin accumulation in soybean cotyledons could be inhibited by an extracellular invertast—a mannan glycoprotein (Ziegler and Albersheim, 1977)—from a compatible *P. megasperma* race, but not by invertase from an incompatible race (Ziegler and Pontzen, 1982).

In the future, investigations on phytoalexins are expected to make further important contributions to basic and applied research. Knowledge about the molecular mechanisms of phytoalexin induction will contribute to the elucidation of the problems of host-parasite recognition and of differential gene activation in higher plants. The finding that certain chemicals can enhance the accumulation of phytoalexins in the host-parasite interaction is probably the most important finding of practical potential, and it could lead to the development of new pesticides which activate the plants' own defense mechanisms (Cartwright et al., 1977). Another possibility of taking advantage of phytoalexins in plant defense could be the breeding of plants with an enhanced genetic potential for phytoalexin synthesis. The external application of elicitors as a means of plant protection has also been suggested (P. Albersheim, personal communication).

The complexity of the problems involved make the combined effort of scientists from different fields strongly desirable.

REFERENCES

Albersheim, P., and Valent, B. (1978). Host-pathogen interactions in plants. Plants, when exposed to oligosaccharides of fungal origin, defend themselves by accumulating antibiotics. J. Cell Bio. 78:627-643.

Anderson, A. J. (1978). Isolation from three species of *Colletotrichum* of glucan-containing polysaccharides that elicit browning and phytoalexin production in bean. Phytopathology 68:189-194.

Anderson-Prouty, A. J., and Albersheim, P. (1975). Host-pathogen interactions. VIII. Isolation of a pathogen-synthesized fraction rich in glucan that elicits a defense response in the pathogen's host. Plant Physiol. 56:286-291.

Ayers, A. R., Ebel, J., Valent, B., and Albersheim, P. (1976). Host-pathogen interactions. X. Fractionation and biological activity of an elicitor isolated from the mycelial walls of *Phytophthora megasperma* var. *sojae*. Plant Physiol. 57:760-765.

Bailey, J. A. (1974). The relationship between symptom expression and phytoalexin concentration in hypocotyls of *Phaseolus vulgaris* infected with *Colletotrichum lindemuthianum*. Physiol. Plant Pathol. 4:477-488.

Bailey, J. A., and Berthier, M. (1981). Phytoalexin accumulation in chloroform-treated cotyledons of *Phaseolus vulgaris*. Phytochemistry 20:187-188.

Bailey, J. A., and Ingham, J. L. (1971). Phaseollin accumulation in bean (*Phaseolus vulgaris*) in response to infection by tobacco necrosis virus and the rust *Uromyces appendiculatus*. Physiol. Plant Pathol. 1:451-456.

Bailey, J. A., Burden, R. S., Mynett, A., and Brown, C. (1977). Metabolism of phaseollin by *Septoria nodorum* and other non-pathogens of *Phaseolus vulgaris*. Phytochemistry 16:1541-1544.

Barz, W. (1978). Microbial degradation of flavonoids, isoflavonoids and isoflavonoid phytoalexins. Bulletin de liaison n° 8 du groupe polyphenols, Narbonne, France. Compte-Rendu des Journées Internationales d'Etude et de l'Assemblée Générale 1978. Nancy 17-19 Mai, pp. 63-89.

Bell, A. A. (1980). The time sequence of defense. In *Plant Disease,* Vol. 5, J. G. Horsfall and E. B. Cowling (Eds.). Academic, New York, pp. 54-73.

Bell, A. A. (1981). Biochemical mechanisms of disease resistance. Annu. Rev. Plant Physiol. 32:21-81.

Börner, H., and Grisebach, H. (1982). Enzyme induction in soybean infected by *Phytophthora megasperma* f. sp. *glycinea*. Arch. Biochem. Biophys. 217: 65-71.

Bostock, R. M., Kuć, J. A., and Laine, R. A. (1981). Eicosapentaenoic and arachidonic acids from *Phytophthora infestans* elicit fungitoxic sesquiterpenes in the potato. Science *212*:67-69.

Bruegger, B. B., and Keen, N. T. (1979). Specific elicitors of glyceollin accumulation in the *Pseudomonas glycinea*-soybean host-parasite system. Physiol. Plant Pathol. *15*:43-51.

Burden, R. S., and Bailey, J. (1975). Structure of the phytoalexin from soybean. Phytochemistry *14*:1389-1390.

Burden, R. S., Bailey, J. A., and Vincent, G. G. (1974). Metabolism of phaseollin by *Colletotrichum lindemuthianum*. Phytochemistry *13*:1789-1791.

Cartwright, D. W., Langcake, P., Pryce, R. J., Leworthy, D. P., and Ride, J. P. (1977). Chemical activation of host defence mechanisms as a basis for crop protection. Nature *267*:511-513.

Cartwright, D. W., Langcake, P., and Ride, J. P. (1980). Phytoalexin production in rice and its enhancement by a dichlorocyclopropane fungicide. Physiol. Plant Pathol. *17*:259-267.

Cartwright, D. W., Langcake, P., Pryce, R. J., Leworthy, D. P., and Ride, J. P. (1981). Isolation and characterization of two phytoalexins from rice as momilactones A and B. Phytochemistry *20*:535-537.

Chalutz, E., and Stahmann, M. A. (1969). Induction of pisatin by ethylene. Phytopathology *59*:1972-1973.

Cline, K., and Albersheim, P. (1981a). Host-pathogen interactions. XVI. Purification and characterization of a beta glucosyl hydrolase-transferase from the walls of suspension-cultured soybean cells. Plant Physiol. *68*: 207-220.

Cline, K., and Albersheim, P. (1981b). Host-pathogen interactions. XVII. Hydrolysis of biologically active fungal glucans by enzymes located from soybean cells. Plant Physiol. *68*:221-228.

Cruickshank, I. A. M., and Perrin, D. R. (1960). Isolation of a phytoalexin from *Pisum sativum* L. Nature *187*:799-800.

Deverall, B. J. (1976). Current perspectives in research on phytoalexins. In *Biochemical Aspects of Plant-Parasite Relationships*, J. Friend and D. R. Threlfall (Eds.). Academic, London, pp. 207-223.

Dewick, P. M. (1975). Pterocarpan biosynthesis: Chalcone and isoflavone precursors of demethylhomopterocarpin and maackiain in *Trifolium pratense*. Phytochemistry *14*:979-982.

Dewick, P. M. (1977). Biosynthesis of pterocarpan phytoalexins in *Trifolium pratense*. Phytochemistry *16*:93-97.

Dewick, P. M., and Martin, M. (1976). Biosynthesis of isoflavonoid phytoalexins in *Medicago sativa* : The biosynthetic relationship between pterocarpans and $2'$-hydroxyisoflavans. J. Chem. Soc. Chem. Commun., pp. 637-638.

Dewick, P. M., Barz, W., and Grisebach, H. (1970). Biosynthesis of coumestrol in *Phaseolus aureus*. Phytochemistry *9*:775-783.

Doke, N., Garas, N. A., and Kuć, J. (1980). Effect on host-hypersensitivity of suppressors released during the germination of *Phytophthora infestans* cytospores. Phytopathology *70*:35-39.

Ebel, J., Ayers, A. R., and Albersheim, P. (1976). Host-pathogen interactions XII. Response of suspension-cultured soybean cells to the elicitor isolated from *Phytophthora megasperma* var. *sojae*, a fungal pathogen of soybean. Plant Physiol. *57*:775-779.

Egin-Bühler, B., Loyal, R., and Ebel, J. (1980). Comparison of acetyl-CoA carboxylases from parsley cell cultures and wheat germ. Arch. Biochem. Biophys. *203*:90-100.

Ende, G., van den, and Müller, K. O. (1964). Zur Kinetik der Phytoalexinbildung. Naturwissenschaften *51*:317.

Flor, H. H. (1942). Inheritance of pathogenicity in a cross between physiologic races 22 and 24 of *Melampsora lini*. Phytopathology *32*:5.

Flor, H. H. (1971). Current status of the gene-for-gene concept. Annu. Rev. Phytopathology *9*:275-296.

Frank, J. A., and Paxton, J. D. (1970). Time sequence for phytoalexin in Harosoy and Harosoy 63 soybean. Phytopathology *60*:315-318.

Fuchs, A., de Vries, F. W., and Platero Sanz, M. (1980). The mechanisms of pisatin degradation by *Fusarium oxysporum* f. sp. *pisi*. Physiol. Plant Pathol. *16*:119-133.

Glazener, J. A., and Van Etten, H. D. (1978). Phytotoxicity of phaseollin to, and alteration of phaseollin by, cell suspension cultures of *Phaseolus vulgaris*. Phytopathology *68*:111-117.

Gnanamanickam, S. S., and Mansfield, J. W. (1981). Selective toxicity of wyerone and other phytoalexins to gram-positive bacteria. Phytochemistry *20*: 997-1000.

Gnanamanickam, S. S., and Smith, D. A. (1980). Selective toxicity of isoflavonoid phytoalexins to gram-positive bacteria. Phytopathology *70*:894-896.

Grisebach, H., and Ebel, J. (1978). Phytoalexins, chemical defense substances of higher plants? Angew, Chem. Int. Ed. Engl. *17*:635-647.

Gross, G. (1977). Phytoalexine und verwandte Pflanzenstoffe. Progr. Chem. Org. Nat. Prod. *34*:187-247.

Hadwiger, L. A., and Beckmann, J. M. (1980). Chitosan as a component of pea-*Fusarium solani* interactions. Plant Physiol. *66*:205-211.

Hahlbrock, K., and Grisebach, H. (1975). Biosynthesis of flavonoids. In *The Flavonoids*, J. B. Harborne, T. J. Mabry, and H. Mabry (Eds.). Chapman and Hall, London, pp. 866-915.

Hahlbrock, K., Lamb, C. J., Purwin, C., Ebel, J., Fautz, E., and Schäfer, E. (1981). Rapid response of suspension-cultures of parsley cells to the elicitor from *Phytophthora megasperma* var. *sojae*. Plant Physiol. *67*:768-773.

Hahn, M. G. (1981). Fragments of plant and fungal cell wall polysaccharides elicit the accumulation of phytoalexins in plants. Ph.D. thesis, University of Colorado.

Hahn, M. G., and Albersheim, P. (1978). Host-pathogen interactions. XIV. Isolation and partial characterization of an elicitor from yeast extract. Plant Physiol. *62*:107-111.

Hahn, M. G., Darvill, A. G., and Albersheim, P. (1981). Host-pathogen interacions. XIX. The endogenous elicitor, a fragment of a plant cell wall polysaccharide that elicits phytoalexin accumulation in soybeans. Plant Physiol. *68*:1161-1169.
Hargreaves, J. A. (1979). Investigations into the mechanism of mercuric chloride stimulated phytoalexin accumulation in *Phaseolus vulgaris* and *Pisum sativum*. Physiol. Plant Pathol. *15*:279-287.
Hargreaves, J. A. (1980). A possible mechanism for the phytotoxicity of the phytoalexin phaseollin. Physiol. Plant Pathol. *16*:351-357.
Hargreaves, J. A. (1981). Accumulation of phytoalexins in cotyledons of french bean (*Phaseolus vulgaris* L.) following treatment with triton (T-octylphenol polyethoxyethanol) surfactants. New Phytol. *87*:733-741.
Hargreaves, J. A., and Bailey, J. A. (1978). Phytoalexin production by hypocotyls of *Phaseolus vulgaris* in response to constitutive metabolites released by damaged bean cells. Physiol. Plant. Pathol. *13*:89-100.
Hargreaves, J. A., Mansfield, J. W., and Rossall, S. (1977). Changes in phytoalexin concentrations in tissues of the broad bean plant (*Vicia faba* L.) following inoculation with species of *Botrytis*. Physiol. Plant Pathol. *11*: 227-242.
Heller, W., and Hahlbrock, K. (1980). Highly purified "flavanone synthase" from parsley catalyses the formation of naringenin chalcone. Arch. Biochem. Biophys. *200*:617-619.
Hess, S. L., Hadwiger, L. A., and Schwochau, M. E. (1971). Studies on biosynthesis of phaseollin in excised pods of *Phaseolus vulgaris*. Phytopathology *61*:79-82.
Heuvel, J. van den, Van Etten, H. D., Serum, J. W., Coffen, D. L., and Williams, T. H. (1974). Identification of 1a-hydroxy phaseolone, a phaseollin metabolite produced by *Fusarium solani*. Phytochemistry *13*:1129-1131.
Higgins, V. J., and Smith, D. G. (1972). Separation and identification of two pterocarpanoid phytoalexins produced by red clover leaves. Phytopahtology *62*:235-238.
Higgins, V. J., Stoessl, A., and Heath, M. C. (1974). Conversion of phaseollin to phaseollinisoflavan by *Stemphylium botyrosum*. Phytopathology *64*:105-107.
Hille, A., Purwin, C., and Ebel, Y. (1982). Induction of enzymes of phytoalexin synthesis in cultured soybean cells by an elicitor from *Phytophythora megasperma* f. sp. *glycinea*. Plant Cell Reports *1*:123-127.
Ingham, J. L. (1976). Fungal modification of pterocarpan phytoalexins from *Melilotus alba* and *Trifolium pratense*. Phytochemistry *15*:1489-1495.
Ingham, J. L. (1978). Isoflavonoid and stilbene phytoalexins of the genus *Trifolium*. Biochem. System. Ecol. *6*:217-223.
Ingham, J. L. (1980). Induced isoflavonoids of *Erythrina sandwicensis*. Z. Naturforsch. *35c*:384-386.
Ingham, J. L., and Harborne, J. B. (1976). Phytoalexin induction as a new dynamic approach to the study of systematic relationships among higher plants. Nature *260*:241-243.

Ingham, J. L., Keen, N. T., Mulheirn, L. J., and Lyne, R. L. (1981). Inducibly-formed isoflavonoids from leaves of soybean. Phytochemistry 20:795-798.

Kaplan, D. T., Keen, N. T., and Thomason, I. J. (1980). Studies on the mode of action of glyceollin in soybean incompatibility to the root knot nematode, *Meloidogyne incognita*. Physiol. Plant Pathol. 16:319-325.

Keen, N. T. (1975). Specific elicitors of plant phytoalexin production: Determinants of race specificity in pathogens. Science 187:74-75.

Keen, N. T., and Legrand, M. (1980). Surface glycoproteins: Evidence that they may funcion as the race specific phytoalexin elicitors of *Phytophthora megasperma* f. sp. *glycinea*. Physiol. Plant Pathol. 17:175-192.

Keen, N. T., Sims, J. J., Erwin, D. C., Rice, E., and Partridge, J. E. (1971). 6a-Hydroxyphaseollin: An antifungal chemical induced in soybean hypocotyls by *Phytophthora megasperma* var. *sojae*. Phytopathology 61:1084-1089.

Keen, N. T., Partridge, J. E., and Zaki, A. I. (1972). Pathogen-induced elicitor of a chemical defense mechanism in soybeans monogenically resistant to *Phytophthora megasperma* var. *sojae*. Phytopathology 62:768.

Király, Z. (1980). Defenses triggered by the invader: Hypersensitivity. In *Plant Disease*, Vol. 5, J. G. Horsfall and E. B. Cowling, (Eds.). Academic New York, pp. 201-224.

Klarman, W. L., and Stanford, J. B. (1968). Isolation and purification of an antifungal principle from infected soybeans. Life Sci. 7:1095-1103.

Kojima, M., and Uritani, I. (1978). Isolation and characterization of factors in sweet potato root which agglutinate germinated spores of *Ceratocystis fimbriata*, black rot fungus. Plant Physiol. 62:751-753.

Kuan, T. L., and Erwin, D. C. (1980). Formae speciales differentiation of *Phytophthora megasperma* isolates from soybean and alfalfa. Phytopathology 70:333-338.

Kuć, J. A. (1976). Phytoalexins. In *Encyclopedia of Plant Physiology*, Vol. 4, R. Heitefuss and P. H. Williams (Eds.). Springer-Verlag, Berlin, pp. 632-652.

Kuć, J., and Caruso, F. L. (1977). Activated coordinated chemical defense against disease in plants. ACS Symp. Ser. 62:78-89.

Kuć, J. A., Currier, W. W., and Shih, M. J. (1976). Terpenoid phytoalexins. In *Biochemical Aspects of Plant-Parasite Relationships*, J. Friend and D. R. Threlfall (Eds.). Academic, London, pp. 225-237.

Kuhn, P. J., and Smith, D. A. (1979). Isolation from *Fusarium solani* f. sp. *phaseoli* of an enzymic system responsible for kievitone and phaseollin detoxification. Physiol. Plant Pathol. 14:179-190.

Kuhn, P. J., Smith, D. A., and Ewing, D. F. (1977). 5,7,2',4'-Tetrahydroxy-8-(3''-methyl-butyl) isoflavanone, a metabolite of kievitone produced by *Fusarium solani* f. sp. *phaseoli*. Phytochemistry 16:296-297.

Langcake, P. (1981). Disease resistance of *Vitis* spp. and the production of the stress metabolites resveratrol, ε-viniferin, α-viniferin and pterostilbene. Physiol. Plant Pathol. 18:213-226.

Langcake, P., and Pryce, R. J. (1977). A new class of phytoalexins from grapevines. Experientia *33*:151-152.
Langcake, P., and Wickins, S. G. A. (1975). Studies on the action of dichlorocylopropanes on the host-parasite relationship in the rice blast disease. Physiol. Plant Pathol. 7:113-126.
Langcake, P., Cornford, C. A., and Pryce, F. (1979). Identification of pterostilbene as a phytoalexin from *Vitis vinifera*. Phytochemistry *18*:1025-1027.
Lappe, U., and Barz, W. (1978). Degradation of pisatin by fungi of the genus *Fusarium*. Z. Naturforsch. *33c*:301-302.
Lawton, M. A., Dixon, R. A., and Lamb, C. J. (1980). Elicitor modulation of the turnover of L-phenylalanine ammonia-lyase in french bean cell suspension cultures. Biochim. Biophys. Acta *633*:162-175.
Lazarovits, G., Stössel, R., and Ward, E. W. B. (1981). Age-related changes in specificity and glyceollin production in the hypocotyl reaction of soybeans to *Phytophthora megasperma* var. *sojae*. Phytopathology 71:94-97.
Lee, S. C., and West, C. A. (1981a). Polygalacturonase from *Rhizopus stolonifer* an elicitor of casbene synthetase activity in castor bean (*Ricinus communis* L.) seedlings. Plant Physiol. *67*:633-639.
Lee, S. C., and West, C. A. (1981b). Properties of *Rhizopus stolonifer* polygalacturonase, an elicitor of casbene synthetase activity in castor bean (*Ricinus communis* L.) seedlings. Plant Physiol. *67*:640-645.
Lisker, N., and Kuć, J. (1977). Elicitors of terpenoid accumulation in potato tuber slices. Phytopathology *67*:1356-1359.
Lyne, R. L., and Mulheirn, L. J. (1978). Minor pterocarpinoids of soybean. Tetrahedron Lett. 3127-3128.
Lyne, R. L., Mulheirn, L. J., and Leworthy, J. C. (1976). New pterocarpinoid phytoalexins of soybean. J. Chem. Soc. Chem. Commun. pp. 497-498.
Mansfield, J. W., Porter, A. E. A., and Smallman, R. V. (1980). Dihydrowyerone derivatives as components of the furanoacetylenic phytoalexin response of tissues of *Vicia faba*. Phytochemistry *19*:1057-1061.
Moesta, P., and Grisebach, H. (1980). Effect of biotic and abiotic elicitors on phytoalexin metabolism in soybean. Nature *286*:710-711.
Moesta, P., and Grisebach, H. (1981a). Investigation of the mechanism of phytoalexin accumulation in soybean induced by glucan or mercuric chloride. Arch. Biochem. Biophys. *211*:39-43.
Moesta, P., and Grisebach, H. (1981b). Investigation of the mechanism of glyceollin accumulation in soybean infected by *Phytophthora megasperma* f. sp. *glycinea*. Arch. Biochem. Biophys. *212*:462-467.
Müller, K. O. (1956). Einige einfache Versuche zum Nachweis von Phytoalexinen. Phytopathol. Z. *27*:237-254.
Müller, K. O., and Börger, H. (1940). Experimentelle Untersuchungen über die Phytophthora Resistenz der Kartoffel. Arb. Biol. Reichsanst. Land Forstwirtsch. Berlin Dahlem *23*:189-231.
Ôba, K., Tatematsu, H., Yamashita, K., and Uritani, I. (1976). Induction of furano-terpene production and formation of the enzyme system from

mevalonate to isopentenyl pyrophosphate in sweet potato root tissue injured by *Ceratocystis fimbriata* and by toxic chemicals. Plant Physiol. *58*: 51-56.
Paradies, I., and Elstner, E. (1980). Wirt-Parasit Beziehungen: Untersuchungen zur Induktion der Aethylenbildung in Höheren Pflanzen und zur Rolle des Aethylens bei der Ausprägung von Krankheitssymptomen und der Einleitung der Abwehrreaktionen. Ber. Dtsch. Bot. Ges. *93*:635-657.
Paradies, I., Hümme, B., Hoppe, H. H., Heitefuss, R., and Elstner, E. (1979). Induction of ethylene formation in bean (*Phaseolus vulgaris*) hypocotyl segments by preparations isolated from germ tube cell walls of *Uromyces phaseoli*. Planta *146*:193-197.
Partridge, J. E., and Keen, N. T. (1977). Soybean phytoalexins: Rates of synthesis are not regulated by activation of initial enzymes in flavonoid biosynthesis. Phytopathology *67*:50-55.
Paxton, J. D., and Chamberlain, D. W. (1969). Phytoalexin production and disease resistance in soybeans as affected by age. Phytopathology *59*:775-777.
Perrin, D. R., and Bottomley, W. (1962). Studies on phytoalexins. V. The structure of pisatin from *Pisum sativum* L. J. Am. Chem. Soc. *84*:1919-1922.
Perrin, D. R., and Cruickshank, I. A. M. (1969). The antifungal activity of pterocarpans toward *Monilinia fructicola*. Phytochemistry *8*:971-978.
Pierre, R. E., and Bateman, D. F. (1967). Induction and distribution of phytoalexins in *Rhizoctonia*-infected bean hypocotyl. Phytopathology *57*:1154-1160.
Planck, van der, J. E. (1975). *Principles of Plant Infection.* Academic, New York.
Reilly, J., and Klarman, W. L. (1972). The soybean phytoalexin, hydroxyphaseollin, induced by fungicides. Phytopathology *62*:1113-1115.
Schlösser, E. W. (1980). Preformed internal chemical defenses. In *Plant Disease,* Vol. V, J. G. Horsfall and E. B. Cowling (Eds.). Academic, New York, pp. 161-177.
Shih, M., and Kuć, (1973). Incorporation of ^{14}C from acetate and mevalonate into rishitin and steroid glycoalkaloids by potato tuber slices inoculated with *Phytophthora infestans*. Phytopathology *63*:826-829.
Sims, J. J., Keen, N. T., and Honwad, V. K. (1972). Hydroxyphaseollin, and induced antifungal compound from soybeans. Phytochemistry *11*:827-828.
Skipp, R. A., Selby, C., and Bailey, J. A. (1977). Toxic effects of phaseollin on plant cells. Physiol. Plant Pathol. *10*:221-227.
Slusarenko, A. J., and Wood, R. K. S. (1981). Differential agglutination of races 1 and 2 of *Pseudomonas phaseolica* by a fraction from cotyledons of *Phaseolus vulgaris* cv. Red Mexican. Physiol. Plant Pathol. *18*:187-193.
Smith, D. A. (1976). Some effects of the phytoalexin, kievitone, on the vegetative growth of *Aphanomyces luteiches, Rhizoctonia solani* and *Fusarium solani* f. sp. *phaseoli*. Physiol. Plant Pathol. *9*:45-55.
Smith, D. A., Kuhn, P. J., Bailey, J. A., and Burden, R. S. (1980). Detoxification of phaseollidin by *Fusarium solani* f. sp. *phaseoli*. Phytochemistry *19*:1673-1675.

Stekoll, M., and West, C. A. (1978). Purification and properties of an elicitor of castor bean phytoalexin from culture filtrates of the fungus *Rhizopus stolonifer*. Plant Physiol. *61*:38-45.

Stoessl, A., Stothers, J. B., and Ward, E. W. B. (1976). Sesquiterpenoid stress compounds of the Solanaceae. Phytochemistry *15*:855-872.

Suzuki, H., Ôba, K., and Uritani, I. (1975). The occurrence and some properties of 3-hydroxy-3-methylglutaryl coenzyme A reductase in sweet potato roots infected by *Ceratocystis fimbriata*. Physiol. Plant Pathol. *7*:265-276.

Uritani, I., and Ôba, K. (1978). The tissue slice system as a model for studies of host-parasite relationships. In *Biochemistry of Wounded Plant Tissues* G. Kahl (Ed.). W. de Gruyter, Berlin, pp. 287-308.

Uritani, I., Ôba, K., Kujima, M., Kim, W., Ohuni, I., and Suzuki, H. (1976). *Biochemistry and Cytology of Plant-Parasite Interactions*. Elsevier, New York, pp. 239-252.

Van Etten, H. D. (1976). Antifungal activity of pterocarpans and other selected isoflavonoids. Phytochemistry *15*:655-659.

Van Etten, H. D., and Pueppke, S. G. (1976). Isoflavonoid phytoalexins. In. *Biochemical Aspects of Plant-Parasite Relationships*, J. Friend and D. R. Threlfall (Eds.). Academic, London, pp. 239-289.

Van Etten, H. D., and Smith, D. A. (1975). Accumulation of antifungal isoflavonoids and 1a-hydroxyphaseollone, a phaseollin metabolite, in bean tissue infected with *Fusarium solani* f.sp. *phaseoli.* Physiol. Plant Pathol. *5*:225-237.

Verhoeff, K. (1972). *In vitro* and *in vivo* production of cell wall degrading enzymes by *Botrytis cinerea* from tomato. Neth. J. Plant Pathol. *78*:179-185.

Wade, M., and Albersheim, P. (1979). Race-specific molecules that protect soybeans from *Phytophthora megasperma* var. *sojae*. Proc. Nat. Acad. Sci. *76*: 4433-4437.

Ward, E. W. B., Lazarovits, G., Unwin, C. H., and Buzzell, R. I. (1979). Hypocotyl reactions and glyceollin in soybean inoculated with zoospores of *Phytophthora megasperma* var. *sojae*. Phytopathology *69*:951-955.

Ward, E. W. B., Lazarovits, G., Stössel, P., Barrie, S. D., and Unwin, C. H. (1980). Glyceollin production associated with control of *Phytophthora* rot of soybeans by the systemic fungicide, metalaxyl. Phytopathology *70*:738-740.

Weinstein, L. I., Hahn, M. G., and Albersheim, P. (1981). Isolation and biological activity of glycinol, a pterocarpan phytoalexin synthesized by soybeans. Plant Physiol. *68*:358-363.

Wit de, P. J. G. M. (1977). A light and scanning-electron microscopic study of infection of tomato plants by virulent and avirulent races of *Cladosporium fulvum*. Neth. J. Plant Pathol. *83*:109-122.

Wit de, P. J. G. M., and Flach, W. (1979). Differential accumulation of phytoalexins in tomato leaves but not in fruits after inoculation with virulent and avirulent races of *Cladosporium fulvum*. Physiol. Plant Pathol. *15*:257-267.

Wit, P. J. G. M. de, and Kodde, E. (1981a). Induction of polyacetylenic phytoalexins in *Lycopersicum esculentum* after inoculation with *Cladosporium fulvum* (syn. *Fulvia fulva*). Physiol. Plant Pathol. *18*:143-148.

Wit, P. J. G. M. de, and Kodde, E. (1981b). Further characterization and cultivar-specificity of glycoprotein elicitors from culture filtrates and cell walls of *Cladosporium fulvum*. Physiol. Plant Pathol. *18*:297-314.

Wit, P. J. G. M. de, and Roseboom, P. H. M. (1980). Isolation, partial characterization and specificity of glycoprotein elicitors from culture filtrates, mycelium and cell walls of *Cladosporium fulvum*. Physiol. Plant Pathol. *16*:391-408.

Wong, E. (1976). Biosynthesis of flavonoids. In *Chemistry and Biochemistry of Plant Pigments*, Vol. 1, T. W. Goodwin (Ed.). Academic, London, p. 464.

Woodward, M. D. (1981). Identification of the biosynthetic precursors of medicarpin in inoculation droplets on white clover. Physiol. Plant Pathol. *18*:33-39.

Wyman, J. G., and Van Etten, H. D. (1978). Antibacterial activity of selected isoflavonoids. Phytopathology *68*:583-589.

Yoshikawa, M. (1978). Diverse mode of action of biotic and abiotic phytoalexin elicitors. Nature *275*:546-547.

Yoshikawa, M., Yamauchi, K., and Masago, H. (1978). Glyceollin: Its role in restricting fungal growth in resistant soybean hypocotyls infected with *Phytophthora megasperma* var. *sojae*. Physiol. Plant Pathol. *12*:73-82.

Yoshikawa, M., Yamauchi, K., and Masago, H. (1979). Biosynthesis and biodegradation of glyceollin by soybean hypocotyls infected with *Phytophthora megasperma* var. *sojae*. Physiol. Plant Pathol. *14*:157-169.

Yoshikawa, M., Matama, M., and Masago, H. (1981). Release of a soluble phytoalexin elicitor from mycelial walls of *Phytophthora megasperma* var. *sojae* by soybean tissue. Plant Physiol. *67*:1032-1035.

Zähringer, U., Ebel, J., and Grisebach, H. (1978). Induction of phytoalexin synthesis in soybean. Elicitor-induced increase in enzyme activities of flavonoid biosynthesis and incorporation of mevalonate into glyceollin. Arch. Biochem. Biophys. *188*:450-455.

Zähringer, U., Ebel, J., Mulheirn, L. J., Lyne, R. L., and Grisebach, H. (1979). Induction of phytoalexin synthesis in soybean. Dimethylallylpyrophosphate: Trihydroxypterocarpan dimethylallyl transferase from elicitor-induced cotyledons. FEBS Lett. *101*:90-92.

Zähringer, U., Schaller, E., and Grisebach, H. (1981). Induction of phytoalexin synthesis in soybean. Structure and reactions of naturally occurring and enzymatically prenylated pterocarpans from elicitor-treated cotyledons and cell cultures of soybean. Z. Naturforsch. *36c*:234-241.

Ziegler, E., and Albersheim, P. (1977). Host-pathogen interactions. VIII. Extracellular invertases secreted by three races of a plant pathogen are glycoproteins which possess different carbohydrate structures. Plant Physiol. *59*:1104-1110.

Ziegler, E., and Pontzen, R. (1982). Specific inhibition of glucan-elicited glyceollin accumulation in soybeans by an extracellular mannan-glycoprotein of *Phytophthora megasperma* f. sp. *glycinea*. Physiol. Plant Pathol. *20*:321-331.

13
Evolution, Ecology, and Mycotoxins: Some Musings

Alex Ciegler / Computer Sciences Corporation, National Space Technology Laboratory Station, Mississippi

Ecology, the dynamic interrelationship between living organisms in an environmental niche, can be regarded as an ever-shifting point on an evolutionary continuum. This concept of a smooth continuum (microevolution) has been challenged over the past few years by rather disparate groups—religious fundamentalists and evolutionists. In both cases, disagreement is with the generally accepted theory that an evolutionary progression occurs as a result of the accumulation of small genetic changes which are then naturally selected for by the environment; this leads smoothly over long periods of time to the origin of new species. Under the circumstances, one should be able to observe in any given species over a time continuum, a variety of forms blending progressively and smoothly into the next closely related species. This is not the case according to the fossil record, where species tend to remain constant over long time spans and then exhibit sudden jumps to new species (macroevolution). The absence of transitional forms is the basis for current objections by both of the aforementioned groups to microevolution, fundamentalists contending it proves the fallibility of the evolutionary theory and macroevolutionists arguing this absence as proof of the "jerk" or "punctuated equilibrium" theory.

A new concept that may help explain the punctuated equilibrium theory suggests that the process of development itself may structurally limit what is feasible with respect to viability. Therefore evolutionary morphological gaps exist not because genes have failed to fill them, or having filled them have been selected against and eliminated, but, rather, that laws of biological form limit the possible options (Lewin, 1981).

As is often the case with seemingly opposite concepts, what we may be observing are two opposed points on a circle. The concept of "punctuated equilibrium" need not rule out change via the gradual accumulation of small genetic perturbations. Even where these represent swings about a mean, an arc of sufficient magnitude may land a given specimen into another species circle.

Among microorganisms this perturbation about a mean was recognized and used to define species by Smith et al. (1952) for the genus *Bacillus* and by Raper and Thom (1949) for the species in the genera *Aspergillus* and *Penicillium*. Both groups of authors developed the concept of overlapping species characteristics which, when drawn as a cartoon, show overlapping rings with some rings overlapping more than once. These latter two genera and their comparative taxonomy as well as their ability to produce secondary metabolites, mycotoxins in particular, have recently gained considerable importance.

As early as 1913, Alsberg and Black (1913), in a classic publication that clearly contained the mycotoxin hypothesis, also posed the possibility of biochemical taxonomy: "It is not impossible that characteristic chemical properties may help to distinguish between species or strains not now sharply separated by morphologists." Among fungal taxonomists this suggestion is never received with open arms; on the contrary, it is usually rejected with relative indifference. And yet, isolated publications (Ciegler et al., 1973, 1981; Frisvad, 1981) have shown the value of secondary metabolites in differentiating among morphologically hard-to-separate fungi. It is recognized that biochemical characteristics change, but, one may argue, morphological and sexual characteristics may also be lost or altered, although, admittedly, not as readily. Does this ability to change characteristics play an important role in the survival of a given fungus during the intense competition that takes place in any given ecological niche? Do secondary metabolites function in the competitive process? The antibiotic nature of some of these metabolites suggest that they may indeed play a role. However, the early arguments based on this antibiotic capacity tended to ignore the fact that antibiotic-producing strains as used in industry (strains that are usually in themselves genetic monstrosities) were cultured under highly artificial conditions conducive to antibiotic synthesis. Does secondary metabolite production to any *effective* concentration take place in nature? This question has been answered, at least to some extent, as a result of the intensive surveys carried out on the potential contamination of foods and feeds by mycotoxins. Mycotoxins have been found in considerable concentrations as natural contaminants in agricultural commodities and many of these substances do have antibiotic properties, for example, patulin, penicillic acid, citrinin, and ochratoxin. However, of greater importance from a teleological standpoint, most mycotoxins have negligible antibiotic activity, for example, aflatoxins, trichothecenes, secalonic acid D, the tremorgens, and zearalenone. Under the circumstances, it would be difficult to argue a simple antimicrobial role for secondary metabolites. Perhaps the problem lies in attempting to categorize all secondary metabolites into one given function. More likely, secondary metabolites play a variety of roles or, perhaps, no role at all in the struggle of a given microorganism for "Lebensraum."

Additionally, the role may be of minimal importance or may change its degree of importance should the environment change. It is recognized that a multiplicity of roles can be assigned a secondary metabolite other than an immediate external function, for example, an energy sink for the elimination of unnecessary or excess metabolized nutrients, metabolic control mechanisms, initiators of differentiation, storage substances, and so on. It should be recognized that human assignment of a function to a metabolite does not preclude that metabolite from performing additional functions, even if unrecognized.

We have observed (Ciegler et al., 1973) that of 17 cultures of *Penicillium viridicatum* group II isolated from vegetable matter (peanuts, beans, wheat), all had the ability to produce the mycotoxins, ochratoxin, and/or citrinin; isolates were culturally stable. In contrast, 13 strains of *P. viridicatum* group III all isolated from meat only produced ochratoxin and were highly unstable in culture—unstable to the point that it was difficult to determine which was the wild type. After repeated transfers on various mycological media instability continued. Does this instability represent, in a teleological sense, an attempt by the fungus to compete in an environment normally foreign to it? If so, then production of ochratoxin per se, assuming that it even occurred in nature, was insufficient to accord it a secure niche. What factors in meat triggered or selected for this instability? Why did the fungus remain unstable on transfer? Would stability have been acquired if repeated transfers were made on grain? Why did none of the group III isolates produce citrinin as did members of group II? We currently have no answers to any of these question, but believe that an investigation into the points raised could provide partial answers to how fungi compete and some insight into the role of secondary metabolites in the competitive process.

Another aspect to the *P. viridicatum* problem cited above was our finding that members of *P. viridicatum* group I, also isolated from grain, grew at a more rapid rate than did members of group II and produced a completely different set of toxins, xanthomegnin and viomellein, instead of ochratoxin and citrinin; all are kidney toxins. Ochratoxin has been implicated in kidney nephroses of humans in the Balkans and in pigs in Denmark. However, in the laboratory group I strains out-compete group II strains. If this is also the case in nature, the toxins that might be of concern are xanthomegnin and viomellein instead of ochratoxin and citrinin. Currently there are no experimental data available from field trials to test this hypothesis.

In a provocatively applied philosophical argument Lillehoj (1980) proposed that secondary metabolites, with particular attention being paid to mycotoxins, may function as chemical signals between species in an ecological niche and that they perform a function in establishing that species or cultivar within a niche. More specifically, he pointed out that insects collected from corn at various geographical locations show a general internal presence of *Aspergillus flavus*, the fungus often being present in pure culture. This dominance may be achieved by aflatoxin elaboration, although, again, it must be noted that aflatoxin per se is a poor antibiotic. Yet the fact remains that *A. flavus* has in many cases established dominance in the insect milieu. We have noted a similar ability to dominate in competitive field experiments involving corn inoculated simultaneously with equal numbers of *Penicillium oxalicum* and *A. flavus*. Although *P. oxalicum* is a fast-growing and aggressive fungus in the laboratory, it was rapidly eliminated from the competitive site (A. Ciegler, unpublished data). However, aflatoxin production was considerably diminished. Whether or not aflatoxin is synthesized in the adult insect gut remains to be established, but it has been shown to occur in *A. flavus*-contaminated larvae of silk worms (Ohtomo et al., 1975).

Lillehoj and his colleagues (1980) have clearly shown that *A. flavus* infections of developing corn kernels are linked to the development of the second-generation European corn borer (*Ostrinia nubilalis*) and that this developmental period coincides with the time span when corn kernels are susceptible to *A. flavus* infection. Lillehoj also contended that current agricultural practices of intense monoculture actually constitutes a system of ecological disequilibrium; this disequilibrium then provides the requisite milieu for the syntheses of toxic-mutagenic interspecies chemical signals, that is, secondary metabolites. Stated more simply, monoculture provides selective conditions for increases in closely associated biota; for example, more corn, more corn insects; more corn insects, more select fungi associated with corn insects; more select fungi, greater production of select secondary metabolites (e.g., mycotoxins) in a rich monoculture milieu; and more secondary metabolites, more trouble for those who are recipients of the monoculture system. The net result is that man has inadvertently directed the formation of ecological niches and guided the evolutionary process in a direction unfavorable to him. In this ecological daisy chain, Lillehoj (1980) proposed that the presence of aflatoxin in a fungal-insect ecological niche is responsible for two major effects:

> (1) advantage for the toxin-producing fungus in competition with other microbes in the insect digestive tract which becomes in the extreme case of elevated production of toxin a causal factor of disease and death in the insect host and (2) advantage for the associated biota including the insect hosts in the introduction of a mutagenic agent that provides an opportunity for initiation of genetic plasticity in an environment of instability and new species selection pressure.

The difficulty with point (1) is both lack of proof that aflatoxin is produced in the insect gut and the minor antimicrobial capability of aflatoxin as an offensive weapon. Point (2) is even more difficult to perceive. One could argue that once a crop monoculture is set up and its associated biota is in place, the system would be intrinsically more stable then an ecological niche containing a large number of contending biological units. Under the circumstances, introduction of a mutagen such as aflatoxin could only tend to disrupt the system. Therefore it would appear that in a system attempting to achieve ecological and evolutionary stability, introduction of extrinsic factors, for example, mutagenic mycotoxins leading to "genetic plasticity" within a econiche, becomes intrinsically undesirable. Furthermore, a monoculture with its associated biota may be regarded from a genetic standpoint as a simpler system than one in which there is a heteroculture. There are a number of convincing instances of an association between the genetic diversity of a population and the ecological complexity of its habitat. Drosophila studies show that there are many more chromosome inversions in complex habitats than in simpler ones (Jones, 1981). Hence, genetic plasticity would be expected to a greater extent in heteroculture than in monoculture. Regardless of these objections, the proposal of Lillehoj remains to date the only known hypothesis attempting to explain the role of mycotoxins in the environment from the standpoint of the producing fungus and its ultimate effect via ecological perturbation on man. It is true that Janzen (1977) had previously argued that toxic fungal secondary metabolites function ecologically in a manner similar to higher plant chemical defences, that is, they render seeds, fresh fruit, and meat objectionable or unusable to larger organisms in the shortest period of time. With respect to mycotoxins, the concept implies that the elaborating microorganisms thereby effect a defense against loss of viability or of resource (substrate). The teleological reasoning implicit in this argument is obvious and should require no further elaboration. Or does this view represent bias on the part of the writer?

Because the title of this volume refers to secondary metabolism and fungal differentiation, perhaps this rambling discourse should bring itself at least momentarily to the subject area. An appropriate encompassing topic is fungal resting bodies such as spores and sclerotia. With respect to the latter, the hyphae of some fungi may become interwoven to form small aggregates which can further coalesce. This coalescence produces sclerotia, bodies that are able to resist adverse conditions for longer periods than ordinary mycelia. The evolutionary origins, morphogenesis, and survival of sclerotia under adverse environmental conditions have been discussed in detail by Willetts (1971, 1972). More uncertain is the status of secondary metabolism, or the occurrence of secondary metabolites in resting fungal bodies. Apparently, fungal spores or spore-like structures such as sclerotia in some cases may have metabolic potential equal to or beyond that of the mycelium. Brief reflection would indicate that a fungal spore can only represent a special case of apparent inactiviy; under certain circumstances it must be ready to respond rapidly to changes in its environment.

While dormancy may appear to afford a survival advantage, in reality it can only confer such an advantage under very limited circumstances, for example, a very hostile environment. Under normal circumstances it must be ready to compete in the same arena with other life forms—the alternative is slow decay and death. This facet has been exploited industrially. Fungal spores have been used as a bed of fixed enzymes for transformation of a variety of compounds (Vezina et al., 1968). Additionally, spores of *Penicillium roqueforti* can carry out a β-oxidation of octanic acid to 2-heptanone, a capacity either not possessed at all, or only minimally, by the mycelium (Gehrig and Knight, 1963). A group of highly toxic epipolythiadioxpiperazines, the sporidesmins, are found primarily in the spores of *Pithomyces chartarum* (Taylor, 1971). Whether or not aflatoxin B_1 occurs in the spores of *A. flavus* or *A. parasiticus* is somewhat more uncertain, since the method of harvest of the spores prior to analysis was somewhat suspect. Sclerotia of *Claviceps* represent a commerical source of the ergot drugs. In general, the occurrence of mycotoxins in fungal spores or resting bodies has not been adequately investigated.

Do these toxic substances have an ecological function such as defense or offense in the struggle for species existence? In view of the substantial energy investment involved in the production of these complex metabolites, one would like to believe that nature is not wasteful (to paraphrase Einstein, God does not play dice with the universe) and that function or purpose can be assigned to their synthesis. Janzen (1973) has hypothesized an analogous situation for secondary metabolites in plants, assigning them a defensive role against herbivores. As an example, he cited the feeding specificty of tropical burchid beetles and attributes the high degree of specialization to the nature and concentration of secondary compounds in the seeds rather than in the seeds' nutritional value. Laying of eggs by the female beetle on a nonhost plant results in death of the larvae when they initiate feeding. Janzen also noted that mammalian herbivores avoid certain plants while grazing, and conjectured that this avoidance mechanism results from the presence of certain secondary metabolites. How does this pertain to the ecological survival of fungi? Certainly, fungal contamination of given agricultural commodities often leads to the rejection of the contaminated food or feed by man or animals. An excellent example is the rejection by swine of corn contaminated with vomitoxin, a secondary metabolite produced by *Fusarium graminearum* (Vesonder et al., 1976). But contrary to conjecture of Janzen (1977), it is difficult to see how this type of refusal aids in the survival of the producing fungus.

An unusual example of a plant-mycotoxin interaction having possible ecological significance is the uptake of trichothecenes via the roots by the Brazilian shrub *Baccharis megapotamica* (Jarvis et al., 1981). In Brazil this shrub colonizes large marshy areas in which the only other higher plants present are native grasses. It would appear that the phytotoxic trichothecenes confer an ecological advantage upon *Baccharis megapotamica*, although there appears to be no ready advantage to the producing fungus.

On a microscale, do toxic secondary metabolites play a role in protecting the fungal mycelium, spore, or sclerotium against micropredators? The use of logic (which in itself may be illogical) would serve to cast doubts on this interpretation, since, as has been repeatedly pointed out, most mycotoxins are poor antibiotics. Where does this leave us? Obviously, in the dark with respect to assigning an ecological function to toxic secondary metabolites, but in the delightful position of needing more research.

Ecological relationships with respect to secondary metabolites became more apparent when one considered fungal-plant communities. The field of plant pathology is based on various extremes of such relationships. However, a facet to which even plant pathologists do not appear to have paid much attention is the possibility of a triangle of interactions, that is,

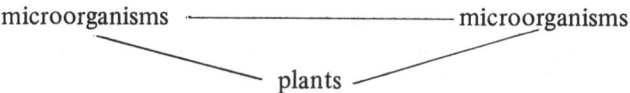

As Jimmy Durante used to say, "Everybody wants to get into the act." This possibility was first brought to our attention in 1977 during an aflatoxin outbreak in Iowa and Illinois. Based on the degree of *A. flavus* present, the levels of aflatoxin found were unusually low. Mycological examination revealed primarily the presence of only two fungi, *A. flavus* and *P. oxalicum* (A. Ciegler, unpublished data). In a similar incident occurring in Louisiana in the same year, we isolated from toxin-contaminated corn three dominant fungi, *P. oxalicum, Cladosporium* sp., and *Fusarium moniliforme* (Ciegler et al., 1980). Horses and pigs eating this corn had died of what appeared to be a mixed mycotoxicosis. Both of the above incidents raised a host of mostly unanswerable but very basic questions: Why was such a limited number of species found on grain that is normally exposed to insult by innumerable microorganisms? Was the metabolism of each fungus affected by the other fungi present, for example, in its growth rate, sporulation, or secondary metabolite synthesis? Did the fungi present play a role—active or passive—in presenting contamination, growth, or invasion by other fungi? Did the host plant participate in a potential fungal selection process? What plant defensive processes were elicited or activated by the fungi? Did the plants' response effect secondary metabolite synthesis by the fungi? If a multiplicity of secondary metabolites were produced by the fungi and/or plants, was synergism, a series of additive effects, or independent toxic activity elicited? How do these various processes function in establishing ecological niches and what is their role in the larger and longer-term evolutionary process? Despite the multiplicity of questions—with relatively few definitive answers—many of them relate directly to one concept: biochemical communication. Some of these questions have already been probed to some extent; for example, Cooke and Whipps (1980) discussed in detail the potential evolution of modes of nutrition in fungi parasitic on terrestial plants. Equally provocative are the lectin studies on the endosymbiotic association of *Rhizobium* spp. with leguminous plants (Wong, 1980; Stacey et al., 1980) and the growth inhibition by lectins of Penicillia and Aspergilli (Barkai-Golan et al., 1978); both involve recognition phenomena between plants and microbes. In the plant host-microbial parasite relationship, the lectins of the two components of the system are often antigenically similar. This may at least partially account for the ability of the parasite to negate the host's defences (Devay, 1976). In a similar vein, biochemical recognition or elicitor function on the part of microorganisms may give rise to phytoalexin responses by plants, some of these phytoalexins exhibiting mycotoxic properties. It is obvious that antigen commonality favors parasite development in this intimate relationship and that it probably arose via evolutionary pressures. Antigen commonality would result in no defensive host action when undergoing parasite attack. It would also imply an element of compatibility that would permit host and parasite to assume a position of commensualism. By this mechanism selective pressures are exerted on microbial populations, facilitating parasite establishment. The same mechanism may operate in recognition between plant stigmatic tissue and pollen germ tubes during pollination (Knox et al., 1972) or, conversely, in the rejection of foreign pollen.

We attempted to answer the competition problem between fungi in a common ecosystem by growing in mixed cultures those fungi found together in natural outbreaks. We used as a model system *A. flavus, P. oxalicum,* and *F. moniliforme.* In laboratory experiments using sterile corn, various combinations of two cultures were inoculated either simultaneously or 2 days apart. The results were clear-cut (Table 1). Mycological analyses indicated no difficulty in two cultures growing together and coexisting, although secondary metabolism was affected. On simultaneous inoculation, mycotoxin synthesis by both cultures was significantly reduced. However, where the two cultures were inoculated 2 days apart, the initial culture produced toxin equivalent to the control, whereas the second culture suffered considerable or complete diminution in mycotoxin synthesis. Field competition experiments using corn on the plant did not repeat laboratory results (Table 2). *Aspergillus flavus* in mixed culture, regardless of inoculaton time, produced aflatoxin levels equivalent to the control, whereas *P. oxalicum* required a lead time in order to effectively compete against *A. flavus*. Furthermore, mycological analysis of field inoculated corn revealed a predominance of *A. flavus* over *P. oxalicum,* even in those situations where *P. oxalicum* had a 4-day inoculation lead. The applied aspects of these results are obvious. Less obvious are the theoretical and ecological implications with respect to the differences between laboratory and field data. The labor-

Table 1 Laboratory Competition Experiments

Culture	Secalonic acid D mg/kg corn	Aflatoxin B$_1$ ug/kg corn
Penicillium oxalicum	2758	–
Aspergillus flavus	–	372
Fusarium moniliforme	0	0
P. oxalicum + A. flavus[a]	101	72
P. oxalicum + F. moniliforme[a]	1950	–
A. flavus + F. moniliforme[a]	–	450
P. oxalicum, A. flavus[b]	2085	17
A. flavus, P. oxalicum[b]	0	502
P. oxalicum, F. moniliforme[b]	2287	0
F. moniliforme, P. oxalicum[b]	0	–

[a]Cultures were inoculated simultaneously.
[b]Second cultures in series was inoculated 2 days after the first culture.

atory work involved a nonviable substrate that had been subjected to an intense heating process, which probably affected its nutritional properties. In the field the corn was viable, at a different physiological and nutritional stage, and it had not been subjected to heat. Hence, although the inocula were the same, the environment niche in the two experimental sets was quite different. Aside from nutrition, did the plant function actively in selecting for one or the other of the competing fungi via defensive mechanisms such as phytoalexin synthesis or reaction with lectins? If mycotoxins functioned as ecological defensive or offensive weapons, one would have expected an increase in the level of these compounds during the struggle for "Lebensraum." No such increase was noted.

Table 2 Field Competition Experiments in Corn

Culture	Secalonic acid D (μg/kg corn)	Aflatoxin B$_1$ (μg/kg corn)
P. oxalicum	37,373	–
A. flavus	–	1,823
P. oxalicum + A. flavus[a]	267	1,137
A. flavus + P. oxalicum[b]	690	1,290
P. oxalicum, A. flavus[b]	21,880	1,983

[a]Cultures were inoculated simultaneously.
[b]Second cultures in series was inoculated 4 days after the first culture.

Therefore the facile assignment of ecological significance to mycotoxin synthesis should be reexamined. It is true that *A. flavus* produces aflatoxin B_1, a compound that is teratogenic, carcinogenic, and mutagenic, and which exhibits acute toxicity against various biological systems; yet the question remains, Are these properties of ecological significance to the producing organism or are we being egocentric and tangentially assigning a mistaken function to a substance because it exhibits properties that affect our personal interests? Perhaps we need fewer publications with idle speculation such as this one and more definitive experiments.

REFERENCES

Alsberg, C. L., and Black, O. F. (1913). Biochemical and toxicological investigations of *Penicillium puberulum* and *Penicillium stoloniferum*. Bulletin 270. Bureau of Plant Industry, U.S. Department of Agriculture, Washington, D.C.

Barkai-Golan, R., Mirelman, D., and Sharon, N. (1978). Studies on growth inhibition by lectins of Penicillia and Aspergilli. Arch. Microbiol. *116*:119-124.

Ciegler, A., Fennell, D. J., Sansing, G. A., Detroy, R. W., and Bennett, G. A. (1973). Mycotoxin-producing strains of *Penicillium viridcatum*: Classification into subgroups. Appl. Microbiol. *26*:271-278.

Ciegler, A., Hayes, A. W., and Vesonder, R. F. (1980). Production and biological activity of secalonic acid D. Appl. Environ. Microbiol. *39*:285-287.

Ciegler, A., Lee, L. S., and Dunn, J. J. (1981). Naphthoquinone production and taxonomy of *Penicillium* viridicatum. Appl. Environ. Microbiol. *42*:446-449.

Cooke, R. C., and Whipps, J. M. (1980). The evolution of modes of nutrition in fungi parasitic on terrestrial plants. Biol. Rev. *55*:341-362.

Devay, J. E. (1976). Protein specificity in plant disease development: Protein sharing between host and parasites. In *Specificity in Plant Diseases*, R. K. S. Wood and A. Granite (Eds.). Plenum, New York, pp. 199-215.

Frisvad, J. C. (1981). Physiological criteria and mycotoxin production as aids in identification of common asymetric Penicillia. Appl. Environ. Microbiol. *41*:568-579.

Gehrig, R. F., and Knight, S. G. (1963). Fatty acid oxidation by spores of *Penicillium roqueforti*. Appl. Microbiol. *11*:166-170.

Janzen, D. H. (1973). Community structure of secondary compounds in plants. Pure Appl. Chem. *34*:529-538.

Janzen, D. H. (1977). Why fruits rot, seeds mold, and meats spoil. Am. Nat. *111*:691-713.

Jarvis, B. B., Midiwo, J. O., and Tuthill, D. (1981). Interaction between the antibiotic trichothecenes and the higher plant *Baccharis megapotamica*. Science *214*:460-462.

Jones, J. G. (1981). Models of speciation—The evidence from *Drosophila*. Nature *289*:743-744.
Knox, R. B., Willing, R. R., and Ashford, A. E. (1972). Role of pollen wall proteins as recognition substances in interspecific incompatibility in poplars. Nature *237*:381-383.
Lewin, R. (1981). Seeds of change in embryonic development. Science *214*:42-44.
Lillehoj, E. B. (1980). Secondary metabolites as chemical signals between species in an ecological niche. *Proceedings of the Sixth International Fermentation Symposium, London, Ontario* (in press).
Lillehoj, E. B., Kwolek, W. F., Horner, E. S., Widstrom, N. W., Josephson, L. M., Franz, A. O., and Catalano, E. A. (1980). Aflatoxin contamination of preharvest corn: Role of *Aspergillus flavus* inoculum and insect damage. Cereal Chem. *57*:255-257.
Ohtomo, T., Murakoshi, S., Sugiyama, J., and Kurata, H. (1975). Detection of aflatoxin B_1 in silkworm larvae attacked by an *Aspergillus flavus* isolated from a sericultural farm. Appl. Microbiol. *30*:1034-1035.
Raper, K. B., and Thom, C. (1949). *A Manual of the Penicillia*. Williams and Wilkins, Baltimore.
Smith, N. R., Gordon, R. E., and Clark, F. E. (1952). Aerobic sporeforming bacteria. U.S. Dep. Agric. Agric. Monogr. *16*:1-48.
Stacey, G., Paau, A. S., and Brill, W. J. (1980). Host recognition in the *Rhizobium*-soybean symbiosis. Plant Physiol. *66*:609-614.
Taylor, A. (1971). The toxicology of sporidesmins and other epipolythiadioxpiperazines. In *Microbial Toxins VII. Algal and Fungal Toxins*, S. Kadis, A. Ciegler, and S. J. Ajl (Eds.). Academic, New York, pp. 337-376.
Vesonder, R. F., Ciegler, A., Jensen, A. H., Rohwedder, W. K., and Weisleder, D. (1976). Co-identity of the refusal and emetic principle from *Fusarium*-infected corn. Appl. Environ. Microbiol. *31*:280-285.
Vezina, C., Sehgal, S. N., and Singh, K. (1968). Transformations of organic compounds by fungal spores. Adv. Appl. Microbiol. *10*:221-268.
Willetts, H. J. (1971). The survival of fungal sclerotia under adverse environmental conditions. Biol. Rev. *46*:387-407.
Willets, H. J. (1972). The morphogenesis and possible evolutionary origins of fungal sclerotia. *47*:515-536.
Wong, P. P. (1980). Interactions between Rhizobia and lectins of lentil, pea, broad beans, and jack bean. Plant Physiol. *65*:1049-1052.

Author Index

Italic numbers give page on which the complete reference is listed.

A

Aasen, A. J., 97, *135*
Abe, K., 254, 255, 258, *259*, *262*, *265*
Abelson, J., 181, *191*
Abraham, F. R. S., 162, *170*
Abrego, V. A., 166, *168*
Acevedo, H.F., 77, *89*
Adams, A., 326, 327, *371*
Adler, J., 279, *300*
Affronti, L. F., 77, *89*
Aharonowitz, Y. A., 13, *25*, 41, 43, 44, *51*, 153, *168*
Aigle, M., 187, *190*
Ailion, D. C., 103, 113, 120, *142*
Ainsworth, G. C., 2, 4, *25*
Aisen, P., 84, *91*
Akagawa, H., 161, *168*
Akers, H. A., 166, *168*
Alasoadura, S. O., 97, *135*
Albersheim, P., 382, 383, 384, 385, 399, 403, 404, 407, 419, *420*, *421*, *422*, *423*, *427*, *428*
Albert, A., 291, *296*
Aldersley, T., 283, *297*
Alexander, N. J., 181, *191*
Alexopoulos, C. J., 3, *25*, *26*
Algeri, A. A., 353, *365*
Alsberg, C. L., 430, *438*
Altman, L. J., 100, *135*

Alvarez, M. I., 113, 131, *139*, 251, *261*
Alwine, J. L., 183, *189*
Amano, T., 83, *92*
Anderson, A. J., 382, *420*
Anderson, J. G., 67, *69*
Anderson, M., 82, *90*
Anderson, S. M., 108, *135*
Anderson-Prouty, A. J., 382, *420*
Andrewes, A. G., 97, *135*
Angerer, R. C., 179, *189*
Anke, H., 155, 163, *168*
Anke, T., 155, 163, *168*
Anwar, M., 108, *135*
Aragón, C. M. G., 113, 114, 116, *135*, *142*, *145*
Aranachalam, T., 243, *259*
Arens, H., 78, 80, 81, *94*
Arita, I., 198, 219, *232*
Arnaud, M., 342, *360*
Arpin, N., 96, 97, *135*, *136*
Arsenault, G. P., 243, *259*, *263*
Ash, L., 100, *135*
Ashford, A. E., 437, *439*
Ashraf, M., 346, *356*
Ashworth, J. M., 3, *26*
Astell, C., 254, *264*, 309, 314, *366*
Aszalos, A., 153, *168*
Atkinson, D. E., 44, *51*
Atkinson, K. D., 348, *356*
Audhya, T. K., 58, *69*

Auger, P., 283, *296*
Aung Than, 101, 103, *136*
Austin, D. J., 131, *136*, 250, *259*
Avers, C. J., 311, *356*
Axelrod, D. E., 88, *91*
Ayers, A. R., 399, 403, 407, *420*, *422*

B

Bach, M.-L., 177, *189*
Backus, B. T., 77, *89*
Bae, M., 97, *136*
Bailey, J. A., 383, 384, 386, 390, 391, 395, 396, 399, *420*, *421*, *423*, *426*
Baird, M. L., 103, 116, 117, 118, 119, *137*, *140*
Baker, B. S., 338, 348, *356*
Baker, P. N., 84, *89*
Baldacci, G., 354, *359*
Baldwin, H. H., 7, *26*
Baldwin, J. E., 162, *168*
Balish, E., 283, 290, 291, *296*, *301*
Ball, C. B., 279, *300*
Ball, S. F., 245, *261*
Ballman, G. E., 285, *296*
Ballou, C. E., 253, *263*, 284, *296*, 323, 324, 326, 328, 329, 333, 338, 348, *356*, *364*, *373*, *374*
Bandoni, R. J., 254, *259*
Banerjee, A. K., 181, *189*
Banerjee, M., 351, *356*
Banerjee, S., 289, *297*
Bangham, A. D., 106, *151*
Baranowska, H., 354, *356*, *368*
Barkai-Golan, R., 436, *438*
Barker, W. C., 331, *359*
Barker, W. G., 105, *136*
Barksdale, A. W., 243, *259*, *263*
Barlow, A. J. E., 283, 284, 288, 289, *297*
Barnard, E. C., 182, 184, *193*
Barnes, D., 83, *89*

Barratt, R. W., 12, *26*, 112, *136*
Barrera, C. R., 278, *296*
Barrie, S. D., 404, 417, *427*
Bartman, C. D., 57, 58, 62, 63, 65, *69*, *70*, *71*
Bartnicki-Garcia, S., 9, *26*, 254, *264*, 270, 272, 273, 274, 275, 284, 286, *296*, *297*, *301*, *304*
Barton, J. K., 343, *356*, *374*
Barz, W., 388, 390, 391, *420*, *422*, *425*
Bass, R., 355, *366*
Bateman, D. F., 395, *426*
Bates, A. L., 107, *148*
Batra, P. P., 126, *142*
Bawnik, N., 205, *234*, *237*
Baxter, J. D., 82, *89*
Baxter, M. G., 204, *236*
Bean, G. A., 22, *29*
Bechmann, H., 312, *361*
Beck, A., 342, *362*
Becker, J. M., 331, *363*
Beckmann, J. M., 382, *422*
Bedbrook, J. R., 127, *136*
Bedell, G. W., 86, *89*, 285, 287, 293, 294, *297*, *305*
Beemsterboer, G., 161, *170*
Behal, V., 45, 47, *52*
Bekhtereva, M. N., 120, 128, 130, *136*, *140*
Bell, A. A., 390, 418, *420*
Bell, G. I., 176, *189*
Belozerskeya, T. A., 131, *143*
Benham, R. W., 280, *297*
Bennett, G. A., 430, 431, *438*
Bennett, J. W., 23, *26*
Bensaude, M., 196, *232*
Bentley, R., 13, *26*
Benton, W. D., 176, 183, *189*
Benveniste, R. E., 167, *169*
Bérdy, J., 36, *52*, 153, *168*
Berezinkov, V. M., 130, *136*
Bergman, K., 121, 122, 124, *136*
Berlespch, K. von, 153, *168*
Bernardi, G., 354, *359*
Berry, D. R., 7, *30*, 338, *373*

Author Index

Berry, R. A., 126, *151*
Berthier, M., 384, *420*
Bespalova, J. D., 318, 323, 331, *369*
Betz, B., 81, *92*
Betz, H., 343, *356*
Betz, R., 251, 253, *259*, *263*, 322, 323, 328, 329, 330, 331, 332, 333, *356, 365*
Bhalerao, U. T., 241, *264*
Bharadwaj, T. P., 283, *303*
Bhattacharya, A., 288, 289, *297*
Bianchi, L., 353, *365*
Biemann, K., 243, *259*, 287, *301*
Bilinski, C. A., 346, *356*
Bindl, E., 114, 126, *136*
Birch, A. J., 17, *26*
Bird, B. A., 55, 62, 63, *69, 70*
Bischoff, R., 23, *28*, 77, 80, *91*
Bishop, R. J., 176, *189*, 288, *297*
Bissell, M. J., 83, *92*
Bissett, Y., 319, *367*
Bistis, G. N., 200, *232*, 256, *259*
Biswas, D. K., 84, *90*
Black, O. F., 430, *438*
Blair, L. C., 309, 337, *362, 363*
Blakeslee, A. F., 246, *259*
Blamire, J., 322, *358*
Blanc, H., 187, *190*
Blanc, P. L., 107, *136*, 187, *190*
Blattner, E. R., 162, *171*
Blumenthal, H. J., 105, 120, 121, *142*
Bobowski, G. C., 105, *136*
Boekelheide, K., 166, *170*
Boell, E. J., 355, *356*
Boettger, H. G., 97, *136*
Böger, P., 127, *148*
Bogorad, L., 127, *136*
Boguslawski, G., 268, *297*
Bohlmann, F., 155, *168*
Bohm, H., 21, *29*, 60, *70*, 74, 79, 80, *90*
Bolotin-Fukuhara, M., 347, *360*
Boltyanskaya, E. B., 130, *136*
Bonner, J. T., 1, 12, *26*

Boorstein, R., 83, *91*
Börger, H., 377, *425*
Borgia, P. T., 271, 273, 277, 278, *297, 298*
Börner, H., 404, 405, 418, *421*
Borrow, A., 58, 59, *69*
Borst, P., 353, *356*
Bostian, M., 153, *168*
Bostock, R. M., 382, *421*
Botstein, D., 177, *189*, 310, 311, *362, 374*
Bottomley, W., 377, 378, *426*
Boucher, F., 108, *136*
Bowers, B., 275, 290, *298*, 326, *357*
Boyd, D. H., 206, *236*
Brackenbury, R., 3, *31*
Bracker, C. E., 199, *232*, 274, 275, *296, 297, 299*
Bradshaw, T. K., 84, *89*
Brain, R. D., 125, *136*
Brambl, R., 175, *189*, 353, *356, 357, 363*
Bramley, P. M., 101, 103, 105, 127, *136, 147, 148*
Brandhorst, B. P., 245, *261*
Brar, S. S., 58, *69*
Braun, P. C., 285, 290, *297*, 327, *357*
Braun, R. J., 347, 348, *361, 368*
Braun, V., 167, *168*
Brawley, J. V., 351, *357*
Brawley, V., 253, *265*, 323, 326, 327, 332, *369*
Brayton, A. R., 282, 289, 293, *306*
Breitenbach, M., 315, 338, 342, 351, 352, 353, 354, *361, 370*
Bremner, D. A., 287, *300*
Brennan, P. J., 16, *26*
Brewer, B. J., 322, 351, 353, *357*
Brian, P. W., 17, 21, *26*, 68, *69*
Briedis, A. V., 104, *145*
Briggs, W. R., 125, *136, 144*
Bright, W., 161, *170*
Brill, W. J., 436, *439*
Britten, R. J., 226, *232*

Britton, G., 97, 100, *136, 137, 150*
Britton, R. J., 179, 180, 182, *189, 190*
Broach, J. R., 242, 254, *262,* 309, *364*
Brock, T. D., 329, *358*
Brody, S., 175, *192*
Brown, A. E., 88, *91*
Brown, C., 391, *420*
Brown, D. T., 24, 25, *26*
Brown, L. A., 293, *297*
Brown, S., 58, *69*
Bruegger, B. B., 383, *421*
Brufani, M., 161, *168*
Brummel, M., 287, 293, 294, *305*
Bruschi, C., 353, *365*
Bryan, R. F., 161, *170*
Buchecker, R., 99, *149*
Bucholtz, M. L., 104, *145*
Buck, J. D., 280, *299*
Bücking-Throm, E., 228, 232, 253, *260,* 322, 330, *357, 359*
Buckley, H. R., 284, 286, *301*
Bulant, V., 43, *53*
Bull, A. T., 7, *26*
Bulla, L. A.,348, 349, *369*
Bu'Lock, J. D., 13, 16, 17, 18, 20, 21, *26, 27,* 38, 43, *52,* 58, 59, *69,* 109, 130, 131, *136, 137,* 155, *169,* 247, 250, 251, *259, 260*
Bünning, E., 122, *137*
Burchall, J. J., 286, *305*
Burden, R. S., 390, 391, 399, *420, 421, 426*
Burgeff, H., 113, *137*
Burke, D. C., 84, *89*
Burke, P. V., 113, *139*
Burnett, J. H., 7, 8, *27,* 107, *137*
Burris, R. H., 10, *27,* 58, *70*
Bussey, H., 324, 331, 337, *368*
Butler, E. E., 199, *232*
Butler, G. M., 7, *27*
Butler, W. L., 125, *145, 146*
Buzzell, R. I., 381, 404, *427*

Byers, B., 317, 320, 323, 324, 327, 340, 341, 346, 355, *357, 359, 370, 374*

C

Cabib, E., 275, 290, *297, 298, 300,* 326, *357*
Caglioti, L., 130, *137,* 245, *260*
Cainelli, G., 245, *260*
Cainelli, T., 130, *137*
Calam, C. T., 23, *27*
Calderón, I. L., 133, *145*
Calderone, R. A., 283, 285, 290, *297, 299,* 327, *357*
Camerino, B., 130, *137,* 245, *260*
Cameron, J. R., 176, *191*
Campbell, C. C., 284, 286, *301*
Campbell, D. A., 323, 328, *357*
Campbell, I. M., 13, *26,* 55, 57, 58, 61, 62, 63, 65, *69, 70, 71*
Cano, R. J., 256, *266*
Cantino, E. C., 3, 11, *27,* 109, 121, 122, *137*
Carbon, J., 176, 177, *189,* 316, *358*
Carlier, C., 167, *169*
Carlile, M. J, 241, 242, *260*
Carlton, W. W., 18, *27*
Carmi, P., 214, *232*
Carpenter, A. T. C., 338, 348, *356*
Carter, B. L. A., 318, 352, *362, 367*
Cartledge, J. L., 246, *259*
Cartwright, D. W., 396, 415, 420, *421*
Caruso, F. L., 418, *424*
Case, M. E., 184, 185, *189, 190*
Casselton, L. A., 9, *27,* 198, 202, *232*
Cassone, A., 282, 284, 285, 286, 289, *297, 302, 304*
Catalano, E. A., 432, *439*
Catt, K. J., *374*
Cattrall, M. E., 117, 119, *137*

Author Index

Cawson, R. A., 282, 286, 287, 289, 293, *303, 305, 306*
Cederberg, E., 105, 106, 132, *137*
Cerdá-Olmedo, E., 113, 114, 115, 116, 121, 122, 124, 131, 133, 134, *135, 136, 138, 139, 142, 144, 145, 150,* 248, *260*
Chacko, S., 23, *28,* 77, 80, *91*
Chaffin, W. L., 282, 285, 289, 293, *296, 297*
Chalutz, E., 385, *421*
Chamberlain, D. W. 381, *426*
Chan, R. K., 253, *260,* 333, *358, 374*
Chapman, D. J., 96, 103, 116, 117, 118, 119, *140, 147*
Chattaway, F. W., 49, *52,* 283, 284, 288, 289, *297, 298*
Chaudhuri, R. K., 217, *233*
Chen, A. W., 343, *358*
Chernysheva, E. K., 131, *143*
Chichester, C. O., 97, 98, 100, 103, 104, 107, 121, 127, 128, 129, 131, 132, *136, 138, 139, 142, 144, 145, 147, 148, 149, 151*
Chiew, Y. Y., 283, 284, 290, 293, *298, 304*
Chilton, M.-D., 162, *169, 171*
Chino, N., 331, *365*
Choih, S. J., 351, *358*
Christensen, C. M., 18, *29*
Chu, I. S., 122, *138*
Chu, M. I., 342, *261*
Chung, C. W., 291, *302*
Ciegler, A., 19, *27,* 130, 131, *138,* 430, 431, 434, 435, *438, 439*
Ciejek, E., 253, *260,* 330, 331, 333, *358*
Cihlar, R. L., 176, *189,* 271, 280, *299*
Cikes, M., 78, *90*
Clancy, M. J., 343, 349, *361*
Clark, F. E., 430, *439*
Clarke, L., 176, 177, *189,* 316, *358*

Clark-Walker, G. D., 273, 276, *298, 304*
Clay, P. T., 16, *26*
Cline, K., 385, *421*
Clutterbuck, A. J., 12, *27,* 64, *70*
Codington, J. F., 330, *358*
Coen, D. M., 127, *136*
Coffen, D. L., 390, *423*
Coggins, C. W., Jr., 103, 104, 105, 129, 130, 131, 133, *138, 142*
Cohen-Bazire, G., 110, *143*
Colarusso, L., 83, *92*
Cole, G. T., 5, *27,* 268, 280, *298*
Coleman, J. R., 84, *90*
Collatz, E., 167, *169*
Collie, J. N., 17, *27*
Colonna, W. J., 342, *358*
Conde, J., 319, 338, 354, *358*
Condit, A., 202, *232*
Conti, S. F., 351, 353, *362*
Cooke, A., 22, *27*
Cooke, R. C., 436, *438*
Cooney, C. L., 37, *53*
Cooney, P. H., 344, *358*
Cooper, R. D. G., 161, *169*
Coppola, C. P., 283, *304*
Cornford, C. A., 411, *425*
Corral, J., 278, *296*
Cortat, M., 327, *358*
Costerton, J. W., 67, *71*
Cota-Robles, E., 274, *296*
Cottrell, S. F., 321, *358*
Courvalin, P., 167, *169*
Cousins, R. J., 83, *90*
Cox, B. S., 348, *372*
Cox, R. P., 77, *90*
Crain, W. R., 179, *189*
Crandall, M. A., 329, *358*
Creanor, J., 352, *367*
Crick, F., 181, *189*
Croes, A. F., 340, 342, *358, 360*
Cruickshank, I. A. M., 377, 397, 398, *421, 426*
Cryer, A., 83, *90*

Cryer, D. R., 322, *358*
Curdova, E., 44, 45, 47, *52*
Curran, J. F., 82, *93*
Currier, W. W., 378, 392, *424*
Curry, G. M., 124, *146*
Curtis, D., 280, *299*
Cutler, J. E., 282, 287, 294, *299*
Cybulska, E. B., 326, *372*
Czeczuga, B., 96, 97, *138*

D

Dabrowa, N., 282, 283, 284, 287, 291, 293, *298, 300*
Da Costa, E. W., 219, *234*
Dahlberg, K. R., 175, 182, *192*
Dailey, R. G., 161, *170*
Dandekar, S., 131, *138*
Darvill, A. G., 384, *423*
Datta, A., 288, 289, *297, 305*
Davidow, L., 346, *359*
Davidson, E. H., 179, 180, 182, *189, 190*, 214, 216, 226, *232, 233*
Davidson, N., 179, *191*
Davies, B. H., 96, 97, 99, 100, 101, 103, 120, 121, 127, 128, 129, 132, *136, 138*
Davies, J. E., 167, *169*
Davies, R. R., 290, *298*
Davis, B. N., 55, *69, 70*
Davis, M. M., 182, *190*
Davis, R. W., 176, 177, 183, *189, 191, 192*
Dawes, I. W., 342, 343, *359, 372, 373*
Dawid, I. B., 181, *191*
Day, A. W., 116, 118, *138, 139*
Day, L. E., 159, *169*
Day, L. L., 116, 118, *138*
Day, P. R., 200, 206, *233*
Day, R. R., 111, 113, 116, *140*
Dayhoff, M. O., 331, *359*
De Fabo, E. C., 122, 123, 124, *139*

DeGennaro, L. J., 176, *189*
Delbrück, M., 103, 108, 113, 114, 115, 120, 121, 122, 123, 124, 125, 132, *139, 142, 143, 145, 146, 147*, 250, *263*
Del Rey, F., 342, *359*
Demain, A. L., 13, 20, 21, 22, 24, *25, 27, 29*, 37, 41, 43, 44, 47, 48, 49, *51, 52, 53, 54*, 55, 58, 60, *70*, 153, 155, 157, 159, *168, 169, 170*
Dembeck, P., 181, *191*
Den Hollander, J. A., 343, *356, 374*
Denning, T. J., 290, *298*
Dennis, D. T., 132, *139*
Depicker, A., 75, *94*
Deppe, C. S., 209, *237*
De Robertis, E. M., 313, *359*
Detory, R. W., 38, 43, *52*
Detroy, R. W., 58, 59, *69*, 430, 431, *438*
Deus, B., 78, 80, 81, *94*
Devay, J. E., 436, *438*
Deverall, B. J., 378, *421*
De Vries, G. J. M. L., 342, *358*
De Vries, O. M. H., 178, 179, 180, *189*
Dewick, P. M., 387, 388, *421, 422*
Dexter, D. L., 84, *90*
De Zamaroczy, M., 354, *359*
Diala, E. S., 353, *359*
Dick, S., 219, *234*
Diekmann, H., 163, *168, 169*
Dixon, R. A., 418, *425*
Dobzhansky, T., 22, *27*
Doerfler, D. L., 55, 57, 58, 62, 63, 65, *69, 70, 71*
Doerschug, M. R., 82, 89, *90*
Doi, S., 329, 338, *359*
Doke, N., 419, *422*
Domek, D. B., 271, 273, 278, *298*
Domer, J. E., 286, *298*
Donavan, F. W., 17, *26*
Dons, H., 216, 217, *238*
Dons, J. J. M., 178, 179, 180, *189*, 214, *233*

Doty, P., 83, *92*
Dougall, D. K., 80, 81, *91*
Dow, J. M., 273, *298*
Dowbenko, D. J., 351, *359*
Doyle, M. E., 75, *92*
Doyon, J. D., 179, 180, *192*, 214, *237*
Drake, D., 130, 131, *136*, *137*, 250, *259*
Drew, S. W., 159, *169*
Driver, C. H., 256, *260*
Droms, K. A., 179, 180, *192*, 214, *237*
Dubovoy, C., 208, 213, *233*
Dulaney, E. L., 42, *52*
Dunkle, L. D., 175, *189*, *192*
Dunn, J. J., 430, *438*
Duntze, W., 228, *232*, 251, 253, *259*, *260*, *263*, *265*, 322, 323, 328, 329, 330, 331, 332, 333, *356*, *357*, *359*, *365*, *370*, *371*, *372*
Duran, A., 275, 290, *298*
Dusenbery, R. L., 214, *233*
Dutcher, S. K., 319, 321, *359*
Dutta, S. K., 214, 217, *233*, *235*

E

Eagle, H., *90*
Ebel, J., 378, 387, 390, 392, 399, 403, 405, 407, 408, 419, *420*, *422*, *423*, *428*
Eberhardt, N. L., 104, *139*
Eddy, E. M., 324, *370*
Edelman, G. M., 332, 333, *364*
Eden, F. C., 179, *189*
Edwards, J. A., 243, *259*
Egashira, T., 107, *148*
Egel, R., 329, *359*
Eggert, W. D.,
Egin-Bühler, B., 387, *422*

Ehmann, C., 343, *373*
Ehrlich, M., 24, *28*
Eijk, G. W., van, 97, 120, *139*, *150*
Eisner, T., 77, *90*
Ejchart, A., 354, *356*
Elahi, M., 128, 129, *139*
Elander, R. P., 49, *52*
El Kousy, S., 61, *70*
Ellingboe, A. H., 202, 204, 215, *233*, *236*, *238*
Elliott, C. G., 255, *260*
El-Shagi, H., 78, 80, 81, *94*
Elstner, E., 385, *426*
Emanuel, J. R., 343, *359*
Emeis, C. C., 351, *360*
Emerson, R. L., 58, *70*, 241, 242, *260*
Ende, G., van den, 389, *422*
Ende, H., van den, 130, *139*, 246, 250, *260*, *263*
Engel, H., 122, *140*
Engels, F. M., 340, *360*
Ennis, H. D., 351, *359*
Ephrussi, B., 352, *360*
Epstein, A., 270, 273, 275, 276, 277, *299*, *303*
Epstein, W. W., 100, *135*
Ernst, L. A., 57, *70*
Erwin, D. C., 381, 395, 398, *424*
Eslava, A. P., 113, 121, 122, 124, 131, *136*, *139*, 251, *261*
Esmon, B., 337, *360*
Esposito, M. S., 338, 339, 342, 344, 345, 346, 347, 348, 353, *356*, *360*, *361*, *362*, *365*, *374*
Esposito, R. E., 338, 339, 342, 344, 345, 346, 347, 348, *356*, *360*, *361*, *362*, *363*, *365*
Esser, K., 10, 12, *28*, 216, *236*
Eugster, C. H., 97, 98, *140*, 155, *169*
Evans, E. G. V., 282, 286, *298*
Ewing, D. F., 390, *424*

F

Failla, M. L., 83, 86, 87, *90*
Falcone, G., 291, *298, 302*
Fangman, W. L., 178, *192,* 321, 322, 351, 353, *357, 366*
Farkas, V., 270, 275, *299*
Faugeron-Fonty, G., 354, *359*
Faulkner, D. J., 153, *169*
Fautz, E., 407, *422*
Fawcett, P., 109, *141*
Federici, B. A., 96, *140*
Feeny, P. P., 74, 75, *94*
Fehrenbacher, G., 322, 323, 328, 333, *360*
Feldon, R. A., 178, *189*
Fennell, D. I., 7, *30*
Fennell, D. J., 430, 431, *438*
Feofilova, E. P., 105, 106, 120, 128, 130, *136, 140*
Fernandez-Pol, J. A., 88, 89, *90*
Ferri, M., 353, *365*
Ferro, A. J., 351, *357, 358*
Fiasson, J. L., 121, *140*
Field, C., 337, *366*
Filner, P., 124, 125, *148*
Fincham, J. R. S., 111, 113, 116, *140*
Fineman, B. C., 280, 287, 290, *299*
Fink, G. R., 177, *190,* 313, 319, 338, 354, *358, 362*
Finkelstein, D., 253, *261*
Finkelstein, D. B., 322, 333, *358, 360, 372*
Firtel, R. A., 181, *190,* 217, *233*
Fischer, S., 84, *90*
Flach, W., 408, 409, *427*
Flegel, T. W., 254, 255, *261*
Flexer, A. S., 200, *234*
Flor, H. H., 380, 419, *422*
Flores-Carreon, A., 275, *301*
Floss, H. G., 51, *52, 54*
Fogel, S., 321, 323, 324, 328, 344, 345, 348, *356, 360, 368, 370*

Fonzi, W. A., 342, *360*
Foppen, F. H., 97, *142*
Forbes, D., 309, *362*
Formanek, H., 106, *151*
Forte, J. W., 277, *299*
Foster, J. W., 13, 21, *28*
Fournier, P., 187, *190*
Fowell, R. R., 309, *361*
Francis, S., 201, *235*
Frank, J. A., 401, *422*
Frank, K. R., 343, *361*
Frankel, C., 215, *233*
Franz, A. O., 432, *439*
Frazer, R. S. S., 352, *367*
Frazier, W. C., 87, *91*
Frederick, J. F., 167, *169*
Free, S. J., 176, *190*
Freedberg, J. A., 125, *136*
Freedman, J. H., 87, *92*
Freer, S. N., 175, 181, 182, *190, 192*
Freese, E., 342, 344, *358, 361*
Freese, E. B., 342, *361*
Freifelder, D., 308, *361*
Frey-Wyssling, A., 106, *140*
Friedenthal, M., 270, 273, 275, *299*
Friederichsen, J., 122, *140*
Friedman, L., 282, 283, 286, 288, 290, 291, 293, *300*
Friedman, M., 182, *190,* 343, *362*
Friedmann, A., 340, *362, 374*
Friend, J., 120, 121, 127, 131, *140*
Friesen, J. D., 182, *192,* 343, *372*
Frisvad, J. C., 430, *439*
Frosch, S., 121, *140*
Fuchs, A., 390, *422*
Fugit, D. R., 348, *364*
Fujimori, E., 107, *144*
Fujino, M., 254, 255, 258, *261, 262, 265*
Fukui, S., 254, 255, 258, *259, 262, 265, 266*
Fuller, M. S., 240, *264*

Author Index

G

Gaber, R. F., 219, *234*
Galau, G. A., 179, 182, *189, 190*
Galbraith, J. C., 7, *31*
Galeazzi, D. R., 111, *146*
Galland, P., 109, 110, *140, 148*
Gallo, M., 38, 40, *52*
Gallowitz, D., 181, *190*
Galston, A. W., 122, *140*
Game, J. C., 315, 347, 348, *361, 368*
Garas, N. A., 419, *422*
Garber, E. D., 103, 116, 117, 118, 119, *137, 140, 141, 151*
Garcea, R. L., 181, *192*
Garcia, J. R., 271, 278, 279, 291, *299*
Garcia-Acha, I., 342, *359*
Gardner, J., 84, *91*
Garland, E., 166, *168*
Garnjobst, L., 112, *141*
Garton, G. A., 121, *141*
Gaucher, G. M., 58, 59, 63, 65, 66, 67, *70, 71*
Gaucher, M., 21, *30*
Gauger, W. L., 251, *261*
Gavin, J. B., 287, *300*
Gehrig, R. F., 434, *439*
Geier, M., 331, *358*
Gelehrter, T. D., 82, *93*
Gelfand, D. H., 176, *189*
Gerbaud, C., 187, *190*
Gerschenson, L. E., 82, *90*
Gershon, H., 287, *301*
Ghazali, H. M., 282, 284, 289, *304*
Giam, C. S., 58, *69*
Giesy, R. M., 200, *233*
Gilbert, L. E., 21, *28*
Giles, N. H., 184, 185, *189, 190, 192*
Gilman, A. G., 83, *92*, 213, *233*
Gingras, G., 108, *136*
Girbardt, M., 199, 201, 202, 203, *233*

Gladstone, P., 212, *233*
Glass, J., 84, *90*
Glass, R. W., 129, *139*
Glazener, J. A., 396, *422*
Gnanamanickam, S. S., 396, *422*
Goetsch, L., 323, 324, 327, 340, 341, 346, *357, 359, 374*
Goff, C. G., 178, *190*
Goldberg, B., 78, *90*
Goldberg, R. B., 178, 179, 180, 181, 182, *191, 192*, 214, *234*
Goldblatt, L. A., 18, *28*
Goldfien, A., 20, *29*
Goldie, A. H., 111, 112, *141, 149*
Goldstrohm, D. D., 107, 122, *141*
Goldthwaite, C. D., 322, *358*
Gong, C.-S., 175, *191*
Gooday, G. W., 9, *28*, 109, 130, *141*, 239, 242, 246, 247, 250, 251, *260, 261, 262*
Goodgal, S. H., 105, *143*
Goodman, H. M., 58, *71*, 75, *94*
Goodwin, T. W., 19, *28*, 96, 97, 100, 101, 103, 113, 120, 121, 127, 130, 131, *137, 138, 140, 141, 150*
Gordee, E. Z., 159, *169*
Gordon, M. P., 162, *171*
Gordon, R. E., 430, *439*
Gorman, M., 49, *52*
Gottlieb, F. J., 57, 58, 65, *70*
Goulden, S. A., 49, *52*
Goursot, R., 354, *359*
Govind, N. S., 130, *141*
Graham, H., 88, *91*
Grange, A., 121, *140*
Granner, D., 82, *93*
Grant, D. W., 57, *70*
Grant, G. A., 268, 273, *302*
Grant, P., 354, *361*
Grappel, S. F., 283, *299*
Green, D., 109, *141*
Green, H., 77, 78, 79, 80, 82, 83, 84, *90, 91, 92*
Green, J. M., 243, *259*
Greenfield, P. C., 355, *356*

Greenwood, K. T., 167, *171*
Gregg, C. T., 355, *366*
Gregg, J. H., 3, *28*
Gregonis, D. G., 100, *135*
Gressel, J., 122, 127, *141*
Gresser, I., 79, 80, *91*
Grewal, N. S., 344, 346, *361*
Gribanovski-Sassu, O., 97, *142*
Grieshaber, E., 106, *140*
Griffiths, L. A., 120, 127, 131, *140*
Griffiths, M., 107, *149*
Grisebach, H., 378, 386, 388, 392, 396, 399, 400, 402, 404, 405, 407, 408, 418, *421, 422, 425, 428*
Grisvard, J., 88, *91*
Grivell, L. A., 313, 353, *356, 365*
Gröger, D., 161, *169*
Grootwassink, J. W. D., 58, 59, 65, *70*
Gross, G., 378, 390, 392, *422*
Gross, P. R., 1, *28*
Grossman, L., 177, *190*
Grove, S. N., 275, *299*
Grunstein, M., 176, *190*
Guardia, M. D., de la, 113, 114, *135, 142*
Guerdoux, J. L., 12, *28*
Guerineau, M., 187, *190*
Guerra, P., 287, *300*
Guilacci, P. L., 284, *300*
Guille, E., 88, *91*
Günther Sillero, M. A., 355, *372*
Gurdon, J. B., 313, *359*
Gutmann, H., 19, *29*
Gutz, H., 351, *360*
Guzman, G., 19, *28*
Gwynne, D. J., 245, *261*
Gyllenberg, H. E., 120, *151*

H

Haag, G., 253, *265,* 331, *370*
Haagen-Smit, A. G., 243, *264*
Haapala, O. K., 214, 215, *233*

Haavik, H. I., 21, *28*
Haber, J. E., 175, *190,* 338, 342, 343, *361, 364, 367, 373*
Hadwiger, L. A., 382, 389, *422, 423*
Hagiya, M., 323, 328, 329, *372, 373, 374*
Hahlbrock, K., 386, 387, 407, *422, 423*
Hahlbrook, K., 81, *92*
Hahn, M. G., 382, 384, 399, *423, 427*
Haid, A., 312, *361*
Haidle, C. W., 270, 275, *299*
Hall, B. D., 176, 182, *190, 191,* 254, *264,* 309, 314, 343, 353, 354, *362, 366*
Hall, N. E. L., 88, *91*
Halvorson, H. O., 321, 338, 342, 343, 348, 349, 351, 353, 354, *360, 361, 362, 363, 364, 368, 369, 370, 372*
Hamada, K., 323, 328, *373*
Hamill, R. L., 49, *52*
Hamilton, D., 13, *27,* 58, 59, *69*
Hamilton, J. G., 286, *298*
Handschin, B., 353, *357*
Hansen, A. M., 18, 19, *28*
Hansen, E. C., 353, *361*
Hanson, D. K., 181, *191*
Hanson, R. S., 75, *92*
Harashima, S., 346, *371*
Harborne, J. B., 378, *423*
Hardie, I. D., 343, *359*
Harding, R. W., 100, 104, 121, 122, 123, 126, 134, *139, 142, 149*
Hardy, T. M., 251, *260*
Hargreaves, J. A., 383, 384, 386, 396, 413, 414, *423*
Harkin, J. C., 286, *298*
Harold, F. M., 242, *261*
Harper, D. B., 88, *91*
Harper, J. F., 343, 349, *361*
Harris, R. C., 130, *150*
Hart, J., 124, 125, *148*
Hartig, A., 315, 338, 351, 352, 353, 354, *361*
Hartmann, K. M., 126, *142*

Hartwell, L. H., 228, *232*, 253, *261*, 267, *299*, 311, 312, 315, 317, 318, 319, 322, 327, 328, 330, 331, 332, 333, 335, 336, 338, 355, *357*, *361*, *362*, *363*, *368*
Harwig, J., 17, *28*
Hashimoto, T., 105, 120, 121, *142*
Hashimoto, T., 351, 353, *362*
Hasilik, A., 288, *299*
Hatch, W. R., 240, 242, *261*
Hautala, J. A., 184, *192*
Hawker, L. E., 3, 10, *28*
Hawthorne, D. C., 310, 338, *366*
Haxo, F. T., 97, 111, 112, 122, 124, 127, *142*, *149*, *150*
Hayashi, M., 80, *91*
Hayashibe, M., 352, *366*
Hayes, A. W., 435, *438*
Hayes, W. A., 12, *28*
Hayman, E. P., 129, *142*
Hazen, K. C., 282, 287, 294, *299*
Heath, M. C., 390, *423*
Heathcote, J. G., 18, *28*
Hedden, D. M., 280, *299*
Hedden, P., 18, *28*
Hegnauer, R., 155, *169*
Heisenberg, M., 113, *142*
Heitefuss, R., 385, *426*
Heller, W., 387, *423*
Hendrie, M. R., 255, *260*
Hendrix, J. W., 63, *70*
Henning, G. L., 103, 129, 133, *138*
Henry, S. A., 338, 342, 348, *356*, *362*, *372*
Herber, R., 127, *142*
Herdson, P. B., 287, *300*
Hereford, L. M., 180, *190*
Herskowitz, I., 253, 254, *263*, *265*, 309, 333, *362*, *371*, *374*
Heslot, A., 187, *190*
Hess, S. L., 389, *423*
Hess, W. M., 7, *31*
Heuvel, J., van den, 390, *423*
Hiatt, W. R., 271, 278, 279, 291, *299*
Hibbert, J. R., 18, *28*

Hicks, J. B., 177, *190*, 242, 254, *262*, 309, 313, 333, *362*, *364*, *371*
Higgins, V. J., 387, 390, *423*
Hilgenberg, W., 121, 127, *148*
Hill, W. B., 293, *300*
Hille, A., 419, *423*
Hinkle, O., 321, *368*
Hinnen, A., 41, *52*, 177, *190*, 313, *362*
Hiragun, A., 83, *91*
Hirschberg, J., 320, *362*
Hirschberg, R., 322, *358*
Ho, K. H., 343, *362*
Hoeksema, H. L., 231, *238*
Hoffman, R. M., 12, *30*, 210, *234*
Hoffmann, A., 161, *169*
Hoffmann-Ostenhof, O., 310, 342, *365*
Hofmann, A., 19, *31*
Hogan, G., 280, *299*
Hoge, J. H. C., 180, *193*, 215, 216, 217, 230, *238*
Hogness, D. S., 176, *190*
Hohn, B., 313, *362*
Holland, K. T., 282, 286, *298*
Holm, P. B., 214, *232*
Holmes, M. R., 284, 288, 289, *297*
Holsters, M., 75, *94*
Holtzer, H., 23, *28*, 77, 80, *91*
Honwad, V. K., 399, *426*
Hopfield, J. J., *374*
Hoppe, H. H., 385, *426*
Hopper, A. K., 182, *190*, 343, *362*, *364*
Hopwood, D. A., 161, 162, *169*, *171*
Horecker, B. L., 87, *93*
Horenstein, G. A., 122, *137*
Horesh, O., 340, *362*
Horesh-Cabilly, O., *374*
Horgen, P. A., 7, *30*, 178, *190*, 244, 245, *261*, *265*
Horne, R. W., 324, *366*
Hornemann, K. M., 51, *54*
Horner, E. S., 432, *439*
Hostalek, Z., 38, 43, 44, 45, 47, *52*,

[Hostalek, Z.], 58, 59, *69*
Hottinguer, H., 352, *360*
Howard, D. H., 283, 291, 293, *298*
Howes, C. D., 126, *142*
Hsu, W.-J., 103, 104, 105, 113, 114, 115, 120, 124, 128, 129, 130, 131, 132, *142, 146, 147*
Hu, H.-Y. Y., 84, *91*
Huang, P. C., 112, *142*
Hubbard, M. J., 293, *303*
Huber, A., 107, 108, *143*
Huber, F. M., 59, *71*
Huber-Willer, A., 125, *148*
Huda, S., 108, *147*
Huddleston, J. A., 162, *170*
Hudspeth, M. E. S., 178, 179, 180, *191*, 214, *234*
Hughes, S. J., 7, *29*
Huisman, J. G., 246, *263*
Hulme, M. A., 13, *27*, 58, 59, *69*
Hümme, B., 385, *426*
Hungate, M. V. G., 112, *143*
Hunt, D. F., 287, *301*
Hunt, L. T., 331, *359*
Hurst, S. K., 283, *301*
Hutchinson, S. A., 255, *263*
Huttunen, E., 201, *237*
Hyatt, M. T., 109, *137*

I

Ikawa, M., 97, *149*
Illingworth, R. F., 342, *362*
Inaba, R., 323, *373*
Inamine, E., 41, *52*
Inderlied, C. B., 271, 278, 279, 280, *299*
Ingham, J. L., 378, 381, 384, 390, 399, *420, 423, 424*
Ininger, G., 61, *70*
Inlow, D., 268, *304*
Ishikawa, T., 213, *238, 374*
Islam, M. S., 256, *261*
Isler, O., 19, *29*

Isogai, A., 254, 255, 258, *261, 262, 265, 266*
Itakura, K., 181, *191*

J

Jackson, R. L., 229, *234*
Jacobson, G., 347, *362*
Jacobson, J. W., 184, *192*
Jaffe, L. F., 7, *29*
Jagadish, M. N., 318, *362*
Jamikorn, M., 127, *141*
Jani, U. K., 131, *138*
Janszen, F. H. A., 211, *234*
Janzen, D. H., 21, 22, *29, 30,* 433, 434, 435, *439*
Jarvik, J., 310, 311, *362*
Jarvis, B. B., 22, *29,* 161, *170,* 435, *439*
Jawets, E., 20, *29*
Jayaram, M., 121, 123, *143*
Jaynes, P. K., 277, *299*
Jeanloz, R. W., 330, *358*
Jechova, V., 45, 47, *52*
Jeffery, K., 293, *303*
Jefferys, E. G., 58, 59, *69*
Jeffs, P. W., 166, *170*
Jensen, A. H., 434, *439*
Jensen, B., 348, *356*
Jensen, R., *374*
Jensen, S. L., 110, *143*
Jerebzoff, S., 109, 120, 121, *143*
Jerebzoff-Quintin, S., 109, 120, 121, *143*
Jersild, R. A., 200, 202, *234, 235*
Jillson, O. F., 287, 290, *300, 302*
Jindras, A., 51, *54*
Jobbagy, A. J., 276, *306*
Johnke, R., 352, *363*
Johnson, B. C., 166, *169*
Johnson, E. A., 98, *143*
Johnson, J. D., 181, *191*
Johnson, P., 181, *191*
Johnson, S. A., 175, *191*

Author Index

Johnson, S. A. M., 280, 282, 286, 287, 290, 291, *300*
Johnston, G. S., 318, *363*
Johnston, L. H., 315, 316, *363*
Joly, J., 283, *296*
Jones, B. E., 109, 130, 131, *137*, 250, 251, *260, 261, 262*
Jones, E. W., 343, *374*
Jones, J. G., 433, *439*
Jones, J. K., 116, *138, 139*
Jones-Brown, Y. R., 337, *363*
Josephson, L. M., 432, *439*
Josephson, M., 353, *357, 363*
Joshi, K. R., 283, 287, *300, 303*
Joyce, B. K., 57, *70*
Julian, G. S., 348, 349, *369*

K

Kalitzer, S., 253, *260*
Kamalay, J. C., 182, *191*
Kamiya, Y., 255, 258, *262*, 333, *363*
Kane, S., 342, *363*
Kaplan, D. T., 397, *424*
Karasawa, K., 127, 128, 132, *144*
Kasimos, J. N., 347, *368*
Kassir, Y., 309, 346, *362, 363, 374*
Katayama, T., 98, *149*
Katayar, S. S., 104, *145*
Kates, M., 111, 112, *143*
Katz, E., 21, 24, *29*, 38, 40, *52*, 159, *170*
Katzir, M., 124, 125, *139*
Kaudewitz, F., 312, *361*
Kaufman, S. H., 82, *92*
Kawanabe, Y., 323, 328, *363, 373*
Kawasaki, G., 315, 321, *363*
Keen, N. T., 381, 383, 395, 397, 399, 404, 419, *421, 424, 426*
Keeping, J. W., 162, *168*
Keller, F. A., 290, *297, 300*
Keller-Schierlein, W., 163, *170*

Kemp, D. J., 183, *189*
Kendrick, B., 5, *27, 29*
Kennel, Y. M., 42, *52*
Kerridge, D., 285, *300*
Kerruish, R. M., 219, *234*
Kessell, R. H. J., 58, 59, *69*
Kessler, G., 284, *300*
Key, J. L., 126, *150*
Keyhani, E., 105, *143*
Keyhani, J., 105, *143*
Keynan, A., 355, *363*
Khan, S. A., 331, *363*
Khokhlov, A. A., 159, *170*
Kielland-Brandt, M. C., 343, *367*
Killick, K. A., 351, *363*
Kiltz, H., 251, *265*, 330, 331, *371*
Kim, W., 418, *427*
Kimura, A., 41, *52*
King, J. C., 77, *90*
King, R. D., 2, *29*
Kinnaird, J. H., 342, *372*
Kinnersly, A. M., 80, 81, *91*
Király, Z., 378, *424*
Kirk, J. T. O., 130, *150*
Kisaki, T., 82, *93*
Kita, H., 331, 332, 333, *365, 371*
Kitada, C., 254, 255, 258, *261, 262, 265*
Kitano, S., 328, *370*
Klapholz, S., 338, 344, 346, *360, 363, 374*
Klapper, B. F., 242, *262*
Klapper, M. H., 242, *262*
Klar, A. J. S., 242, 254, *262*, 309, 338, 343, *363, 364, 372*
Klarman, W. L., 395, 415, 417, *424, 426*
Klebs, G., 11, *29*
Klein, W. H., 182, *190*
Kleinerova, E., 43, *53*
Kleinig, H., 129, *143*
Klemm-Wolfgramm, E., 125, *146*
Klotz, L. C., 176, 178, 179, *191*
Kniep, H., 196, *234*

Knight, R. H., 175, *192*
Knight, S. G., 87, *91*, 434, *439*
Knights, B. A., 255, *260*
Knizley, H., 101, 127, 133, *146*
Knox, R. B., 437, *439*
Ko, W. H., 256, *262*
Kobayashi, G. S., 268, 273, 284, *300, 301, 302*
Koch, K. S., 83, *91*
Kodde, E., 378, 382, 395, 409, *428*
Koehrn, F. J., 129, *146*
Koffler, H., 58, *70*, 87, *91*
Kogane, F., 62, 63, *72*
Kohorn, B. D., 214, *237*
Kojima, M., 418, *424*
Kokontis, J. M., 116, *141*
Kolat, A. I., 348, *356*
Kolbitzer, S., 330, *359*
Kolthoum, I., 155, *168*
Koltin, Y., 200, 204, 205, 206, 209, 211, 214, 215, 222, *232, 234, 236, 237, 238*
Kominek, L. A., 41, *52*
Kondoh, H. C., 279, *300*
Konieczny, S. F., 84, *90*
Konigsberg, I. R., 80, *91*
Kopecka, M., 342, *364*
Korf, R. P., 97, *143*
Kornberg, R. D., 178, *191*
Kot, E. J., 290, *300*
Kowerski, R. C., 100, *135*
Kozinn, P. J., 286, *305*
Kozlova, Y. I., 120, 128, *140*
Kraig, E., 343, *364*
Kreger-Van Rij, N. J. W., 340, 342, 349, 350, 351, *364*
Kremen, A., 44, *52*
Krinsky, N. I., 107, 108, *135, 143, 144*
Kristkii, M. S., 131, *143*
Krogh, P., 17, *29*
Kruger, K., 75, *94*
Krumlauf, R., 178, 179, 180, *191*
Krupinski, V. M., 51, *52*
Kuan, T. L., 398, *424*

Kuć, J. A., 378, 382, 392, 418, 419, *421, 424, 425, 426*
Kuehn, H. H.,97, *151*
Kuenen, R., 10, *28*
Küenzi, M. T., 354, *364, 372*
Kuhn, J., 226, 227, *234*
Kuhn, P. J., 390, *424, 426*
Kujima, M., 418, *427*
Kulikova, A., 120, *151*
Kumagai, T., 108, *143*
Kumar, B. V., 268, *301*
Kumar, V. B., 268, 273, *302*
Kundu, S. C., *374*
Kupchan, S. M., 161, *170*
Kurata, H., 432, *439*
Kuri-Harcuch, W., 83, *91*
Kurjan, J., *374*
Kurylowicz, W., 57, *70*, 155, *170*
Kusaka, I., 254, *259*
Kushner, P., 309, *362*
Kushner, S. R., 184, 185, *189, 192*
Kushwaha, S. C., 111, 112, *143*
Kuvshinova, V. I., 130, *136*
Kwolek, W. F., 432, *439*
Kwon-Chung, K. J., 293, *300*
Kybal, J., 43, *53*

L

Laatsch, H., 155, *168*
Lacroute, F., 177, *189*
Lafer, E. M., 24, *30*
Lai, E. Y., 175, *190*
Laine, R. A., 382, *421*
Lam, K. B., 322, *358*
Lamb, C.J., 408, 418, *422, 425*
Lambowitz, A. M., 268, 273, *302*
Lamie, F., 313, *365*
Lampen, J. O., 326, *372*
Land, G. A., 282, 283, 286, 288, 290,291, 293, *300*
Land, M. D., 82, *92*
Landau, J. W., 282, 284, 287, 293,

Author Index

[Landau, J. W.], 298, 300
Lane, P. L., 202, 235
Lang, W., 114, 126, 136, 143
Langcake, P., 394, 395, 396, 411, 415, 420, 421, 424, 425
Lange, V. I., 200, 219, 224, 234
Langeron, M., 287, 300
Lang-Feulner, J., 124, 143
Lansbergen, J. C., 126, 144
Lappe, U., 390, 425
Larsen, A. D., 271, 276, 278, 300, 301
Larson, B. R., 100, 135
Lauer, G. D., 176, 178, 179, 191
Laurence, J. B., 84, 90
Lawellin, D. W., 57, 70
Lawrence, L. M., 329, 358
Lawton, M. A., 418, 425
Lazaridis, I., 166, 170
Lazarovits, G., 381, 404, 417, 425, 427
Leclerc, R. F., 182, 184, 193
Lee, K. L., 284, 286, 301
Lee, L. H., 321, 358
Lee, L. S., 430, 438
Lee, S. C., 382, 384, 425
Lee, S. G., 157, 170
Lee, T.-C., 100, 127, 128, 129, 132, 139, 144
Lee, T. H., 97, 100, 127, 128, 129, 132, 136, 139, 144
Lee, T. M., 343, 356
Leffert, H. L., 83, 91
Legrand, M., 383, 419, 424
Leibowitz, M. J., 331, 337, 364, 373
Leith, W. H., 250, 261
Lembach, K. J., 280, 301
Lemontt, J. F., 348, 364
Leonard, T. J., 219, 234
Leong, T.-Y., 125, 144
Leutwiler, L., 123, 143
Levi, J. D., 251, 262, 330, 364
Levin, D. A., 75, 91
Lewin, B., 179, 191

Lewin, R., 429, 439
Lewis, K. E., 256, 262, 264
Lewis, M. J., 98, 143
Leworthy, D. P., 415, 420, 421, 425
Leworthy, J. C., 399, 425
Liaaen-Jensen, G., 97, 98, 144
Liaaen-Jensen, S., 96, 97, 99, 135, 136, 144
Liao, H., 253, 262, 333, 364
Light, R. J., 17, 29, 58, 59, 70
Lightly, A. C., 114, 115, 132, 146
Lijinsky, W., 121, 141
Lillehoj, E. B., 432, 439
Lilly, V. G., 107, 122, 138, 141
Lincoln, R. E., 100, 110, 147
Lindemann, I., 122, 147
Lingappa, B. T., 287, 301
Lingappa, Y., 287, 301
Link, G., 127, 136
Linnane, A. W., 354, 366
Lipke, P. N., 253, 263, 323, 324, 326, 328, 333, 364
Lipmann, F., 157, 170
Lipson, E. D., 123, 124, 125, 144
Lisker, N., 382, 425
Littau, V., 157, 170
Liu, C. M., 47, 50, 53
Lizlova, L. V., 318, 323, 331, 369
Lloyd, E. C., 58, 59, 69
Lloyd, P. B., 58, 59, 69
Lodish, H. F., 181, 190
Loker, W. J., 87, 92
Long, E. O., 181, 191
Long, I., 283, 304
López-Días, I., 122, 124, 133, 134, 144, 145
Lopez-Romero, E., 275, 304
Loras, P., 47, 48, 50, 53
Losel, D. M., 16, 26
Löser, G., 125, 144
Losick, R., 355, 364
Loumaye, E., 374
Louria, D. B., 283, 305
Lovelace, E., 82, 93
Lovett, J. S., 11, 27, 29, 175, 191

Loyal, R., 387, *422*
Lubben, A., 19, *30*
Luckner, M., 21, *29*, 60, 61, *70, 71,* 157, 159, *170*
Ludescher, R. D., 321, *366*
Ludvik, J., 59, *72*
Ludwig, G. D., 105, *145*
Luke, R. K. J., 167, *171*
Lusnak, K., 340, *368*
Lutkon, A., 121, *138*
Luxton, M., 22, *27, 29*
Lyne, R. L., 399, 407, 408, *424, 425, 428*
Lyons, J., 84, *90*

M

Mabe, J. A., 49, *52*
McAlear, J. H., 199, *235*
McAlister, L., 333, *360*
McCann, P. A., 40, *53*
McClary, D. O., 280, 282, 283, 287, 290, *300, 301*
McCorkindale, N. J., 255, *263*
McCullough, J., 253, *263*
McCusker, J. H., 342, *364*
McDaniel, L. E., 47, 50, *53*
McDonald, W. C., 282, 283, 286, 288, 290, 291, 293, *300*
McDonough, J. P., 277, *299*
Machlis, L., 240, 241, 242, *260, 263, 264*
Mackall, J. C., 82, *92*
MacKay, V. L., 253, *259, 263,* 323, 325, 326, 327, 333, 334, 336, 348, *356, 359, 364, 372*
Mackinney, G., 121, *138, 144, 145*
Mackinnon, J. E., 280, *301*
Macko, V., 287, *301*
MacLaughlin, A., 343, *356*
McLaughlin, C. S., 18, *32*
Macmillan, 18, *28*
MacMillan, J. D., 18, *28,* 107, *145*
McMorris, T. C., 243, 244, 245, *259, 263*

McMurrough, I., 275, *301*
McNitt, K. L., 153, *168*
Magee, P. T., 182, *190,* 338, 342, 343, 349, *358, 359, 361, 362, 364*
Magill, J. M., 111, 112, *151*
Magni, G., 287, *301*
Mahlar, H. R., 181, *191*
Mahler, H. R., 277, *299*
Majumdar, M. K., 41, *53*
Majumdar, S. K., 41, *53*
Malek, I., 36, *53*
Manabe, K., 125, *144*
Maness, P. F., 332, 333, *364*
Mangin, M., 354, *359*
Mankowski, Z., 282, 283, 284, 286, 287, 291, 293, *302*
Manney, T. R., 228, *232,* 251, 253, *259, 263,* 322, 323, 328, 329, 330, 331, 332, 334, 335, 336, *356, 357, 359, 364, 365*
Manning, J. E., 179, *191*
Manning, M., 273, 284, 286, 292, 293, *301*
Mansfield, J. W., 396, 413, 414, *422, 423, 425*
Marchant, R., 200, 211, *235, 238*
Marco, R., 355, *372*
Mardon, D. N., 283, 290, *301*
Maresca, B., 268, 273, *301, 302*
Margulis, L., 201, *235*
Marian, B., 340, *365*
Markowitz, O., 108, *148*
Marmiroli, N., 353, *365*
Marmur, J., 322, *358*
Marnati, M. P., 157, *171*
Marotta, R., 354, *359*
Marriott, M. S., 285, *302*
Marrs, B., 111, *151*
Martin, D., Jr., 82, *93*
Martin, J. F., 21, *29,* 44, 47, 48, 50, *53,* 58, 60, *70,* 155, 157, 159, *170*
Martin, M., 388, *421*
Martin, R., 285, 286, *304*
Marzluf, G. A., 178, 179, 180, *191, 193*

Masago, H., 385, 400, 401, 402, 404, *428*
Mase, Y., 121, *144*
Massee, G., 11, *29*
Masui, Y., 331, *365*
Masurekar, P. S., 49, *53*
Matama, M., 385, *428*
Mathew-Roth, M. M., 107, *144*
Matile, P., 327, *358, 365*
Matrone, G., 86, *92*
Matsumoto, K., *374*
Matsumoto, T., 81, 82, *91, 93*
Matteo, C. C., 37, *53*
Mattia, E., 282, 289, *302*
Maudinas, B., 104, 127, *142, 145*
Maugh, T. H., 20, *29*
Maurer, R., *374*
Maxwell, W. A., 107, *138, 145*
Mayama, M., 181, *190*
Mayfield, J. E., 200, *235*
Meade, J., 323, 330, *364*
Medine, V. J., 280, *304*
Medoff, G., 268, 273, *301, 302*
Medoff, J., 268, *301*
Meissner, G., 113, *145*, 250, *263*
Merkel, G. J., 331, *363*
Merker, H. J., 355, *366*
Merrick, M. J., 162, *169*
Merza, A. P., 103, 116, 117, *141*
Mesland, D. A. M., 246, *263*
Metha, B., 130, *141*
Metzenberg, R. L., 176, *190*
Meuth, M., 84, *90*
Meyers, F. H., 20, *29*
Michaelis, G., 353, *368*
Midiwo, J. O., 22, *29*, 435, *439*
Miller, A. L., 48, *53*
Miller, C. O., 82, 89, *90*
Miller, J. J., 310, 342, 343, 344, 346, 349, *356, 358, 361, 362, 365, 370, 371*
Miller, R. D., 12, *30*
Mills, D., 343, *361, 365*
Mirelman, D., 436, *438*

Mirocha, C. J., 18, *29*, 256, *266*
Misawa, M., 80, *91*
Mishkin, S., 200, *234*
Mitchell, A. P., 343, *374*
Mitchell, H. K., 123, *142*
Mitchell, P., 84, *90*
Mitchell, T. G., 273, 284, 286, 292, 293, *301*
Mitchison, J. M., 352, *367*
Mitsui, H., 83, *91*
Mitzka, U., 106, *145*
Mitzka-Schanbel, U., 100, 104, 105, 106, 126, *145*
Mize, P. D., 166, *170*
Modi, V. V., 130, 131, *138, 141, 147*
Moens, P. B., 182, *192*, 340, 341, 343, 344, 346, 347, *360, 365, 372, 374*
Moesta, P., 386, 400, 402, 407, *425*
Mohan, Ram, H. Y., 3, *31*
Mohr, H., 121, 125, *140, 145*
Moldave, K., 177, *190*
Moller, A., 24, *30*
Molson, J., 82, *90*
Mondelli, R., 130, *137*, 245, *260*
Montague, M., van, 75, *94*
Montelone, B. A., 316, *365*
Mooney, D. T., 272, 276, 287, *302*
Moor, H., 327, *365*
Moore, R. T., 199, *235*
Moore, S. K., 88, *91*
Moore, T. A., 124, *149*
Moorman, A. F. M., 313, *365*
Moran, T., 83, *91*
Morehouse, L. G., 155, *171*
Morgan, E. H., 84, *94*
Morrill, R. C., 276, *305*
Morris, P. W., 3, *30*, 77, 79, 84, 85, 88, *92*
Morris, R. N., 178, *189, 191*
Morris, S. A. C., 107, *145*
Morser, J., 84, *89*
Mortimer, R. K., 310, 338, 341, *366*
Moss, G. P., 97, *145*
Moss, M. O., 19, *30*

Mothes, K., 60, 61, *70, 91*
Mucke, E., 351, 352, 353, 354, *361*
Mühlethaler, K., 106, *140*
Mulheirn, L. J., 399, 407, 408, *424, 425, 428*
Müller, K. O.,377, 389, *422, 425*
Muller, W., 63, 66, *71*
Mullins, J. T., 245, *263*
Mummery, R. S., 97, 101, 108, 120, 122, 125, 127, 128, 129, *139, 150*
Munim-al-Shakarchi, A., 38, 43, *52*, 58, 59, *69*
Munoz, A., 213, *233*
Muñoz, V., 125, *145*
Murai, N., 75, *92*
Murakami, T., 331, *365, 371*
Murakoshi, S., 432, *439*
Murillo, F. J., 113, 114, 115, 116, 129, 130, 131, 133, *135, 142, 145, 150*
Murray, N. N., 111, *146*
Murthy, V. S., 47, *54*
Murygina, V. P., 130, *136*
Muscio, F., 100, *135*
Musgrave, A., 245, *263*
Mynett, A., 391, *420*

N

Nagano, M., 351, 353, *369*
Nagley, P., 354, *366*
Naharro, G., 50, *53*
Naider, F., 331, *363*
Nair, N. G., 12, *28*
Nakayama, H., 283, 288, *302*
Nakayama, T. O. M., 97, 100, 103, 104, 121, 128, 131, *138, 144, 145, 147, 148, 151*
Nandi, S., 166, *170*
Narita, K., 331, *371*
Nash, C. H., 59, *71*

Nasmyth, K. A., 254, *264*, 309, 313, 314, 315, 316, *363, 366, 374*
Naylor, H. B., 351, 353, *362*
Needham, J., 3, *30*
Nelson, N., 274, *296*
Nester, E. W., 162, *171*
Neuberth, D., 355, *366*
Neujahr, H. Y., 105, 106, 132, *137*
Neupert, W., 105, *145*
Neway, J., 63, 66, *71*
Newcomer, V. D., 282, 284, 287, *298, 300*
Newlon, C. S., 321, *366*
Newlon, M. C., 353, 354, *366*
Nickerson, W. J., 270, 272, 273, 274, 282, 283, 284, 286, 287, 290, 291, 293, *296, 298, 300, 302, 304*
Niederpruem, D. J., 12, *30*, 200, 201, 202, 211, *234, 235, 238*
Nielsen, R. I., 251, *264*
Nienstedt, I., 214, 215, *233*
Nieuwenhuis, D., 245, *263*
Niimi, K., 283, 288, *302*
Niimi, M., 283, 288, *302*
Nilsson-Tillgren, T., 343, *367*
Ninet, L., 103,120, 128. 129, 132, *146*
Ninnemann, H., 125, *146*
Nirenberg, M. W., 83, *92*
Nishi, K., 352, *366*
Nishida, K., 81, *91*
Nishioka, Y., 282, 283, 293, *302*
Nixon, I. S., 58, 59, *69*
Noguchi, M., 81, 82, *91, 93*
Nombela, C., 342, *359*
Nordheim, A., 24, *30*
Northcote, D. H., 324, 327, *366, 370*
Nover, L., 21, *29*, 60, 61, 63, 66, *70, 71*
Novick, P., 327, 337, *360, 366*
Nozawa, Y., 268, 280, *298*
Nuesch, J., 41, *52*
Nulton, C. P., 57, 58, 61, 65, *70, 71*

Nunez, M. T., 84, *90*
Nunheimer, T. D., 49, *54*
Nurse, P., 318, 319, *366, 367*
Nutter, R., 162, *171*
Nutting, W. H., 241, *263, 264*

O

Ôba, K., 386, 392, 393, 418, *425, 426, 427*
O'Connell, B. T., 278, *302*
Oda, Y., 108, *143*
O'Day, D. H., 7, *30*, 239, 256, *262, 264*
Odds, F. C., 268, 280, 282, 286, 288, *297, 298, 303*
O'Farrell, P. H., 58, *71*, 314, *367*
O'Farrell, P. Z., 58, *71*
Ogata, W. N., 112, *136*
Ogawa, H., 41, *54*
Ogden, R., 181, *191*
Oguchi, T., 351, 353, *369*
Ohja, M., 214, *235*
Ohta, S., 80, 81, *92*
Ohtomo, T., 432, *439*
Ohuni, I., 418, *427*
Oishi, S., 344, *374*
Ojha, M. H., 242, 243, *266*
Okanishi, M., 161, *168*
Okigaki, T., 82, *90*
Olaiya, A. G., 293, *303*
Olson, J. A., 101, 127, 133, *146*
Olson, L. W., 338, 340, 345, *367, 374*
Olson, M. V., 176, *191*
Olson, V. L., 290, *300*
Omerud, W., 201, *235*
Ootaki, T., 113, 114, 115, 132, *146*
Opheim, D. J., 342, *360, 367*
Orcutt, B. C., 331, *359*
O'Reilly, J., 283, 288, 289, *297, 298*

Orlowski, M., 271, 275, 276, 277, 278, *299, 303*
Orr, H. T., 229, *235*
Orr, W. C., 180, 182, 184, *192, 193*
Orton, W. L., 329, *371*
Osagie, A. U., 130, *137*, 247, *260*
Oshima, Y., 346, *371*
Osman, M., 125, *146, 150*
O'Sullivan, J., 162, *170*
Osumi, M., 326, 351, 353, *367, 369*
Otte, C. A., *374*
Overmeer, W. P. J., 109, *150*
Owens, A. E., 118, *140*

P

Paau, A. S., 436, *439*
Packter, N. M., 16, 18, *30*, 43, *54*
Padilla, G. M., 352, *363, 367*
Page, R. M., 124, *146*
Paietta, J., 124, *146*
Pall, M. L., 276, 288, *303, 305*
Palleroni, N. J., 351, *367*
Pandey, K. K., 227, *235*
Papastephanou, C., 104, *145*
Paradies, I., 385, *426*
Parag, Y., 203, 205, 226, 227, *234, 235*
Pardo, M., 77, *89*
Pardou, M. L., 24, *30*
Parker, W., 255, *260*
Parn, P., 132, *146*
Parr, D. N., 287, *300*
Partridge, J. E., 381, 395, 404, *424, 426*
Passeron, S., 270, 273, 275, 276, 277, *299, 303*
Pateman, J. A., 12, *31*
Pathres, S. V., 18, *29*
Paveto, C., 276, 277, *303*
Paxton, J. D., 381, 401, *422, 426*
Paznokas, J. L., 271, 273, 276, 277, 278, *302, 303*

Pazoutova, S., 67, *71*
Peace, J. N., 63, *69*, *71*
Pearson, N. J., 343, *367*
Pearson, W. R., 179, *189*
Peat, A., 60, *71*
Peberdy, J. F., 9, *30*, 162, *170*
Pederson, T., 88, *92*
Pedrosa, R. O., 87, *93*
Peisach, J., 87, *92*
Pelaez, M. I., 251, *261*
Perkins, C. M., 87, *92*
Perkins, D. D., 111, *146*
Perlman, D., 58, *70*
Perlman, P. S., 181, *191*
Perrin, D. R., 377, 378, 397, 398, *421*, *426*
Perry, K., 322, 323, 328, 333, *360*
Peters, J., 271, 278, 279, 291, *299*, *303*
Petersen, J. G. L., 338, 343, *367*
Petes, T. D., 178, 181, *192*, 313, *367*
Pfeiffer, E., 61, *70*
Phaff, H. J., 97, 98, 121, *135*, *143*, *145*, 284, *303*
Phillips, A. W., 283, 290, *301*
Phillips, R. L., 111, 112, *151*
Phinney, B. O., 18, *28*
Pickart, L., 87, *92*
Pictet, R. L., 3, *30*, 77, 79, 84, 85, 88, *92*
Pierre, R. E., 395, *426*
Piggot, P. J., 307, 310, 312, *367*
Piñon, R., 340, 343, 344, 347, *362*, *367*, *371*
Pirt, S. J., 38, *53*, 59, 60, *71*
Pivovarova, T. M., 105, 106, *140*
Planck, J. E., van der, 378, *426*
Platero Sanz, M., 390, *422*
Plattner, J. J., 241, *264*
Ploegh, H. L., 229, *235*
Poff, K. L., 124, 125, *144*, *146*, *148*
Pogell, B. M., 40, *53*
Pohl, U., 109, *148*
Pokorny, V., 67, *71*

Polak, A., 284, *303*
Poling, S. M., 128, 129, *142*, *146*
Pollack, J. H., 105, 120, 121, *142*
Pollak, J. K., 355, *368*
Pommerville, J., 240, 241, 242, *264*
Pontecorvo, G., 110, *146*
Pontremoli, S., 87, *93*
Pontzen, R., 419, *428*
Popplestone, C. R., 244, *264*
Porter, A. E. A., 413, *425*
Porter, J. W., 100, 101, 104, 110, *145*, *147*
Postan, I., 82, *93*
Postle, K., 162, *171*
Pouchet, G. R., 77, *89*
Poulter, R., 293, *303*
Powell, A., 162, *171*
Powell, A. J., 13, *27*, 58, 59, *69*
Prakash, L., 316, 321, *365*, *368*
Prakash, S., 316, 321, *365*, *368*
Prasad, K., 124, *149*
Prasad, M., 287, *301*
Pratje, E., 353, *368*
Prazmo, W., 354, *368*
Prebble, J., 105, 106, 108, *135*, *147*
Prelog, V., 163, *170*
Presti, D., 108, 121, 122, 123, 124, 125, *139*, *143*, *144*, *147*
Preston, J. F., 166, *169*
Prevost, G., 201, 227, *235*
Prézelin, B. B., 124, *149*
Price, M. F., 282, 293, *305*, *306*
Prieto, A., 245, *260*
Prieto, R., 130, *137*
Pringle, J. R., 228, *238*, 311, 315, 317, 318, 326, 327, 331, 338, 355, *363*, *368*, *371*, *373*
Pryce, R. J., 411, 415, 420, *421*, *425*
Pueppke, S. G., 378, 390, 391, 397, *427*
Pugh, D., 287, *303*
Puglisi, P. P., 353, *365*, *368*
Purohit, B. C., 283, *303*
Pursey, B. A., 255, *263*
Purwin, C., 407, 419, *422*, *423*

Author Index

Putrament, A., 354, *356, 368*

Q

Quackenbush, F. W., 121, *144*
Quarrie, S. A., 109, 131, *137*
Quilico, A., 130, *137,* 245, *260*

R

Rabie, C. J., 19, *30*
Rabourn, W. J, 121, *144*
Radford, A., 111, 113, 116, *140*
Radin, D. N., 323, 324, 328, 330, 338, *368, 370*
Radu, M., 215, *236*
Radunz, A., 106, *147*
Ragan, M. A., 96, *147*
Rajasingham, K. C., 286, *303*
Ram, S. P., 283, 284, 293, *304*
Ramacek, Z., 67, *71*
Ramadan-Talib, Z., 105, 106, 108, *147*
Ramakrishnan, T., 288, *303*
Ramedo, I. N., 283, *303*
Rao, G. R., 288, *303*
Rao, S., 130, 131, *147*
Raper, C. A., 12, *30,* 198, 204, 206, 208, 209, 219, 222, 227, 230, *236*
Raper, J. A., 64, *70*
Raper, J. R., 9, 12, *30,* 198, 201, 204, 206, 208, 209, 210, 211, 216, 222, 224, 227, *234, 236, 237, 238,* 239, 243, *264*
Raper, K. B., 7, *30,* 430, *439*
Rapoport, H., 241, *263, 264*
Rapport, E., 340, 341, *365*
Rasmussen, S. W., 214, *232*
Rast, D., 211, *237*
Rau, U., 124, *143*
Rau, W., 100, 104, 105, 106, 114, 121, 122, 124, 125, 126, 127, 132, *136, 143, 145, 147, 148, 149, 150*

Raudaskoski, M., 200, 201, 205, 206, 210, *236, 237*
Rau-Hund, S., 122, *147*
Raven, P. H., 21, *28*
Redshaw, P. A., 40, *53,* 351, *368*
Reed, B. C., 82, *92*
Reed, S. I., 314, 315, 316, *366, 368*
Rees, A. F., 101, 103, *136, 138*
Rehacek, Z., 59, *72*
Reichard, P., 88, *93*
Reid, B. J., 318, 327, *368*
Reid, I. D., 254, *264*
Reilly, J., 415, 417, *426*
Remaley, A. T., 55, 62, 63, *69, 70*
Renaud, R. L, 111, 112, 126, *143, 144*
Renaut, J., 103, 120, 128, 129, 132, *146*
Renwick, J. A. A., 287, *301*
Resnick, M. A., 347, 348, *361, 368*
Reyes, E., 273, 274, 275, *296*
Reyes, P., 128, 131, *147*
Rhoads, D. D., 293, *304*
Rice, E., 381, 395, *424*
Rice, P. W., 176, *190*
Rich, A., 24, *30,* 127, *136*
Rich, M. A., 209, *237*
Richardson, M. D., 282, 286, *298*
Rickenberg, H. V., 288, *304*
Ride, J. P., 396, 415, 420, *421*
Righelato, R. C., 38, *53,* 59, 60, *71*
Riley, G. J. P., 105, *147*
Rillema, J. A., 83, *92*
Rilling, H. C., 100, 104, *135, 139, 147*
Rine, J., 254, *265,* 309, *362*
Rippon, J. W., 268, 280, *304*
Robbers, J. E., 51, *52, 54*
Robbins, E., 88, *92*
Roberts, T. M., 176, 178, 179, *191*
Robertson, L. W., 51, *54*
Robinow, C. F., 327, *365*
Robinson, S. H., 84, *90*
Rodgers, M. A., 107, *148*
Rodriguez, D. B., 127, 128, 132, *144*
Roeymans, H. J., 97, 120, *139, 150*

Rogers, D. T., 324, 331, 337, *368*
Rogers, R. J., 273, 276, *304*
Rohwedder, W. K., 434, *439*
Rolewic, L. J., 290, *300*
Romano, A. H., 280, 291, *302, 304*
Roos, W., 61, *70, 71*
Rosato, L. M., 55, *69, 70*
Rosbash, M., 180, *190*
Rose, A. H., 342, *362*
Roseboom, P. H. M., 381, 409, *428*
Rosenberg, H., 163, *170*
Rosenthal, G. A., 21, *30*
Ross, I. K., 201, *237*
Rossall, S., 413, 414, *423*
Roth, J., 82, *93*
Roth, R. M., 338, 340, 342, 344, 345, 347, 348, *360, 361, 363, 368*
Rothwell, A., 58, *69*
Rothwell, B., 58, *69*
Rottem, S., 108, *148*
Rousseau, G. G., 82, *89*
Rousseau, P., 348, 349, 351, 353, *369*
Rowe, J. R., 282, *300*
Rozek, C. E., 180, *192*, 245, *264*
Rubaclava, B., 83, *91*
Rubery, P. H., 273, *298*
Ruddat, M, 103, 116, 117, *141, 151*
Ruiz-Herrera, J., 274, 275, *296, 297, 304*
Rusch, H. P., 7, *26*
Russell, D. H., 280, *304*
Russell, D. W., 58, *69*
Russell, P. J., 182, *192*
Russell, T. R., 84, *92*
Russo, G. M., 175, 182, *192*
Russo, V. E. A., 109, 110, *140, 148*
Rutter, W. J., 3, *30*, 77, 79, 84, 85, 88, *92*, 176, 181, *189, 192*

S

Sage, J., 214, *232*
Sakagami, Y., 254, 255, 258, *261,*
[Sakagami, Y.], *262, 265, 266*
Sakai, K., 323, 328, 329, *369*
Sakakibara, S., 331, *365*
Sakato, K., 80, *91*
Sakurai, A., 251, 255, 258, *262, 265,* 322, 323, 328, 329, 330, 331, 333, *363, 369, 370, 373*
Saleh, F. K., 126, *151*
Saltarelli, C. G., 283, 287, *304*
Salts, Y., 340, 344, *367, 371*
Salvatori, T., 130, *137*, 245, *260*
Samejima, H., 80, *91*
Samokhin, G. P., 318, 323, 331, *369*
Samson, R. A., 5, *27*
Samuels, H. H., 82, *93*
San-Blas, F., 268, *304*
San-Blas, G., 268, *304*
Sanders, M. M., 178, *189*
Sandler, L., 338, 348, *356*
Sandmann, G., 121, 127, *148*
Sando, N., 344, 351, 353, *369, 374*
Sanford, B. H., 330, *358*
Sankaran, L., 40, *53*
Sansing, G. A., 430, 431, *438*
Sansome, E., 255, *260*
Santos, T., 342, *359*
Sarachek, A., 293, *304*
Sargent, M. L., 107, 108, 124, *136, 146, 150*
Sato, G., 83, *89*
Sato, M., 83, *91*
Satoh, A., 41, *54*
Satomura, Y., 41, *54*
Saunders, R. M., 329, *358*
Savarese, J. J., 349, 351, *369*
Saville, D., 331, 337, *368*
Schaar, G., 283, *304*
Schäfer, E., 125, *144*, 407, *422*
Schaffner, C. P., 47, 50, *53*
Schaller, E., 399, 408, *428*
Schatz, G., 312, *361*
Schekman, R., 253, *265*, 323, 326, 327, 332, 337, *360, 366, 369*
Schell, J., 75, *94*
Scherer, G., 253, *265*, 331, *370*

Scherer, S., 177, *192*
Scherwitz, C., 285, 286, *304*
Schild, D., 310, 320, 341, 346, *366, 370*
Schipper, M. A. A., 251, *265*
Schlösser, E. W., 378, 418, *426*
Schmid, G. H., 106, *147*
Schmidt, C. W., 179, *191*
Schmidt, W., 124, 125, *148*
Schmit, J. C., 175, *192*
Schnierer, S., 353, *368*
Schroder, J., 81, *92*
Schroeder, R., 342, 351, 352, 353, 354, *361, 370*
Schrott, E. L., 107, 108, 122, 123, 125, 126, 127, *143, 148*
Schulz, R., 353, *368*
Schwartz, F. J., 86, *92*
Schwartz, R. M., 331, *359*
Schwarz, R. I., 83, *92*
Schwarzhoff, R. H., 293, *304*
Schweizer, M., 184, 185, *189*
Schweyen, R. J., 312, *361*
Schwochau, M. E., 389, *423*
Scott, W. T., 255, *263*
Seddon, B., 166, *170*
Seeds, N. W., 83, *92*
Sehgal, S. N., 434, *439*
Seidel, R., 181, *190*
Seigel, J. L., 343, *370*
Sekiguchi, J., 21, *30,* 67, *71*
Selby, C., 396, *426*
Sell, S., 83, *91*
Selva, A., 130, *137,* 245, *260*
Sena, E. P., 321, 323, 324, 328, *370*
Sentandreu, R., 327, *370*
Sergeeva, L. N., 130, *136*
Serum, J. W., 390, *423*
Seshadri, R., 243, *259, 263*
Seviour, R. J, 132, *146*
Shaffer, M., 84, *90*
Shanley, M., 342, *360*
Shantha, T., 47, *54*

Shapiro, B. M., 324, *370*
Shapiro, S. K., 351, *358*
Sharma, M., 130, *141*
Sharon, N., 436, *438*
Shaw, G., 109, *141*
Shechter, Y., 293, *298*
Sheng, G., 111, *148*
Sheng, T.C., 111, *148*
Shepherd, D., 13, *27,* 58, 59, *69*
Shepherd, M. G., 282, 283, 284, 285, 286, 287, 289, 290, 293, *298, 303, 304,305*
Sherwood, W. A., 255, *265*
Shields, R., 280, *299*
Shih, M. J., 378, 392, *424, 426*
Shilo, B., 318, 320, *370*
Shilo, V, 318, 320, *370*
Shimizu, M., 107, *148*
Shimoda, C., 251, *265,* 322, 323, 326, 328, 329, 330, 331, 333, 352, *366, 367, 369, 370*
Shio, I., 80, 81, *92*
Shizuri, Y., 161, *170*
Shropshire, W., Jr., 121, 122, 123, 134, *139, 142, 148, 151*
Shulman, R. G., 343, *356, 374*
Shumard, D. S., 180, 181, *192*
Shuster, J. R., *374*
Sietsma, J. H., 211, *237*
Silagi, S., 84, 85, *92*
Silva-Hutner, M., 282, 283, 293, *302*
Silver, J. C., 178, *190,* 245, *265*
Simchen, G., 204, *234,* 311, 312, 318, 320, 340, 344, 346, *362, 363, 367, 370, 371, 374*
Simonetti, N., 282, 284, 285, 286, 289, *297, 304*
Simpson, K. L., 97, 98, 103, 121, 127, 128, 129, 132, *139, 142, 144, 148, 149*
Sims, J. J., 381, 395, 399, *424, 426*
Singh, B. R., 289, *305*
Singh, K., 434, *439*
Singh, P. D., 162, *168*

Sinha, R. N., 22, *30*
Sirsi, M., 288, *303*
Sissoeff, I., 88, *91*
Sistrom, W. R., 107, *149*
Skinner, C. E., 280, 290, *305*
Skipp, R. A., 396, *426*
Skoog, F., 75, *92*
Slifkin, M., 77, *89*
Sloat, B. F., 326, 327, *371*
Slusarenko, A. J., 418, *426*
Smalley, H. M., 13, 17, *27*, 58, 59, *69*
Smallman, R. V., 413, *425*
Smirnov, V. N., 318, 323, 331, *369*
Smith, D. A., 390, 396, 397, *422, 424, 426, 427*
Smith, D. G., 387, *423*
Smith, G. N., 13, 17, *27*, 58, 59, *69*
Smith, H. G., 16, *26*
Smith, J. E., 7, 12, *30, 31*, 67, *69*
Smith, J. K., 283, *305*
Smith, M., 254, *264*, 309, 314, *366*
Smith, M. R., 111, *146*
Smith, N. R., 430, *439*
Sneden, A. T., 161, *170*
Snider, P. J., 201, 224, *237*
Snyder, S. H., 280, *304*
Sogin, S. J., 282, 285, 289, 293, *297, 303*
Sokolovskii, V. Y., 131, *143*
Solioz, M., 312, *361*
Soll, D. R., 86, *89*, 285, 287, 293, 294, *297, 305*
Solms, U., 19, *29*
Somerson, N. L., 49, *54*
Song, P. S., 96, 124, *149*
Sowder, L., 317, *357*
Spalla, C., 157, *171*
Specht, C. A., 179, 180, *192*, 214, *237*
Spiegelmann, B. M., 80, 83, *92*
Sprague, G., Jr., 309, *362*
Sprague, G. F., 254, *265*, 374
Spurgeon, S. L., 100, 104, 126, *147, 149*

Srivastava, P. K., 346, *371*
Stacey, G., 436, *439*
Stadler, D. R., 112, *149*
Stadler, P. A., 19, *31*
Stahmann, M. A., 385, *421*
Stalpers, J. A., 251, *265*
Stamberg, J., 204, 205, 206, 209, 222, *234, 237*
Stanford, J. B., 395, *424*
Stanier, R. Y., 107, 110, *143, 149*
Staples, R. C., 287, *301*
Stark, G. R., 183, *189*
Starr, M. P., 97, *135*
Stasi, M., 293, *305*
Steed, J. R., 293, *303*
Steele, S. D., 349, *371*
Steglich, W., 155, *168, 171*
Steijns, J. M. J. M., 342, *358*
Steinlauf, R., 215, *236*
Stekoll, M., 382, *427*
Stemerding, D., 231, *238*
Stenkamp, R. E., 87, *92*
Sterjna, S. M. S. M., 342, *358*
Stetler, D. A., 268, *297*
Steward, F. C., 3, *31*
Stewart, P. R., 273, 276, *304*
Stich, H. F., 88, *93*
Stinchcomb, D. T., 177, *192*
Stjernholm, R. L., 282, 283, 286, 288, 290, 291, 293, *300*
Stockdale, F. E., 83, *93*
Stockigt, J., 78, 80, 81, *94*
Stoessl, A., 378, 390, *423, 427*
Stoll, A., 19, *31*
Stollar, D. D., 24, *30*
Storck, R., 270, 275, 276, *299, 305*
Stork, R., 177, *192*
Storm, E. M., 322, 348, *356, 358*
Stössel, P., 404, 417, *427*
Stössel, R., 381, 404, *425*
Stothers, J. B., 378, *427*
Stötzler, D., 251, 253, *260, 265*, 330, 331, *359, 371*
Strathern, N. J., 242, 254, *262*, 309, *362, 364, 371*

Author Index

Strausberg, S., 253, *261*, 333, *360*
Street, H. E., 3, *31*
Strippoli, V., 282, 284, 285, 286, 289, *297, 304*
Strominger, J. L., 229, *235*
Struhl, K., 177, *192*
Stuart, S. J., 167, *171*
Student, A. K., 82, *92*
Sturzenegger, V., 99, *149*
Stutz, P., 19, *31*
Subden, R. E., 105, 107, 111, 112, 126, *136, 141, 143, 144, 145, 149*
Suda, K., *374*
Sugiyama, J., 432, *439*
Sullivan, P. A., 282, 283, 284, 285, 286, 287, 289, 290, 293, *298, 303, 304, 305*
Sundaram, S., 285, *305*
Sundeen, J., 243, *259*
Sures, I., 181, *190*
Sussman, M., 3, *31*
Suto, T., 344, *374*
Sutter, R. P., 122, 131, *149*, 248, 250, 258, *265, 266*
Sutton, R., 355, *368*
Suzuki, A., 254, 255, 258, *261, 262, 265, 266*
Suzuki, H., 82, *93*, 392, 418, *427*
Svihla, G., 291, *296*
Swain, T., 21, *31*, 74, 77, *93*
Swait, J. C., 58, *69*
Swatek, F. E., 268, *305*
Swinburne, T. R., 88, *91*
Sypherd, P. S., 176, *189*, 271, 272, 273, 275, 276, 277, 278, 279, 280, 287, 291, *297, 299, 300, 301, 302, 303*
Szczech, G. M., 18, *27*
Szulmajster, J., 355, *371*

T

Tabata, M., 79, 80, 81, 82, *93*
Taber, W. A., 7, *31*, 58, *69*

Takahama, U., 107, *148*
Takahashi, N., 255, 258, *262, 265*
Takano, A., *374*
Takeo, K., 276, *305*
Tamarkin, A., 205, 206, *234, 237*
Tamura, S., 251, 254, 255, 258, *261, 262, 265*, 322, 323, 328, 329, 330, 331, 333, *369, 370, 373*
Tanaka, H., 80, *91*
Tanaka, T., 331, 332, 333, *365, 371*
Tanaka, Y., 98, *149*
Tanaki, E., 81, *91*
Taschdjian, C. L., 286, *305*
Tashjian, A. H., Jr., 84, *90*
Tassi, F., 353, *365*
Tatchell, K., 254, *264*, 309, 314, *366*
Tatematsu, H., 393, *425, 426*
Tatum, E. L., 112, *141*
Tavlitzki, J., 352, *360*
Taxer, S. S. S., 283, 291, *298*
Taylor, A., 253, *263*, 323, 324, 326, 328, 333, *364, 434, 439*
Taylor, D., 109, 131, *137*
Taylor, N. W., 329, *371*
Taylor, R. F., 97, 100, *138, 149*
Tejwani, G. A., 87, *93*
Tekamp, P. A., 181, *192*
Terenzi, H. F., 270, 275, 276, *305, 306*
Thaler, M. M., 87, *92*
Theimer, R. R., 106, 124, 132, *149, 150, 151*
Thelander, L., 88, *93*
Thiel, J., 293, *305*
Thom, C., 7, *30*, 430, *439*
Thomas, D. M., 130, 131, *150*
Thomas, P. S., 314, *372*
Thomas, S. A., 107, 108, *150*
Thomashow, M. F., 162, *171*
Thomason, I. J., 397, *424*
Thomopoulus, P., 82, *93*
Thompson, E. B., 82, *93*
Thompson, S. N., 96, *140*
Thomson, R. H., 155, *171*
Thorner, J. 253, 260, 262, 322, 323,

[Thorner, J.], 328, 331, 332, 333, 337, 338, *358, 360, 363, 364, 372, 374*
Threlkeld, S. F. H., 112, *149*
Throm, E., 322, 333, *372*
Timberlake, W. E., 178, 179, 180, 181, 182, 183, 184, *191, 192, 193,* 214, *234, 237,* 245, *264, 266*
Tingle, M. A., 338, 354, *364, 372*
Tissier, R., 103, 128, 129, *146*
Titov, M. I., 318, 323, 331, *369*
Tkacz, J. S., 325, 326, 327, *372*
Todaro, G. J., 77, 79, 80, 82, *90*
Tohoyama, H., 329, *372*
Tokunaga, J., 283, 288, *302*
Tomkins, G. M., 82, *93*
Topper, Y. J., 83, *93*
Torres-Martínez, S., 113, 115, 116, 124, *138, 145, 150*
Toselli, M., 109, *148*
Tovarova, I. I., 159, *170*
Travis, R. L., 126, *150*
Trevillyan, J. M., 276, *305*
Trew, B. J., 182, *192,* 343, *372*
Trewavas, A., 239, *266*
Trinci, A. P. J., 7, *26, 27, 31,* 60, *71*
Trouilloud, M., 121, *140*
Tsuboi, M., 342, *372*
Tsuchiya, E., 254, 255, 258, *262, 265, 266*
Tubaki, K., 5, *31*
Tuite, M. F., 348, *372*
Turian, G., 12, *31,* 111, 112, 126, 127, *149, 150,* 214, *235,* 240, 242, 243, *266*
Turler, H., 214, *235*
Turner, G., 10, *31*
Turner, R. V., 100, 104, 126, 134, *142, 149*
Turner, W. B., 2, 13, 14, 16, 17, 19, *31*
Tuthill, D., 22, *29,* 435, *439*
Tuveson, R. W., 107, 108, *136, 150*

U

Uchida, A., *374*
Udden, M. M., 322, *372*
Ueberberg, H., 285, 286, *304*
Ullrich, R. C., 179, 180, *192,* 214, 216, 226, 227, *237*
Umezawa, H., 155, 161, *168, 171*
Unger, M. W., 318, *362*
Uno, I., 213, *238, 374*
Unrau, A. M., 244, *264*
Unwin, C. H., 381, 404, 417, *427*
Upper, C. D., 132, *139*
Uritani, I., 386, 392, 393, 418, *424, 425, 426, 427*

V

Valadon, L. R. G., 96, 97, 99, 101, 108, 120, 122, 125, 126, 127, 128, 129, *139, 146, 150*
Valent, B., 382, 403, *420*
Valenzuela, P., 176, *189*
Vallejo, C. A., 162, *168*
Vallejo, C. G., 355, *372*
Vandamme, E. J., 159, *171*
van der Putte, T. M. J. A., 342, *358*
Van der Rest, M., 108, *136*
VanDerWatt, J. J., 19, *30*
Vanek, Z., 44, *52*
Van Etten, H. D., 378, 390, 391, 396, 397, 398, *422, 423, 427, 428*
Van Etten, J. L., 175, 181, 182, *189, 190, 192,* 353, *357*
Van Rensburg, S. J., 19, *30*
Van Rij, N. J. W., 282, 283, 290, *302*
Van Zon, A. Q., 109, *150*
Vapnek, D., 184, *192*
Vaughn, V., 87, *93*
Vaughn, V. J., 282, 284, *305*
Vecher, A. S., 120, *151*
Veerman, A., 109, *150*

Author Index

Verhoeff, K., 413, *427*
Vesonder, R. F., 434, 435, *438, 439*
Vezina, C., 434, *439*
Vezinhet, F., 342, *372*
Vierstra, R. D., 125, *144*
Vigfusson, N. V., 256, *266*
Villa, T. G., 98, *143*
Villanueva, J. R., 50, *53*
Villoutreix, J., 103, 127, *138, 142*
Vincent, G. G., 390, *421*
Visser, S. A., 97, *135*
Vogelmann, H., 78, *93*
Volpi, L., 109, *148*
Vorisek, J., 59, *72*
Vries, F. W., de, 390, *422*
Vries, O. M. H., de, 214, 215, 216, 217, 219, 230, *233, 238*
Vuori, A. T., 120, *151*

W

Wade, M., 383, 399, 404, *427*
Wade, N., 18, *31*
Wagner, F., 78, *93*
Wagnière, G., 99, *149*
Wagstaff, J. E., *374*
Wain, W. H., 282, 284, 289, 293, *303, 305, 306*
Walker, J. B., 13, *31*, 38, 39, 48, *53, 54*
Walker, M. S., 48, *54*
Walter, S. K., 321, *366*
Wang, C. S., 216, *238*
Wang, S. S., 111, 112, *151*
Wang, Y.-H., 24, *28*
Wanner, G., 106, *151*
Ward, A. C., 43, *54*
Ward, E. W. B., 378, 381, 404, 417, *425, 427*
Watkins, W. M., 330, *372*
Watrud, L. S., 202, *238*
Watson, D. C., 338, *373*
Weber, D. J., 7, *31*
Weedon, B. C. L., 97, *145, 151*

Weeks, O. B., 126, *151*
Wei, D., 18, *32*
Wei, L., 88, *93*
Weiche, G. R., 243, *263*
Weidman, F. D., 295, *306*
Weiler, E. W., 78, 80, 81, *94*
Weinberg, E. D., 20, 21, *32*, 44, *54*, 75, 79, 85, 86, 87, 88, 89, *90, 93, 94*, 155, *171*, 282, 284, *305*
Weinstein, B., 87, *92*
Weinstein, L. I., 399, *427*
Weisleder, D., 434, *439*
Weiss, C. V., 125, *136*
Weiss, L. M., 116, *140*
Wejksnora, P. J., 175, *190*, 343, *373*
Welch, J. W., 321, *370*
Welch, S. K., 182, *190*, 343, *362*
Weld, J. T., 290, *306*
Went, F. A. F. C., 122, *151*
Werczberger, R., 205, *234*
Werkman, T. A., 250, *266*
Werth, A., 293, *297*
Wessels, J. G. H., 12, *30*, 178, 179, 180, *189, 193*, 200, 210, 211, 214, 215, 216, 217, 219, 230, 231, *233, 234, 235, 237, 238*
West, C. A., 132, *139*, 382, 384, *425, 427*
Wheeler, H. E., 256, *260*
Wheeler, P. R., 283, 288, 289, *298*
Wheeler, R., 255, *263*
Whelan, W. L., 348, *373*
Whipps, J. M., 436, *438*
Whitaker, B. D., 123, *151*
Whitaker, J. P., 250, *266*
White, R. M., 244, 245, *263*
Whiteman, P. F., 344, *358*
Whiting, R. F., 88, *93*
Whittaker, R. H., 73, 74, 75, *94*
Wickins, S. G. A., 415, *425*
Wickner, R. B., 331, 337, 338, 348, *364, 373*
Widra, A., 283, *304*
Widstrom, N. W., 432, *439*
Wiemken, A., 327, *358*

Wilkerson, W. M., 182, *192*
Wilkie, E., 353, *359*
Wilkinson, L. E., 228, *238*, 331, *373*
Will, O. H., III, 116, *141*, *151*
Willetts, H. J., 433, *439*, *440*
William, M. W., 241, *263*
Williams, P. H., 167, *171*
Williams, R. J. H., 103, *138*
Williams, T. H., 390, *423*
Williamson, I. P., 250, 251, *262*
Willing, R. R., 437, *439*
Willmer, J. S., 120, 127, *141*
Wilson, B. J., 17, *32*
Wilson, C. M., 241, 242, *260*
Wilson, T., 107, *144*
Winskill, N., 109, 130, 131, *137*, 250, *260*
Winstanley, D. J., 59, *69*, 130, 131, *137*
Wintersberger, U., 340, *365*
Wirahadikusumah, M., 126, *151*
Wit, P. J. G. M., de, 378, 381, 382, 395, 408, 409, *427*, *428*
Wold, B. J., 182, *190*
Wolf, D. H., 343, *373*
Wolf, J. C., 256, *266*
Wong, E., 386, *428*
Wong, L.-J. C., 180, *193*
Wong, P. P., *440*
Wood, R. K. S., 418, *426*
Woodruff, H. B., 21, *32*, 77, *94*
Woods, V., 332, 335, *365*
Woodward, M. D., 388, *428*
Wright, B. E., 11, *32*
Wright, J. F., 343, *373*
Wright, L. F., 161, *171*
Wu, R., 177, *193*
Wygal, D. D., 175, *190*
Wyllie, T. D., 155, *171*
Wyman, J. G., 396, *428*

Y

Yagisawa, M., 344, *374*
Yamaguchi, H., 282, 283, 293, *306*
Yamamoto, H. Y., 106, *151*
Yamashita, K., 393, *425*, *426*
Yamauchi, K., 400, 401, 402, 404, *428*
Yanagishima, N., 251, 253, *265*, *266*, 322, 323, 326, 328, 329, 330, 331, 333, *363*, *367*, *369*, *370*, *372*, *373*, *374*
Yanagita, T., 62, 63, *72*, 344, *374*
Yen, H.-C., 111, *151*
Yen, P. H., 329, *374*
Yeoh, G. C. T., 84, *94*
Yokoyama, H., 97, 100, 103, 104, 105, 121, 128, 129, 130, 131, 133, *136*, *138*, *142*, *146*, *151*
Yoshida, K., 253, *266*, 322, 323, 328, 329, *363*, *372*, *373*, *374*
Yoshida, M., 254, *262*, *265*, *266*
Yoshikawa, M., 381, 385, 386, 400, 401, 402, 404, 407, *428*
Yoshimura, M., 329, 338, *359*
Young, I. G., 163, *170*

Z

Zähner, H., 68, *72*, 160, *171*
Zähringer, U., 399, 405, 407, 408, *428*
Zajic, J. E., 97, *151*
Zaki, A. I., *424*
Zalokar, M., 105, 111, *151*
Zamb, T. J., 347, 348, *361*
Zambryski, P., 75, *94*
Zantinge, B., 180, *193*, 216, 217, *238*
Zechmeister, L., 99, *151*

Author Index

Zenk, M. H., 78, 80, 81, *94*
Zennaro, E., 353, *368*
Zerahn, K., 291, *302*
Zickler, D., 214, *232*, 338, 340, *367, 374*
Zickler, H., 256, *266*

Ziegler, E., 419, *428*
Zimmerman, C. R., 182, 184, *193*
Zimmerman, F. K., 340, 345, *367*
Zinker, S., 322, *358*
Zorzopulos, J.,276, *306*
Zubenko, G. S., 343, *374*

Subject Index

A

Abscisic acid, 131, 258
Action spectrum, 122
　distortion, 123
Alpha factors, 251, 252, 322
Agglutination, 323
Agglutinins, 322, 328, 332
　a and α, 328
Alkaloid, 19, 60, 61, 64, 66
α factor, 251, 252, 322, 330
　degradation, 333
　precursor processing, 336
　proteolytic cleavage, 333
　secretion, 331
Aminopyridine, 129
Animal cells
　kinetics, 78
　secondary metabolism in, 82-85
　secondary metabolism, problems, 79
　secondary metabolites of, 73-76
Antheridiol, 243-258
Antibiotics
　β-lactam, 59
　resistance information transfer, 167
Apomixis, 344, 346
Apocarotenoids, 95
Ascocarps, 8
Ascomycetes, 96, 97
Ascospores, 307

Ascus, 307
　two-spored, 346
　wall, 342
Asperphenamate, 63, 64, 66

B

Bacteriostats, 107
Basidiocarps, 8
Basidiomycetes, 96, 97
Benzodiazapine, 60
　alkaloids, 61
　in cyclopenase, 61
　formation, 66
β-Ionone, 128, 129, 131
Between cultures, 66
Bikaverin production, 38
Biochemical taxonomy, 430
Biphasic fluence, 123
Branched pathways
　feedback regulation of, 40-50
　candicidin, 50
　penicillin, 49
Brevianamides A and B, 63, 64
5-Bromodeoxyuridine alteration of differentiation, 84-85
Budding pathways, 307, 317, 351
Bud emergence, 317
Bud scars, 307, 326

C

cAMP, 250, 333
  in differentiation, 212-214
  in dimorphism, 276-278
  glucose repression, 41
  in dikaryon morphogenesis, 212-214
Candicidin production
  branched pathway and feedback regulation of, 50
Carotene, 247, 249
  biosynthesis, 110
  α-carotene, 100
  β-carotene, 109, 131, 132, 133, 134
  cyclization, 103, 127
  short life, 124
  singlet state, 124
  synthesis, 121
  turnovers, 120
Carotenogenesis, 112, 131
  enzyme model, 110
Carotenogenic, enzyme, 104, 126
  complex, 117
Carotenoids, 97
  accumulation, 120
  deficient white mutants, 107
  distribution, 120
  extraction, isolation, and identification,
  membrane stabilizers, 108
  in mitochondrial fraction, 105
  as photoreceptors, 108
  production, 120
  protecting, 107
  quantitative determination, 99
  stimulation (increase), 130
  synthesized in ER, 106
  triplet state, 124
"Cassette" mechanism, 242, 351
Catabolite regulation
  inhibition, 41
  nitrogen, 42, 43
  repression, 40
  streptomycin production, 42

Cell cycle, 253, 307
Cell differentiation, 56
Cell fusion, 322
Cell preparation, homogenous batch mode in, 66
Cellular differentiation, 56
  monitoring, 57
Cellulase, 245
Centriole, 317
Cephalosporin synthesis, repression of, 41
Chemotropism, 241, 244, 246
Chitin, 9, 253, 254, 326
  synthetase, 326, 332
Chlamydospores; ergot alkaloids in *Claviceps purpurea*, 59
Chlorotetracycline production
  energy charge in, 45
  phosphate regulation of, 44, 45
Cholestanol, 255
Chromatographic technology radiogas/radioliquid, 57
Chromophores, 108, 122
Chromosome
  chromatin structure, 178
  dyads, 346
  loss, 321
  organization, 178
Circadian rhythm, 125
Cladosporium fluvum, elicitor from, 409-410
Cloning, *Neurospora crassa*, 184
Compactin, 63
Competition, fungi in, 437-438
Complementation, 114, 310
Conidiation, 63, 66, 125
  induction of, 59
Conjugation, 310
Copper in secondary metabolism and differentiation, 87
Cryptochrome, 122
Culture instability in environmental competition, 431
Cytochrome *c*, 116
  overproduction, 119
Cytoductants, 324

Cytokinesis, 317
Cytoskeleton, 327

D

Dehydrogenase; complex, 103
  multimeric, 117
Dehydrogenations, 110, 127
Dehydrojuvabione, 241, 258
2-Deoxy-D-glucose, 326
Deoxyribose nucleic acid (DNA), 24
  2 μm, 313
  organization, 179
  reiterated, 179
  synthesis, 253, 255
  synthesis (premeiotic), 340, 345
  unique, 199
Deuteromycetes, 4, 96
Dichlorocyclopropanes in phytoalexin production, 415
Differentiation
  definition of, 2
  solvent effects on, 84
Dikaryon morphogenesis, 196-203
  cytology of, 199-201
  definition of, 195
  enzymes associated with, 209-212
  general scheme of, 198
  genetic control of, 203-209
  regulation of nuclear behavior in, 218
  time course of, 201-203
Dimorphism
  cAMP in, 276-278, 288-289
  in *Candida albicans*, 280
  cell membranes in, 284-286
  cell walls in, 273-275
  factors affecting, 272-273, 282-284
  hexose metabolism in, 278
  in *Mucor*, 268
  physiological basis, 273
  protein synthesis in, 278-280
  respiration in, 275-276

[Dimorphism]
  Y and M cell walls in, 284-286
  in yeasts and molds, 267
Diploidization, 309
DNA (*See* Deoxyribose nucleic acid)

E

Ecdysone, 258
Ergosterol, 18, 63, 64
Ergot alkaloids, 19, 58
Evolution, theory of, 258, 429
Execution point, 310, 319
Extrachromosomal inheritance, 10
Extranuclear genomes, 321

F

Factors (*See* A Factor, α Factor)
Feedback regulation, 49
Fungicides, 107
Furanoterpenoids, biosynthesis of, 392-394

G

G_1 period, 318
Gamete, 322
Gene derepression, 131
Genetic engineering, 313
Genome, size of the haploid fungal genome, 177
Genomic library, 176
Germination, 308, 321, 349
Germ tube, 349
Gibberellic acid, 18
  fermentation, 58, 387
  production by *Gibberella fujikuroi*, 58, 59
Glucan, 253, 254
Glucanase, 324

Gluconeogenesis, 342
Glucosamine, 348
Glyceollins
  biosynthesis of, 405-406, 408
  as phytoalexins, 401-404
  production stimulation of, 417
Gonadotropin synthesis, 77
Griseofulvin, 63

H

Herbicides, 107, 129
Heterokaryosis
  definition of, 196
  mutual analysis in, 205-208
Heterothallic, 310
Histone, 178
  acetylation, 245
Homothallic, 309
Hyphae, 63
  aerial, 64
  growth, 327

I

Idiolite, 13, 35
Idiophase, 13, 35, 37, 58, 60, 61, 66
Incompatibility factor, β, 218-225
Induction of secondary metabolism, 50-51
Inhibitors of protein synthesis, 59
Introns, 312
Invertebrate secondary metabolites, 77
Isoprene units, 18, 95

J

Juvabione, 241, 258

K

Kanamycin synthesis, repression of, 41
Karyogamy, 319
Killer Toxins, 331, 337

L

λ Phage, 183
Light wavelength, effect on plant secondary metabolites, 81
Lipids
  floating, 106
  storage (flow from ER to spherosomes), 106
Luxury molecules, 23, 77
Lycopene, 100
  accumulation, 133

M

Mannan, 325
Mass spectrometry, 57
Master switch gene characteristics, 225-232
Mating, 307
Mating-Type Locus (MAT), 308
Maturation of the spores, 320
Mediator substance, 109
Meiosis, 244, 307
Meiotic recombination, 310, 341, 345
Messenger RNA sequence sets, 182
Metabolism of sporulating cells, 342
Metalaxyl, stimulation of phytoalexin by,
Metal metabolism
  iron in, 167
  synergism in, 88

Subject Index

Metals, effect of on secondary
  metabolism, 86-88
Methylation of lipids, 348
6-Methylsalicyclic Acid; production
  by *Penicillium urticae,* 58
  nonproducing state of *Penicillium
  patulum,* 59, 63, 64, 66
Mevalonic acid, 100
Microcycle condiation, 67
Microtubuli, 317
Mitochondria, 321
  differentiation, 352
  in protein synthesis, 352
Mitotic recombination, 345
Morphogenesis, 4, 7, 307
  amino acid effects on, 291-293
  autoinhibitors of, 287
  exogenous substances affecting,
    286-287
  chitin synthesis in, 289-290
  cysteine in, 290-291
  factors affecting, 294
  nucleic acid synthesis in, 293
  physiological concomitants of,
    209-214
  protein profiles during, 215-216
  protein synthesis in, 215
  respiration in, 290
Multienzyme, 104
Mutagenesis, 334
Mutations,
  albino, 111
  APO, 344
  bar, 337
  β-carotene, 133
  car, 115, 116, 133
  con, 344
  conditioned, 66
  double, 311
  flavin-deficient, 115, 124
  Ger, 352
  Kar, 354
  Kex, 337
  mei, 344

[Mutations]
  meiosis/sporulation, 344
  mit, 353
  nonmating, 334
  nonsense, 312
  nonspecific sterile, 334
  pep, 343
  rad, 347
  rho, 352
  sec, 338
  specific sterile, 334
  spo, 344
  tup, 348
  yellow (*ylo-1*),
Mycelium
  aerial, 63, 64
  in *Candida albicans,* 287
  formation, 2
Mycophenolic acid, 58, 63, 64, 66,
  67
Myogenesis, 80

N

Neomycin, glucose inhibition of, 41
Neurosporene, 100
Novobiocin production, 41
Nuclear genome, 214-215
  pathway, 317
Nucleolus, 339

O

Oogonia, 258
Oogoniol, 243
"Over" culture, 63

P

Pachytene, 340
Parasexual cycle, 116

Patulin, 58, 63, 64
Penicillin
  branched pathway and feedback regulation of, 49
  fermentations, 58
  glucose inhibiton of, 41
  synthesis, 38-39
Penicillus formation, 63
Peptides, 251-258
Petites, 321
Phenylalanine ammonia-lyase as phytoalexin, 404
Pheromone secretion, 337
Phosphate regulation of streptomycin synthesis, 48
Photocaroteneogenesis, 122, 124, 125
Photodynamic activation, 107
  inactivation, 107
Photoinduction, 124, 133, 134, 121
Photoreceptor, sporangiophore initiation, 109
Photoregulation, 122, 134
Photosensitizers, 107
Photostimulation, 122
Phototropism, 125
Phytoalexins
  antibacterial activity of, 396
  antifungal activity of, 396
  bioassays of, 395-396
  biological action spectrum of, 395
  biosynthesis of, 386-389
  definition, 378
  endogenous elicitors in production of, 384-385
  host resistance in grapevines, 411-413
  induction by biotic elicitors, 381-385
  induction by infection, 380-381
  isoflavonoid grouping in, 386-389
  metabolism by fungi, 390-391
  mode of action of, 397
  in plant-parasite interactions, 398-405

[Phytoalexins]
  production enhancement by fungicides, 415-417
  properties of, 377
  of soybeans, 399-400
  structure and distribution, 378-380
  theories in host defense, 417-420
Phytochrome, 125
Phytoecdysones, 75
Phytoene
  dehydrogenation, 100
  synthetase, 127
Phytofluene, 100
Phytopathogen resistance by grapevine, 411-413
Plant cell, secondary metabolites of, 73-76
Plant cell culture
  callus culture in, 77, 80
  compound transformation, 79
  kinetics of, 78
  secondary metabolism, 77-82
  suspension culture, 77
Plant cell differentiation, hormonal effects on, 83
Plant cell metabolites; environmental effects on production of, 82
Plant-parasite interaction phytoalexins in, 398-405
Plant secondary metabolism; morphological differentiation, effect on, 80
Plant secondary metabolites, 74
  categories, 74
Plasmids
  bacterial, 75
  pBr 322, 184
  RNA, 337
Polycomplex body, 339
Polyketides, 17
Posttranscriptional modification, 181
Prenols, synthesis, 130
Prenyltransferase, 104, 126, 127, 132

Subject Index

Prephytoene, 100
Primary metabolism, 13
  definition, 155
Primary and secondary metabolism
  comparison of, 155-157
  interrelationship between, 157, 158
Protein, 245
  light-induced synthesis, 126
  phosphorylation, 333
  proteinases, 343
Protoplasts, 9, 342
Pterocarpans, structure-activity relationships of, 397
Puromycin repression, 50

Q

Quinic acid catabolic pathway, 184
Quinoline, 60
  alkaloids, 61

R

Radiation sensitivity, 347
Receptor α factor, 332
  blue light, 122
  photochromic, 125
Recombination, 119, 320
  intermediates, 347
Regulation, carbon catabolite, 40
Repression
  carbon catabolite, 354
  glucose, 342
  nitrogen, 342
Response
  blue light, 124, 125
  curve, 123
Retinol, 110
Retinal, 247, 249, 258
Rhodotorucine, 252, 254, 258
Riboflavin, 124

Ribose nucleic acid (RNA), 25
  in differentiation, 216-218
  mRNA, 180, 245
  rRNA, 245
  synthesis, 59
Rishitin, biosynthesis of, 392-394
RNA (*See* Ribose nucleic acid)

S

Sclerotium formation, 7, 67
Secondary metabolism, 13-14
  biochemical playground, 159, 160
  cell differentiation, 22, 56
  in colony development, 55
  conditions for, 160, 161
  in evolution, 153-168
  genetics of, 161-163
  monitoring status, 57
  regulation of, 157, 159
  regulation mechanisms of, 159
  role, 55
  in sclerotia, 433-434
  in submerged liquid culture, 65
  transfer of genetic information in, 166, 167
Secondary metabolites, 2, 15
  from amino acids, 19
  chemical signals, 432
  from citric acid cycle, 19
  from fatty acids, 16
  function, 20
  ecological relationships, 435-436
  production in nature, 430-431
  production variability, 155, 156
  structure, 14-15
  taxonomic value, 155
Secretory pathway, 337
  vesicles, 327
Sensitizing pigment
  excited triplet, 107
  singlet oxygen, 107
Sex factors, 330
  hormone, 109

Sexual agglutination, 322
　agglutinins, 253
　cycle, 116
　interaction, 133
　spore germination, 67
"Shmoo," 251, 318, 323, 332
Shunt metabolites, 75
Siomycin, glucose inhibition of, 41
Sirenin, 240, 258
Solid culture, 62, 63
Spectrophotometry, 99
Spheroplasts, 313
Spherosomes, 105, 106
Spindle pole body (SPB), 317
　in meiosis, 339
　satellite formation, 320
Spore, 4
　arthrospore, 6
　blastospores, 6
　chlamydospores, 6, 59
　commitment to sporulation, 317, 343
　conidiospores, 5
　germination, 348
　mitochondria, 351
　phialospores, 6
　porospores, 6
　prospores, 347, 349
　purification, 351
　radulaspores, 6
　zygospore, 251
Sporopollenin, 109, 251
Sporulation, 11, 307, 320, 338
　genes, 182
　medium, 338
　synchronous, 338
Start (G_0), 317, 318, 322
Sterol, 243, 255, 258
　synthesis, 130
Succinimide, 129, 132
Suppressors, 312
Synaptonomal complex, 339, 340, 344
Synthetic hormones, effect on callus culture, 80-81

T

Taxonomy, secondary metabolite value in, 155
Terpenoids, 15, 18, 240
Tetrad dissection, 310
Tonoplast membrane, disruption of, 396
Toxins, role in ecology, 434-435
Transcription, 181, 245
Transformation, 185, 313
Transition metabolism, secondary to primary, 163, 166
Tremerogen, 252, 254
Trisporic acid, 9, 109, 130, 131, 132, 245
Trophophase, 39, 35, 37, 58, 60, 62

U

Ubiquinones, 130

V

Vegetative phenotype, 64
Viniferin, 412-413
Vitamin A, 131, 133

X

Xanthophylls, 96, 97

Z

Zearalenone, 256
Zinc metalloenzymes, 87
　in secondary metabolism and differentiation, 86-87
Zygomycetes, 96, 109
Zygote, 307, 324